Neural Control of Speech

Neural Control of Speech

Frank H. Guenther

The MIT Press
Cambridge, Massachusetts
London, England

© 2016 Massachusetts Institute of Technology

All rights reserved. No part of this book may be reproduced in any form by any electronic or mechanical means (including photocopying, recording, or information storage and retrieval) without permission in writing from the publisher.

This book was set in Syntax and Times New Roman by Toppan Best-set Premedia Limited. Printed and bound in the United States of America.

Library of Congress Cataloging-in-Publication Data

Names: Guenther, Frank H., author.
Title: Neural control of speech / Frank H. Guenther.
Description: Cambridge, MA : The MIT Press, [2015] | Includes bibliographical references and index.
Identifiers: LCCN 2015044773 | ISBN 9780262034715 (hardcover : alk. paper)
Subjects: | MESH: Speech—physiology | Brain—physiology | Speech Disorders | Vocal Cords—physiology | Models, Neurological
Classification: LCC QP306 | NLM WV 501 | DDC 612.7/8—dc23 LC record available at
    http://lccn.loc.gov/2015044773

10 9 8 7 6 5 4 3 2 1

To those who struggle to express themselves through word of mouth.

Contents

Preface xi
Acknowledgments xiii

1 **Introduction** **1**
    1.1 Situating This Book 1
    1.2 Reference Frames for Speech Motor Control 6
    1.3 Methods for Investigating Speech 9
    1.4 Summary 29

2 **Neural Structures Involved in Speech Production** **37**
    2.1 The Primate Vocalization Circuit 37
    2.2 Brain Stem Structures 39
    2.3 Cerebellum, Basal Ganglia, and Thalamus 44
    2.4 Cerebral Cortex 57
    2.5 Summary 80

3 **Overview of Speech Motor Control** **93**
    3.1 Segmental and Prosodic Components of Speech 93
    3.2 The Control Task of Speech Production 94
    3.3 The DIVA Model 99
    3.4 Speech Motor Development 109
    3.5 Summary 114

4 **The Targets of Speech** **121**
    4.1 Definition of a Speech Target 121
    4.2 Roles of Different Sensory Systems in Speech 123
    4.3 Motor Equivalence in Speech 125
    4.4 Target Regions 129
    4.5 The Neural Bases of Speech Targets 140
    4.6 Summary 145

5 Auditory Feedback Control   153
5.1   The Auditory Feedback Control Subsystem   153
5.2   Behavioral Studies of Auditory Feedback Control in Speech   155
5.3   Neural Circuitry Underlying Auditory Feedback Control   163
5.4   Summary   170

6 Somatosensory Feedback Control   177
6.1   The Somatosensory Feedback Control Subsystem   178
6.2   Behavioral Studies of Somatosensory Feedback Control in Speech   179
6.3   Neural Circuitry Underlying Somatosensory Feedback Control   183
6.4   Summary   188

7 Feedforward Control   193
7.1   The Feedforward Control System   193
7.2   Behavioral Studies of Feedforward Control Mechanisms   197
7.3   Neural Circuits Underlying Feedforward Control of Speech   206
7.4   Summary   214

8 Sequencing of Speech Sounds   221
8.1   Models of Serial Behavior   221
8.2   Frame-Content Models of Phonological Representation   224
8.3   The Baddeley and Hitch Model of Working Memory   226
8.4   Phonological Working Memory in Speech Production   227
8.5   The Neural Bases of Speech Sound Sequencing   229
8.6   Summary   242

9 Prosody   251
9.1   Behavioral Studies of Prosodic Control   252
9.2   Neural Bases of Prosodic Control   257
9.3   Summary   267

10 Neurological Disorders of Speech Production   273
10.1   Dysarthria   275
10.2   Apraxia of Speech   288
10.3   Medial Premotor Syndromes   293
10.4   Stuttering   294
10.5   Future Directions   306
10.6   Summary   307

Appendix A   Articulator Meta-analyses   321
    A.1   Methods   321
    A.2   Results   324
    A.3   Discussion   327

Appendix B   Cortical Parcellation Scheme   339
Appendix C   Speech Network Cortical Connectivity Maps   345
        C.1   Rolandic Cortex   346
        C.2   Medial Frontal Cortex   356
        C.3   Inferior Parietal Cortex   356
        C.4   Inferior Frontal Cortex   356
        C.5   Insular Cortex   369
        C.6   Superior Temporal Cortex   369
Appendix D   DIVA Brain Activity Simulations   389
        D.1   Cortical Components   390
        D.2   Subcortical Components   394
Index   397

Preface

The primary purpose of this book is to provide a comprehensive, unified account of the neural computations underlying speech production. We focus herein on speech motor control, that is, the production of phoneme strings with minimal regard for linguistic content. More specifically, we focus on the brain mechanisms responsible for commanding the musculature of the vocal tract to produce articulations that lead to an acoustic signal that conveys a desired string of syllables. Our treatment includes neuroanatomical and neurophysiological descriptions of the primary brain structures involved in speech production, with particular emphasis on the cerebral cortex and its interactions with the cerebellum and basal ganglia. Furthermore, we use basic control theory concepts to explore the computations believed to be performed by these brain regions. Since many otherwise interested readers may not be versed in control theory or mathematical modeling, control theory descriptions include nontechnical accounts in addition to any technical details or equations.

In order to facilitate a *unified* account, the book lays out in detail a theoretical framework that has been developed over the past 20+ years at Boston University and the Massachusetts Institute of Technology to account for a broad range of behavioral and neurological data concerning the production of speech. We believe this framework to be the most detailed and comprehensive account of the neural computations underlying speech production to date, and it provides a unified account in the following two senses: (1) it accounts for both the behavioral and neurological literatures on speech, which have to date been very largely distinct, and (2) it is entirely self-consistent—for example, the portions of the framework dealing with syllable sequencing are completely integrated with the portions dealing with speech motor control at the single-syllable level.

The theoretical framework has been implemented mathematically as an artificial neural network, with equations describing the electrical activities and synaptic strengths of neurons in each of the brain regions responsible for speech. Computer simulations have verified the theory's ability to account for a wide range of existing neurophysiological, acoustic, and kinematic data concerning speech, and a number of experimental tests have been performed that verify specific predictions of the model and identify shortcomings, thus leading to improvements in the theory. Nonetheless the reader should not obediently accept our theoretical account as undisputed fact, as many of the hypotheses embodied by the model are still being actively investigated and debated in the speech, motor control, linguistics, and neuroscience literatures. Indeed, a number of alternative proposals regarding speech production are discussed in the chapters that follow in order to provide a more comprehensive coverage of the current state of the field.

Because the material in this book is highly interdisciplinary, the target audience also spans several fields. The text is written at a level appropriate for graduate students and researchers in these fields, as well as motivated undergraduates interested in speech and brain mechanisms. The first part of the target audience consists of individuals studying the speech sciences from a behavioral perspective and/or a neurological perspective. The theoretical account of speech presented herein is intended to provide these researchers with a detailed bridge between the neurophysiological and behavioral literatures on speech production. The second group consists of individuals investigating communication disorders, particularly motor speech disorders. The book provides these individuals with a detailed account of how different parts of the brain are involved in different aspects of speech, including an account of what can go awry when particular brain regions are damaged. A third group consists of graduate students and researchers in the cognitive and computational neurosciences who are interested in speech, language, or motor control. Finally, control engineers and other researchers in non-speech motor control will find that the book provides one of the most detailed treatments currently available of the brain mechanisms underlying the control of a complex motor act.

In order to accommodate this interdisciplinary audience, chapter 1 provides background information that may be familiar to more advanced readers who study speech neuroscience but not those from other domains. Chapter 2 provides a detailed treatment of the brain structures involved in speech production that serves as the foundation for later chapters describing the neural computations performed by these structures. The book then turns to issues of control, with chapter 3 providing an overview of the speech motor control problem and the DIVA neurocomputational model of speech production. Chapter 4 tackles the issue of what exactly the goals are for the neural controller of speech, along with how such goals are learned by the developing infant. Chapters 5 through 7 detail the central components of the neural controller responsible for learning and producing an individual speech sound "chunk," such as a phoneme or syllable, including the auditory feedback control subsystem (chapter 5), somatosensory feedback control subsystem (chapter 6), and feedforward control subsystem (chapter 7). Chapters 8 and 9 address the neural mechanisms involved in producing longer utterances, consisting of multiple speech sounds, including mechanisms for buffering and sequencing through the individual speech sounds (chapter 8) and for generating the global rhythm and intonation patterns, or *prosody*, of speech (chapter 9). Finally, chapter 10 addresses a number of disorders of the speech production system that are neurological in origin, with reference to the theoretical framework developed throughout the book. Additional information that may prove useful for researchers in speech neuroscience is provided in a set of appendixes.

Acknowledgments

This book would not have been completed without the efforts of several wonderful longtime colleagues in the Boston University Speech Lab. The neuroanatomical content has benefited greatly from the expert input of Jason Tourville, who has taught me much of what I know about the subject. Alfonso Nieto-Castanon's invaluable contributions span computational modeling, statistical analysis, and data visualization; I am in awe of his abilities in these areas. Bobbie Holland contributed substantially to the editing process for this book and, more generally, has been the administrative force keeping our lab afloat and productive. I am truly blessed to have these three individuals as colleagues.

The research described herein involved important contributions from a number of excellent students and postdocs in the lab, including (in alphabetical order) Heather Ames, Deryk Beal, Jay Bohland, Maya Peeva Brainard, Jon Brumberg, Shanqing Cai, Oren Civier, Ayoub Daliri, Satra Ghosh, Elisa Golfinopoulos, Michelle Hampson, Fatima Husain, Dave Johnson, Rob Law, Edwin Maas, Daniele Micci Barreca, Carrie Niziolek, Misha Panko, Kevin Reilly, Jenn Segawa, Emily Stephen, Hayo Terband, Emir Turkes, Virgilio Villacorta, and Majid Zandipour. It has been a privilege and delight to work with so many bright young minds.

Fellow faculty members and research scientists at Boston University and the Massachusetts Institute of Technology also played important roles in the development of the theoretical framework that lies at the heart of this book. Dan Bullock's knowledge regarding the motor regions of the brain is truly immense, and he has strongly influenced my thinking for over two decades, as have the neural modeling techniques developed by Steve Grossberg and Gail Carpenter. Joseph Perkell's knowledge of speech production has been another important influence, often through lively meetings at MIT's RLE Speech Communication Group that included Margaret Denny, Harlan Lane, Melanie Matthies, Lucie Ménard, Pascal Perrier, Mark Tiede, Jennell Vick, Reiner Wilhelms-Tricarico, and numerous bright research assistants and postdocs. Cara Stepp's quick mind, scientific enthusiasm, and knowledge of all things voice have also made their mark on this volume.

I am also thankful to the external collaborators who have shaped the work herein, including Xavier Alario, Kirrie Ballard, Anne Blood, Suzanne Boyce, Dan Callan, Carol

Espy-Wilson, Robert Hillman, Philip Kennedy, Psyche Loui, Christy Ludlow, Ben Maassen, Dara Manoach, Ludo Max, Rupal Patel, Don Robin, Gottfried Schlaug, Kristie Spencer, and Steve Tasko. I am indebted to Bruce Rosen and the Athinoula A. Martinos Center for Biomedical Imaging at Massachusetts General Hospital for giving me the opportunity to learn and practice the art of brain imaging at such a fine institution. Financial support for the research summarized in this volume came from the National Institutes of Health (particularly National Institute on Deafness and Other Communication Disorders grants R01 DC002852 and R01 DC007683), the National Science Foundation, and Sargent College of Health & Rehabilitation Sciences at Boston University.

Finally, my heartfelt appreciation goes out to my wonderful wife, Chris; mother, Nicole; sister, Carolyn; and writing companion, Petey, for patiently supporting me throughout the decade-long odyssey that culminated in this book.

1

Introduction

The production of speech requires integration of diverse information sources in order to generate the intricate patterns of muscle activation required for fluency. These sources include auditory, somatosensory, and motor representations in the temporal, parietal, and frontal cortical lobes, respectively, in addition to linguistic information regarding the message to be conveyed. Accordingly, a large portion of the cerebral cortex, along with associated subcortical structures, is involved in even the simplest speech task, such as reading a single word (e.g., Petersen et al., 1988; Turkeltaub et al., 2002) or meaningless syllable (Sörös et al., 2006; Ghosh, Tourville, & Guenther, 2008).

The primary goal of this book is to provide an integrated account of these brain processes, viewed from three different perspectives: (1) a *neurobiological perspective*, concerned with the regions of the brain responsible for speech and their interconnections; (2) a *behavioral perspective*, concerned with acoustic and articulatory measures of speech; and (3) a *computational perspective*, concerned with mathematical characterization of the computations performed by the brain in order to produce speech. In the process, we will detail a unified theory of how the brain performs the complex transformation of discrete phonological units (phonemes and/or syllables) into properly timed activations of the muscles of the vocal tract that lead to the spoken word.

This introductory chapter first situates the current book within the existing scientific literature. This will be followed by definitions of some of the terms that will be used to describe the different types of information important for speech and how they are encoded in different regions of the brain. The chapter closes with descriptions of many of the techniques for measuring and characterizing brain function and speech performance that have been used to carry out the studies of speech addressed in the remainder of the book.

1.1 Situating This Book

The contents of this book represent the intersection of several different scientific disciplines. In the following paragraphs we situate the current work within three domains of study: motor control, linguistics, and neuroscience.

Motor Control

Speech production is the most complex motor skill that is routinely performed by humans, and among the most skilled motor acts performed by any species. An average speaker of English can easily say the 11-phoneme word "dilapidated" in less than 1 second. Approximately 100 different muscles are located in the respiratory, laryngeal, and supralaryngeal vocal tract, and each muscle contains approximately 100 different *motor units* that consist of a motor neuron and associated muscle fibers (Darley, Aronson, & Brown, 1975; Duffy, 1995). If we assume that each phoneme requires activation of 1/10 of these motor units and each motor unit requires only one command per phoneme (the actual situation is likely more complex than this), then we arrive at an estimate of 11,000 motor command signals that need to be generated in less than 1 second. To further complicate matters, the neural controller of articulation integrates a complex array of auditory and somatosensory information when generating commands to the musculature, and the system is highly adaptive; that is, the neural controller is constantly retuning itself to deal with changes in the morphology of the vocal tract and/or hearing status that occur as we grow, age, and are subjected to strange experimental situations.

Despite the enormous complexity of the problem, recent years have seen a great deal of progress in explicating the neural control processes underlying speech, as documented in the remainder of this book. The same is true of other complex motor acts such as eye movements (Kaminski & Leigh, 2002) and reaching (e.g., Shadmehr & Wise, 2005), though those literatures will not be treated in detail here.

The control of movement, both biological and robotic, is often divided into two subproblems: *inverse kinematics* and *inverse dynamics*. Inverse kinematics involves the transformation of a desired movement specified in terms of a task-based spatial representation (e.g., the desired spatial position of the fingertip for a reaching movement or a particular acoustic target for speech production) into movement trajectories of body parts, typically specified by joint angles or muscle lengths. Inverse dynamics concerns the transformation of these kinematic trajectories of the effectors into muscle forces or motor neuron activation levels, which depend on things like gravity, limb mass, and inertia. In the current work we will concentrate primarily on inverse kinematics, though many of the principles and neural substrates are expected to apply to inverse dynamics as well. Even within the domain of speech motor control, our focus will be somewhat limited; in particular we will only superficially address respiratory control in favor of control of the larynx and supralaryngeal speech articulators such as the tongue, lips, and jaw, and we will avoid detailed treatment of the vocal tract musculature, instead focusing on a simpler, lower-dimensional articulator-based view. Treatments of the respiratory system and its neural control, as well as detailed treatments of the anatomy and musculature of the vocal tract, can be found elsewhere (e.g., Barlow, 1999; Duffy, 1995; Kent, 1997; Zemlin, 1998).

Linguistics

Speech is commonly defined as the act of expressing concepts, feelings, perceptions, and so forth through the articulation of words. In this sense, speech production is intimately linked with language. However, the words *speech* and *language*, as used in the scientific literature, are not synonymous.

Bloom and Lahey (1978) have described language as consisting of three components: content (*semantics*), form (*grammar*), and use (*pragmatics*). Grammar is further subdivided into *syntax*, *morphology*, and *phonology*. Syntax describes the rules for combining words into sentences and phrases. Morphology describes the rules for combining words and meaningful word "parts" (*morphemes*) to form larger words; for example, "walk" + "ed" = "walked." Phonology describes the sound system of a language and the rules for combining sounds to form words; for example, "pelf" is a legal sound combination in English, whereas "pfefl" is not.

Speech is typically broken into two processes: *production* and *perception*. Speech production concerns the process of taking a linguistically well-formed utterance and articulating it with the vocal tract. Speech perception concerns the process of transforming auditory (and, when available, visual) information generated by a speaker into phonemes, syllables, and/or words in the brain of the listener. Most studies of speech per se are only peripherally concerned with the meaning (semantics) of the utterance, if they are concerned with meaning at all. This is not to claim that speech perception, for example, is carried out in the brain completely independently of word meaning; to the contrary, it is clear that semantics helps disambiguate poorly articulated or degraded speech (e.g., Warren & Sherman, 1974). Instead the underlying assumption is that speech can be studied fruitfully without fully addressing the complexities of language. The valuable insights regarding speech perception and production that have been gained from hundreds of studies of this type testify to the appropriateness of this assumption.

Studies of speech perception and production inherently involve *phonetics*, which is the study of the sounds of speech, including studies of their acoustic properties (*acoustic phonetics*) and the manner in which they are articulated (*articulatory phonetics*). Although the terms phonetics and phonology are sometimes used interchangeably (and some studies can be considered to be examples of both phonetics and phonology), the two terms have somewhat different meanings. Phonology is concerned with the rules that govern the combination of sounds into larger units such as syllables or words. Phonetics is concerned with the details of the acoustic signals and articulations that convey phonemes, including things like the particular tongue shape used to articulate a sound and the resulting frequency makeup of the acoustic signal.

Our primary concern here is speech production, with a secondary emphasis on speech perception. Though the material in this book touches on many issues in linguistics (especially phonology), we will typically sidestep these issues in order to maintain our focus on the motor control of speech. For example, no attempt is made to explain why humans do not produce arbitrary phoneme strings but instead follow certain rules that determine which sounds can be produced in sequence. When addressing phoneme string production, we will typically just assume that only appropriately structured strings will be sent to the speech production mechanism. Likewise, many issues concerning the development of language in children are touched on but not directly addressed. Instead, attention is paid only to those aspects of infant development relevant to the acquisition of the motor skills necessary for the production of speech sounds independent of any underlying linguistic meaning. This approach should not be taken as a conviction that speech and language are completely independent entities. Rather, this simplified approach is used to keep the problems addressed herein tractable, with the belief that this account of speech motor control will eventually mesh seamlessly with descriptions of the higher-level linguistic processes underlying language (e.g., Levelt, 1989).

The subject matter of the current book can be placed within the larger context of theories of language production. Perhaps the most widely used model of lexical access and word production is that of Levelt and colleagues (e.g., Levelt, Roelofs, & Meyer, 1999), schematized in figure 1.1. At the highest level of the model is the *conceptual preparation* stage,

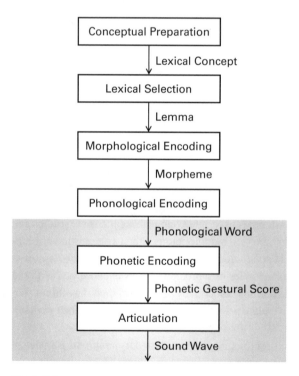

Figure 1.1
Stages of word production according to the model of Levelt and colleagues. The current book focuses primarily on the *phonetic encoding* and *articulation* stages (blue shading).

where the intended meaning of the utterance is formulated. This is followed by *lexical selection*, where one of the candidate *lemmas* (which can be thought of as canonical forms of words, as would be used for a dictionary entry) for expressing the intended concept is chosen. This lemma then undergoes *morphological encoding*, where the lemma is translated into one or more morphemes, which are the smallest meaningful units of a language. For example, the lemma "escort" may be transformed into the morphemes "escort" and "ing" to indicate an ongoing act. The chosen morphemes are then translated into syllables, which are the favored units for speech motor programming according to the Levelt scheme, in the *phonological encoding* stage. Here "escorting" becomes "e-SCOR-ting," with capitalization indicating emphatic stress. The resulting *phonological word* (which may consist of multiple words but just one emphasized syllable) then undergoes *phonetic encoding*, resulting in a *phonetic gestural score* that specifies, at a somewhat abstract level, the articulatory task that will produce the phonological words, and finally *articulation*, in which the phonetic gestural score is transformed into the muscle activations that produce the word. Our focus herein is on the phonetic encoding and articulation stages, which are only briefly sketched out in the Levelt model.

Neuroscience
The field of neuroscience consists of a number of subfields, including *molecular neuroscience*, which is concerned with the molecular processes underlying the function of neurons; *developmental neuroscience*, concerned with how the nervous system's structure and function develop during the prenatal period and in early life; *systems/cognitive neuroscience*, concerned with how large-scale networks in the brain mediate perceptions, thoughts, and actions; and *computational neuroscience*, concerned with the mathematical (or other computational) characterization of neural processes. In the current work we are primarily concerned with systems neuroscience studies of motor control in primates and cognitive neuroscience studies of speech in humans, with an additional focus on the neural computations involved. Molecular and developmental neuroscience issues related to speech will not be addressed in detail herein, except in a few instances where they relate closely to system-level behavior.

The neuroscience literature on motor control consists of both animal and human studies. Animal studies can provide a highly detailed description of the electrical spiking patterns of individual neurons or relatively small groups of neurons. The close correspondence between many monkey and human brain regions, such as the motor and visual cortices, allows us to infer the function of neurons in many parts of the human nervous system from monkey studies. However, monkey studies are limited to nonspeech studies for obvious reasons, and considerable differences exist between the parts of the human brain that are responsible for language compared to the closest homologues (if any) in the monkey brain. Thus animal studies must be supplemented with studies of human brain function in order to understand the neural bases of speech. Human studies, with rare exception, do not allow measurements of neural functioning at this fine a grain, instead requiring techniques that measure the combined effects of thousands or millions of neurons acting at the same time, as detailed in the last section of this chapter.

One challenge for neuroscience is to bridge these diverse measures of nervous system function. The remaining chapters of this book will detail neural models that attempt to bridge the gap between animal single-cell recording studies and human functional neuroimaging studies while also accounting for behavioral observations of speech movements and acoustics.

1.2 Reference Frames for Speech Motor Control

Controlling speech movements requires a complex integration of many different types of information by the brain, including phonetic, auditory, tactile, proprioceptive, and motor representations. The term *reference frames* will be used herein to refer to the different types of information represented in different brain regions. Consider, for example, a hypothetical, highly simplified brain region consisting of three neurons that collectively represent the position of a visual object in three-dimensional (3-D) space, characterized by a 3-D Cartesian coordinate system consisting of X, Y, and Z dimensions. Assume that the firing rate, or *activity level*, of one neuron represents (or *encodes*) the X-coordinate of the object in 3-D space, and the other two neurons encode the Y- and Z-coordinates of the object in a similar fashion. Together, the activities of these neurons fully specify the position of the object in 3-D Cartesian space; in other words, they represent the position of the visual object in a 3-D Cartesian reference frame. A somewhat more realistic scenario would be a brain region consisting of hundreds of neurons where each encodes a single Cartesian dimension, thereby providing a redundant representation of the object's position. Even more realistic would be a case where some neurons represent combinations of dimensions, such as $X + Y$, rather than just the position along a single dimension. Despite these complexities, the term "3-D Cartesian reference frame" would still be used to refer to these neural representations since it provides the simplest characterization of the represented information.

The remainder of this section defines a number of important reference frames for the planning of speech movements. For the sake of definiteness and clarity, discussions in later chapters will adhere to the terms defined here.

Motor Reference Frames

The *muscle length reference frame* describes the lengths and/or shortening velocities of the muscles that move the speech articulators. Activities of alpha motor neurons (which are primarily responsible for the contractile state of the muscle) constitute the *muscle activation reference frame*. This reference frame differs from a muscle length reference frame since the muscle activation commands are load dependent; that is, the muscle activation needed to achieve a particular muscle length will depend on the weight of the articulator, its inertia, and any external forces (e.g., from other muscles) acting on the articulator.

As described above, control of the vocal tract involves some 100 muscles whose commands must be precisely timed to produce fluent speech. Detailed treatments of the musculature involved in speech production can be found in Barlow (1999) and Zemlin (1998). This book will typically take a simplified view of the articulatory musculature, which we will refer to as an *articulatory reference frame*. The coordinates of this frame roughly correspond to the primary movement degrees of freedom of the speech articulators (e.g., Mermelstein, 1973; Rubin, Baer, & Mermelstein, 1981; Maeda, 1990). Although it is clear that the primary movement degrees of freedom are closely related to the musculature, the articulatory reference frame is assumed to be of lower dimensionality than the muscle reference frame. For example, several muscles may move together in a synergy that effectively controls a single articulatory movement degree of freedom. Such a representation may be utilized, for example, at the level of primary motor cortex and primary somatosensory cortex. Within this view, the corticobulbar tract projections from motor cortex to cranial nuclei in the brain stem perform an articulatory-to-muscular transformation, and projections from the muscle spindles to the primary somatosensory cortex via the cranial nerve nuclei and thalamus perform a muscular-to-articulatory transformation.

For the purposes of this book, the distinction between the articulatory reference frame and muscle length or activity reference frames is relatively unimportant aside from noting that the articulatory reference frame has a lower dimensionality. For this reason, the term *motor reference frame* will often be used as an umbrella term to describe reference frames that contain articulatory and/or muscle representations. The distinction between articulatory and muscle reference frames becomes far more important for detailed modeling of the dynamics of the speech articulators (e.g., Perrier, Ostry, & Laboissière, 1996; Ostry, Gribble, & Gracco, 1996; Wilhelms-Tricarico, 1995, 1996), a topic not treated in depth herein.

Several researchers have proposed reference frames for speech production whose coordinates describe the locations and degrees of key constrictions in the vocal tract (e.g., Browman & Goldstein, 1990; Coker, 1976; Guenther, 1994, 1995; Saltzman & Munhall, 1989). Such a *constriction reference frame* includes dimensions corresponding to the location and degree of the tongue body constriction, tongue tip constriction, and lip constriction, among others. The relationship between the constriction frame and the articulatory frame is one to many; that is, a given set of constriction locations and degrees can be reached by an infinite number of different articulator configurations or muscle length configurations. In the case of a vowel, for example, the same target tongue body constriction could be reached with the mandible high and the tongue body low relative to the mandible under normal conditions, or with the mandible lower and the tongue body higher if a bite block is present. This one-to-many relationship makes it possible for a movement controller that uses invariant constriction targets and an appropriate mapping between the constriction and articulator frames to overcome constraints on the articulators (such as a bite block) by utilizing different articulator configurations to produce the same constrictions (e.g., Saltzman & Munhall, 1989; Guenther, 1992, 1994, 1995). This ability to use different movements to reach the same goal under different conditions, called *motor equivalence*, is a ubiquitous property of biological motor systems and is addressed within the context of speech production in later chapters.

Sensory Reference Frames

Sensory information about muscle length and shortening velocity is available to the central nervous system via muscle spindles, though this information is not pure muscle length information since it depends on the descending motor command (via the gamma motor system) and is subject to habituation during maintained postures. The term *proprioceptive* will be used to describe sensory feedback concerning muscle lengths and shortening velocities along with other forms of information about articulator positions and motions. The term *tactile* will be used to describe the states of pressure receptors (mechanoreceptors) on the surfaces of the speech articulators. Tactile mechanoreceptors provide important information about articulator positions when contact between articulators is made (e.g., when producing the precise constrictions necessary for fricative consonants) but provide little or no information when contact is absent. The term *somatosensory reference frame* will be used as an umbrella term that refers to the combination of tactile and proprioceptive information from the vocal tract.

The *acoustic reference frame* describes important properties of the acoustic signal produced by the vocal tract. This reference frame describes the primary signal that is emitted by the speaker and received by the listener, although face-to-face conversations also involve a variety of visual signals as well (see Massaro, 1997, for a detailed treatment of visual speech perception). The most commonly used descriptors of the acoustic signal are the *fundamental frequency*, which is the frequency of vocal fold vibration and is perceived as voice pitch, and the *formant frequencies*, which are the peaks of the acoustic spectrum and are related to the overall shape of the vocal tract and important for differentiating phonemes. The fundamental frequency is often denoted as *F0*, and the formant frequencies are labeled (from lowest to highest) *F1, F2, F3*, and so on. The first three formant frequencies are the most important for speech perception.

The central nervous system has access to acoustic signals only after transduction by the auditory system. The term *auditory perceptual* is sometimes used to refer to the transduced version of the acoustic signal (cf. Miller, 1989; Savariaux et al., 1999) as represented in auditory cortical areas. Here the term *auditory reference frame* will be used to refer generally to the representation of sounds in the central nervous system, particularly in the auditory cortex, and it will typically be assumed that the dimensions of this auditory reference frame correspond to perceptually salient aspects of the speech signal such as the fundamental frequency and the first three formant frequencies. (In actuality there are many different representations of auditory signals in the auditory cortical areas; when necessary, we will use more specific terminology to describe these auditory representations in later chapters.) It will be further assumed that the auditory parameters are *talker normalized*, that is, they take into account differences in the overall sizes of vocal tracts across men, women, and children. The topic of talker-normalized representations is a rich one, and different proposals have been put forth for transforming raw formant frequencies into talker-normalized coordinates (e.g., Syrdal & Gopal, 1986; Miller, 1989; Traunmüller, 1994). This book will typically sidestep this issue and simply assume talker-normalized formant frequencies unless otherwise specified.

Phonological Reference Frames
At higher levels of the nervous system, speech is processed in terms of a discrete set of sound units, called *phonemes*, for a particular language. Phonemes are combined into larger sound units referred to as *syllables*. The syllable appears to be a particularly important unit for speech motor control (e.g., Levelt, Roelofs, & Meyer, 1999; Guenther, Ghosh, & Tourville, 2006). Many researchers also refer to a *featural representation* of speech (e.g., Chomsky & Halle, 1968; Fant, 1973; Jakobson & Halle, 1956; Stevens, 1998). Within this view, each phoneme is comprised of a bundle of *distinctive features* that each take on one of a finite set of values (usually binary). For example, the vowel[1] /i/ can be characterized as having the feature *voiced* and the tongue body position features *high* and *front*. In this view the smallest discretized unit of speech is the feature rather than the phoneme.

Within our theoretical view, the brain contains learned *motor programs* for producing both phonemes and syllables from the native language, as well as for producing some commonly used multisyllabic words. We will use the term *speech sounds* to refer to a more general set of discrete sound "chunks" where *each has its own distinct motor program*, a concept that will be developed throughout this book.

1.3 Methods for Investigating Speech

Since Paul Broca's mid-nineteenth-century discovery that damage to the left inferior frontal gyrus of the cerebral cortex results in the language output deficit now known as *Broca's aphasia*, a large number of experimental techniques have been used to study how the brain produces speech. The following paragraphs present an overview of some of the most relevant techniques, and subsequent chapters attempt to synthesize the knowledge gained using these techniques into a unified and detailed account of the neural processes underlying speech.

Behavioral Measures

From the perspective of speech science, movements of the speech articulators differ from other movements of the body in two important ways. First, one of the most important speech articulators, the tongue, is not visible to the naked eye, and thus its position can only be measured by an experimenter using relatively sophisticated techniques such as those described below. Second, whereas most movements of the body are meant to achieve goals in the 3-D world (e.g., reaching to a cup in a particular spatial position relative to the body), the goal of speech is to create a desired acoustic signal that will be perceived by the listener as a string of speech sounds. In other words the goals of speech are acoustic or auditory rather than spatial (see however Browman & Goldstein, 1990, and Fowler, 1986, for alternative views).

As a result of these factors, along with the fact that measuring the acoustic signal is inexpensive and easy to do, behavioral studies of speech usually focus on the acoustic signal that is produced by the speech articulators rather than on the spatial positions of the articulators themselves. Excellent treatments of speech acoustics and phonetics are provided elsewhere (e.g., Lieberman & Blumstein, 1988; Stevens, 1998). For the purposes of the current book it will suffice to note a few key properties of speech acoustics here, with more detailed treatments provided as needed in later chapters.

The frequency of vibration of the vocal folds, F0, is perceived as *pitch*, with males having relatively low-pitched voices due to relatively slow vibration of the vocal folds while females and children have respectively higher-pitched voices due to faster vibrations of the vocal folds. These differing speeds of vocal fold vibration are in large part related to differences in vocal fold mass, with lower mass leading to a higher frequency of vibration, though actively controlled changes in F0 are heavily utilized in speech, particularly for the generation of *intonation*, which provides linguistic information such as lexical stress and whether a sentence is a statement or a question. Intonation is a major component of *prosody*, which refers to aspects of the acoustic signal that do not distinguish words but nonetheless convey communicative information.

At any given point in time, the 3-D shape of the vocal tract (i.e., the shape of the "air tube" that exists from the vocal folds to the lips) is determined by the locations of speech articulators such as the tongue, lips, jaw, larynx, and soft palate (velum) that separates the oral and nasal cavities. This shape is manifested in the speech signal as a set of formant frequencies, often referred to simply as *formants*. Roughly speaking, F1 is related to the degree of constriction formed by the tongue in the vocal tract. A lower F1 value corresponds to a tighter constriction formed by a higher tongue position. F2 is related to the location of the tongue constriction along the length of the vocal tract, with higher F2 values corresponding to constrictions closer to the lips. It should be noted that these characterizations are rather crude "rules of thumb" that do not apply to all vocal tract configurations and leave out many important aspects of the relationship between speech articulation and acoustics. The interested reader is referred to Stevens (1998) for a thorough treatment of these topics. Vowels, which are relatively static sounds produced with relatively large constrictions, can be identified based on the first three formants. The fourth and higher formants contribute relatively little to the perception of speech, and thus most studies focus on only the first two or three formants. Changes in the formant frequencies over time, or *formant transitions*, are important cues for identifying consonants, which are more dynamic in nature and involve tighter constrictions than vowels. Together, F0 and the first three formants can be thought of as a first approximation to the dimensions of the auditory reference frame defined previously, though the dimensions of the auditory reference frame as identified in neurophysiological studies of mammalian auditory cortex appear to be significantly more complex than this characterization (see, e.g., Depireux et al., 2001; Gaese & Ostwald, 2003; Wang et al., 1995), and for humans the precise dimensions of the auditory representation depend in part on the native language.

Panel A of figure 1.2 shows a speech waveform corresponding to the utterance "red socks." Panel B illustrates a *speech spectrogram*, which characterizes the sound energy of the acoustic signal as a function of frequency (on the vertical axis) and time (on the horizontal axis). Dark bands indicate formant frequencies, which are particularly clear during vowels. Panel C shows the acoustic spectrum of the portion of the utterance indicated by the black rectangular outline in panel B (corresponding to the vowel in "socks"), with the first three formants indicated.

Some studies focus on the phonemic level of the acoustic signal rather than on acoustic details such as the formant frequencies or fundamental frequency. For example, a number of studies have been performed to investigate speech errors at the phonemic level, focusing on deletions, insertions, and substitutions of phonemes in running speech in both neurotypical (e.g., Postma & Kolk, 1992; Shattuck-Hufnagel, 1992) and disordered speakers (e.g., Peach & Tonkovich, 2004; Tanji et al., 2001). Such studies provide us with many insights concerning the sequencing of speech sounds, a topic we will return to in chapter 8.

Although the acoustic signal is clearly crucial for speech research, more direct measures of articulator movements are also important for defining and evaluating theories of how the brain controls speech movements. Movements of the lips can be measured relatively easily with video equipment. However, movements of the tongue, as well as most of the other speech articulators, require more sophisticated measurement techniques. As described in the review by Honda (2002), the earliest study of the positions of the tongue during vowels was performed by Meyer (1911) using x-ray technology that had been invented 15 years earlier by Roentgen, and a landmark study of vowels using this technique was published by Chiba and Kajiyama (1941). *Cineradiography* can provide an x-ray motion picture of the vocal tract during speech and was used in several important speech studies of the 1960s and 1970s (e.g., Moll, 1960; Stevens & Öhman, 1963; Perkell, 1969; Kent, 1972; Wood, 1979). However, the relatively high doses of radiation required with this technique limited its use in favor of less potentially dangerous methods developed in later years. The *X-ray*

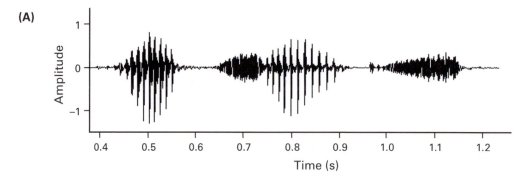
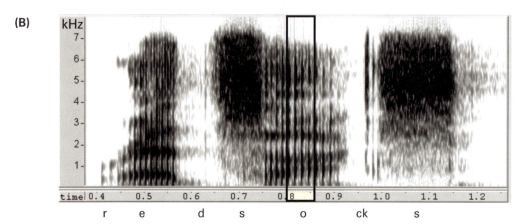
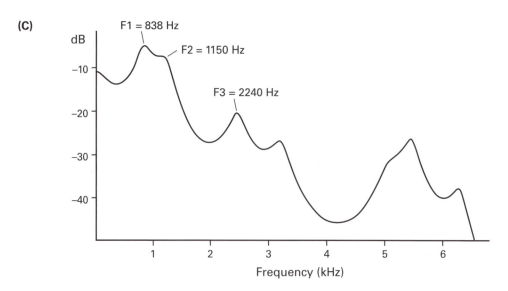

Figure 1.2
(A) Speech waveform of the utterance "red socks." (B) Speech spectrogram of the same utterance. Dark portions of the spectrogram correspond to frequencies (*y*-axis) with high energy in the acoustic signal at particular points in time (*x*-axis). The fundamental frequency is evident from the vertical "stripes" during voiced periods. Each stripe indicates an individual glottal pulse, and the reciprocal of the time between two consecutive stripes is the fundamental frequency. (C) The acoustic spectrum for a brief time frame during the vowel in "socks" (indicated by the black rectangular outline on the spectrogram in panel B). The peaks of the spectrum (corresponding to the dark horizontal bands on the spectrogram) are the formant frequencies that define the vowel. The locations of these peaks (and thus the values of the formants) are dependent upon the shape of the vocal tract, especially the positions of the jaw, tongue, and lips.

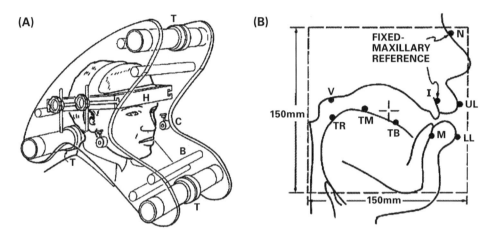

Figure 1.3
(A) Depiction of a subject in an electromagnetic midsagittal articulometry system. The system utilizes three transmitter coils (labeled T) to induce magnetic fields that are measured by small receiver coils (transducers) glued to the speech articulators. (B) Typical locations of the transducers on the articulators. A triangulation technique is used to determine the positions of the receiver coils during speech with high spatial and temporal resolution. I, upper incisor; LL, lower lip; M, maxilla; N, nose; TB, tongue blade; TM, tongue middle; TR, tongue rear; UL, upper lip; V, velum. Images provided by Joseph Perkell.

microbeam technique developed by Kiritani, Itoh, and Fujimura (1975) uses lower radiation doses to track the positions of pellets mounted on the tongue and other speech articulators whereas *electromagnetic articulometry* (*EMA*; Schönle et al., 1987; also called *EMMA* for electromagnetic midsagittal articulometry, Perkell et al., 1992) utilizes completely safe magnetic fields to localize the positions of transducers mounted on the speech articulators (see figure 1.3). *Ultrasound* also provides a safe means for measuring the oral articulators (e.g., Minifie et al., 1970; Shawker & Sonies, 1984). *Electromyography* (*EMG*) involves the measurement of electrical signals from the muscles, via either surface-mounted electrodes or small wire electrodes inserted into the muscle. According to our earlier reference frame definitions, x-ray cineradiography, x-ray microbeam, and EMMA systems provide information regarding the articulatory reference frame whereas EMG provides information in a muscle activation reference frame.

The large majority of behavioral studies of speech provide no specific information concerning the particular regions of the nervous system involved in the task under study; instead they provide evidence regarding the functionality of the nervous system as a whole. One of the goals of this book is to synthesize the results of these behavioral studies with the results of neurobiological studies, such as those described in the next subsection, to provide a more unified account of the brain function underlying speech.

Lesion Studies

The earliest convincing demonstrations of localization of speech and language function in the brain came from studies of *aphasias*, or language difficulties arising from cortical lesions, by Paul Broca, Carl Wernicke, and their colleagues in the nineteenth century. The French surgeon Paul Broca's 1861 paper is often cited as a crucial landmark in the history of aphasia, providing some of the earliest evidence of localization of language function. Broca noted that damage to the inferior frontal gyrus of the cerebral cortex (labeled *Broca's Area* in figure 1.4) was associated with impaired speech output. In a second classic paper, Broca (1865) reported that lesions of the left hemisphere, but not the right hemisphere, interfered with speech, with only a small number of exceptions. An earlier unpublished paper presented by Marc Dax in 1836 had also noted that loss of speech was associated with lesions to the left hemisphere. In keeping with its proposed role in speech production, Broca's area lies immediately anterior to the motor cortical representation of the speech articulators in the ventral portion of the precentral gyrus.

The Broca studies primarily concerned loss of speech output, and patients with Broca's aphasia showed a relatively spared ability to perceive speech, though more recent studies have noted that individuals with Broca's area damage often have difficulty understanding sentences with complex grammatical structure (e.g., Nadeau, 1988). In 1874, German neurologist Carl Wernicke identified a different brain region associated with a second type of aphasia, *sensory aphasia* (also called *Wernicke's aphasia*), which is primarily characterized by poor speech comprehension with relatively fluent (though often nonsensical) speech output. Wernicke identified the posterior portion of the superior temporal gyrus in the left cerebral hemisphere, labeled *Wernicke's Area* in figure 1.4, as the crucial lesion site for sensory aphasia. This region is immediately posterior to the primary auditory cortex, which is located in Heschl's gyrus in the superior temporal lobe within the Sylvian fissure and is considered to be a higher-order auditory cortical area important for perceptual processing of speech.

Broca's and Wernicke's areas involve relatively large expanses of cortex and are often inconsistently defined in the literature. For example, portions of the supramarginal gyrus, angular gyrus, and/or middle temporal gyrus are sometimes included in the definition of Wernicke's area (Penfield & Roberts, 1959) while other researchers limit Wernicke's area to the posterior superior temporal gyrus and planum temporale (Goodglass, 1993; Martin, 1996; Kuehn, Lemme, & Baumgartner, 1989). Similarly, Broca's area is sometimes

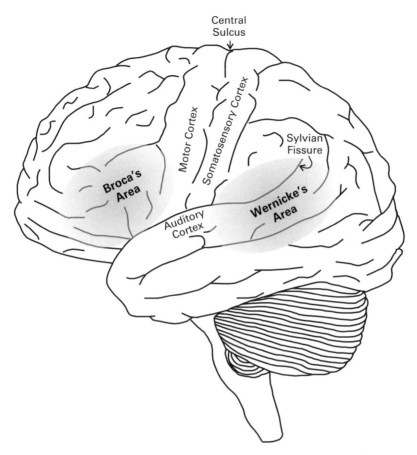

Figure 1.4
Approximate locations of Broca's and Wernicke's areas in the left cerebral hemisphere.

limited to the opercular portion of the inferior frontal gyrus (Martin, 1996) while other definitions also include the more anterior triangular portion (Duvernoy, 1999; Goodglass, 1993). For this reason, the use of more precise anatomical and/or functional descriptions instead of (or in addition to) these terms will be used in later chapters.

Shortly after the discoveries of Broca and Wernicke, German physician Ludwig Lichtheim (1885) formulated a model of language function in the brain that is now referred to as the *Wernicke-Lichtheim model* (e.g., Goodglass, 1993). As illustrated in figure 1.5, this model identifies seven possible lesion sites in or between three major centers of language: an auditory center corresponding to Wernicke's area, a motor center corresponding to Broca's area, and a conceptual center whose location in the brain was not specified. This model has been modified over the years, most notably by Geschwind (1965), who

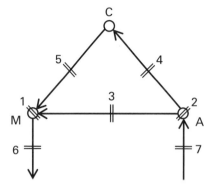

Figure 1.5
Schematic of the Wernicke-Lichtheim model of the interactions between auditory (A), motor (M), and conceptual (C) centers in the brain. Numbers indicate hypothesized lesion sites corresponding to various speech and language deficits. Lesion Site 1, Broca's aphasia; 2, Wernicke's aphasia; 3, conduction aphasia; 4, transcortical sensory aphasia; 5, transcortical motor aphasia; 6, speech motor impairment; 7, auditory system impairment.

proposed a role for the supramarginal and angular gyri in the inferior parietal cortex in associating auditory word forms for an object with visual and/or somesthetic representations of the object's properties, thus forming part of the conceptual system.

Since the time of Broca, Wernicke, and Lichtheim, a tremendous number of lesion studies have led to many important contributions regarding the brain functions underlying speech and language that go far beyond the Wernicke-Lichtheim model (for reviews, see Dronkers, Pinker, & Damasio, 2000; Goodglass, 1993; Kuehn, Lemme, & Baumgartner, 1989). In subsequent chapters we will address lesion studies that bear directly on the neural bases of speech production. It is important to note, however, that such studies are notoriously difficult to interpret. There are several reasons for this. First, rarely do different patients have lesions in precisely the same region of the cortex. Second, most lesions span several cortical areas and thus affect a number of neural systems, not just the system involved in a particular task such as speech production. Third, spared portions of cortex will often take over functions formerly performed by the damaged portion. Finally, considerable variability may exist in the location of a particular brain function in different individuals, particularly for higher-level (nonprimary) regions of cortex (e.g., Caplan, 2001). For these reasons, later chapters will supplement lesion evidence with additional sources of information regarding the functionality of specific brain regions in speech production.

Intracranial Electrical Stimulation
Not long after humankind mastered the fundamentals of electricity, the first studies of electrical stimulation to the exposed brain surface were performed. One of the earliest such

studies was performed by Fritsch and Hitzig (1870), who evoked movement of the limbs in dogs by electrically stimulating the cerebral cortex. A number of pioneering neurophysiologists refined these techniques in the following decades, including Cushing (1909), who demonstrated phantom somatic sensation in human patients when stimulating the postcentral gyrus (see Penfield & Roberts, 1959, for a historical review).

In the 1930s, 1940s, and 1950s, Wilder Penfield and colleagues at the Montreal Neurological Institute collected a large body of crucial neurological data regarding the effects of direct electrical stimulation of the cerebral cortex on speech and language function. In these studies, which were performed prior to cortical excision surgery in focal cerebral epileptics, the researchers used an electrode to deliver brief bursts of electrical stimulation (frequencies of 30–60 Hz, durations of 0.2–0.5 ms, amplitude of 2–4 volts) to carefully documented locations on the cerebral cortical hemispheres in conscious human patients (e.g., Penfield & Rasmussen, 1950; Penfield & Roberts, 1959). These landmark studies uncovered a number of fundamental properties regarding the organization of speech and language function in cerebral cortex.[2]

One such property is the somatotopic organization of the body surface, including the vocal tract organs, in the primary somatosensory and motor cortices. When the postcentral gyrus was stimulated, patients generally reported tingling, numbness, or pulling sensations in some portion of the body,[3] sometimes accompanied by movements in that body region. Panel A of figure 1.6, adapted from Penfield and Rasmussen (1950), illustrates the somatotopic organization of the primary somatosensory cortex along the postcentral gyrus as constructed from the results of these studies. When the precentral gyrus was stimulated instead, simple movements of the body were produced, identifying this region as primary motor cortex. The somatotopic organization of primary motor cortex, summarized in panel B of figure 1.6, roughly mirrors the somatotopic organization of primary somatosensory cortex, which is not surprising (at least in retrospect) since these areas abut each other and are heavily interconnected. The most notable findings for our purposes concern the speech articulators, including the lips, tongue, jaw, laryngeal system, and respiratory system. With the exception of the respiratory system, which later studies locate in the dorsolateral portion of the cerebral cortex, the speech-related motor and sensory organs are represented in the ventral half of the lateral surface of the postcentral (somatosensory) and precentral (motor) gyri.

Stimulation of the ventral precentral gyrus produces grunts, groans, or vowel-like vocalizations, sometimes with a consonant component. These responses can occur with stimulation to either hemisphere. More recent work has noted that, although the production of entire words is very rarely or never produced with cortical stimulation, stimulation of the language-dominant anterior thalamus or head of the caudate nucleus often produces whole-word vocalizations such as phrases and short sentences (Schaltenbrand, 1965, 1975; Van Buren, 1963, 1966; see Crosson, 1992, for a review). Whereas these productions typically

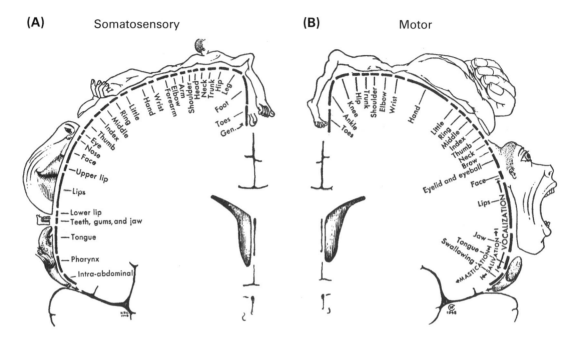

Figure 1.6
Schematized coronal views of the somatotopic organization of (A) the primary somatosensory cortex and (B) the primary motor cortex, located in the postcentral and precentral gyri, respectively. Each cerebral hemisphere includes a motor and somatosensory map that primarily (but not exclusively) represents the opposite side of the body. From Penfield /Rasmussen. *THE CEREBRAL CORTEX OF MAN.* © 1950 Gale, a part of Cengage Learning, Inc. Reproduced by permission. www.cengage.com/permissions.

occur at the onset of stimulation in the thalamus, it is the *offset* of caudate stimulation that induces vocalizations (Crosson, 1992).

Figure 1.7, adapted from Penfield and Roberts (1959), summarizes the locations where stimulation of the left hemisphere cerebral cortex interferes with speech.[4] In addition to the primary somatosensory and motor cortical areas just described, we see that stimulation in a number of frontal, temporal, and parietal lobe sites can cause interference with normal speech function, including the language areas identified by Broca and Wernicke.

Large-Scale Electrical and Magnetic Field Recordings

Electrophysiology is the general term used to describe techniques that measure electrical activity in the body, especially from the brain. The most commonly used electrophysiological technique for human studies is *electroencephalography* (*EEG*), which involves the measurement of electrical activity through electrodes placed on the scalp. The term *electroencephalography* was coined by German physiologist Hans Berger, who collected some of the earliest human EEG recordings in the 1920s. Modern EEG systems for neuroscience

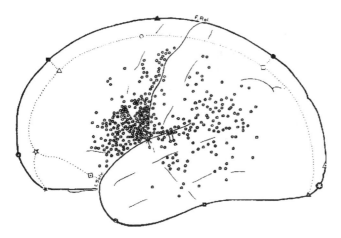

Figure 1.7
Locations on the lateral surface of the left cerebral hemisphere where electrical stimulation interferes with speech in some manner. f. Rol., Rolandic fissure (central sulcus). Adapted from Penfield and Roberts, 1959, figure VIII-3. Reproduced by permission from Princeton University Press.

research record from as many as 200 electrodes simultaneously, each positioned over a different part of the head.

The signals measured by EEG represent the combined electrical activity of tens of thousands of neurons in the brain. The contribution of a neuron to the EEG signal depends on its orientation relative to the electrode and distance from the electrode, among other factors. For this reason the contributions from neurons that are randomly oriented in the cortical tissue, such as stellate cells, tend to cancel each other out, while the contributions of neurons that are aligned with each other, such as pyramidal cells, tend to sum together and produce relatively large contributions to the EEG signal. When aligned cells fire synchronously in an oscillatory fashion, a rhythmic EEG pattern results; these patterns (referred to as *waves* or *rhythms*) are often characterized by their primary frequency, for example as *delta* (<4 Hz), *theta* (4–8 Hz), *alpha* (8–12 Hz), *beta* (12–30 Hz), or *gamma* (>30 Hz) waves.[5] These rhythms relate in complex ways to different states of consciousness and action. With regard to motor control, rhythmic activity in the motor cortex is often suppressed when movement is initiated, especially in the beta frequency range. It has been noted that this pattern of suppression is abnormal in people who stutter (Salmelin et al., 2000), a topic we will return to in chapter 10.

Like EEG, *magnetoencephalography* (*MEG*) measures a signal produced by the combined electrical activity of a large number of neurons in cortex, but MEG uses extremely sensitive magnetometers called *superconducting quantum interference devices* (*SQUIDs*) to measure the magnetic field produced by this activity rather than the electric currents measured by EEG. Modern MEG systems can collect data from 300 or more sensors simultaneously. The magnetic field produced by a current is in a direction orthogonal to the current direction, a relationship characterized by the "right-hand rule" of physics. As a result, whereas EEG is most sensitive to electrical activity in gyri (where the pyramidal cells are oriented roughly perpendicular to the skull surface), MEG is most sensitive to activity in sulci (where the pyramidal cells are oriented roughly parallel to the skull surface).

A number of different techniques are used to analyze EEG and MEG data. One commonly used technique involves the identification and characterization of *event-related potentials* (*ERPs*), which are large spikes in the EEG/MEG waveform that occur at specific times relative to the onset of a sensory stimulus or a motor event. For example, the M100 is a magnetic evoked potential arising in auditory cortex approximately 100 ms after the onset of an acoustic signal. Many studies identify changes in the magnitude or timing of these ERPs in different tasks, thus providing information that relates perception or behavior to ERPs in the brain.

A major advantage of EEG and MEG over functional neuroimaging techniques based on *regional cerebral blood flow* (*rCBF*, discussed below) is that EEG and MEG can differentiate neural events occurring very closely together in time (i.e., on a millisecond timescale or faster) whereas rCBF-based techniques have much worse temporal resolution, on the order of 1 s or more, due to the relatively sluggish nature of the hemodynamic response. The major disadvantages of EEG and MEG are (1) they are limited to measuring an average of activity over a large number of neurons, (2) they are primarily limited to measurements of cerebral cortex, and (3) it is difficult to accurately localize the sources of the measured activity in the brain. Regarding (3), EEG localization is usually limited to noting the position of the sensor(s) that provide the strongest readings. More sophisticated techniques have been developed to provide more accurate localization of EEG and MEG data (e.g., Lin et al., 2006; Mattout et al., 2006; Rodriguez-Rivera et al., 2006)—for example, by using magnetic resonance imaging data from the same subject to help constrain the localization of the EEG/MEG sources in the cortical surface. At the current time, however, the reliability of these methods is variable, at least for relatively complex tasks such as speaking. An additional limitation of EEG and MEG specific to speech production is the fact that these techniques are susceptible to large artifacts when the muscles of the face are contracted, though some studies have overcome this issue (e.g., Houde et al., 2002; Salmelin et al., 2000).

The recent scientific literature has seen a rapid rise in the number of publications that utilize *electrocorticography* (*ECoG*) to study the functioning human brain. ECoG is similar to EEG but involves electrodes placed on the surface of the cerebral cortex (either subdural or epidural), thus providing much higher signal-to-noise ratio and spatial resolution. This technique is currently limited almost exclusively to neurosurgery patients undergoing presurgical functional mapping to guide the surgical intervention. Despite the relatively limited opportunities for ECoG use, ECoG during speech production has already been used to investigate temporal lobe activity in response to one's own voice (Creutzfeldt et al., 1989), to compare spoken and sign language (Crone et al., 2001), to identify the time course of activity across the speech network during overt (Pei et al., 2011; Leuthardt et al., 2012) and covert (Pei et al., 2011) word repetition, and to identify the spatiotemporal pattern of activity in articulatory representations in sensorimotor cortex (Bouchard et al., 2013). This trend is likely to continue because of ECoG's unique combination of high spatial and temporal resolution. ECoG is not without limitations, however. One limitation is that ECoG signals can only be measured from the outer surface of the cortex; this means activity within the brain's sulci (a large proportion of the cerebral cortex) as well as subcortical structures cannot be measured. A second limitation is that current ECoG methods do not have the spatial resolution to record from individual neurons.

Early EEG studies identified correlations between mental state or behavior and brain rhythms (e.g., alpha rhythm, delta rhythm, etc.) that can be identified by peaks in the frequency spectrum of the recorded signals. The higher spatial resolution afforded by MEG and ECoG compared to EEG has led to the development of computational methods aimed at analyzing interactions between large-scale brain rhythms in different cortical areas or in different frequency bands within the same cortical area. Among other things, these techniques can identify functional connectivity between different brain regions or functional coupling across frequency bands within a brain region. An increasing number of studies are reporting relationships between brain rhythms and speech processing (e.g., Giraud et al., 2007; Luo & Poeppel, 2007; Leuthardt et al., 2011).

Microelectrode Recordings

Another form of electrophysiology involves the use of microelectrodes to measure the electrical activity of single neurons (*single-unit recordings*), the combined activities of relatively small local groups of neurons (*multi-unit recordings*), or *local field potentials* (*LFPs*) that sum the contributions of many neurons in a region of the brain. These techniques have provided a wealth of information concerning neuronal activity during a wide range of tasks in a wide range of animals. Of particular interest for speech are studies of sensory and motor tasks in nonhuman primates. As detailed in the next chapter and throughout the remainder of the book, the primary motor cortex was a major early focus of these studies, and more recently the premotor and auditory cortical areas have been studied in great detail. These studies provide insight into the likely functioning of the close human homologues of these areas, though such extrapolations are on weaker footing with regard to higher-level human areas such as Broca's or Wernicke's areas, which differ substantially from their closest monkey homologues. Although rare, some microelectrode recordings have been done in humans, either during presurgical mapping (e.g., Sahin et al., 2009) or in profoundly paralyzed individuals implanted in an attempt to restore motor function via a brain-computer interface (e.g., Guenther et al., 2009).

Single-cell recordings provide the most detailed information regarding the firing properties of individual neurons. One shortcoming of these studies has been that they are typically limited to studying one or two brain areas (as opposed to functional neuroimaging techniques that measure activity in the entire brain), and cells are typically recorded sequentially (i.e., after one neuron has been studied, the experimenter moves on to another) rather than simultaneously. However, an increasing number of recent studies have used simultaneous recording from many neurons in one or more brain regions during a task (see Brown, Kass, & Mitra, 2004, for a review).

Structural Neuroimaging
Although not a direct indicator of brain function, structural neuroimaging can provide important information concerning the locations of lesions in living patients and can improve the anatomical localization of the results of functional neuroimaging studies. Structural imaging can also provide important information regarding connectivity patterns in the brain that can be used to evaluate theories regarding the neural circuitry involved in speech and language.

In the 1890s, Wilhelm Conrad Roentgen produced the first x-ray images of the human body. Unfortunately, however, the brain is nearly invisible in standard x-ray images because it consists of soft tissue and is encased in a bath of cerebrospinal fluid. In 1918, the American neurosurgeon Walter Dandy developed a technique in which cerebrospinal fluid was replaced with air, oxygen, or helium in order to produce contrasts indicating brain structure on x-ray images (Dandy, 1918). Although a risky and painful procedure, this technique, referred to as *pneumoencephalography*, was used until the advent of more sophisticated structural imaging techniques in the 1970s.

A more elaborate x-ray technique, *computed tomography* (*CT*, also known as *computed axial tomography*, or *CAT*), was invented independently by British engineer Godfrey Newbold Hounsfield and Tufts University professor Allan Cormack, who shared the Nobel Prize in 1979 for their research on this technology. Early CT machines were capable of producing an anatomical image of a brain slice consisting of an 80×80 matrix of pixels. Current CT machines are capable of 512×512 matrix images.

Magnetic resonance imaging (*MRI*), another technology for obtaining structural images of the brain and other body parts that emerged during the 1970s, is based on pioneering research in nuclear magnetic resonance in the 1940s by Félix Bloch and Edward Purcell, who were awarded the 1952 Nobel Prize in Physics for this work. Like CT, structural MRI is capable of producing high-resolution 3-D images of anatomical structure in the brain. Magnetic resonance imaging is now by far the most common technique for characterizing brain structure in living humans.

A more recently developed magnetic resonance imaging technology called *diffusion-weighted imaging* (*DWI*) can be used to image white matter pathways in the brain. Often the more specific term *diffusion-tensor imaging* (*DTI*) is used to describe this technology.

These techniques work by identifying the direction of water molecule diffusion in a region of the brain, which is indicative of the direction of axonal projections in that region. Many early diffusion MRI studies focused on speech and language circuits (e.g., Catani, Jones, & ffytche, 2005; Golestani et al., 2006; Henry et al., 2004; Sommer et al., 2002), and the number of studies of this type continues to grow rapidly.

Functional Neuroimaging

By allowing noninvasive collection of high-resolution images of the functioning human brain, technologies for measuring brain activity as represented by changes in rCBF have led to a revolution in our understanding of the neural bases of speech and language. A relationship between cerebral blood flow and neural activity was first noted by Roy and Sherrington (1890). Later studies verified that neural activity utilizes oxygen in the blood, which in turn leads to increased blood flow to the active region. An increase in blood flow to a region, or *hemodynamic response*, while a subject performs a particular task thus provides evidence of that region's involvement in the task.

The first technology to measure local blood flow in the entire brain while a subject was performing an experimental task was *positron emission tomography* (*PET*), which is based on positron annihilation radiation research from the 1950s by Gordon Brownell and William Sweet at the Massachusetts General Hospital. The technology was steadily developed over the next 25 years, and the first commercial PET scanners were produced in the 1970s. PET is capable of imaging metabolic processes through the use of radioactive tracers that are injected into the subject (typically into the bloodstream) prior to scanning. Different tracers highlight different aspects of metabolic function. One major use of PET is in the detection of malignant cancers using the tracer *fluorodeoxyglucose* (*FDG*), which highlights areas of high metabolic activity. Another technique using radioactive tracers is *single photon emission computed tomography* (*SPECT*). Among other things, SPECT can be used to image dopamine transport in the brain. Dopamine is a major neurotransmitter known to be involved in motor function (particularly in the basal ganglia), and damage to the dopamine system characterizes Parkinson's disease. It has also been suggested that stuttering may arise from abnormal function in the dopamine system of the basal ganglia (see Alm, 2004, and chapter 10 in this volume for further details).

The most common use of PET for studies of speech involves measurement of rCBF while subjects perform speech tasks after injection of the tracer $H_2^{15}O$. In one of the earliest such studies, Petersen et al. (1988) measured rCBF while subjects produced single words presented either auditorily or visually. Brain areas involved in both the visual and auditory production tasks included motor and somatosensory areas along the ventral portion of the central sulcus, the superior temporal gyrus, and the supplementary motor area, in keeping with earlier studies of the effects of cortical stimulation on speech (e.g., Penfield & Roberts, 1959).

More recently, MRI technology has been applied to the problem of rCBF measurement (e.g., Ogawa et al., 1990; Belliveau et al., 1991). This technique, referred to as *functional magnetic resonance imaging* (*fMRI*), provides improved spatial resolution over PET and does not require radioactive tracers, though some early fMRI studies used such tracers (e.g., Belliveau et al., 1991). The most commonly used fMRI approach uses *blood-oxygen-level-dependent* (*BOLD*) contrasts that capitalize on differences in magnetic susceptibility for hemoglobin in its oxygenated and deoxygenated state (Ogawa et al., 1990). A large number of fMRI studies of speech and language have been performed since the early 1990s, as documented throughout the remainder of this book. Figure 1.8 illustrates brain activity in the speech production network as measured by BOLD fMRI in 116 adult subjects. The active brain regions in this figure will be addressed in the chapters that follow.

Like other brain measurement techniques, PET and fMRI have limitations. First, even at the highest resolutions available with fMRI—a typical fMRI study using a 3 tesla scanner will have a spatial resolution of approximately $2 \times 2 \times 2$ mm—the measured activities represent the average activation of a large number of neurons. Furthermore, when a brain region becomes electrically active for a brief period, it takes several seconds before the hemodynamic response reaches a peak, and the blood flow remains at a higher than baseline level for several seconds after the electrical activity subsides (e.g., Logothetis et al., 2001). In other words, the hemodynamic response operates on a slower timescale than the electrical activity. This means that fMRI and PET inherently have poor temporal resolution compared to imaging technologies that more directly measure electrical activity (EEG, ECoG) or the magnetic field induced by electrical activity (MEG) of pyramidal cells in the cerebral cortex. Whereas the temporal resolution of fMRI is on the order of seconds (for standard fMRI techniques) or at best tenths of seconds (for specially designed event-related fMRI techniques), the temporal resolution of EEG and MEG is on the order of a millisecond. Conversely, fMRI has relatively good spatial resolution compared to EEG and MEG.

Transcranial Stimulation

Transcranial magnetic stimulation (*TMS*) allows the experimenter to noninvasively interrupt the function of a particular brain region with a high-energy pulse of a focalized magnetic field aimed at that region through the skull (see Cowey & Walsh, 2001, for a review). This has the advantage over neuroimaging techniques of directly testing whether the brain region in question is required for a particular task. For example, if we see fMRI activation in the inferior parietal lobule during an auditory spatial localization task, it is possible that this activation is only a side effect of other brain activations that are involved in identifying the spatial location of a sound source, rather than a direct contributor to the computation of spatial location. Greater confidence about the importance of the inferior parietal lobule for spatial localization can be gained if the fMRI experiment is supplemented by a TMS

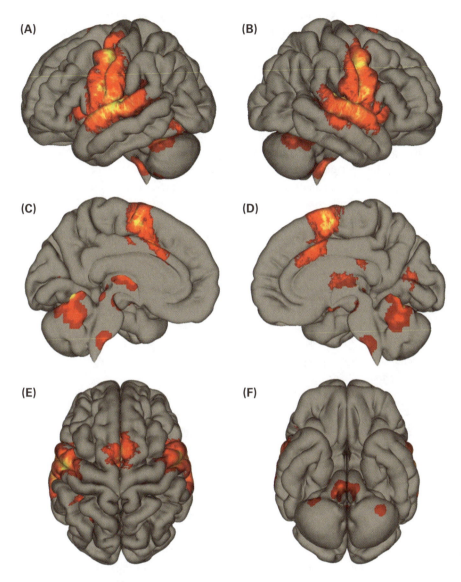

Figure 1.8
Active brain regions during reading aloud of monosyllabic and bisyllabic utterances contrasted with a baseline task of passively viewing letters. Top panels show activation on the (A) left and (B) right lateral surface of the brain. Middle panels show (C) left and (D) right medial views. Bottom panels show (E) superior and (F) inferior views. Data are from 116 subjects. Cortical surface vertices were thresholded at $p < 0.001$ uncorrected, followed by a cluster-level analysis with a family-wise error threshold of $p < 0.05$. Subcortical voxels were thresholded at a false discovery rate of $p < 1 \times 10^{-6}$. No smoothing of functional data was performed. Activity is shown using a "hot" colormap, starting with red for relatively low activity levels and ranging to bright yellow for the highest activity levels.

experiment that involves brief inactivation of the inferior parietal lobule while a subject is trying to localize a sound source. If this inactivation leads to poor spatial localization performance, one can more definitively conclude that normal inferior parietal lobule function is important for computing spatial localization. TMS has been used in a number of studies of speech and language (e.g., Fadiga et al., 2002; Gentilucci et al., 2006; Ito, Kimura, & Gomi, 2005; van Rijn et al., 2005; Watkins & Paus, 2004), as discussed in later chapters.

Transcranial current stimulation (*tCS*), a technique that has gained popularity in recent years for studying the functioning brain, acts in a manner similar to TMS but utilizes electrical stimulation at low current levels rather than magnetic stimulation (see Ruffini et al., 2013, for an overview). The technique often utilizes direct (rather than alternating) current and is thus often referred to as *transcranial direct current stimulation* (*tDCS*). The concept of stimulating the brain electrically has been around for well over a century, but stimulation was usually applied for clinical rather than scientific purposes until the 2000s, as guidelines for safe use of the technique in humans became more clearly established. In addition to using tCS to disrupt normal (e.g., Berryhill et al., 2010) or abnormal (e.g., Pedron et al., 2014) functioning of a portion of the brain, the ability of certain tCS protocols to *improve* function of a stimulated brain region has also been demonstrated (e.g., Boggio et al., 2006), similar to findings that have been reported in the TMS literature (e.g., Pascual-Leone et al., 1996; Luber & Lisanby, 2014).

Computational Modeling

The modeling of physical systems by mathematical equations that can be simulated on digital computers has become a crucially important scientific tool in the past half century, as data sets regarding task performance and brain function become increasingly large and complex. Speech science is no exception, as a number of important studies have used computational models to account for the physics and neural processes underlying speech.

Articulatory synthesizers are models that transform the positions of simulated speech articulators into acoustic signals (e.g., Coker, 1976; Flanagan et al., 1970; Maeda, 1990; Rubin, Baer, & Mermelstein, 1981). These models consist of two main components: (1) a *source* or *excitation signal* that characterizes the sound arising from vocal fold vibration and/or acoustic noises created by air turbulence at any tight constrictions within the vocal tract and (2) a *vocal tract filter* derived from the overall shape of the vocal tract that amplifies certain sound source frequencies and attenuates others (see Stevens, 1998, for a mathematical treatment of this topic, often referred to as *source-filter theory*).

Biomechanical models of the vocal tract (e.g., Gerard et al., 2005; Payan & Perrier, 1997; Wilhelms-Tricarico, 1995, 1996) provide detailed characterizations of the musculature of the vocal tract and how muscle commands translate into articulator positions, typically using finite element computational methods to model soft articulators such as the tongue. The resulting articulator positions and vocal tract shapes can be transformed into acoustic signals using the same techniques used for articulatory synthesis.

Articulatory synthesizers and biomechanical models do not, by themselves, address the neural mechanisms involved in controlling movements of the vocal tract. One of the earliest attempts to apply engineering techniques to study speech motor control was Fairbanks's (1954) *servosystem model* of speech production. This model characterized speech production as an auditory feedback control problem. Although useful as an approximation to the auditory feedback control component of speech, this model by itself cannot account for many key characteristics of speech, such as its rapid rate (e.g., Neilson & Neilson, 1987). As detailed in chapter 3, speech motor control involves feedforward control mechanisms in addition to auditory and somatosensory feedback control contributions.

One of the first models of speech motor control to be implemented on a computer was Henke's (1966) *dynamic articulatory model*, which was capable of producing strings of vowels and stop consonants with articulator movements that were formed using a simple set of rules to move between phoneme targets. In order to account for known characteristics of speech movements, vowels and consonants had different rule sets in this model. The *task dynamic model* of Saltzman and Munhall (1989) provides a detailed account of computations used to transform a *gestural score* (Browman & Goldstein, 1990), consisting of a time series of constriction targets for the vocal tract, into movements of a set of speech articulators that could then be synthesized into an acoustic signal using the Rubin, Baer, and Mermelstein (1981) articulatory synthesizer. Within this theoretical view, articulatory gestures constitute the phonological units of speech (see Browman & Goldstein, 1990, for details), and therefore no auditory targets are utilized for speech movement planning in the task dynamic model.

The computational models described above have contributed a tremendous amount of information regarding speech production. However, they provide little information about the neural mechanisms underlying speech. Over the past two decades our laboratory has applied *artificial neural networks* (*ANNs*), often referred to simply as *neural networks*, to computationally model the neural computations underlying speech production (for related approaches, see Dell, 1986; Fagg & Arbib, 1998; Roelofs, 1997; Horwitz, Friston, & Taylor, 2000; Husain et al., 2004; Tagametz & Horwitz, 1999; Kröger et al., 2006; Garagnani & Pulvermüller, 2013). Neural models[6] developed in our laboratory will be used to organize and interpret a wide range of experimental findings concerning speech in the remainder of this book. Particular emphasis will be placed on the *DIVA model* of speech acquisition and production (Guenther, 1992, 1994, 1995; Guenther et al., 1998, 2006), which will be introduced in chapter 3 and discussed in more depth in later chapters.

The field of ANN research started with the publication of a seminal paper by McCulloch and Pitts (1943) describing a simple computational model of the electrical state of a neuron based on its inputs, which project to it through synapses that scale, or *weight*, the incoming input signal (see figure 1.9, panel A). Roughly speaking, a small synaptic weight attenuates the signal arriving via the synapse, and a large synaptic weight amplifies it. A neural network is simply a collection of abstracted neurons of this sort, referred to as *nodes* herein to avoid conflation with real neurons in the brain, that are connected to each other through weighted synapses (figure 1.9, panel B). The synapses can be excitatory (represented by arrowheads in figure 1.9) or inhibitory (represented by circles in figure 1.9), and they can be either fixed or adaptive. In the latter case, the process of adjusting the synaptic weights is referred to as *learning* as it is an approximation of learning and memory formation in the brain.

The term *map* is often used to denote a set of nodes in a neural model that collectively represent a particular type of information, similar to the somatotopic maps of the body in the pre- and postcentral gyri identified by Penfield and colleagues. In many cases such a map will constitute a single layer of neurons in a multilayer ANN, such as the top row of neurons in the three-layer ANN illustrated in panel B of figure 1.9. Panel C provides a simplified depiction of the three-layer network in panel B in which each map of nodes is represented by a single box and synaptic projections between layers are schematized as a single projection. Later chapters will rely primarily on this type of simplified schematic to depict neural models.

A large number of studies have been done to identify useful algorithms for adjusting synaptic weights in ANNs (e.g., Grossberg, 1976; Rosenblatt, 1962; Rumelhart, Hinton, & Williams, 1986; Widrow & Hoff, 1960); such algorithms are often referred to as *learning laws*. Additional studies provide detailed characterizations of the (often complex) dynamics of electrical activity in networks of interconnected neurons (e.g., Grossberg,

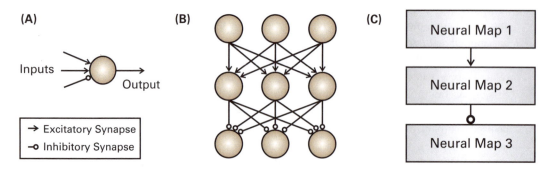

Figure 1.9
(A) Schematic of a single neuron, or node, in a neural network model. This node receives two excitatory synaptic inputs (represented as arrowheads) and one inhibitory synaptic input (represented as a circle).
(B) A three-layer neural network with excitatory synaptic projections from the first layer to the second layer and inhibitory projections from the second layer to the third layer. (C) Simplified depiction of the neural network in (B).

1973). The neural models described in the current book rely primarily on biologically motivated equations for node activities and synaptic weights similar to those developed by Grossberg and colleagues since the early 1970s. Most of the mathematical details of these models will be avoided herein in favor of pointers to journal articles that provide relevant equations.

There are several motivations for using neural modeling as a theoretical framework to account for the neural control of speech. First, because the models are computational, they can be simulated on computers to verify their properties. A realistic account of speech motor control necessarily involves reasonably complex interactions between a number of different systems and sensory modalities; only by formulating and simulating computational models can the performance of such complex systems be verified. Second, like other computational models, detailed comparisons can be made between model simulations and experimental data, thereby providing clear mechanistic accounts of the findings of speech experiments and making clear, testable predictions for future experiments. Third, the models provide a unified account of behavioral findings and neurobiological findings. Fourth, because these models are able to tune themselves by adjusting synaptic weights according to biologically based learning laws, they are able to address issues related to the development of speaking skills in infants and the continued tuning of the speech motor system throughout life. Finally, because model components can be "damaged" to simulate the effects of neurological damage, these models are capable of providing unique insights into the possible roots of communication disorders, a topic addressed in chapter 10.

1.4 Summary

The material in this book is situated at the intersection of three major disciplines: neuroscience, linguistics, and motor control. Our specific focus is on the cortical and subcortical mechanisms involved in moving the speech articulators to produce speech sounds, with little or no emphasis on the meaning of the utterance. Furthermore, we focus primarily on motor control at the kinematic level (i.e., the level describing the movements of the speech articulators without regard to muscle forces) rather than the dynamic level (the level describing the forces needed to carry out the articulator movements). Because the focus is on neural control, detailed treatments of speech and hearing anatomy and acoustic properties of speech will also be avoided (see Zemlin, 1998, and Stevens, 1998, for thorough treatments of these topics).

The various neural structures involved in speech represent information within a number of different reference frames. Of particular importance to the neural control of speech are (1) an auditory reference frame that characterizes sound representations in the auditory cortical areas, (2) an articulatory reference frame that represents the positions and velocities of the speech articulators, (3) a somatosensory reference frame that represents tactile and proprioceptive information from the vocal tract, (4) a phonemic reference frame that represents the discrete phonemes of a language, and (5) a speech sound reference frame that represents sound "chunks" (which can be phonemes, syllables, or words) that each have their own optimized speech motor program for production.

The studies described in this book involve a large number of experimental and modeling techniques for studying speech. Experimental studies include behavioral studies of speech acoustics and movement kinematics; lesion studies that investigate the effects of damage to different brain regions; structural imaging studies that investigate gross neuroanatomy in normal and disordered populations; electrophysiology studies that measure electrical activity in individual neurons or small groups of neurons; studies of large-scale electrical and magnetic fields using EEG, MEG, or ECoG; and neuroimaging techniques that measure cerebral blood flow associated with neural activity such as PET and fMRI. Each technique has its own strengths and shortcomings; as a result, a complete picture of the neural bases of speech will require the synthesis of data collected using many different experimental techniques. Computational models, in particular adaptive neural network models, will be used in this book to provide a unified account of this broad range of behavioral and neurological findings.

Notes

1. We will use slashes to indicate phonemes or phoneme strings that are depicted using *international phonetic alphabet* (*IPA*) symbols. For simplicity we will omit diacritic marks from these symbols.

2. Care must be taken when interpreting the results of these stimulation studies because of the unnaturally large voltages applied (on the order of volts, compared to the natural membrane potential range of just over 100 millivolts for neurons) and the large numbers of cells that are simultaneously affected by the electrical pulse. Nonetheless, disruptions of behavior or perception induced by such stimulations provide important insights into the gross-level functions of different locations in the cerebral cortex.

3. The patient never believes these sensations are caused by external stimuli, as with ordinary sensations, instead considering them to be artifactual.

4. Interference with speech as defined by Penfield and Roberts (1959) included the following phenomena: involuntary vocalizations, speech arrest, hesitation, slurring, sound distortions, repetition, inability to name objects, misnaming, perseveration, and confusion of numbers while counting.

5. The frequency ranges listed here are approximate; there is no definite consensus on the exact ranges for each rhythm.

6. The term *neural models* will be used herein to refer to ANNs that are intended to directly model the function of specific biological neural systems such as the speech motor control system, in contrast to ANNs used for technological applications rather than as a means of directly modeling brain function.

References

Alm, P. A. (2004). Stuttering and the basal ganglia circuits: a critical review of possible relations. *Journal of Communication Disorders*, *37*, 325–369.

Barlow, S. M. (1999). *Handbook of clinical speech physiology*. San Diego: Singular.

Belliveau, J., Kennedy, D., McKinstry, R., Buchbinder, B., Weisskoff, R., Cohen, M., et al. (1991). Functional mapping of the human visual cortex by magnetic resonance imaging. *Science*, *254*, 716–719.

Berryhill, M. E., Wencil, E. B., Coslett, H. B., & Olson, I. R. (2010). A selective working memory impairment after transcranial direct current stimulation in right parietal lobe. *Neuroscience Letters*, *479*, 312–316.

Bloom, L., & Lahey, M. (1978). *Language development and language disorders*. New York: Wiley.

Boggio, P. S., Ferrucci, R., Rigonatti, S. P., Covre, P., Nitsche, M., Pascual-Leone, A., et al. (2006). Effects of transcranial direct current stimulation on working memory in patients with Parkinson's disease. *Journal of the Neurological Sciences, 249*, 31–38.

Bouchard, K. E., Mesgarani, N., Johnson, J., & Chang, E. F. (2013). Functional organization of human sensorimotor cortex for speech articulation. *Nature, 495*, 327–332.

Broca, P. (1861). Perte de la parole: Ramollissement chronique et destructions partielle du lobe antérior gauche du cerveau. *Bulletin de la Société d'Anthropologie, 2*, 235–238.

Broca, P. (1865). Sur la faculté du langage articulé. *Bulletin de la Société d'Anthropologie, 6*, 337–393.

Browman, C. P., & Goldstein, L. (1990). Gestural specification using dynamically-defined articulatory structures. *Journal of Phonetics, 18*, 299–320.

Brown, E. N., Kass, R. E., & Mitra, P. (2004). Multiple neural spike train data analysis: state-of-the-art and future challenges. *Nature Neuroscience, 7*, 456–461.

Caplan, D. (2001). Functional neuroimaging studies of syntactic processing. *Journal of Psycholinguistic Research, 30*, 297–320.

Catani, M., Jones, D. K., & ffytche, D. H. (2005). Perisylvian language networks of the human brain. *Annals of Neurology, 57*, 8–16.

Chiba, T., & Kajiyama, M. (1941). *The vowel: its nature and structure*. Tokyo: Tokyo-Kaiseikan.

Chomsky, N., & Halle, M. (1968). *The sound pattern of English*. New York: Harper & Row.

Coker, C. H. (1976). A model of articulatory dynamics and control. *Proceedings of the IEEE, 64*, 452–460.

Cowey, A., & Walsh, V. (2001). Tickling the brain: studying visual sensation, perception and cognition by transcranial magnetic stimulation. *Progress in Brain Research, 134*, 411–425.

Creutzfeldt, O., Ojemann, G., & Lettich, E. (1989). Neuronal activity in the human lateral temporal lobe: I. Responses to speech. *Experimental Brain Research, 77*, 451–475.

Crone, N. E., Hao, L., Hart, J., Jr., Boatman, D., Lesser, R. P., Irizarry, R., et al. (2001). Electrocorticographic gamma activity during word production in spoken and sign language. *Neurology, 57*, 2045–2053.

Crosson, B. (1992). *Subcortical functions in language and memory*. New York: Guilford Press.

Cushing, H. (1909). A note upon the faradic stimulation of the postcentral gyrus in conscious patients. *Brain, 23*, 44–53.

Dandy, W. (1918). Ventriculography following the injection of air into the cerebral ventricles. *Annals of Surgery, Philadelphia, 68*, 5–11.

Darley, F. L., Aronson, A. E., & Brown, J. R. (1975). *Motor speech disorders*. Philadelphia: Saunders.

Dax, M. (1836). Lésions de la moitié gauche de l'encéphale soincidant avec l'oubli des signes de la pensée. *Congrès méridional tenu à Montpellier*.

Dell, G. S. (1986). A spreading activation theory of retrieval and sentence production. *Psychological Review, 93*, 283–321.

Depireux, D. A., Simon, J. Z., Klein, D. J., & Shamma, S. A. (2001). Spectro-temporal response field characterization with dynamic ripples in ferret primary auditory cortex. *Journal of Neurophysiology, 85*, 1220–1234.

Dronkers, N. F., Pinker, S., & Damasio, A. (2000). Language and the aphasias. In E. R. Kandel, J. H. Schwartz, & T. M. Jessell (Eds.), *Principles of neural science* (pp. 1169–1187). New York: McGraw-Hill.

Duffy, J. R. (1995). *Motor speech disorders: substrates, differential diagnosis, and management*. St. Louis: Mosby.

Duvernoy, H. M. (1999). *The human brain: surface, blood supply, and three-dimensional anatomy* (2nd ed.). New York: Springer-Verlag Wien.

Fadiga, L., Craighero, L., Buccino, G., & Rizzolatti, G. (2002). Speech listening specifically modulates the excitability of tongue muscles: a TMS study. *European Journal of Neuroscience, 15*, 399–402.

Fagg, A. H., & Arbib, M. A. (1998). Modeling parietal-premotor interactions in primate control of grasping. *Neural Networks, 11*, 1277–1303.

Fairbanks, G. (1954). Systematic research in experimental phonetics: I. A theory of the speech mechanism as a servosystem. *Journal of Speech and Hearing Disorders*, *19*, 133–139.

Fant, G. (1973). *Speech sounds and features*. Cambridge, MA: MIT Press.

Flanagan, J. L., Coker, C. H., Rabiner, L. R., Schafer, R. W., & Umeda, N. (1970). Synthetic voices for computers. *IEEE Spectrum*, *7*, 22–45.

Fowler, C. A. (1986). An event approach to the study of speech perception from a direct-realist perspective. *Journal of Phonetics*, *14*, 3–28.

Fritsch, G., & Hitzig, E. (1870). Über die elektrische Erregbarkeit des Grosshirns. *Archiv für anatomie, physiologie und wissenschaftliche medizin*, *37*, 300–332.

Gaese, B. H., & Ostwald, J. (2003). Complexity and temporal dynamics of frequency coding in the awake rat auditory cortex. *European Journal of Neuroscience*, *18*, 2638–2652.

Garagnani, M., & Pulvermüller, F. (2013). Neuronal correlates of decisions to speak and act: spontaneous emergence and dynamic topographies in a computational model of frontal and temporal areas. *Brain and Language*, *127*, 75–85.

Gentilucci, M., Bernardis, P., Crisi, G., & Dalla Volta, R. (2006). Repetitive transcranial magnetic stimulation of Broca's area affects verbal responses to gesture observation. *Journal of Cognitive Neuroscience*, *18*, 1059–1074.

Gerard, J. M., Ohayon, J., Luboz, V., Perrier, P., & Payan, Y. (2005). Non-linear elastic properties of the lingual and facial tissues assessed by indentation technique: application to the biomechanics of speech production. *Medical Engineering & Physics*, *27*, 884–892.

Geschwind, N. (1965). Disconnexion syndromes in animals and man: II. *Brain*, *88*, 585–644.

Ghosh, S. S., Tourville, J. A., & Guenther, F. H. (2008). A neuroimaging study of premotor lateralization and cerebellar involvement in the production of phonemes and syllables. *Journal of Speech, Language, and Hearing Research*, *51*, 1183–1202.

Giraud, A. L., Kleinschmidt, A., Poeppel, D., Lund, T. E., Frackowiak, R. S., & Laufs, H. (2007). Endogenous cortical rhythms determine cerebral specialization for speech perception and production. *Neuron*, *56*, 1127–1134.

Golestani, N., Molko, N., Dehaene, S., Lebihan, D., & Pallier, C. (2006). Brain structure predicts the learning of foreign speech sounds. *Cerebral Cortex*, *17*, 575–582.

Goodglass, H. (1993). *Understanding aphasia*. New York: Academic Press.

Grossberg, S. (1973). Contour enhancement, short term memory, and constancies in reverberating neural networks. *Studies in Applied Mathematics*, *LII*, 213–257.

Grossberg, S. (1976). Adaptive pattern classification and universal recoding: I. Parallel development and coding of neural feature detectors. *Biological Cybernetics*, *23*, 121–134.

Guenther, F. H. (1992). *Neural models of adaptive sensory-motor control for flexible reaching and speaking*. PhD dissertation, Boston University.

Guenther, F. H. (1994). A neural network model of speech acquisition and motor equivalent speech production. *Biological Cybernetics*, *72*, 43–53.

Guenther, F. H. (1995). Speech sound acquisition, coarticulation, and rate effects in a neural network model of speech production. *Psychological Review*, *102*, 594–621.

Guenther, F. H., Brumberg, J. S., Wright, E. J., Nieto-Castanon, A., Tourville, J. A., Panko, M., et al. (2009). A wireless brain-machine interface for real-time speech synthesis. *PLoS One*, *4*, e8218.

Guenther, F. H., Ghosh, S. S., & Tourville, J. A. (2006). Neural modeling and imaging of the cortical interactions underlying syllable production. *Brain and Language*, *96*, 280–301.

Guenther, F. H., Hampson, M., & Johnson, D. (1998). A theoretical investigation of reference frames for the planning of speech movements. *Psychological Review*, *105*, 611–633.

Henke, W. (1966). *Dynamic articulatory model of speech production using computer simulation*. PhD dissertation, Massachusetts Institute of Technology.

Henry, R. G., Berman, J. I., Nagarajan, S. S., Mukherjee, P., & Berger, M. S. (2004). Subcortical pathways serving cortical language sites: initial experience with diffusion tensor imaging fiber tracking combined with intraoperative language mapping. *NeuroImage*, *21*, 616–622.

Honda, K. (2002). Evolution of vowel production studies and observation techniques. *Acoustical Science and Technology, 23*, 189–194.

Horwitz, B., Friston, K. J., & Taylor, J. G. (2000). Neural modeling and functional brain imaging: an overview. *Neural Networks, 13*, 829–846.

Houde, J. F., Nagarajan, S. S., Sekihara, K., & Merzenich, M. M. (2002). Modulation of the auditory cortex during speech: an MEG study. *Journal of Cognitive Neuroscience, 14*, 1125–1138.

Husain, F. T., Tagamets, M. A., Fromm, S. J., Braun, A. R., & Horwitz, B. (2004). Relating neuronal dynamics for auditory object processing to neuroimaging activity: a computational modeling and an fMRI study. *NeuroImage, 21*, 1701–1720.

Ito, T., Kimura, T., & Gomi, H. (2005). The motor cortex is involved in reflexive compensatory adjustment of speech articulation. *Neuroreport, 16*, 1791–1794.

Jakobson, R., & Halle, M. (1956). *Fundamentals of language*. The Hague: Mouton.

Kaminski, H. J., & Leigh, J. R. (2002). *Neurobiology of eye movements: from molecules to behavior*. New York: New York Academy of Sciences.

Kent, R. D. (1972). Some considerations in the cineradiographic analysis of tongue movements during speech. *Phonetica, 26*, 293–306.

Kent, R. D. (1997). *The speech sciences*. San Diego: Singular.

Kiritani, S., Itoh, K., & Fujimura, O. (1975). Tongue-pellet tracking by a computer-controlled x-ray microbeam system. *Journal of the Acoustical Society of America, 57*, 1516–1520.

Kröger, B. J., Birkholz, P., Kannampuzha, J., & Neuschaefer-Rube, C. (2006). Modeling sensory-to-motor mappings using neural nets and a 3D articulatory synthesizer. *Proceedings of the 9th International Conference on Spoken Language Processing* (pp. 565–568).

Kuehn, D. P., Lemme, M. L., & Baumgartner, J. M. (1989). *Neural bases of speech, hearing, and language*. Austin: Pro-Ed.

Leuthardt, E. C., Gaona, C., Sharma, M., Szrama, N., Roland, J., Freudenbergy, Z., et al. (2011). Using the electrocorticographic speech network to control a brain-computer interface in humans. *Journal of Neural Engineering, 8*, 036004.

Leuthardt, E. C., Pei, X. M., Breshears, J., Gaona, C., Sharma, M., Freduenberg, Z., et al. (2012). Temporal evolution of gamma activity in human cortex during an overt and covert word repetition task. *Frontiers in Human Neuroscience, 6*, 99.

Levelt, W. (1989). *Speaking: from intention to articulation*. Cambridge, MA: MIT Press.

Levelt, W. J. M., Roelofs, A., & Meyer, A. S. (1999). A theory of lexical access in speech production. *Behavioral and Brain Sciences, 22*, 1–75.

Lichtheim, L. (1885). On aphasia. *Brain, 7*, 433–484.

Lieberman, P., & Blumstein, S. E. (1988). *Speech physiology, speech perception, and acoustic phonetics*. Cambridge, UK: Cambridge University Press.

Lin, F. H., Witzel, T., Ahlfors, S. P., Stufflebeam, S. M., Belliveau, J. W., & Hamalainen, M. S. (2006). Assessing and improving the spatial accuracy in MEG source localization by depth-weighted minimum-norm estimates. *NeuroImage, 31*, 160–171.

Logothetis, N. K., Pauls, J., Augath, M., Trinath, T., & Oeltermann, A. (2001). Neurophysiological investigation of the basis of the fMRI signal. *Nature, 412*, 150–157.

Luber, B., & Lisanby, S. H. (2014). Enhancement of human cognitive performance using transcranial magnetic stimulation (TMS). *NeuroImage, 85*, 961–970.

Luo, H., & Poeppel, D. (2007). Phase patterns of neuronal responses reliably discriminate speech in human auditory cortex. *Neuron, 54*, 1001–1010.

Maeda, S. (1990). Compensatory articulation during speech: evidence from the analysis and synthesis of vocal tract shapes using an articulatory model. In W. J. Hardcastle & A. Marchal (Eds.), *Speech production and speech modelling* (pp. 131–149). Boston: Kluwer Academic.

Martin, J. H. (1996). *Neuroanatomy: text and atlas* (2nd ed.). Stamford, CT: Appleton & Lange.

Massaro, D. W. (1997). *Perceiving talking faces: from speech perception to a behavioral principle.* Cambridge, MA: MIT Press.

Mattout, J., Phillips, C., Penny, W. D., Rugg, M. D., & Friston, K. J. (2006). MEG source localization under multiple constraints: an extended Bayesian framework. *NeuroImage, 30,* 753–767.

McCulloch, W. S., & Pitts, W. (1943). A logical calculus of the ideas immanent in nervous activity. *Bulletin of Mathematical Biophysics, 5,* 115–133.

Mermelstein, P. (1973). Articulatory model for the study of speech production. *Journal of the Acoustical Society of America, 53,* 1070–1082.

Meyer, E. A. (1911). *Untersuchungen über lautbildung: experimentalphonetische untersuchungen über die vokalbildung im deutschen, holländischen, englischen, schwedischen, norwegischen, französischen und italienischen.* Marburg, Germany: Elwert.

Miller, J. D. (1989). Auditory-perceptual interpretation of the vowel. *Journal of the Acoustical Society of America, 85,* 2114–2134.

Minifie, F. D., Hixon, T. J., Kelsey, C. A., & Woodhouse, J. (1970). Lateral pharyngeal wall movement during speech production. *Journal of Speech and Hearing Research, 13,* 584–594.

Moll, K. (1960). Cinefluorographic techniques in speech research. *Journal of Speech and Hearing Research, 3,* 227–241.

Nadeau, S. E. (1988). Impaired grammar with normal fluency and phonology: implications for Broca's aphasia. *Brain, 111,* 1111–1137.

Neilson, M. D., & Neilson, P. D. (1987). Speech motor control and stuttering: a computational model of adaptive sensory-motor processing. *Speech Communication, 6,* 325–333.

Ogawa, S., Lee, T., Kay, A., & Tank, D. (1990). Brain magnetic resonance imaging with contrast dependent on blood oxygenation. *Proceedings of the National Academy of Sciences of the United States of America, 87,* 9869–9872.

Ostry, D. J., Gribble, P. L., & Gracco, V. L. (1996). Coarticulation of jaw movements in speech production: is context sensitivity in speech kinematics centrally planned? *Journal of Neuroscience, 16,* 1570–1579.

Pascual-Leone, A., Rubio, B., Pallardó, F., & Catalá, M. D. (1996). Rapid-rate transcranial magnetic stimulation of left dorsolateral prefrontal cortex in drug-resistant depression. *Lancet, 348,* 233–237.

Payan, Y., & Perrier, P. (1997). Synthesis of V–V sequences with a 2D biomechanical tongue model controlled by the equilibrium point hypothesis. *Speech Communication, 22,* 185–205.

Peach, R. K., & Tonkovich, J. D. (2004). Phonemic characteristics of apraxia of speech resulting from subcortical hemorrhage. *Journal of Communication Disorders, 37,* 77–90.

Pedron, S., Monnin, J., Haffen, E., Sechter, D., & Van Waes, V. (2014). Repeated transcranial direct current stimulation prevents abnormal behaviors associated with abstinence from chronic nicotine consumption. *Neuropsychopharmacology, 39,* 981–988.

Pei, X., Leuthardt, E. C., Gaona, C. M., Brunner, P., Wolpaw, J. R., & Schalk, G. (2011). Spatiotemporal dynamics of electrocorticographic high gamma activity during overt and covert word repetition. *NeuroImage, 54,* 2960–2972.

Penfield, W., & Rasmussen, T. (1950). *The cerebral cortex of man: a clinical study of localization of function.* New York: MacMillan.

Penfield, W., & Roberts, L. (1959). *Speech and brain mechanisms.* Princeton, NJ: Princeton University Press.

Perkell, J. S. (1969). *Physiology of speech production: results and implications of a quantitative cineradiographic study.* Cambridge, MA: MIT Press.

Perkell, J. S., Cohen, M., Svirsky, M., Matthies, M., Garabieta, I., & Jackson, M. (1992). Electromagnetic midsagittal articulometer (EMMA) systems for transducing speech articulatory movements. *Journal of the Acoustical Society of America, 92,* 3078–3096.

Perrier, P., Ostry, D. J., & Laboissière, R. (1996). The equilibrium point hypothesis and its application to speech motor control. *Journal of Speech and Hearing Research, 39,* 365–377.

Petersen, S. E., Fox, P. T., Posner, M. I., Mintun, M., & Raichle, M. E. (1988). Positron emission tomographic studies of the cortical anatomy of single-word processing. *Nature, 331,* 585–589.

Postma, A., & Kolk, H. (1992). The effects of noise masking and required accuracy on speech errors, disfluencies, and self-repairs. *Journal of Speech and Hearing Research, 35*, 537–544.

Rodriguez-Rivera, A., Baryshnikov, B. V., Van Veen, B. D., & Wakai, R. T. (2006). MEG and EEG source localization in beamspace. *IEEE Transactions on Biomedical Engineering, 53*, 430–441.

Roelofs, A. (1997). The WEAVER model of word-form encoding in speech production. *Cognition, 64*, 249–284.

Rosenblatt, F. (1962). *Principles of neurodynamics: perceptrons and the theory of brain mechanisms*. Washington, DC: Spartan.

Roy, C., & Sherrington, S. (1890). On the regulation of the blood supply of the brain. *Journal of Physiology, 11*, 85–108.

Rubin, P., Baer, T., & Mermelstein, P. (1981). An articulatory synthesizer for perceptual research. *Journal of the Acoustical Society of America, 70*, 321–328.

Ruffini, G., Wendling, F., Merlet, I., Molaee-Ardekani, B., Mekonnen, A., Salvador, R., et al. (2013). Transcranial current brain stimulation (tCS): models and technologies. *IEEE Transactions on Neural Systems and Rehabilitation Engineering, 21*, 333–345.

Rumelhart, D. E., Hinton, G. E., & Williams, R. J. (1986). Learning representations by back-propagating errors. *Nature, 323*, 533–536.

Sahin, N. T., Pinker, S., Cash, S. S., Schomer, D., & Halgren, E. (2009). Sequential processing of lexical, grammatical, and phonological information within Broca's area. *Science, 326*, 445–449.

Salmelin, R., Schnitzler, A., Schmitz, F., & Freund, H.-J. (2000). Single word reading in developmental stutterers and fluent speakers. *Brain, 123*, 1184–1202.

Saltzman, E. L., & Munhall, K. G. (1989). A dynamical approach to gestural patterning in speech production. *Ecological Psychology, 1*, 333–382.

Savariaux, C., Perrier, P., Orliaguet, J. P., & Schwartz, J. L. (1999). Compensation strategies for the perturbation of French [u] using a lip tube: II. Perceptual analysis. *Journal of the Acoustical Society of America, 106*, 381–393.

Schaltenbrand, G. (1965). The effects of stereotactical electrical stimulation in the depth of the brain. *Brain, 88*, 835–840.

Schaltenbrand, G. (1975). The effects on speech and language of stereotactical stimulation in the thalamus and corpus callosum. *Brain and Language, 2*, 70–77.

Schönle, P., Grabe, K., Wenig, P., Schrader, J., & Conrad, B. (1987). Electromagnetic articulography: use of alternating magnetic fields for tracking movements of multiple points inside and outside the vocal tract. *Brain and Language, 31*, 26–35.

Shadmehr, R., & Wise, S. P. (2005). *The computational neurobiology of reaching: a foundation for motor learning*. Cambridge, MA: MIT Press.

Shattuck-Hufnagel, S. (1992). The role of word structure in segmental serial ordering. *Cognition, 42*, 213–259.

Shawker, T. H., & Sonies, B. C. (1984). Tongue movement during speech: a real-time ultrasound evaluation. *Journal of Clinical Ultrasound, 12*, 125–133.

Sommer, M., Koch, M. A., Paulus, W., Weiller, C., & Buchel, C. (2002). Disconnection of speech-relevant brain areas in persistent developmental stuttering. *Lancet, 360*, 380–383.

Sörös, P., Sokoloff, L. G., Bose, A., McIntosh, A. R., Graham, S. J., & Stuss, D. T. (2006). Clustered functional MRI of overt speech production. *NeuroImage, 32*, 376–387.

Stevens, K. N. (1998). *Acoustic phonetics*. Cambridge, MA: MIT Press.

Stevens, K. N., & Öhman, S. E. G. (1963). Cineradiographic studies of speech. *Quarterly Progress and Status Report, 4*(2), 9–11.

Syrdal, A. K., & Gopal, H. S. (1986). A perceptual model of vowel recognition based on the auditory representation of American English vowels. *Journal of the Acoustical Society of America, 79*, 1086–1100.

Tagametz, M. A., & Horwitz, B. (1999). Functional brain imaging and modeling of brain disorders. In J. A. Reggia, E. Ruppin, & D. L. Glanzman (Eds.), *Disorders of brain, behavior, and cognition: the neurocomputational perspective* (pp. 185–202). New York: Elsevier.

Tanji, K., Suzuki, K., Yamadori, A., Tabuchi, M., Endo, K., Fujii, T., et al. (2001). Pure anarthria with predominantly sequencing errors in phoneme articulation: a case report. *Cortex, 37*, 671–678.

Traunmüller, H. (1994). Conventional, biological, and environmental factors in speech communication: a modulation theory. *Phonetica, 51*, 170–183.

Turkeltaub, P. E., Eden, G. F., Jones, K. M., & Zeffiro, T. A. (2002). Meta-analysis of the functional neuroanatomy of single-word reading: method and validation. *NeuroImage, 16*, 765–780.

Van Buren, J. M. (1963). Confusion and disturbance of speech from stimulation in the vicinity of the head of the caudate nucleus. *Journal of Neurosurgery, 20*, 148–157.

Van Buren, J. M. (1966). Evidence regarding a more precise localization of the frontal-caudate arrest response in man. *Journal of Neurosurgery, 24*, 416–417.

van Rijn, S., Aleman, A., van Diessen, E., Berckmoes, C., Vingerhoets, G., & Kahn, R. S. (2005). What is said or how it is said makes a difference: role of the right fronto-parietal operculum in emotional prosody as revealed by repetitive TMS. *European Journal of Neuroscience, 21*, 3195–3200.

Wang, X., Merzenich, M. M., Beitel, R., & Schreiner, C. E. (1995). Representation of a species-specific vocalization in the primary auditory cortex of the common marmoset: temporal and spectral characteristics. *Journal of Neurophysiology, 74*, 2685–2706.

Warren, R. M., & Sherman, G. L. (1974). Phonemic restorations based on subsequent context. *Perception & Psychophysics, 16*, 150–156.

Watkins, K., & Paus, T. (2004). Modulation of motor excitability during speech perception: the role of Broca's area. *Journal of Cognitive Neuroscience, 16*, 978–987.

Wernicke, C. (1874). *Der aphasische symptomencomplex.* Breslau: Cohn and Weigert.

Widrow, B., & Hoff, M. E., Jr. (1960). Adaptive switching circuits. In *1960 IRE WESCON convention record, IV* (pp. 96–104). New York: Institute of Radio Engineers.

Wilhelms-Tricarico, R. (1995). Physiological modeling of speech production: methods for modeling soft-tissue articulators. *Journal of the Acoustical Society of America, 97*, 3085–3098.

Wilhelms-Tricarico, R. (1996). A biomechanical and physiologically-based vocal tract model and its control. *Journal of Phonetics, 24*, 23–38.

Wood, S. W. (1979). A radiographic analysis of constriction locations for vowels. *Journal of Phonetics, 7*, 25–43.

Zemlin, W. R. (1998). *Speech and hearing science: anatomy and physiology* (4th ed.). Needham Heights, MA: Allyn & Bacon.

2

Neural Structures Involved in Speech Production

As mentioned in the Introduction, producing even the simplest speech utterance involves a large number of cortical and subcortical regions of the brain. In this chapter we will provide an overview of these regions, with later chapters providing more detailed descriptions of their hypothesized functions within the neural control system for speech. This discussion will utilize standard terminology for relative anatomical locations; these terms and the corresponding anatomical directions are illustrated in figure 2.1.

2.1 The Primate Vocalization Circuit

Control of speech production builds on neural circuits used for other purposes by our evolutionary predecessors, most notably the production of learned voluntary vocalizations. The study of the neural mechanisms of speech thus naturally begins with a treatment of the neural systems underlying these vocalizations in our closest evolutionary relatives, nonhuman primates.

The nonhuman primate vocalization system has been more precisely characterized than the human system because of the availability of invasive techniques that are not suitable for use in humans, including single-unit electrophysiology, focal lesioning, and axonal tracers. Figure 2.2, adapted from Jürgens (2009), schematizes the brain regions and axonal tracts responsible for the production of learned vocalizations in primates, based largely on studies of the squirrel monkey. The remainder of this section follows the Jürgens model presented in this figure.

The *reticular formation* acts as the final convergence zone of projections from higher-level brain areas involved in vocalization and is partially responsible for coordinating movements of the various muscles involved in vocalization, including respiratory, laryngeal, and orofacial muscles. The coordinated commands are sent to the motoneuron pools in the brain stem that control these muscles. Activation of the reticular formation produces full coordinated vocalizations rather than isolated components of vocalization, supporting the assertion of a role for the reticular formation in coordination across muscle systems (Jürgens & Richter, 1986).

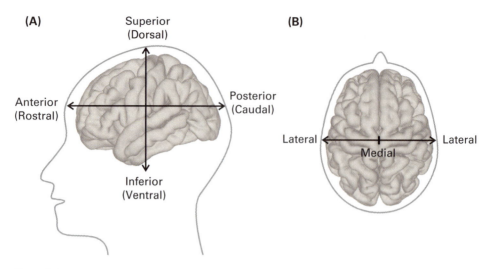

Figure 2.1
Anatomical terms of location. (A) Lateral view. (B) Superior view.

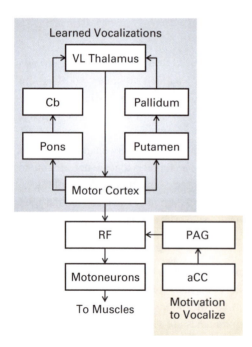

Figure 2.2
Primate vocalization model proposed by Jürgens (2009). aCC, anterior cingulate cortex; Cb, cerebellum; PAG, periaqueductal gray matter; RF, reticular formation; VL, ventral lateral nucleus.

The reticular formation receives projections from two distinct pathways from the cerebral cortex: a *limbic pathway* involving the *anterior cingulate cortex* (*aCC*) and *peri-aqueductal gray matter* (*PAG*) and a *motor cortical pathway* emanating from the primary motor cortex.

The limbic pathway is not heavily involved in motor coordination but instead serves a gating function that allows commands from the cerebral cortex to reach the motor periphery via the reticular formation. This pathway appears to be involved in controlling the intensity, but not the muscle patterning, of the vocalization. The gating signal from the limbic system is believed to represent the *motivation* or *readiness* to vocalize. In keeping with this characterization, neural activity in PAG correlates with global loudness but is not sensitive to specific acoustic patterns within the vocalization (Düsterhöft, Häusler, & Jürgens, 2004).

The motor cortical pathway is responsible for the production of learned vocalizations, including higher-level coordination of the muscles involved in vocalization (in concert with the lower-level coordination provided by the reticular formation). Signals in motor cortex are under the influence of two reentrant loops with subcortical structures: a loop through the *pontine nuclei* (or *pons*), *cerebellum*, and *thalamus* (*cortico-cerebellar loop*) and a loop through the *putamen*, *pallidum* (also called the *globus pallidus*), and *thalamus* (*cortico–basal ganglia loop*). The roles of these loops will be discussed within the context of the speech motor system in later sections.

The remainder of this chapter details the neural structures and pathways that make up the human speech motor system, starting from the motor periphery and moving upward in the nervous system to the cerebral cortex. These circuits include the primate vocalization circuit described by the Jürgens model as well as additional neural structures, particularly in the cerebral cortex.

2.2 Brain Stem Structures

Cranial Nerve Nuclei

The most peripheral neurons of the nervous system that are involved in speech are the *cranial nerve nuclei* located in the *medulla* and *pons*, which, along with the *midbrain*, constitute the *brain stem*. The twelve *cranial nerves* are typically identified by Roman numerals along with a name. Figure 2.3 schematizes the cranial nerves involved in speech, along with their associated cranial nuclei and the projections between these nuclei and the cerebral cortex. The cranial nuclei include sensory nuclei, illustrated on the left half of the brain stem in figure 2.3, and motor nuclei, illustrated on the right half of the brain stem.[1] Some nuclei and nerves have both motor and sensory components while others are primarily motor or primarily sensory. The cranial nerves most important for speech are nerves V and VII–XII; these nerves and their associated nuclei are discussed in the following paragraphs.

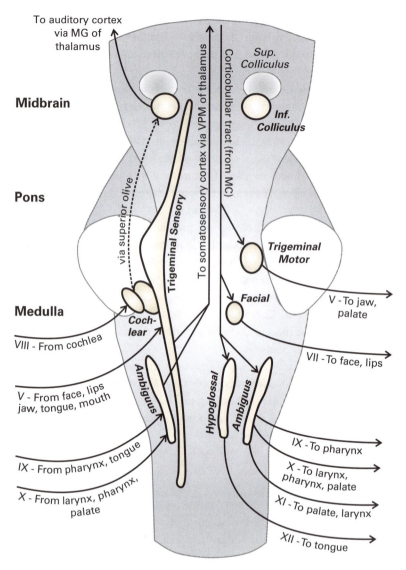

Figure 2.3
Dorsal view of the brain stem (medulla, pons, and midbrain) with cerebellum removed (white patches represent cerebellar peduncles) to illustrate the cranial nerves and associated nuclei involved in speech, including direct connections to cerebral cortex and to the periphery. Connections to the cranial nerve nuclei from the reticular formation and red nucleus are not shown. Sensory pathways are shown on the left; motor projections to muscles on the right. Nerve locations are approximate. Roman numerals indicate cranial nerve number. Inf., inferior; MC, motor cortex; MG, medial geniculate nucleus; Sup., superior; VPM, ventral posterior medial nucleus.

The *Vth nerve* (also called the *trigeminal nerve*) carries tactile and proprioceptive information from the oral cavity (except the posterior third of the tongue), nasal cavity, face, lips, jaw, and pharynx to the *trigeminal sensory nucleus* in the medulla. The motor portion of the trigeminal nerve originates in the *trigeminal motor nucleus* and innervates jaw and soft palate muscles. Somatosensory information from the trigeminal and other somatosensory cranial nerve nuclei projects to the primary somatosensory cortex via the *ventral posterior medial* (*VPM*) *nucleus* of the thalamus. Damage to the trigeminal nerve can cause numbness on the side of the lesion, difficulty chewing, and loss of muscle tone in the floor of the mouth (Zemlin, 1998). Unilateral lesions of the trigeminal nerve have minimal effects on speech, but bilateral damage is devastating for speech since the jaw hangs open (Duffy, 1995).

The *VIIth nerve*, or *facial nerve*, is a combined sensory and motor nerve, but only the motor aspect appears to be heavily involved in speech (Duffy, 1995). This nerve innervates muscles of the face and lips, and unilateral damage can result in paralysis of the facial muscles on the side of the damage, fasciculations (small, local quivering of muscle fibers), and/or atrophy of facial muscles resulting in facial asymmetry (Zemlin, 1998).

The *VIIIth nerve* (also called the *cochleovestibular nerve*, *cochlear nerve*, or *auditory nerve*) is a sensory nerve that carries auditory information from the cochlea to the *dorsal* and *ventral cochlear nuclei* in the medulla. From there, this information projects to the *superior olive* in both hemispheres and continues up to the *inferior colliculus* via a pathway called the *lateral lemniscus*. Auditory information then passes through the *medial geniculate* (*MG*) *nucleus* of the thalamus before arriving at the primary auditory cortex. Unilateral lesions of the VIIIth nerve typically result in partial to full deafness in the ipsilateral ear and/or *tinnitus* (or ringing in the ear), possibly with some facial pain or numbness (Zemlin, 1998). Bilateral lesions can result in total deafness.

The *nucleus ambiguus* in the medulla gives rise to several cranial nerves involved in speech. The *IXth nerve*, or *glossopharyngeal nerve*, innervates the stylopharyngeal muscle of the pharynx and receives somatosensory information from the pharynx and tongue. Damage to this nerve results in difficulty swallowing and loss of sensation and taste in the posterior third of the tongue (Zemlin, 1998). The IXth nerve is also thought to play a role in the gag reflex. Its role in speech motor control appears to be minor. The *Xth nerve*, or *vagus nerve*, innervates muscles in the larynx, pharynx, and soft palate and receives somatosensory information from these same areas. The vagus nerve plays a central role in speech production, and lesions to this nerve can result in severe voicing and swallowing abnormalities as well as weakness in the soft palate, pharynx, and larynx. The *XIth nerve*, or *accessory nerve*, is a motor nerve that innervates muscles of the soft palate and larynx and is intermingled with the vagus nerve. Lesions of the accessory nerve can result in difficulties with head and shoulder movements and cause a variety of voicing problems.

The *XIIth nerve*, or *hypoglossal nerve*, is primarily a motor nerve[2] (Zemlin, 1998) that originates in the *hypoglossal nucleus* and innervates almost all of the muscles of the tongue.

Damage to the hypoglossal nucleus can cause paralysis, weakness, or fasciculations of the tongue on the side of the lesion.

The cranial motor nuclei contain *lower motor neurons* which connect directly with muscles and thus control the contractile state of the muscle. These neurons are often distinguished from *upper motor neurons*, which are primarily located in the motor cortex[3] and can only affect the muscles via the lower motor neurons. In later chapters we will use the terms *motor neuron* or *motoneuron* to refer to the lower motor neurons, and we will refrain from using the somewhat ambiguous term *upper motor neuron* in favor of more precise terms. Furthermore, we will not distinguish between *alpha* and *gamma motor neuron* types, nor will we explore the considerable complexity of the interactions between interneurons, motoneurons, and sensory afferents at the level of the spinal cord and cranial nerve nuclei (see Brooks, 1986, and Duffy, 1995, for treatments of these topics).

Since there are multiple pathways by which activity in the lower motor neurons can be affected (as described below), these neurons are sometimes referred to as the *final common pathway* for neural signals traveling to the muscles. Because of this organization, damage to the lower motor neurons affects all types of movements (voluntary, reflex, or automatic) of the associated musculature.

Reticular Formation and Red Nucleus
As indicated in figure 2.4, descending projections from the cerebral cortex (especially motor cortex) reach the cranial nerve nuclei via direct projections as well as indirectly via subcortical nuclei in the midbrain and brain stem. The *direct pathway* is a phylogenetically newer pathway[4] that involves the *corticobulbar tract*, which, along with the *corticospinal tract*, forms the descending *pyramidal system*.[5] The corticospinal tract is primarily involved in controlling the limbs and trunk. Its only role in speech concerns respiratory function, which will not be covered in detail here; see Barlow (1999) for a treatment of the neural organization of the respiratory system.

The direct pathway plays an essential role in the control of speech. For the speech articulators, these projections are largely bilateral (i.e., each motor cortex cell projects to lower motor neurons on both sides of the body), with the exception of projections to the facial and hypoglossal nuclei, which are mostly to the contralateral side. As a result, unilateral damage to the corticobulbar tract in humans usually has relatively little effect on speech, with only some weakness of the tongue and lower face on the side contralateral to the damage, whereas bilateral damage can have a devastating effect on speech (Duffy, 1995). The pyramidal projections for most other motor systems are primarily to motor neurons on the contralateral side of the body, and thus unilateral damage to the pyramidal tract causes severe weakness or paralysis of the contralateral side in these systems.

The *indirect pathway* refers to projections from cortex that involve intermediate nuclei, specifically the *red nucleus*, located in the midbrain, and the *reticular formation*, which spans the midbrain and medulla (Darley, Aronson, & Brown, 1975; Duffy, 1995). These

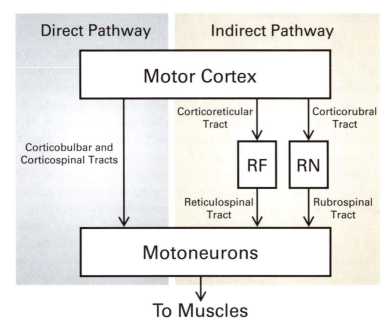

Figure 2.4
Direct and indirect pathways from the cerebral cortex to motoneurons in the cranial nerve nuclei and spinal cord. The direct pathway is formed by the corticobulbar and corticospinal tracts. The indirect pathway has two components: one passing through the reticular formation (RF) via the corticoreticular and reticulospinal tracts and the other passing through the red nucleus (RN) via the corticorubral and rubrospinal tracts.

nuclei receive cortical input via the *corticorubral tract* and *corticoreticular tract*, respectively, as well as input from the cerebellum and thalamus. They project to the motor periphery via the *reticulospinal tract* (which includes projections to the cranial nerve nuclei) and the *rubrospinal tract*, respectively. The indirect pathway's role in speech is poorly understood and appears to be more modulatory than control-oriented, with lesions to the indirect pathway generally affecting muscle tone and reflexes (Duffy, 1995) rather than the patterning of speech movements. However, Jürgens (2009) posits that the reticular formation is responsible for generation of innate vocal patterns, including nonverbal emotional vocal utterances of humans, and it may play a role in coordinating vocal fold vibrations during speech. In chapter 9 we speculate that the reticular formation may also play a role in affective prosody during speech.

It should be noted that the terms *direct* and *indirect pathway* have different meanings when discussed with regard to basal ganglia or cerebellar function, as we will see below. When not obvious from context, we will use care to specify which usage of the terms is intended throughout this book.

Periaqueductal Gray Matter

Less is understood about the PAG in humans than in monkeys. This midbrain structure, which surrounds the cerebral aqueduct, has been known to play a role in vocalization since Brown (1915) demonstrated that stimulation of PAG produced laughter in chimpanzees (for reviews, see Behbehani, 1995; Larson, 2004; Jürgens, 2009). In the Jürgens (2009) primate vocalization model (figure 2.2), PAG plays two key roles: (1) it is responsible for generating motor commands for innate vocalizations such as laughing and crying, and (2) it acts as a gating signal that modulates, rather than generates, the precisely timed motor commands of learned vocalizations which arrive from the motor cortex. This modulation may occur at the reticular formation, where the PAG and motor cortical signals come together. The relatively sparse human literature on PAG involvement in learned vocalizations appears to be compatible with the role it plays in the primate model of Jürgens; that is, as a modulator of descending commands from motor cortical areas rather than a generator of the detailed motor commands needed to coordinate the articulators during speech. However, PAG may coordinate some aspects of learned vocalizations, such as respiratory and laryngeal interactions (Larson, 2004), in concert with the commands generated from motor cortex. The view of PAG as a gating signal is compatible with the observation that bilateral damage to the PAG can lead to complete loss of voluntary vocalization, or *mutism*, in humans (Esposito et al., 1999) as well as animals (Behbehani, 1995; Jürgens, 2009; Larson, 2004) since bilateral damage to PAG would eliminate the gating signal that is needed to read out motor commands for learned vocalization. PAG neural activities also correlate with overall intensity of vocalization but not to acoustic details of the utterance. The notion of PAG as a gating signal for learned vocalizations is not completely settled, however, as studies in squirrel monkeys have shown that bilateral PAG lesions do not significantly affect signals from motor cortex to the laryngeal musculature (Jürgens, 2009), seemingly arguing against a role for PAG in gating of motor cortex commands for learned vocalizations.

2.3 Cerebellum, Basal Ganglia, and Thalamus

Although they are crucial for activating articulatory muscles, the subcortical regions discussed above are not thought to play a major role in the neural computations underlying the finely timed movements that constitute fluent speech. In this section we will treat subcortical structures that have a major influence on the descending motor commands via reentrant loops with the cerebral cortex: the thalamus, cerebellum, and basal ganglia. These structures are schematized in figure 2.5.

Thalamus

The thalamus is situated above the midbrain and below the basal ganglia, which in turn lie below cerebral cortex. In addition to playing roles in arousal and consciousness, the

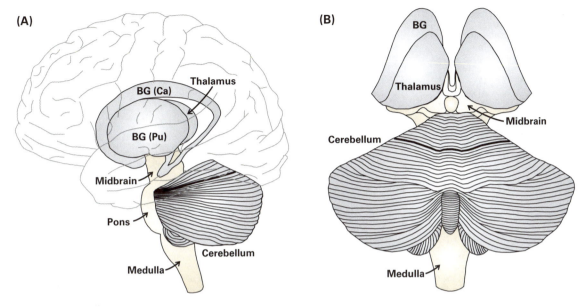

Figure 2.5
(A) Sagittal view showing the location of subcortical structures relative to the cerebral cortex. (B) Dorsal view of subcortical structures with cerebral cortex removed. Bold lines indicate location of the primary fissure of the cerebellum, which separates the anterior and posterior lobes of the cerebellar cortex. BG, basal ganglia; Ca, caudate nucleus; Pu, putamen.

thalamus acts as a massive relay station[6] for information entering or leaving the cerebral cortex. This includes sensory information from the periphery, motor information to the periphery, and outputs of the basal ganglia and cerebellum. The thalamus is divided into a number of distinct nuclei that are connected with different regions of the cerebral cortex. These cortico-thalamic connections are generally bidirectional; that is, each thalamic nucleus sends axons to a particular region of cortex, and that cortical region sends axons back to that same thalamic nucleus.

Electrical stimulation studies have implicated the thalamus, particularly the left *ventral lateral (VL) nucleus*, in speech motor control (see Johnson & Ojemann, 2000, for a review). Schaltenbrand (1975) reported that stimulation of VL in the language-dominant hemisphere can give rise to compulsory speech (monosyllabic yells and exclamations). Other effects of thalamic stimulation noted by Schaltenbrand (1975) include increasing or decreasing the loudness or rate of speech and *speech arrest*, or sudden loss of ability to speak. Mateer (1978) noted increased utterance duration with slurring and other articulatory distortions when the left VL was stimulated but not the right.

Four thalamic nuclei play important roles in speech motor control; these nuclei and their cortical connection zones are highlighted in figure 2.6. VPM projects to the ventral portion

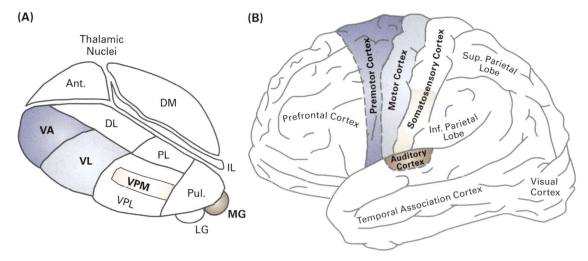

Figure 2.6
(A) Schematic of the thalamus with the nuclei involved in speech motor control highlighted. (B) Cortical projection targets of the thalamic nuclei involved in speech motor control, color coded according to the corresponding thalamic nucleus in panel A. Ant., anterior nucleus; DL, dorsal lateral nucleus; DM, dorsal medial nucleus; IL, intralaminar nucleus; inf., inferior; LG, lateral geniculate nucleus; MG, medial geniculate nucleus; PL, posterior lateral nucleus; Pul., pulvinar; sup., superior; VA, ventral anterior nucleus; VL, ventral lateral nucleus; VPL, ventral posterior lateral nucleus; VPM, ventral posterior medial nucleus.

of the somatosensory cortex, which contains the representation of the speech articulators. MG relays information from auditory brain stem structures to auditory cortex. VL is heavily connected with the primary motor cortex, consistent with the fact that VL is the thalamic nucleus that most frequently affects ongoing articulation when electrically stimulated. The *ventral anterior (VA) nucleus* is connected with premotor cortex. VL and VA form part of two reentrant loops with the cerebral cortex. The cortico–basal ganglia loop involves projections from the cortex to the basal ganglia, then to the thalamus, and then back to cerebral cortex. The cortico-cerebellar loop involves projections from cortex to the pons, then to the cerebellum, then to the thalamus, then back to cerebral cortex. Within VL and VA, neurons connected with the cerebellum are largely segregated from those connected with the basal ganglia (Sakai, Inase, & Tanji, 2002). These loops will be discussed briefly in the following subsections, and further detail of their involvement in speech production will be provided in later chapters.

Basal Ganglia

The basal ganglia lie beneath the cerebral cortex and are heavily interconnected with the frontal cortex via multiple cortico–basal ganglia loops (Alexander, DeLong, & Strick, 1986; Alexander & Crutcher, 1990; Middleton & Strick, 2000). The architecture of the basal ganglia make them suitable for selectively enabling one output from a set of competing alternatives (Mink & Thach, 1993; Mink, 1996; Kropotov & Etlinger, 1999), a property evident in several computational models of basal ganglia function (e.g., Redgrave, Prescott, & Gurney, 1999; Brown, Bullock, & Grossberg, 2004; Prescott et al., 2006). Damage or electrical stimulation to the basal ganglia can cause several disturbances in speech. For example, damage to the basal ganglia can result in inaccuracies in articulation (Pickett et al., 1998), and electrical stimulation of the basal ganglia in the language-dominant hemisphere can evoke word production and cause other speech disturbances, including speech arrest (Van Buren, 1963; Gil Robles et al., 2005).

The basal ganglia consist of four distinct nuclei—the *striatum*, *globus pallidus*, *substantia nigra*, and *subthalamic nucleus* (*ST*)—which in turn contain distinct subdivisions. The striatum is the primary input recipient for the basal ganglia. It is typically separated into three subdivisions: the *caudate nucleus* (or simply *caudate*), *putamen*, and *ventral striatum* (which includes the *nucleus accumbens*). The caudate and putamen are similar in terms of microstructure and function; they are considered separate entities anatomically because they are separated by the *internal capsule*, a major white matter projection between the cerebral cortex and brain stem. The striatum receives projections from a vast expanse of the cerebral cortex. Roughly speaking, prefrontal cortical areas project mostly to the caudate and sensorimotor cortical areas to the putamen. The ventral striatum is heavily interconnected with the *limbic system*, which plays important roles in emotion, motivation, and memory. Functions involving the ventral striatum include olfaction and reward processing; it does not appear to play a significant role in speech motor control.

The globus pallidus receives input from the striatum and subthalamic nucleus. There are two distinct subregions of the globus pallidus: an *internal segment* (*GPi*) and an *external segment* (*GPe*). GPi contains many of the output neurons of the basal ganglia, which project to the thalamus. In chapter 7 we will discuss the different functional roles played by GPi and GPe in the cortico–basal ganglia loop.

The substantia nigra contains two functionally distinct subregions: the *pars compacta* (*SNc*) and the *pars reticulata* (*SNr*). SNr plays a functionally similar role to GPi whereas SNc consists of *dopaminergic neurons* that supply the neurotransmitter dopamine to the striatum via the *nigrostriatal pathway*. Among other roles, these dopaminergic projections are important for the learning of motor behaviors, acting as a sort of "teaching signal" that strengthens rewarding actions. This topic will be addressed in more detail in chapter 7.

ST is connected to both GPi and GPe, and it is the source of the only excitatory projections of the basal ganglia, which impinge on the globus pallidus. ST also receives input from the cerebral cortex, particularly the motor and premotor cortices (Zemlin, 1998).

The cortico–basal ganglia loop involves a substantial "funneling" or fan-in of information from cortical sources to basal ganglia outputs. For the rat, Zheng and Wilson (2002) estimate the number of neurons in cortex that project to the striatum to be about 10 times the number of striatal neurons that receive these projections. Furthermore, there are approximately 100 times as many striatal neurons as basal ganglia output neurons in the GPi and SNr (Oorschot, 1996), indicating a ratio of basal ganglia afferents to efferents of roughly 1,000:1, with tens of millions of cortical neurons channeled to tens of thousands of basal ganglia output neurons. Despite this massive funneling, distinct information channels are maintained (Middleton & Strick, 2000). Brown, Bullock, and Grossberg (2004) hypothesize that there is a substantial fan-out of projections from the basal ganglia output channels back to cerebral cortex via the thalamus. Given the small number of basal ganglia output channels, it is highly unlikely that the basal ganglia are responsible for generating the precise motor commands needed for skilled movement. Instead the structure of the cortico–basal ganglia loops seems suited to choosing between alternative motor programs and sending *gating signals* to activate the cortical neurons responsible for carrying out the chosen motor program and inhibit cortical neurons related to competing motor programs (e.g., Mink, 1996; Brown, Bullock, & Grossberg, 2004).

The most important cortico–basal ganglia loop for speech motor control is the *motor circuit*. This circuit, illustrated in figure 2.7, includes several cortical areas that will be described in later sections, including primary motor cortex, premotor cortex, primary somatosensory cortex, and the supplementary motor area, as well as the putamen, globus pallidus, and VL nucleus of the thalamus (Alexander, DeLong, & Strick, 1986). Neuroimaging reveals that the components of the motor circuit are active for speech tasks as simple as production of a single syllable (Ghosh, Tourville, & Guenther, 2008). As discussed further with regard to the supplementary motor area below, this circuit is likely involved in the initiation of speech motor programs but less involved in choosing the precise muscle patterns that make up the motor programs. The motor circuit will be discussed in further detail in chapter 7.

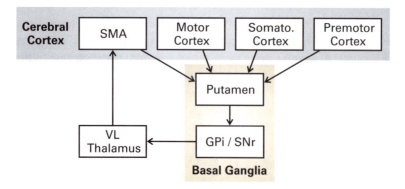

Figure 2.7
The cortico-basal ganglia motor circuit as originally proposed by Alexander, DeLong, and Strick (1986). GPi, internal segment of the globus pallidus; SMA, supplementary motor area; SNr, substantia nigra pars reticulata; Somato., somatosensory; VL, ventral lateral nucleus.

Two additional cortico–basal ganglia loops likely play roles in speech production. The *prefrontal circuit* may be involved in the buffering and sequencing of speech sounds in multisyllabic utterances (Bohland, Bullock, & Guenther, 2010). This topic is addressed in chapter 8. The *cingulate circuit* may be involved in the "will to speak" (discussed further in section 2.4), as well as affective prosody (discussed further in chapter 9).

A number of neuroimaging studies have noted activity in the basal ganglia during simple speech production tasks (e.g., Wildgruber, Ackermann, & Grodd, 2001; Bohland & Guenther, 2006; Riecker et al., 2006; Sörös et al., 2006) as well as during simple non-speech movements of the speech articulators. Appendix A describes a set of *activation likelihood estimate* (*ALE*) *meta-analyses* of 36 neuroimaging studies involving simple movements of individual speech articulators, including the jaw, larynx, lips, respiratory system, and tongue. Also included is an analysis of regions of *high articulatory convergence*, that is, regions where meta-analyses of three or more speech articulators identified activation foci.

The mean locations of high articulatory convergence in the basal ganglia and thalamus are illustrated in panels A–C of figure 2.8. Panel D of figure 2.8 compares the locations of high articulatory convergence to speech-related activity measured with fMRI in 92 subjects reading monosyllabic and bisyllabic utterances contrasted with a silent baseline (Guenther, Tourville, & Bohland, 2015). Significant speech-related activity is seen in the thalamus bilaterally, with the strongest response in VL ($y = -16$ in figure 2.8) very near the thalamic locations of high articulatory convergence (white circles). Speech is also accompanied by bilateral activity throughout much of the basal ganglia, with a distinct peak response in the globus pallidus which is marginally stronger in the left hemisphere. This peak is slightly posterior and inferior to the basal ganglia locations of high articulatory convergence (white circles at $y = -4$). Speech-related activity in the caudate is primarily in the anterior region, or *head of the caudate*, and an additional speech-related peak appears in the substantia nigra ($y = -22$).

Cerebellum

The cerebellum has long been known to play an important role in motor learning and fine motor control, including the control of articulation (Holmes, 1917). The cerebellum has a very regular structure, consisting of the *cerebellar cortex*[7] surrounding a set of *deep cerebellar nuclei*. Cerebellar afferents target both the cerebellar cortex and the deep cerebellar nuclei, whereas cerebellum efferents arise primarily from the deep cerebellar nuclei.

Figure 2.9 provides a schematic of the cortico-cerebellar loop through which the cerebellum exerts its effects on the cerebral cortex. Afferents to the cerebellum come from a wide expanse of the cerebral cortex via the *pons*, a region that looks like a bulge in the brain stem situated immediately anterior to the cerebellum. The pons receives input from ipsilateral cortical regions and projects to the contralateral cerebellar hemisphere, indicative of the crossed relationship between the cerebral cortex and the cerebellum wherein

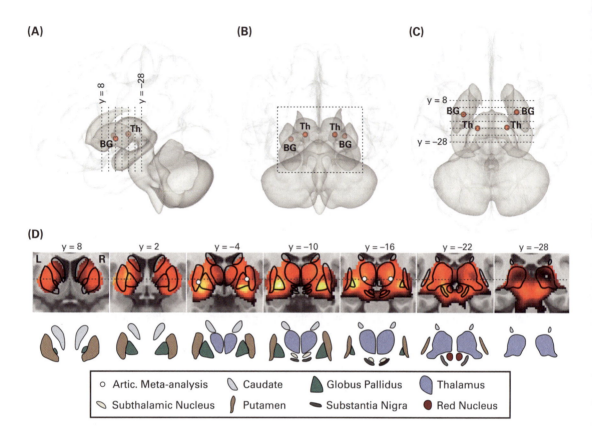

Figure 2.8
(A) Left, (B) posterior, and (C) superior views of basal ganglia and thalamus foci of high articulatory convergence (red spheres) identified from meta-analyses of simple nonspeech articulator movements (see appendix A) viewed through a transparent brain (subcortical structures shown in darker shade). Dashed lines indicate locations of slices in panel D. (D) Coronal slices of speech-related activity in the basal ganglia and thalamus derived from functional magnetic resonance imaging data collected from 92 subjects reading monosyllabic and bisyllabic utterances contrasted with a silent baseline (Guenther, Tourville, & Bohland, 2015). White circles in slices at $y = -4$ and $y = -16$ indicate locations of high articulatory convergence (red spheres in panels A–C). Slice locations are identified by y-coordinate in Montreal Neurological Institute stereotactic space; dashed horizontal line represents $z = 0$. Below each slice is a color-coded guide to the regions outlined in black on the slice. Artic., articulatory; BG, basal ganglia; L, left hemisphere; R, right hemisphere; Th, thalamus.

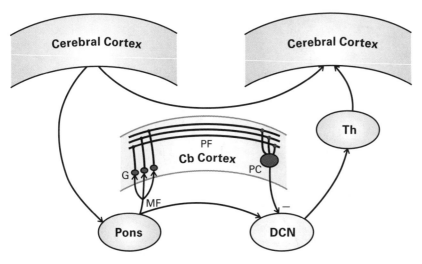

Figure 2.9
Schematic of the cortico-cerebellar loop. The loop consists of an excitatory direct pathway through the pons, deep cerebellar nuclei, and thalamus, as well as a tunable inhibitory side loop from the pons to the cerebellar cortex and back down to the deep cerebellar nuclei. Parallel fiber to Purkinje cell synapses, which are believed to play a central role in cerebellar learning, are indicated by red dots. The components of the cerebellum (beige) are in the hemisphere opposite the other components of the cortico-cerebellar loop (blue). Not shown for clarity are inhibitory interneurons, including Golgi cells that inhibit granule cells as well as stellate and basket cells that inhibit Purkinje cells. Also not shown are climbing fiber projections from the inferior olive to the deep cerebellar nuclei and Purkinje cells. Cb, cerebellum; DCN, deep cerebellar nuclei; G, granule cell; MF, mossy fibers; PC, Purkinje cell; PF, parallel fibers; Th, thalamus.

left cerebral cortex connects primarily to right cerebellum and vice versa. Brodal and Bjaalie (1992) estimate that, in the human brain, approximately 38 million cortical cells impinge on approximately 20 million pontine cells, a ratio of about 2:1. The number of cortical projections to the pons is about 20 times larger than the number of projections from cortex to the brain stem and spinal cord (Brodal & Bjaalie, 1992; Colin, Ris, & Godaux, 2002). As illustrated in figure 2.10, a large proportion of cortico-pontine projections arise from the motor and premotor cortex, with additional projections from somatosensory and auditory cortical areas in the postcentral and superior temporal gyri, respectively, and from parietal and prefrontal cortical areas (Schmahmann & Pandya, 1997).

Pontine neurons send *mossy fibers* to the cerebellar cortex through three massive bilateral fiber bundles: the *inferior, middle, and superior peduncles*. These fibers impinge on *granule cells* in the cerebellar cortex[8] as well as on deep cerebellar nuclei. Colin, Ris, and Godaux (2002) estimate that there are 10–100 billion granule cells in the cerebellum, a quantity that exceeds the number of neurons in the rest of the brain combined. There is thus a massive fan-out from the pons to the cerebellar cortex, with a ratio of one pontine cell for every 10,000–100,000 granule cells. This fan-out has been hypothesized to result in *sparse coding* of sensory, motor, and cognitive information by the granule cells, wherein each

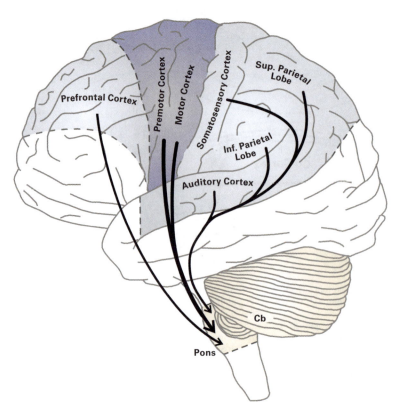

Figure 2.10
Schematic of regions of the cerebral cortex that provide major input (dark blue) or other substantial input (light blue) to the cerebellum via the pons. This schematic is based primarily on findings from the rhesus monkey (Schmahmann & Pandya, 1997). Cb, cerebellum; Inf., inferior; Sup., superior.

granule cell responds to a very specific set of cortical inputs, and only a small percentage of granule cells are active at any given time. A number of models of cerebellar function propose that this sparse code facilitates learning by enhancing pattern discrimination capacity and learning speed (e.g., Marr, 1969; Albus, 1971). Cerebellar *Golgi cells* (not shown in figure 2.9) that receive input from granule cells and mossy fibers may contribute to the sparseness of this code through inhibition of granule cells.

Each granule cell sends an axon to the surface of the cerebellar cortex, where it branches into two fibers that extend approximately 1.5 mm in opposite directions. The fibers from different granule cells run parallel to each other, giving rise to the name *parallel fibers*. The parallel fibers impinge on the approximately planar dendritic trees of *Purkinje cells*, which are large inhibitory neurons that send outputs from the cerebellar cortex to the deep cerebellar nuclei. The human cerebellum contains about 15 million Purkinje cells (Palay &

Chan-Palay, 1974). The Purkinje cell dendritic trees are oriented roughly orthogonally to the parallel fibers, allowing each parallel fiber to synapse on a large number of Purkinje cells and each Purkinje cell to receive synaptic inputs from a large number of parallel fibers. Purkinje cell dendrites also receive input from inhibitory interneurons called *stellate cells* and *basket cells* (not shown in figure 2.9).

In addition to parallel fiber input, Purkinje cells receive projections via *climbing fibers* that arise from cells in the *inferior olivary nucleus* (or simply *inferior olive*) in the medulla. In humans, the inferior olive contains about 1.5 million neurons, and it receives inputs ipsilaterally from all levels of the brain, including the spinal cord, brain stem, and cerebral cortex, with cortical inputs originating only from motor cortex (Colin, Ris, & Godaux, 2002). Each Purkinje cell receives strong input from a single climbing fiber originating in the contralateral inferior olive. These inputs are crucial for motor learning and have been hypothesized to signal movement errors, among other things.

The deep cerebellar nuclei, composed of the *dentate*, *interpositus*, and *fastigial nuclei*, act as the output channels of the cerebellum. They receive inhibitory projections from Purkinje cells in the cerebellar cortex as well as excitatory input from the contralateral pontine nuclei and inferior olive. The ratio of Purkinje cells to deep cerebellar cells has been estimated at 26:1 (Ito, 1984) whereas the ratio of total cerebellar afferents to efferents has been estimated at 40:1 (Colin, Ris, & Godaux, 2002). This ratio is much lower than the approximately 1,000:1 ratio of basal ganglia afferents to efferents. Furthermore, the number of deep cerebellar output neurons is on the order of millions (Andersen, Korbo, & Pakkenberg, 1992, estimate about 5 million neurons in the dentate nucleus alone), compared to tens of thousands of basal ganglia output channels. These numbers, combined with the sparse coding of sensorimotor context information provided by the massive number of granule cells in the cerebellar cortex, highlight why the cerebellum is far more suited to learning and generating precise motor commands than the basal ganglia.

Panel A of figure 2.11 illustrates the superior surface of the cerebellar cortex as well as the projections from the cerebellar cortex to the deep cerebellar nuclei. The portion of the cerebellar cortex near the midline is called the *vermis* and projects to the fastigial nucleus, which in turn sends projections bilaterally to the vestibular nuclei and reticular formation in the brain stem (Colin, Ris, & Godaux, 2002). The vermis and fastigial nucleus are heavily involved in control of eye movements, equilibrium, body stance/posture, and gait (Bastian & Thach, 2002); they are not known to play a major role in speech motor control, though vermal activity is often found in fMRI studies of speech production (see figure 2.12, which shows cerebellar activity during simple speech utterances), perhaps due to reading-related eye movements induced by the orthographic stimuli used in most speech fMRI studies.

The *paravermal cerebellar cortex* (sometimes called the *intermediate cerebellar cortex*) projects to both the interpositus nucleus and the dentate nucleus; these nuclei in turn project to the thalamus, and from there to the regions of the cerebral cortex indicated in panel B of

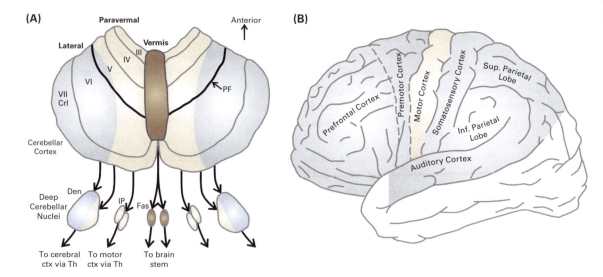

Figure 2.11
(A) Superior view of the cerebellar cortex illustrating the primary cerebellar output channels through the deep cerebellar nuclei. Roman numerals indicate the different lobules of the cerebellar cortex (Schmahmann et al., 2000). (B) Regions of the cerebral cortex receiving projections from the dentate nucleus (blue) and interpositus nucleus (beige) via the thalamus. Motor cortex receives input from both interpositus and dentate nuclei. This schematic is based primarily on monkey data (Dum & Strick, 2003). CrI, crus I; ctx, cerebral cortex; Den, dentate nucleus; Fas, fastigial nucleus; Inf., inferior; IP, interpositus nuclei; PF, primary fissure; Sup., superior; Th, thalamus.

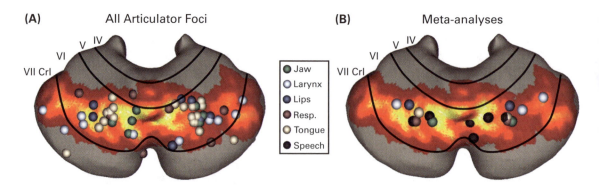

Figure 2.12
(A) Comparison of functional magnetic resonance imaging (fMRI) activity during speech collected from 116 subjects (hot colormap) with activity foci from 36 neuroimaging studies of nonspeech movements of individual speech articulators (spheres color coded by articulator) projected onto the superior surface of the cerebellar cortex. (See figure 1.8 caption for details regarding statistical analysis of speech fMRI data.) Roman numerals indicate lobules of the cerebellar cortex (Schmahmann et al., 2000), and black lines indicate approximate lobule boundaries. (B) Comparison of speech fMRI activity with foci resulting from meta-analyses of the same 36 studies (one meta-analysis per articulator). See appendix A for a list of the studies included and additional details regarding the meta-analyses. Black spheres indicate loci from the meta-analysis of single word reading by Turkeltaub et al. (2002) and the meta-analysis of word and sentence reading by Brown et al. (2005). CrI, crus I; Resp., respiratory system.

figure 2.11. The paravermal cerebellar cortex is strongly interconnected with motor cortex via the interpositus nucleus and is heavily involved in speech motor control. The *lateral cerebellar cortex* is connected to higher levels of the cerebral cortex via the dentate nucleus. The dentate nucleus, which also plays an important role in speech (Duffy, 1995), projects to a large portion of the cerebral cortex, including motor, premotor, auditory, and somatosensory areas as well as prefrontal and parietal association areas.[9] Across species, the size of the dentate nucleus scales with the size of the cerebral cortex; in humans this nucleus contains 90% of the deep cerebellar neurons (Colin, Ris, & Godaux, 2002). The lateral cerebellar cortex appears to be involved in higher-order cognitive processes, including language and working memory (e.g., Desmond & Fiez, 1998; Schmahmann & Pandya, 1997; see Schmahmann, 1997, for a detailed treatment), though it also appears to play a role in speech motor control, as suggested by activity in this area during many speech neuroimaging studies.

The classical view of cerebellar involvement in speech motor control is in regulating fine temporal organization of the motor commands necessary to produce smooth, coordinated productions of words and sentences, particularly during rapid speech (see Ackermann, 2008, for a review). This view receives strong support from cerebellar lesion studies. Speech deficits resulting from cerebellar stroke usually occur with damage to the superior cerebellar artery (Ackermann et al., 1992). This type of infarct often leads to *ataxic dysarthria*, a motor disorder that results in inaccurate articulation, prosodic excess, and phonatory-prosodic insufficiency (Darley, Aronson, & Brown, 1975). Cerebellar damage additionally results in increased duration of sentences, words, syllables, and phonemes (Kent et al., 1997; Ackermann & Hertrich, 1994). As discussed in chapter 10, ataxic dysarthria is most frequently associated with lesions to the superior paravermal cerebellar cortex (Ackermann et al., 1992; Urban et al., 2003; Schoch et al., 2006). Activity in this region is very commonly reported in neuroimaging studies of speech, particularly in cerebellar lobule VI (see figure 2.12). As discussed in chapter 7, Guenther, Ghosh, and Tourville (2006) hypothesize that adaptive timing mechanisms in these portions of the superior cerebellar cortex are involved in motor learning and feedforward control in speech production. The superior cerebellar cortex has also been implicated in predicting sensory consequences of movements (e.g., Blakemore, Frith, & Wolpert, 2001), a topic that will be discussed further in chapters 5 and 6.

A second cerebellar activation locus in the inferior cortex (in/near lobules VIIB and VIII) has been noted in some neuroimaging studies of speech (e.g., Bohland & Guenther, 2006; Tourville, Reilly, & Guenther, 2008; Golfinopoulos et al., 2011), and speech-related activity in this general location is visible in panel F of figure 1.8. Presurgical electrical stimulation of the cerebellum has also revealed distinct superior and inferior paravermal representations of the face (Mottolese et al., 2013). Bohland and Guenther (2006) found that this inferior region, unlike the superior region, did not show sensitivity to syllable complexity (e.g., "stra vs. ta"). However, it did show increased activity for utterances containing three different syllables compared to uttering the same syllable three times, suggesting involvement in speech sequencing at the suprasyllabic level without regard for the complexity of the individual syllable "chunks." Increased activity in cerebellar lobule VIII has also been noted in response to auditory (Tourville, Reilly, & Guenther, 2008) and somatosensory (Golfinopoulos et al., 2011) feedback perturbations during speech, suggesting a role in sensory feedback control (discussed further in chapters 5 and 6). Finally, this activity may be related to verbal working memory, which has been shown to be impaired in cases of cerebellar damage (Silveri et al., 1998; Chiricozzi et al., 2008). Neuroimaging studies by Desmond et al. (1997) and Chen and Desmond (2005) implicate right inferior cerebellar lobule VIIB in verbal working memory tasks.

Also shown in figure 2.12 are activation foci from the 36 neuroimaging studies included in the meta-analyses of simple nonspeech articulator movements reported in appendix A. Panel A shows all foci from all studies, and panel B shows foci derived from ALE meta-analyses[10] of these studies. Also shown in panel B are foci from the Turkeltaub et al. (2002) and Brown et al. (2005) meta-analyses of speech production.

Overall, the data in figure 2.12 support the following conclusions. First, speech-related activity in the superior cerebellar cortex spans from the vermis to the lateral edges of the hemispheres in lobule VI, with additional activity in lobule V in/near the vermis and in lateral portions of lobule VII crus I. Second, there is very high convergence of articulator representations in the paravermal portion of lobule VI, and speech-related foci from the Turkeltaub et al. (2002) and Brown et al. (2005) meta-analyses are also located here, attesting to the crucial role of this region in speech motor control. Third, the representations of individual articulators overlap substantially, though the larynx representation appears to be more lateral on average than the other articulators. Fourth, speech-related activity encompasses nearly all of the foci from the nonspeech articulator movement studies. Finally, speech appears to involve the vermis more heavily than nonspeech articulator movements.

Cerebellar involvement in speech is not limited to controlling ongoing movements; cerebellar activity also arises during *covert speech* (also called *inner speech* or *internal speech*; Riecker et al., 2000a; Callan et al., 2006). Findings such as these led Ackermann (2008) to posit a role for the cerebellum in sequencing of a prearticulatory verbal code. A role in inner speech is compatible with cerebellar involvement in verbal working memory, which is thought to involve subvocal articulatory rehearsal (e.g., Baddeley & Hitch, 1974).

Finally, although the cerebellum is undoubtedly involved in speech motor control, it is interesting to note that nearly normal speech is possible in the complete absence of the cerebellum if the condition is congenital; Richter et al. (2005) describe the case of a woman born without a cerebellum (*cerebellar agenesis*) who exhibited only minor motor impairments and nearly normal speech.

2.4 Cerebral Cortex

The *cerebral cortex*, or simply *cortex*, constitutes the highest processing stage in the brain, and its dramatically increased size in humans compared to other primates is one of the key evolutionary changes that made speech and language possible. There are three distinct types of cerebral cortex: (1) *allocortex*, the phylogenetically oldest part of cortex which contains three distinct cellular layers;[11] (2) *mesocortex* (also called *paleocortex*) containing four to five distinct layers; and (3) *neocortex* (also called *isocortex*) which contains six distinct layers and is the phylogenetically newest form of cortex. In humans approximately 90% of cortex is neocortex, and it is neocortex that plays the most important role in speech and language.

The cortex takes the form of a folded sheet, as illustrated in panel A of figure 2.13. The "bumps" in the sheet are called *gyri* (*gyrus* in singular form) and the grooves are called *fissures* or *sulci* (*sulcus* in singular form). The highest part of a gyrus is called the *crown*, and the deepest part of a sulcus is called the *fundus*. In general we will define a gyrus as extending from the fundus on one side of the gyrus to the fundus on the other side. At times, the location of a functional region is more naturally described as lying within a sulcus. The sides of the sulcus are referred to as *banks*, as illustrated in panel A of figure 2.13. Note that the banks of a sulcus are also part of the gyri on either side of the sulcus.

The *interhemispheric fissure* (also called the *longitudinal fissure*) splits the cortex into left and right hemispheres. Each hemisphere can be broken into four *lobes*, as illustrated in panel B of figure 2.13. Also shown are two major sulci that partially separate the cortical lobes: the *central sulcus* and the *Sylvian fissure* (also known as the *lateral sulcus*). Roughly speaking, the *occipital lobe* contains the cortical components of the visual system; the *temporal lobe* contains the cortical components of the auditory system and is crucial for speech perception; the *parietal lobe* contains somatosensory representations of the body and is crucial for spatial representation; and the *frontal lobe* contains motor representations and is crucial for high-level cognitive tasks such as decision-making and language production. This characterization is highly schematic as each lobe is involved in a variety of tasks. Speech motor control involves portions of the frontal, parietal, and temporal lobes, as described in the following subsections.

Much of the cortex is buried within sulci and therefore not visible in panel B of figure 2.13. This is particularly true within the Sylvian fissure, which separates the temporal lobe from the frontal and parietal lobes. Panel C of figure 2.13 illustrates this situation. Buried deep within the Sylvian fissure is an island of cortex called the *insula*.[12] It is not visible in panel B of figure 2.13 because it is completely covered by the *operculum*,[13] which forms part of the frontal and parietal lobes, and the *supratemporal plane* (also known as the *temporal operculum*), which is part of the temporal lobe. It will be advantageous to visualize the insula and other regions of cortex that lie within sulci when presenting neuroimaging results. We will therefore use an *inflated cortical surface*, as illustrated in panel D of

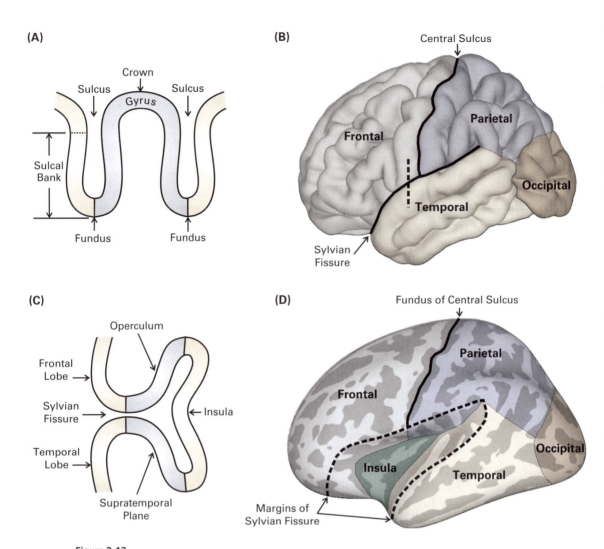

Figure 2.13
(A) Schematic of the folded cortical sheet. (B) Lateral view of the cerebral cortex indicating the four major lobes. Solid black lines indicate the central sulcus and Sylvian fissure; the dashed line indicates slice location for the schematic in panel C. (C) Schematic of a partial coronal slice (see the dashed line in panel B for location) illustrating cortical regions within the Sylvian fissure. (D) Lateral view of an inflated cortical surface showing the lobes of cortex as well as the insula, which is not visible in the uninflated view of panel B. Darker shaded areas represent sulci and lighter shaded areas represent gyri. The solid black line indicates the fundus of the central sulcus; dashed black lines represent the margins of the Sylvian fissure, which are now spread apart to reveal the intra-Sylvian region.

figure 2.13. In this inflated representation, darker shaded areas represent sulci and lighter shaded areas represent gyri. For example, the fundus of the central sulcus is indicated by a black line, and the darker shaded regions on either side of this line represent the portions of cortex located on the banks of the central sulcus. Furthermore, the Sylvian fissure is "spread open" to reveal the insula as well as the operculum and supratemporal plane. The upper margin of the Sylvian fissure (corresponding to the border between the frontal lobe and the operculum in panel C) is indicated by the upper dashed line in panel D, and the lower margin (corresponding to the border between the temporal lobe and the supratemporal plane in panel C) by the lower dashed line. We will use inflated cortical surfaces to illustrate the cerebral cortex throughout this book.

Given the large variety of functions performed even within a single cortical lobe, it is necessary to divide the cerebral cortex into smaller *regions of interest* (*ROIs*). In figures throughout this book, the cortical sheet will be parcellated into a number of anatomically defined ROIs based on the locations of landmarks visible on MRI scans, particularly sulci and gyri. Appendix B provides a detailed description of these ROIs and the anatomical landmarks that define them. The ROI boundaries are indicated by black outlines on inflated cortical surface figures, and abbreviated ROI names will be indicated by white text labels within relevant ROIs (see, e.g., figure 2.14). Although this ROI scheme will be used consistently in figures herein, it is important to note that this is only one of many possible ways to parcellate the cortical sheet, and furthermore these ROIs still involve somewhat large expanses of cortex. It will thus be necessary at times to use alternative region definitions to (1) describe results from studies that do not utilize our parcellation scheme and (2) describe subregions within these ROIs. Such regions will typically be indicated with yellow text labels on inflated cortical surfaces to distinguish them from the parcellation scheme described in appendix B.

One widely used alternative scheme for parcellating the cerebral cortex is based on regional differences in cellular makeup, relative thickness of cortical layers, and other microscopic details of the cortical tissue referred to collectively as *cytoarchitecture*. The most commonly used cytoarchitectonic parcellation scheme is that of German neurologist Korbinian Brodmann (Brodmann, 1909). Within this scheme, the cortex is broken into 52 regions with different cytoarchitecture, referred to as *Brodmann areas* (abbreviated as *BA*), and each region is assigned a unique Brodmann number (e.g., BA 4, which corresponds to primary motor cortex). For completeness, Brodmann areas will often be included in the descriptions that follow. Cytoarchitectonic parcellations might be expected to relate more directly to brain function than parcellations based on sulci and gyri since the cellular makeup of a region affects neural computations within the region, though this issue remains largely unresolved at present. Furthermore, it is not generally possible to identify cytoarchitectural details in human neuroimaging studies, so identification of Brodmann's areas in human studies is only approximate as their localization relative to visible features such as sulci and gyri varies somewhat from individual to individual.

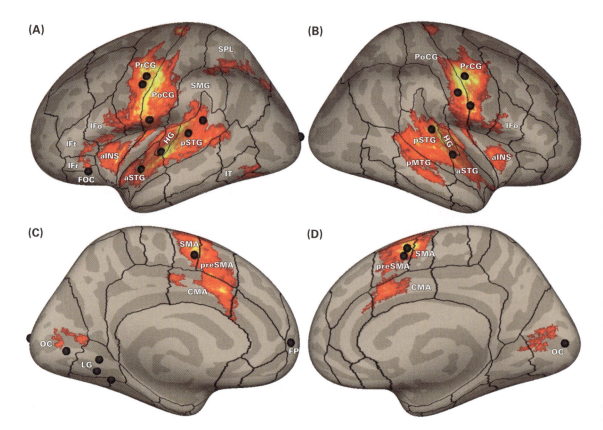

Figure 2.14
Cortical activity measured with fMRI in 116 subjects while reading monosyllabic and bisyllabic utterances (see the caption of figure 1.8 for details) plotted on inflated cortical surfaces. Black spheres represent the locations of foci from the Turkeltaub et al. (2002) and Brown et al. (2005) meta-analyses of speech neuroimaging studies. The top panels show (A) left and (B) right hemisphere views of the lateral cortical surface. The bottom panels show (C) left and (D) right hemisphere views of the medial cortical surface. Boundaries between cortical regions are indicated by black outlines; see appendix B for details regarding cortical parcellation. aINS, anterior insula; aSTG, anterior superior temporal gyrus; CMA, cingulate motor area; FOC, frontal orbital cortex; FP, frontal pole; HG, Heschl's gyrus; IFo, inferior frontal gyrus pars opercularis; IFr, inferior frontal gyrus pars orbitalis; IFt, inferior frontal gyrus pars triangularis; ITO, inferior temporo-occipital junction; LG, lingual gyrus; OC, occipital cortex; pMTG, posterior middle temporal gyrus; PoCG, postcentral gyrus; PrCG, precentral gyrus; preSMA, pre-supplementary motor area; pSTG, posterior superior temporal gyrus; SMA, supplementary motor area; SMG, supramarginal gyrus; SPL, superior parietal lobule.

Figure 2.14 illustrates cortical activity measured with fMRI in 116 subjects producing monosyllabic and bisyllabic utterances on an inflated cortical surface (see the caption of figure 1.8 for details regarding statistical analysis), along with activity foci identified by the Turkeltaub et al. (2002) and Brown et al. (2005) speech production meta-analyses (black spheres). In general, the meta-analysis foci accord well with the displayed fMRI activity although the ALE meta-analyses, which are relatively conservative, omit some areas with significant fMRI activity, most notably the *anterior insula* (*aINS*), *cingulate motor area* (*CMA*), and *inferior frontal gyrus pars opercularis* (*IFo*).

The remainder of this section will treat the ROIs that contain significant activity in figure 2.14, with a few exceptions. Activity and meta-analysis foci in the *occipital cortex* (*OC*) and adjoining *lingual gyrus* (*LG*) are highly likely to be related to viewing the letters that form the words or pseudowords that subjects read aloud. Because these regions are not believed to play a significant role in speech motor control, we will not discuss them in the remainder of this book. Activity in the left *superior parietal lobule* (*SPL*) and left *inferior temporal-occipital junction* (*ITO*) are also likely related to reading rather than speech motor control and will be treated only briefly herein (for reviews of reading-related brain activity, see Dehaene, 2009; Indefrey & Levelt, 2004; and Pugh et al., 2001). Finally, the left *frontal pole* (*FP*), which contains one focus from the speech meta-analyses, is not generally considered to be involved in speech motor control and does not show speech-related activity in the fMRI studies performed in our laboratory. For these reasons it will not be treated herein.

Appendix C provides detailed maps of structural and functional connectivity for the cortical regions that make up the speech network. The discussion that follows will reference these connectivity maps at relevant junctures.

Rolandic Cortex (Precentral and Postcentral Gyri)

Rolandic cortex consists of the *precentral gyrus* (*PrCG*) and *postcentral gyrus* (*PoCG*), which lie immediately anterior and posterior to the *central sulcus* (also called the *Rolandic fissure*), respectively. The central sulcus divides the frontal and parietal lobes. It also divides the motor cortical regions of the PrCG (including *primary motor cortex*, BA 4, in the caudal portion of the precentral gyrus, and *premotor cortex*, BA 6, in the rostral portion) from somatosensory regions in the PoCG (BA 1, 2, and 3, which collectively make up the *primary somatosensory cortex*). The portion of the operculum beneath the pre- and postcentral gyri is called the *Rolandic* or *central operculum* (BA 43); this region is included in the PrCG and PoCG ROIs in figure 2.14.

The primary somatosensory cortex receives projections from peripheral receptors (most notably proprioceptive information from muscle spindles and tactile information from mechanoreceptors) in the speech articulators on the contralateral side of the body via the cranial nerve nuclei and VPM nucleus of the thalamus (see section 2.2 for details). The primary motor cortex sends projections to motoneurons via the corticospinal and corticobulbar tracts as described in section 2.2. The primary motor and somatosensory cortices are both heavily interconnected with the striatum of the basal ganglia (especially the putamen) and with the cerebellum via the pons. Within the cortex, the primary somatosensory and motor cortices are heavily interconnected with each other as well as with the posterior parietal cortex, premotor cortex, and the supplementary motor area (Krubitzer & Kaas, 1990; see figures C.1 through C.11 in appendix C).

Rolandic cortex has long been known to play a role in movement control, beginning with Fritsch and Hitzig's (1870) seminal study demonstrating that electrical stimulation near the central sulcus of an anaesthetized dog caused movement in the contralateral limb. Bartholow (1874) provided the first demonstration of movement via brain stimulation in a human (see Harris & Almerigi, 2009, for a detailed scientific and historical treatment of Bartholow's controversial study). By the 1930s, Wilder Penfield and colleagues were using electrical stimulation prior to epilepsy surgery to locate the regions of the cerebral cortex involved in speech and other functions that can be seriously impacted by cortical excision (referred to as *eloquent cortex*) prior to removal of portions of the cortex to eliminate the epileptogenic focus. In the process, the first maps of the body's representation in the motor and somatosensory cortices were produced (Penfield & Rasmussen, 1950) as discussed in chapter 1 and illustrated in figure 1.6.

Functional neuroimaging studies of overt articulation show extremely reliable activation of the precentral and postcentral gyri bilaterally. In our own fMRI studies of speech production, the ventral portion of PrCG is typically the most strongly active region for any overt speaking task compared to a passive baseline task, a finding consistent with the Turkeltaub et al. (2002) meta-analysis of PET and fMRI studies involving overt single-word reading, where peaks along the precentral gyrus bilaterally had the highest statistical significance. Activity in the ventral PrCG has also been reported in neuroimaging studies that involve covert speech tasks rather than overt speech (e.g., Wildgruber et al., 1996; Riecker et al., 2000a), though less activity is found for covert than overt speech, and the activity is more left lateralized in covert speech.

Lesion/Excision Studies Excision of the precentral gyrus in one hemisphere results in paralysis of skilled and delicate movements of the opposite hand and foot, with minimal effect on somatic sensation (Penfield & Rasmussen, 1950). In contrast, voluntary movement of the proximal joints of the arm and leg are not lost, and only minor disruption of face and mouth movements occurs (Penfield & Roberts, 1959). The latter observation is most likely due to the largely bilateral nature of the direct and indirect pathways projecting from motor cortex to the cranial nerve nuclei.

Unilateral excision of the postcentral gyrus results in a loss of somatic sensation (particularly the sense of movement and position in space) in the contralateral extremities without lasting paralysis in monkeys (Kruger & Porter, 1958) and humans (Penfield &

Rasmussen, 1950). The reduction in somatic sensation of the human face area with unilateral excision of the postcentral gyrus is only partial and improves over time, leaving no noticeable permanent disabilities (Penfield & Rasmussen, 1950).

Although these lesion studies support a separation between motor function in the precentral gyrus and somatosensory function in the postcentral gyrus, the separation is not absolute. For example, stimulation of the postcentral gyrus can elicit muscle contractions similar to those elicited by stimulation of the precentral gyrus (Woolsey et al., 1953), and excitation of peripheral sensory afferents can evoke activity in the precentral gyrus as well as the postcentral gyrus (Malis, Pribram, & Kruger, 1953). Kruger and Porter (1958) found that when both the precentral and postcentral gyri are unilaterally ablated in rhesus monkeys, motor and somatosensory deficits on the contralateral side are more profound and permanent than when only one of the gyri is ablated, presumably because in the latter case the nonablated gyrus takes over some processing from the ablated one.

When the precentral gyrus is ablated bilaterally in a monkey, the motor deficits seen for unilateral ablation arise in both sides of the body (Kruger & Porter, 1958). Bilateral ablation does not result in permanent paralysis; many motor abilities are regained over time, though with increased clumsiness (Kruger & Porter, 1958). Analogously, bilateral postcentral gyrus lesions result in bilateral sensory discrimination deficits similar to those seen contralaterally for unilateral lesions, and the ability to perform sensory discrimination tasks improves somewhat with time (Kruger & Porter, 1958). Reports of bilateral damage or excision of precentral or postcentral gyrus in humans are exceedingly rare and generally involve tissue damage beyond the Rolandic cortex, limiting the interpretability of these rare cases.

Somatotopy As illustrated in figure 2.15, adapted from Penfield and Roberts (1959), the motor and somatosensory cortical representations of the speech articulators lie primarily in the ventral portion of the Rolandic cortex, ranging approximately from the Sylvian fissure up to the midpoint of the lateral surface of the cortex along the precentral gyrus. In the Penfield and Roberts map, the articulators are laid out in the following order, starting dorsally and moving downward toward the Sylvian fissure: vocalization (larynx), lips, jaw, tongue, and throat. More recent finer-grained microstimulation studies in monkeys have identified a "fractured" nature to the somatotopic representation of the body in Rolandic cortex, in which representations of individual muscles that are frequently used together are overlapped with each other along the surface of the primary motor cortex, and most stimulation sites result in the activation of multiple muscles rather than individual muscles (e.g., Humphrey & Tanji, 1991). Notably, Penfield's somatotopic maps identified voicing with a wide expanse of cortex that spanned the lip, jaw, and tongue representations, consistent with overlapping representations of different body parts.

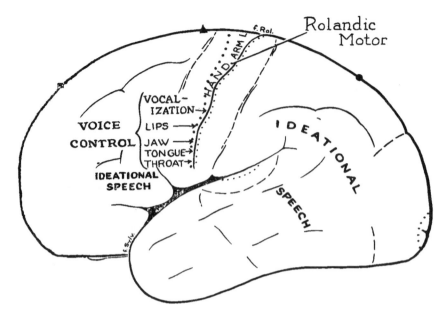

Figure 2.15
Somatotopy of the speech articulators in left hemisphere Rolandic cortex identified by Penfield and colleagues from electrical stimulation prior to epilepsy surgery. f. Rol., Rolandic fissure; f. Sylv., Sylvian fissure. From Penfield and Roberts, 1959, figure X-3. Reproduced by permission from Princeton University Press.

Neuroimaging studies have shed further light on the representation of the speech articulators in Rolandic cortex. Consistent with Penfield's characterization, the fMRI activity map during speech illustrated in figure 2.14 indicates activity in the ventral half of PrCG and PoCG. The somatotopy of the speech articulators within this expanse of cortex has been investigated in a number of fMRI, PET, and ECoG studies. Panels A–D of figure 2.16 illustrate all of the Rolandic cortex activation foci from the set of imaging studies of simple articulator movements that were included in the meta-analyses of individual articulator movements reported in appendix A. Panels E and F indicate the foci resulting from ALE meta-analyses of these studies (see also Takai, Brown, & Liotti, 2010), along with the Rolandic foci from the speech meta-analyses of Turkeltaub et al. (2002) and Brown et al. (2005).

The meta-analyses of appendix A (see also Takai, Brown, & Liotti, 2010) paint a more complex picture regarding somatotopy of the speech articulators. These meta-analyses find some support for a rough dorsal to ventral ordering of respiration,[14] larynx, lips, jaw,[15] tongue, and throat (the latter not included in figure 2.16; see Takai, Brown, & Liotti, 2010). However, they also identified multiple locations for each articulator along the Rolandic cortex, with a very large amount of overlap in the locations of different articulators except for a dorsal region primarily activated by respiration. This high degree of overlap likely

All Articulator Foci

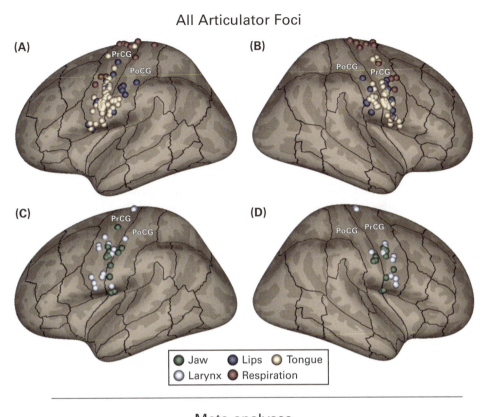

Meta-analyses

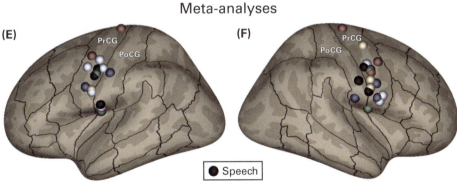

Figure 2.16
(A, B) Activation foci (spheres color coded by articulator) in (A) left- and (B) right-hemisphere Rolandic cortex from studies of simple nonspeech movements of the tongue, lips, and respiratory musculature. (C, D) Activation foci from studies of simple nonspeech movements of the jaw and larynx. (E, F) Results of ALE meta-analyses of the articulator foci shown in panels A–D (see appendix A for details), along with Rolandic cortex foci from the Turkeltaub et al. (2002) and Brown et al. (2005) speech production meta-analyses (black spheres). Black outlines represent cortical region boundaries (see appendix B for details). ALE, activation likelihood estimate; PoCG, postcentral gyrus; PrCG, precentral gyrus.

contributes to the precise interarticulator coordination present in fluent speech. Particularly high articulatory overlap is found bilaterally in the portion of Rolandic cortex immediately posterior to the inferior frontal sulcus. The mean locations of the foci within these three regions are very near the Rolandic cortex activation foci found in the Turkeltaub et al. (2002) and Brown et al. (2005) meta-analyses of speech production, indicating that the portions of Rolandic cortex used for speech also control simple nonspeech movements of multiple speech articulators. In other words, there may not be "speech specific" regions of Rolandic cortex that are involved in speech but not in other articulator movements, at least when viewed at the macroscopic level.

The notion of a rough somatotopy of the speech articulators with a large degree of overlap and multiple locations for the same articulator is also supported by ECoG data collected during evaluations performed prior to epilepsy surgery. Bouchard et al. (2013) used ECoG to identify articulatory representations in the ventral Rolandic cortex during speech production. Again, a rough dorsoventral ordering of larynx, lip, jaw, and tongue was found, along with a second larynx representation ventral to the tongue representation. Although some degree of consistency was found across the three subjects studied, there was also a large degree of variation in location of the articulators across subjects and substantial overlap of the representations of different articulators within subjects. Based on ECoG data from 70 patients, Farrell et al. (2007) also identified a large degree of individual variability in the locations of speech articulators between subjects, leading the authors to conclude that the notion of a regular, point-to-point somatotopy in Rolandic cortex should be replaced by a *functional mosaicism* view.

In summary, several lines of evidence support the following characteristics of the speech articulator representations in Rolandic cortex:

- A rough dorsoventral ordering of respiration, larynx, lip, jaw, and tongue representations.
- Multiple representations for each articulator.
- Substantial overlap of articulatory representations, especially in key speech production areas in the middle and ventral Rolandic cortex.
- Substantial variation in locations of articulatory representations across individuals.

Single Neuron Properties Microelectrode recordings from monkey motor cortex have demonstrated that the firing rates of most pyramidal tract neurons (the subset of motor cortical neurons that project to the motor periphery) are more closely related to movement force rather than displacement. However, a subset of pyramidal tract neurons is related more closely to movement displacement than force (Evarts, 1968). Cheney and Fetz (1980) found that motor cortical cells which synapse onto motoneurons in the spinal cord (or *corticomotoneurons*) respond similarly in an isometric task involving no hand displacement and a movement task that does displace the hand, providing further support for the view that these cells primarily code force (or torque) parameters of the motor act rather than displacement. They also noted that during a movement the neuronal activation pattern of all corticomotoneurons, as well as many other motor cortical cells, fell into one of four types (panels A–D of figure 2.17): (1) *tonic cells* (28% of their sample) whose activity remained at a near-constant level before and after the movement takes place, with a smooth transition between the before-movement and after-movement levels of activation during the movement; (2) *phasic-tonic cells* (59%) whose activity was near zero prior to movement, then showed a brief burst of activity just before and during the movement, then reducing to a near-constant nonzero level during the "hold period" after the movement (while hand position was being

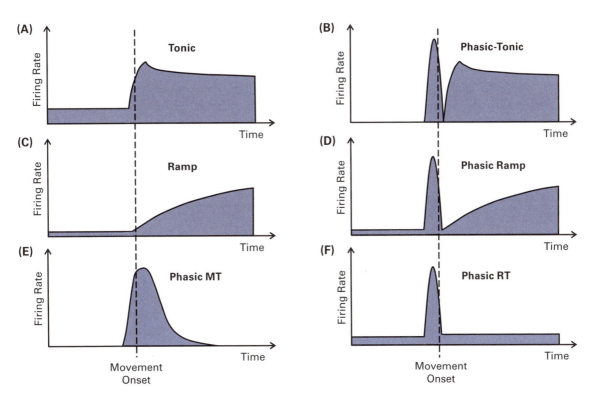

Figure 2.17
Schematic firing patterns for six cell types found in primary motor cortex. (A–D) Cell types reported by Cheney and Fetz (1980) during an isometric task in which monkeys had to apply and maintain a certain torque level on a manipulandum. (E, F) Additional cell types reported by Kalaska et al. (1989) in a task that required monkeys to move a manipulandum held in the hand (with or without an external load applied) to a target location in a 2-D space. Kalaska et al. noted that cell types with a tonic component were more load sensitive than those with only a phasic component. RT, reaction time; MT, movement time.

maintained); (3) *ramp cells* (5%) whose activity was near zero prior to movement and increased steadily throughout the movement and continuing into the hold time; and (4) *phasic-ramp cells* (8%) which showed a phasic burst of activity starting just prior to movement onset, followed by a steady increase in activity during the remainder of the movement and hold period. Cheney and Fetz (1980) noted that the tonic aspects of the activation patterns are likely involved in static aspects of motor control (e.g., maintaining the muscle lengths that determine arm positioning) whereas the phasic aspects are related to the dynamics of the movement, such as muscle shortening velocities. In a later study of motor cortex not limited to corticomotoneurons (Kalaska et al., 1989), additional populations of *phasic cells* were identified for both the movement time and reaction time portions of a cued arm movement (*phasic MT* and *phasic RT* in panels E and F of figure 2.17). The authors note that phasic cells were less sensitive to load than cells with a tonic component to their activity.

In the early 1980s, Georgopoulos and colleagues studied the relationship between motor cortical cells and the direction of a multijoint arm movement (e.g., Georgopoulos et al., 1982, 1984). They noted that most cells had a *preferred direction* of movement, with the firing rate falling off for movements further away from this preferred direction. This firing pattern could be captured by a *cosine-shaped tuning curve*; that is, the cell's activity was roughly proportional to the cosine of the angle between the actual movement direction and the preferred direction. The preferred directions of these cells appear to be roughly uniformly distributed in 3-D space for arm movements in free space (Schwartz, Kettner, & Georgopoulos, 1988). Georgopoulos et al. (1984) noted that the actual movement direction of the arm could be approximated by a *population vector* formed by summing up the preferred direction of task-related motor cortical cells scaled by their activity level during the movement compared to their baseline activity.

The Georgopoulos et al. studies have often been interpreted as indicating that motor cortical cells code spatial aspects of movements (such as 3-D position and velocity of the hand) rather than simply coding the muscle forces required to carry out the movements. Subsequent studies have determined that both spatial and muscle parameters are represented in motor cortex. Alexander and Crutcher (1990) dissociated movement direction from muscle force direction by having monkeys perform a cursor movement task, in which the monkey used a computer mouse to move a cursor from a central position on a video display to a target location at the right or left of the display, as well as a second task in which the cursor movement was reversed relative to the movement of the mouse (e.g., to move the cursor to a target on the right, the mouse had to be moved to the left). They found motor cortical cells whose activity related to limb direction regardless of cursor movement direction and other cells whose activity related to cursor direction regardless of limb direction. Cells coding limb direction tended to be located more caudally in motor cortex (near the central sulcus, which contains many corticomotoneurons) while cells located rostrally tended to encode cursor direction. Other studies have suggested a continuum of neurons between the spatial and motor coding extremes. For example, by dissociating load direction from movement direction, Kalaska et al. (1989) found a wide range of cell types, including cells that encoded movement direction regardless of load direction, cells that had both movement- and load-dependent tuning curves, and cells that were highly sensitive to load direction regardless of movement direction.

Together, these studies support the notion of a caudorostral gradient spanning motor and premotor cortex in which neurons in caudal primary motor cortex tend to encode relatively low-level, motoric aspects of movements such as muscle forces, lengths, and shortening velocities whereas neurons located more rostrally in premotor cortex tend to represent higher-level aspects of the movement, such as arm spatial kinematics or direction of movement of an external object/cursor in 2-D or 3-D space independent of limb movement (Alexander & Crutcher, 1990; Kalaska & Crammond, 1992). According to the DIVA model (described in chapter 3), a similar caudorostral gradient exists for speech production wherein caudal portions of the motor cortex encode articulatory musculature parameters and rostral portions of the premotor cortex represent acoustic or auditory properties of speech sounds.

Although microelectrode recordings from human motor cortex are exceedingly rare, a study by Guenther et al. (2009) verified the existence of an auditory representation related to formant frequencies of an intended speech signal in the precentral gyrus of an individual suffering from locked-in syndrome who was implanted with electrodes in a speech-related region of motor cortex in an attempt to restore speech capabilities via a brain-computer interface. Panel A of figure 2.18 indicates the directional preference (or *tuning curve*) in F1/F2 space of one neural unit from this study. This unit fires maximally for movements that increase F1 and decrease F2 (corresponding to the circled arrow on the *x*-axis). Panel B indicates the distribution of preferred directions for all neural units with statistically significant preferred directions; this distribution spans the entire F1/F2 direction space, analogous to the approximately uniform distribution of preferred spatial directions of arm-related motor cortical units.

Proprioception-related cells in the primary somatosensory cortex have many similar properties to motor cortical cells. Prud'homme and Kalaska (1994) noted a continuum from purely phasic cells to purely tonic cells, with most cells showing a combination of phasic and tonic properties (cf. phasic-tonic cells in figure 2.17). As with motor cortical cells, most cells had a roughly cosine-shaped tuning curve centered at a preferred movement direction, with similar properties for load direction. Overall, somatosensory cortical cells were less load-sensitive than motor cortical cells but more load-sensitive than cells in the parietal cortex.

Supplementary Motor Areas
The supplementary motor areas lie in a portion of the premotor cortex (BA 6) located on the medial surface of the cortex rostral to the precentral sulcus. They consist of at least two

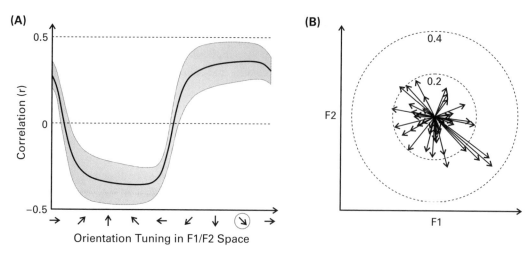

Figure 2.18
Formant frequency representation in speech motor cortex. (A) Tuning curve (black line) of a speech motor cortex neuron recorded during attempted production of a vowel sequence by an individual suffering from locked-in syndrome. Shaded region represents 95% confidence interval. This unit fires preferentially for speech movements that involve increasing F1 while decreasing F2, indicated by the circled arrow on the *x*-axis. (B) Distribution of preferred directions in F1/F2 space for all units showing a directional preference. Dashed circles indicate r values of correlations between firing rates and preferred directions. Adapted from Guenther et al. (2009).

subregions that can be distinguished on the basis of cytoarchitecture, connectivity, and function: the *supplementary motor area* (*SMA*) proper,[16] which is located caudal to the vertical line passing through the anterior commissure, and the *pre-supplementary motor area* (*preSMA*), which lies rostral to this line (Picard & Strick, 1996). As indicated in figure 2.14, both SMA and preSMA are reliably active during simple speech utterances. Activity is also reliably found in the supplementary motor areas for simple movements of individual speech articulators (see the meta-analyses in appendix A). Microstimulation of the supplementary motor areas in humans can yield vocalization, repetitions of words or syllables, speech arrest, slowing of speech, or hesitancy (Penfield & Welch, 1951; Fried et al., 1991).

A number of speech deficits in patients with lesions of the supplementary motor areas have been described in the literature (Jonas, 1981, 1987; Mochizuki & Saito, 1990; Ziegler, Kilian, & Deger, 1997; Pai, 1999; Ulu et al., 2008). If the lesion/excision is unilateral, near-complete recovery of speech usually occurs within months (Penfield & Roberts, 1959; Laplane et al., 1977). Although case studies of speech disturbances after right-hemisphere damage have been reported (e.g., Gelmers, 1983; Mendez, 2004), a review of 57 cases of speech disturbances accompanying lesions to the supplementary motor areas by Jonas (1981) indicated that the percentage of these cases associated with right-hemisphere damage is approximately the same as the incidence of right-hemisphere language dominance in the general population, leading the author to conclude that the language-nondominant (usually right) supplementary motor areas do not play a major role in speech (see also Krainik et al., 2003). However, neuroimaging studies reliably indicate bilateral activity in the supplementary motor areas, suggesting that the right supplementary areas are normally involved in speech production, though they may not be essential for it.

Speech deficits due to lesions of the supplementary motor areas involve a transient period of total mutism, after which patients may suffer from a decline in propositional (self-initiated) speech with nonpropositional speech (automatic speech; e.g., counting, repeating words) nearly intact. Such a deficit is often termed *transcortical motor aphasia* (Freedman, Alexander, & Naeser, 1984). Other problems include involuntary vocalizations, repetitions, paraphasias, echolalia, lack of prosodic variation, stuttering-like behavior, and variable speech rate, with only rare occurrences of distorted articulations. These outcomes are suggestive of roles in sequencing, initiating, suppressing, and timing of speech output, but likely not in providing detailed motor commands to the articulators. Based largely on the lesion literature, Jonas (1987) and Ziegler, Kilian, & Deger (1997) arrived at similar conclusions regarding the role of the supplementary motor areas in speech production, suggesting that they aid in sequencing and initiating speech sounds, but probably not in determining their phonemic content.

Primate neurophysiological studies have suggested that SMA and preSMA are differentially involved in the sequencing and initiation of movements, with preSMA acting at a more abstract level and SMA at a more motoric level (Matsuzaka, Aizawa, & Tanji, 1992; Shima et al., 1996; Shima & Tanji, 1998a, 2000; Tanji, 2001; Tanji & Shima, 1994). SMA and preSMA also have distinct patterns of connectivity with cortical and subcortical areas in monkeys (Jürgens, 1984; Luppino et al., 1993; see figures C.12 and C.13 in appendix C), a finding supported in humans using diffusion-tensor imaging (Johansen-Berg et al., 2004). Specifically, preSMA is heavily connected with the prefrontal cortex and the caudate nucleus, whereas SMA is more heavily connected with the motor cortex and the putamen, again suggesting a functional distinction with preSMA involved in higher-level motor planning and SMA with motor execution.

The roles of SMA and preSMA in speech production, particularly in feedforward control and speech sound sequencing, are discussed further in chapters 7 and 8.

Cingulate Motor Area

The CMA lies in the cingulate cortex in the medial frontal lobe, immediately inferior to motor cortex and the supplementary motor areas. This region is generally considered part of the aCC, which in turn is part of the *paralimbic system* that integrates neocortical function with the emotional/motivational functions of the *limbic system*. In monkeys, the CMA is typically divided into three separate cytoarchitectonic fields which differ in connectivity and function (Picard & Strick, 1996): a *rostral* field (*CMAr*) in the dorsal bank of the *cingulate sulcus* beneath the preSMA, and two caudal fields below SMA and motor cortex, one in the *dorsal* bank of the cingulate sulcus (*CMAd*) and one in the *ventral* bank (*CMAv*). Electrical stimulation of CMAd and CMAv causes motor responses whereas stimulation of CMAr generally does not (Luppino et al., 1993). Using electrophysiological recordings along with reversible chemical inactivation of CMAr during a task that involved choosing between motor actions with different reward levels, Shima and Tanji (1998b) determined that this field is involved in processing reward information for motor selection, operating at a higher level in the motor hierarchy than CMAd and CMAv. In humans there is no widely accepted division of subregions within CMA, though cytoarchitecture suggests a dorsal region located primarily within the cingulate sulcus (BA 32) and a ventral region extending downward from the cingulate sulcus onto the convexity of the cingulate gyrus (BA 24). The speech-related fMRI activity displayed in figure 2.14 primarily involves the dorsal portion of CMA bilaterally, with a larger expanse of activity in the left hemisphere. This location coincides quite closely with speech areas of CMA identified by Shima and Tanji (1998b) and Paus et al. (1993) and appears to more heavily involve BA 32 than BA 24. Both the dorsal and ventral regions of the CMA are heavily interconnected with the supplementary motor areas structurally and functionally, and they both display strong functional connectivity with the insula bilaterally (see figures C.14 and C.15 in appendix C).

Reports of speech deficits after unilateral damage to aCC are uncommon. Although very rare, instances of bilateral damage to aCC can result in *akinetic mutism*, characterized by a near-complete lack of speech and general lack of movement despite normal arousal levels and intact sensory, motor, and cognitive abilities (Rosenbek, 2004). Any speech that is generated is generally grammatically correct and properly articulated (Rubens, 1975; Jürgens, 2009). Minagar and David (1999) reviewed the neurological literature from 1930 to 1999 and noted eight cases of bilateral infarcts of the anterior cerebral artery, which supplies aCC and the supplementary motor areas (as well as other structures), seven of which resulted in mutism. Nicolai et al. (2001) reports an eighth such case due to bilateral anterior cerebral artery infarction. Infarcts of the anterior cerebral artery do not necessarily imply damage to the aCC; however, a handful of additional studies have noted akinetic mutism after bilateral anterior cingulate damage. Németh, Hegedus, and Molnár (1988) report three such cases, and two cases of akinetic mutism with lesions almost entirely limited to the anterior cingulate gyrus have been described by Nielsen and Jacobs (1951) and Barris and Schuman (1953), indicating that akinetic mutism can occur in the absence of damage to SMA or the corpus callosum. Based largely on the lesion literature, Paus (2001) suggests that aCC is involved in the *willed* control of actions. A similar account was provided by Jürgens (2009), who posits that aCC is involved in controlling the "readiness to express oneself vocally" (p. 9).

In a case of bilateral aCC damage reported by von Cramon and Jürgens (1983), the patient was not permanently mute (perhaps because of incomplete damage of the aCC) but showed a permanent lack of emotional expression in his speech, prompting the authors to conclude that aCC plays a key role in the vocal expression of emotions (also known as *affect*). This view receives further support from the observation that electrical stimulation of aCC in monkeys and other nonhuman mammals can elicit species-specific emotive vocalizations, likely via projections to the PAG as schematized in figure 2.2 (Devinsky, Morrell, & Vogt, 1995; Jürgens, 2009). The role of the aCC in affective aspects of speech is addressed in more detail in chapter 9.

Supramarginal Gyrus

The *supramarginal gyrus* (*SMG*), BA 40, lies immediately posterior to the ventral postcentral gyrus in the inferior parietal lobule, which also includes the *angular gyrus* (BA 39) located posterior to SMG. SMG is strongly interconnected with the Rolandic cortex and anterior insula bilaterally, with the posterior portion also showing strong connectivity with the posterior superior temporal sulcus (see figures C.16 through C.18 in appendix C), and it has been implicated in several language processes, including verbal working memory and reading as well as speech motor control.

The most influential model of working memory is that of Baddeley and colleagues (e.g., Baddeley, 1992). Baddeley's model posits a *phonological loop* (also called the *articulatory loop*) that is responsible for maintaining verbal material in working memory. Briefly, the model states that the speech motor system subvocally repeats verbal material (such as the digits of a phone number), with each production activating a *phonological store* that maintains the corresponding phonological information for 1 to 2 seconds, until the next repetition. Verbal material can be kept in working memory indefinitely by repeating this process over and over. (See chapter 8 for further treatment of the Baddeley model.) Later neuroimaging studies have localized the phonological store to SMG and/or the neighboring angular gyrus (e.g., Paulesu, Frith, & Frackowiak, 1993; Jonides et al., 1998). Relatedly, Henson, Burgess, and Frith (2000) found activity in SMG when comparing a delayed matching task involving letters to the same task involving nonverbal symbols, suggesting that SMG participates in phonological recoding of visually presented verbal materials. Crottaz-Herbette, Anagnoson, and Menon (2004) found more activity in the left SMG and adjoining intraparietal sulcus (which separates the superior and inferior parietal lobules) for a verbal working memory task involving visually presented stimuli compared to auditorily presented stimuli; activity in this region can also be seen in our mega-analysis of simple speech utterances (figure 2.14), which involved visually presented verbal material. The review of Pugh et al. (2001) implicates SMG along with the angular gyrus and posterior superior temporal gyrus in the integration of orthographic, phonological, and lexical-semantic dimensions during reading. They distinguish these regions from an inferior temporo-occipital region (*ITO* in figure 2.14) involved in word form processing; whereas the ITO word form area shows increased activation for word reading compared to pseudoword reading, SMG shows the opposite pattern of activation. The role of SMG in verbal working memory processes is treated in further detail in chapter 8.

SMG has also been implicated in somatosensory feedback-based control of speech. For example, Golfinopoulos et al. (2011) found increased activation in ventral somatosensory cortex and supramarginal gyrus during speech when the jaw was unexpectedly perturbed compared to unperturbed speech. According to the DIVA model, the supramarginal gyrus contains *somatosensory error cells* that represent the difference between expected and actual somatic sensation during speech movements. The role of SMG in somatosensory feedback control is addressed in detail in chapter 6.

The regions discussed thus far have all been implicated in nonspeech motor tasks involving the speech articulators. This is illustrated in figure 2.19, which shows all activation foci (blue spheres) reported in the 36 studies of simple nonspeech articulator movements included in the meta-analyses described in appendix A, plotted on inflated cortical surfaces that also show the speech-related activity from figure 2.14. It is noteworthy that nonspeech articulator movements cover nearly the entire range of speech-related activity in the Rolandic cortex, SMA, preSMA, and CMA. Much of the speech-related activity in the superior temporal gyrus, which consists largely of auditory cortex, is also found for simple nonspeech articulator movements, primarily of the larynx. Regions involved in speech but not simple articulator movements include left IFo, *inferior frontal gyrus pars triangularis* (*IFt*), and *pars orbitalis* (*IFr*), bilateral aINS, and portions of the left *anterior superior temporal gyrus* (*aSTG*), left *posterior superior temporal gyrus* (*pSTG*), and right *posterior middle temporal gyrus* (*pMTG*). These regions are discussed in the following subsections.

Inferior Frontal Gyrus

The *inferior frontal gyrus* is classically subdivided into three regions as illustrated in figure 2.19: IFo (BA 44), IFt (BA 45), and IFr (BA 47). Left-hemisphere posterior inferior frontal gyrus, comprising IFo and IFt, is often referred to as *Broca's area* because of the landmark study by neurologist Paul Broca that identified this region as crucial for language production based on the location of lesions in aphasic patients (Broca, 1861; see also Dronkers et al., 2007, for a new analysis of Broca's patients' lesions). Left inferior frontal gyrus also has been implicated in semantic processing (Fiez, 1997; Poldrack et al., 1999, Vigneau et al., 2006), grammatical aspects of language comprehension (Heim, Opitz, & Friederici, 2003; Sahin, Pinker, & Halgren, 2006), reading (Pugh et al., 2001), and a variety of tasks outside of the domain of speech and language production (see, e.g., Fadiga, Craighero, & D'Ausilio, 2009). IFo, IFt, and IFr are strongly interconnected with each other as well as with anterior insula and the supplementary motor areas, both structurally and functionally (see figures C.19 through C.27 in appendix C). Substantial connectivity is also seen

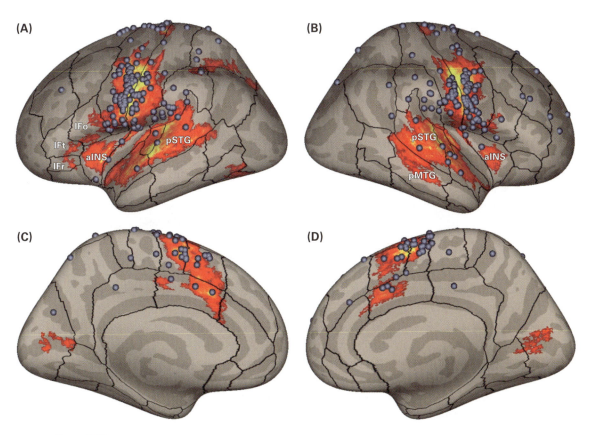

Figure 2.19
All activity foci from the 36 studies considered in the meta-analyses of simple articulator movements in appendix A plotted as blue spheres on inflated cortical surfaces showing speech-related fMRI activity from 116 subjects (hot colormap) in (A) left lateral, (B) right lateral, (C) left medial, and (D) right medial cortex (see the caption of figure 1.8 for details concerning statistical analysis of fMRI data). Areas with significant speech-related activity but little or no activity during simple nonspeech articulator movements are labeled; regions of activity in the occipital and parietal lobes that are likely related to reading of the speech stimuli rather than speech motor control are not labeled. aINS, anterior insula; aSTG, anterior superior temporal gyrus; IFo, inferior frontal gyrus pars opercularis; IFr, inferior frontal gyrus pars orbitalis; IFt, inferior frontal gyrus pars triangularis; pMTG, posterior middle temporal gyrus; pSTG, posterior superior temporal gyrus.

between the inferior frontal gyrus and auditory cortical regions of the superior temporal gyrus and sulcus.

Although more commonly associated with higher-level language tasks, the inferior frontal gyrus is also involved in speech motor control. Evidence for this comes from electrical stimulation, lesion, and neuroimaging studies. Penfield and Roberts (1959) reported that electrical stimulation of the inferior frontal gyrus could give rise to speech arrest, offering direct evidence for its role in speech motor output. Groswasser, Korn, and Groswasser-Reider (1988) report three cases of sustained mutism in the absence of other language problems after bilateral damage to the posterior inferior frontal gyrus. All three individuals suffered from *buccofacial apraxia* (indicated by an impaired ability to produce gestures such as sticking out the tongue, kissing, and chattering the teeth), a condition which may have eliminated the ability to perform the basic articulatory gestures underlying speech and thus may have been responsible for mutism. The lesion study of Hillis et al. (2004) associates damage to Broca's area with *apraxia of speech* (*AOS*), a disorder of speech motor programming in the absence of weakness of the speech articulators (see chapter 10 for detailed treatment).

The neuroimaging study of Ghosh, Tourville, and Guenther (2008) identified activity in left IFo and IFt (particularly the ventral portions) even for utterances consisting of a single nonsense syllable. Similarly, speech activity is evident in figure 2.14 in IFo and IFt, as well as in IFr. Based on differences in cytoarchitecture, Amunts and Zilles (2012) differentiate the opercular portions of IFo and IFt from the rest of these regions. Specifically, they identify a cytoarchitectonic area *op8* that occupies the operculum below BA 44, and another area *op9* that occupies the operculum below BA 45. Within this scheme, the activity shown in figure 2.14 lies primarily in op8 and op9 rather than BA 44 and BA 45, although some activity is evident more dorsally in the posterior-most portion of IFo (BA 44) bilaterally.

As we will see in later chapters, the DIVA model posits the existence of a *speech sound map* in left ventral premotor cortex, including portions of the posterior inferior frontal gyrus, that acts as a link between phonology and speech motor control. Activation of nodes in this map leads to the readout of motor programs that articulate the corresponding speech sounds. Damage to these nodes would therefore result in AOS, consistent with the findings of Hillis et al. (2004) and others. The model also posits that right hemisphere IFo and IFt play a role in the generation of corrective movements for perceived inaccuracies in speech articulation; this topic is covered in chapters 5 and 6. Right hemisphere inferior frontal gyrus may also play a role in prosody, as discussed in chapter 9.

Insula

The *insula* borders several key speech and language areas and has long been considered a candidate for language function (e.g., Wernicke, 1874; Freud, 1891; Dejerine, 1914; see Ardila, 1999, for a review). In the anterior superior direction, the insula borders speech motor areas in the frontal opercular region of the inferior frontal gyrus and Rolandic operculum. In the posterior superior direction, the insula borders the parietal operculum. Inferiorly, the insula borders the auditory cortical areas of the supratemporal plane. In addition to speech, a wide range of functions including memory, drive, affect, gustation, and olfaction have been attributed to the insula in various studies (Türe et al., 1999). The insula has strong functional and structural connectivity with surrounding portions of the inferior frontal gyrus, Rolandic cortex, and superior temporal gyrus, as well as strong functional connectivity with the preSMA and cingulate motor area (see figures C.28 and C.29 in appendix C), suggestive of a role in integrating a wide range of cognitive, motor, somatosensory, and auditory information.

The insula comprises at least three different cytoarchitectonic regions: a rostro-ventral *agranular* region, a transitional *dysgranular* region, and a posterior *granular* region (Augustine, 1996; Ogar & Gorno-Tempini, 2007). "Granularity" in this context refers to the distinctness of cortical layer IV, which is called the granular layer. According to Amunts and Zilles (2012), in agranular cortex layer IV is essentially nonexistent; primary motor cortex (BA 4) has this characteristic. Dysgranular cortex contains a thin granular layer IV that is not clearly distinct from layers III and V because of intermingling of cell types between the layers; BA 44 is an example of dysgranular cortex. Granular cortex contains a distinct layer IV; BA 45 is an example of granular cortex.

Despite much investigation and discussion of the topic in both the neuroimaging and neurology literatures, the role of the insula in speech and language remains remarkably opaque (Ardila, 1999; Türe et al., 1999; Ackermann & Riecker, 2004; Price, 2012). A large number of neuroimaging studies of speech production have found insula activation (e.g., Wise et al., 1999; Riecker et al., 2000a; Sakurai et al., 2001; Bohland & Guenther, 2006; Sörös et al., 2006; Ghosh, Tourville, & Guenther, 2008). Figure 2.14 also shows substantial activity in the aINS bilaterally, though with a left-hemisphere bias. The location of this activity is primarily limited to dysgranular portions of the insula (cf. Morel et al., 2013), suggesting it may share some characteristics with other dysgranular cortical regions, most notably premotor cortical areas. Most of the active area contains six cortical layers (cf. Morel et al., 2013) and is thus neocortex, compared to other portions of the insula (most notably the agranular portion) that contain fewer than six layers and are phylogenetically older mesocortex or allocortex.

Although many speech neuroimaging studies find insula activation, others do not, at least for some speech conditions (Lotze et al., 2000; Riecker et al., 2000b; Soros et al., 2006). Soros et al. (2006) suggest that the difference may lie in whether the speaking condition involves repeated production of the same word/syllable or it involves production of different syllables/words on each trial, a proposal also made by Nota and Honda (2003). It has also been suggested that insula recruitment may hinge on the automaticity of the speaking task, with very simple tasks or overlearned speech tasks not necessitating its involvement, whereas more complex sequential production tasks may recruit insular cortex during or prior to articulation (Ackermann & Riecker, 2004; Baldo et al., 2011). Again, however, conflicting reports can be found. For example, Ghosh et al. (2008) found bilateral insula activation for even very simple consonant-vowel monosyllabic utterances.

Ardila (1999) and Benson (1979) have argued that the insula is likely involved in motivational and affective aspects of speech and language. This view is in keeping with the characterization, by some authors, of the insula as part of the paralimbic system as well as with studies showing that insular damage can lead to deficits in speech initiation (Shuren, 1993) and motivation to speak (Habib et al., 1995). As noted above, however, much of the human insula is neocortex, which is not usually classified as part of the paralimbic system.

At least two distinct subregions of the anterior insula have been associated with speech motor programming. Dronkers (1996) found that the *superior precentral gyrus of the insula* (*SPGI*) in the left hemisphere was the only common site of overlap in a group of patients diagnosed with AOS but was preserved in a second group with aphasia but not AOS. This region, therefore, was suggested to be critical for motor programming, or translating a planned speech sound into articulatory action. This conclusion has been questioned by Hillis et al. (2004), who found (using different analytical methods) no association between AOS and lesions of left insula, instead finding that AOS is associated with damage to the left posterior inferior frontal gyrus. A second, more rostral portion of the anterior insula that abuts the frontal operculum has also been implicated in the generation of speech movements (e.g., Bohland & Guenther, 2006; Ackermann & Riecker, 2010). The fMRI study of Bohland and Guenther (2006) found that increased complexity of an utterance was associated with an increase in activity in the rostral insula location and the adjoining frontal operculum, but not in SPGI.

Superior Temporal Cortex
The superior portion of the temporal lobe contains the auditory cortical areas. The *primary auditory cortex* (BA 41)[17] is located in *Heschl's gyrus* (*HG*, also called the *transverse temporal gyrus*) on the supratemporal plane within the Sylvian fissure. Primary auditory cortex receives input from the auditory periphery via the MG nucleus of the thalamus, with substantially more subcortical processing occurring in the ascending auditory system compared to other sensory systems (see Pickles, 1988, for a review). Unilateral excision of HG (and thus primary auditory cortex) has little effect on hearing or speech (Penfield & Roberts, 1959), likely due to the fact that auditory cortex in each hemisphere receives peripheral input from both ears.

Primary auditory cortex projects to surrounding higher-order auditory cortical areas (BA 42, 22) of the *anterior* and *posterior superior temporal gyrus* (*aSTG* and *pSTG* in figure 2.14), including the planum temporale and planum polare within the supratemporal plane, and laterally to the superior temporal sulcus, possibly including dorsal portions of the *posterior middle temporal gyrus* (*pMTG*).[18] HG and pSTG are strongly structurally and functionally connected with Rolandic cortex and the insula, and pSTG also has strong functional connectivity with the inferior frontal gyrus (see figures C.30 through C.34 in appendix C). aSTG has strong functional and structural connectivity with the insula and inferior frontal gyrus (figures C.35 through C.38). Left pSTG is sometimes referred to as *Wernicke's area* after German neurologist Carl Wernicke, who discovered lesions in this region in patients suffering from *receptive aphasia* (also called *sensory aphasia* or *Wernicke's aphasia*), characterized by impaired ability to understand language with relatively spared ability to produce syllables and words, as opposed to the impairment of language production that characterizes Broca's aphasia. The modern view is that lesions confined to left pSTG typically do not result in lasting receptive aphasia; additional damage to white matter pathways underlying the superior temporal gyrus and supramarginal gyrus is also required.

The acoustic signal produced by a talker impinges on the talker's ears (referred to as *sidetone*, as discussed in chapter 5), where it is transduced by the cochlea and eventually transmitted to auditory cortex. Although much of the activity in auditory cortical areas during speech production can be attributed directly to hearing one's own voice while speaking, it has been demonstrated with MEG and fMRI that covert speech also activates auditory cortex (e.g., Paulesu, Frith, & Frackowiak, 1993; Numminen & Curio, 1999; Hickok et al., 2000; Okada et al., 2003; Okada & Hickok, 2006), indicative of *efference copies* of motor commands projecting from motor cortical regions to auditory cortex. Relatedly, the DIVA model hypothesizes that higher-order auditory cortical areas play a crucial role in auditory feedback control of speech (Guenther, Ghosh, & Tourville, 2006), that is, the generation of motor commands to correct for perceived errors in the acoustic signal. Auditory feedback control is particularly important when initially learning to produce the speech sounds of a language. The model posits the existence of an *auditory error map* that represents mismatches between the expected acoustic signal (arriving via the efference copy pathways) and the actual acoustic signal during speech. In support of this prediction, increased auditory cortical activation has been identified in fMRI studies of speech production with delayed auditory feedback (Hashimoto & Sakai, 2003) and auditory feedback that has been modified in real-time by shifting the first formant frequency (Tourville, Reilly, & Guenther, 2008). Both of these manipulations result in a mismatch between produced speech (as heard through the headphones after modification) and auditory expectations for the produced sounds. A related phenomenon, referred to as *speech-induced suppression*, is observed in studies using MEG or EEG. In this effect, activation in the auditory cortex (measured by the peak amplitude of a stereotyped potential approximately 100 ms after sound onset) is reduced for self-produced speech relative to hearing the speech of others or hearing a recorded version of one's own voice (Houde et al., 2002; Curio et al., 2000; Numminen & Curio, 1999). The topics of efference copies, speech-induced suppression, auditory error maps, and auditory feedback control of speech are treated in further detail in chapter 5.

Lateralization of Speech and Language Processing

As evident in figure 2.8, figure 2.12, and figure 2.14, activity during simple speech tasks is largely bilateral in both cortical and subcortical regions. This differs from activity in higher-level language tasks, which is generally left lateralized in cerebral cortex.[19] In subcortical areas, a minor left-hemisphere bias can be seen in the putamen and thalamus. The largest deviations from bilaterally symmetrical activity for speech and language are found in the cerebral cortex, most notably the inferior frontal gyrus and aINS. Left-hemisphere lesions in these areas result in AOS, whereas right-hemisphere lesions have a much smaller impact, if any, on speaking skills. Accordingly, the DIVA model posits that the key interface between phonological representations and speech motor programs (including the model's speech sound map) is localized to left-hemisphere ventral premotor cortical areas, including IFo, IFt, rostral PrCG, and aINS. Sequencing of speech sounds also appears to preferentially involve left-hemisphere frontal cortex, as discussed in chapter 8.

Left lateralization for language tasks is also found in the temporal and parietal lobes. Again, this is less true for speech motor control compared to language, though there is evidence indicating that right-hemisphere auditory cortical areas are more specialized for processing more slowly changing aspects of the speech signal related to prosody, while the left auditory cortical areas are more specialized for processing the rapid acoustic changes that differentiate phonemes (Poeppel, 2003; Abrams et al., 2008; see chapter 9 for further discussion). This is in keeping with findings of impaired prosodic processing in individuals with right-hemisphere damage (e.g., Weintraub, Mesulam, & Kramer, 1981; Bryan, 1988; Pell, 1999). Perhaps relatedly, neuroimaging studies have identified right-lateralized brain activity during singing (Wildgruber et al., 1996; Riecker et al., 2000a), which imposes intonational control requirements that are similar in some ways to prosodic control.

Although brain activity is found bilaterally in the supplementary motor areas (SMA and preSMA) during speech, studies of transcortical motor aphasia mentioned earlier suggest that the left- and right-hemisphere supplementary motor areas play somewhat different roles during speech, with the left hemisphere more essential for initiating speech. It is likely that additional bilaterally active regions play different roles in the two hemispheres, including the basal ganglia (Gil Robles et al., 2005) and thalamus (Schaltenbrand, 1975).

2.5 Summary

The discovery that the left inferior frontal cortex is specialized for language production by French physician Paul Broca in the mid-nineteenth century provided the first clear evidence that different regions of the brain perform different functions. Since that time a wealth of knowledge has accumulated regarding how the brain carries out speech and language functions. Because invasive technologies such as microelectrode recordings cannot be used in humans except in rare circumstances, the advent of noninvasive functional neuroimaging techniques in the latter half of the twentieth century was particularly important for studies of speech and language. Although the core of the speech production system is organized in a manner similar to the vocalization circuit of nonhuman primates, evolution has elaborated this network in a number of important ways, most notably in the cerebral cortex.

The cranial nerve nuclei in the medulla, pons, and midbrain form the lowest level of the neural system underlying speech. Sensory information from the speech articulators arrives via several cranial nerves. Tactile and proprioceptive information from the face, lips, jaw, tongue, and mouth projects to the trigeminal sensory nucleus via the Vth nerve. Sensory information from the pharynx, tongue, larynx, and palate arrives at the nucleus ambiguus via the IXth and Xth nerves. This somatosensory information then projects to the primary somatosensory cortex via the ventral posterior medial nucleus of the thalamus. Auditory information from the cochlea arrives at the cochlear nucleus via the VIIIth nerve and projects to auditory cortex via the superior olive, inferior colliculus, and medial geniculate nucleus of the thalamus. The trigeminal motor nucleus projects to muscles of the jaw and palate via the Vth nerve. The facial nucleus projects to muscles of the face and lips via the VIIth nerve. The nucleus ambiguus projects to muscles in the pharynx, larynx, and palate via the IXth, Xth, and XIth nerves, and the hypoglossal nucleus projects to the tongue via the XIIth nerve.

The corticobulbar tract carries projections from motor cortex to the cranial nerve nuclei. These direct connections are supplemented by indirect projections that pass through the red nucleus and reticular formation. The indirect projections appear to play a more modulatory role and are likely not heavily involved in generating the intricately patterned muscle activations that underlie speech movements.

The motor cortical commands for speech are shaped by two reentrant loops through subcortical structures: a cortico–basal ganglia loop and a cortico-cerebellar loop. Both of these loops pass through the thalamus on their way back to the cerebral cortex, and both actually consist of many somewhat independent circuits involving different parts of the cerebral cortex. The cortico–basal ganglia loop plays a key role in selecting and initiating the appropriate motor programs for speech whereas the cortico-cerebellar loop is important for generating the finely timed muscle activations needed for rapid speech. Within the cerebellum, the superior paravermal region of cerebellar cortex (specifically lobule VI) appears to play the largest role in controlling movements of the speech articulators. Damage to this region results in ataxic dysarthria, which is characterized by poorly timed and coordinated speech articulations. In addition to key roles in motor control, both the cortico–basal ganglia loop and the cortico-cerebellar loop appear to play roles in higher-level language processes.

The most sophisticated computations underlying speech occur in the cerebral cortex. The Rolandic cortex, consisting of the postcentral gyrus (which includes primary somatosensory cortex) and precentral gyrus (which contains primary motor cortex and part of premotor cortex), constitutes the lowest level of cortical processing for speech production. In concert with the cortico-cerebellar loop, the ventral half of the Rolandic cortex is crucial for integrating somatosensory and motor representations and for generating the motor commands that are sent to the articulatory musculature via the cranial nerve nuclei. Representations of the individual speech articulators follow a coarse dorsal-to-ventral ordering across Rolandic cortex, with the respiratory system represented most dorsally, followed by larynx, lip, jaw, and tongue representations. With the exception of a dorsal region that appears to mainly contain a respiratory representation, the articulator representations are highly overlapped, in keeping with the need for highly coordinated movements of multiple articulators during speech. Individual neurons in the precentral gyrus lying near the central sulcus (primary motor cortex) project to motoneurons in the brain stem and spinal cord. As one moves rostrally toward the precentral sulcus, the firing properties of motor cortical neurons become more abstracted from the musculature, culminating in premotor cortical neurons near the precentral sulcus that represent task space variables such as 3-D space for reaching movements or acoustic/auditory space for speech, though more investigation is necessary to verify this organization for speech.

Several portions of the medial surface of the frontal cortex play important roles in speech production. The supplementary motor area plays a major role in movement initiation, likely in concert with the basal ganglia via the cortico–basal ganglia motor circuit. The pre-supplementary motor area may play a role in sequencing of speech sounds in longer utterances. Damage to the supplementary motor areas in the left hemisphere can result in transcortical motor aphasia, characterized by a general lack of self-generated speech, though normal speech usually returns unless the damage is bilateral. The cingulate motor area appears to play a role in the motivation to speak; bilateral damage to this region can lead to akinetic mutism, characterized by a complete or near-complete lack of spontaneous speech despite intact muscle tone and articulatory control.

In addition to playing a crucial role in speech perception, the auditory cortical areas of the superior temporal gyrus are involved in auditory feedback control of speech, and they play a particularly important role during the development of speaking skills, as discussed in later chapters.

The highest levels of speech motor planning appear to be carried out by the left-hemisphere frontal cortex. Damage to the left inferior frontal gyrus and/or anterior insula results in apraxia of speech, which can be coarsely characterized as the inability to generate the motor programs for speech sounds. The right-hemisphere homologues of these regions appear to be involved in sensory feedback control, that is, the generation of corrective motor commands for perceived errors in articulation. The lateralized nature of processing in the inferior frontal cortex differs from most other areas of the speech motor system, which appear to be largely bilaterally symmetrical, though some additional subtle differences between the left and right hemispheres may exist, such as preferential processing of rapid sound changes (indicative of phonological content) in left-hemisphere auditory cortex and slower sound changes (indicative of prosodic content) in right-hemisphere auditory cortex. Differences in the roles of left- and right-hemisphere subcortical structures, most notably the basal ganglia and thalamus, in the generation of speech output have also been noted.

Notes

1. Note that all nuclei and nerves in figure 2.3 are bilaterally represented in the nervous system.

2. Note, however, that tactile information received by the hypoglossal nucleus via the nucleus of the solitary tract and trigeminal sensory nucleus is important for speech as well as chewing, swallowing, and sucking (Duffy, 1995).

3. Although commonly equated to motor cortical neurons, the term *upper motor neuron* is sometimes used to refer to other neurons in the motor system, such as those in the red nucleus and reticular formation, which can affect activity of the lower motor neurons (e.g., Duffy, 1995).

4. For example, it has been noted that squirrel monkeys, unlike humans, do not possess direct projections from motor cortex to the laryngeal motoneurons in the nucleus ambiguus (Jürgens, 2009).

5. The pyramidal tracts are so named because they pass through the medullary pyramids before crossing to the opposite side of the body en route to the motor periphery. Within this context, the *extrapyramidal system* is the portion of the motor system that does not pass through the pyramids, including the reentrant cortico–basal ganglia loops through the striatum, pallidum, and thalamus (discussed further below), along with projections from the pallidum to the reticular formation. Some authors also consider reentrant cortico-cerebellar loops through the cerebellum via the pons and thalamus (also discussed below) to be part of the extrapyramidal system whereas others use the term to refer only to the basal ganglia circuit.

6. Although it provides a first-order account of thalamic function that serves our current purposes, the characterization of thalamus as a relay station is a considerable simplification of its many roles in sensation, cognition, and action. For example, Llano (2013) proposes that the thalamus may serve as a tunable filter that enhances or suppresses the transfer of particular types of information depending on the task.

7. The existence of both a cerebral cortex and a cerebellar cortex can lead to confusion. In this book, the term *cortex*, when not preceded by *cerebellar*, will refer to the cerebral cortex.

8. The cerebellar cortex is typically divided into three layers. These are (ordered from inner to outer surface): the *granule cell layer*, the *Purkinje cell layer*, and the *molecular layer* which contains the parallel fibers and Purkinje cell dendrites as well as inhibitory basket and stellate interneurons.

9. Association areas of cortex process multiple sensory and motor signals (as opposed to the primary and secondary areas of cortex, which primarily process information from a single sensory or motor modality). Association cortical areas are believed to be the highest-level processing centers in the brain.

10. Foci for respiration are based on an average of foci from the individual studies rather than an ALE meta-analysis—see appendix A for details.

11. Each layer of cortex courses parallel to the cortical surface. The outermost layer of cortex is layer I, with each subsequent layer lying immediately below the previous layer in the cortical sheet.

12. The insula is often considered to be a fifth lobe of the cerebral cortex (the *insular lobe*).

13. The Latin word *operculum* means little cover or lid. In neuroanatomy, the term is used to refer to the portions of cortex located on the banks of the Sylvian fissure that cover the insula, especially the superior bank.

14. Penfield's somatotopic maps did not include a representation of the respiratory musculature. The Takai et al. (2010) and appendix A meta-analyses found a dorsal activation focus for respiration in Rolandic cortex, in the vicinity of the dorsal-most activity cluster in PoCG and PrCG of figure 2.14. These meta-analyses also found a more ventral respiration focus located in the vicinity of the other speech articulator representations.

15. Not included in the Takai et al. (2010) study.

16. Throughout this book, the acronym SMA will be used to refer to SMA proper. Many studies, particularly those occurring before the subregions of the supplementary motor areas were identified, used the acronym SMA to refer to SMA and preSMA collectively; when describing the results of these studies, the term *supplementary motor areas*, without abbreviation, will be used rather than SMA.

17. Some authors consider BA 42, which lies immediately posterior to BA 41 on Heschl's gyrus, to be part of primary auditory cortex along with BA 41. Recent histological studies indicate that there are many more than Brodmann's original set of three cytoarchitectonic subdivisions of auditory cortex (BA 41, 42, 22), and many modern authors (particularly in the nonhuman primate literature) utilize the terms *core*, *belt*, and *parabelt* to describe primary, secondary, and higher-order auditory areas, respectively.

18. This assertion is based partly on the finding of speech-related activity in pMTG in our mega-analysis of simple speech utterances (figure 2.14), particularly in the right hemisphere.

19. It has been estimated that language is left lateralized in approximately 90% of the population, with a somewhat higher prevalence of left-lateralized language in right-handed individuals compared to left-handers. In a sample of 326 healthy individuals, Knecht et al. (2000) found left language lateralization in 96% of strongly right-handed individuals, 85% in ambidextrous individuals, and 73% in strongly left-handed individuals. Throughout this book we will often use the term "left hemisphere" as a synonym for "language dominant hemisphere" with the understanding that a small percentage of the population has either bilateral or right-lateralized speech and language representation.

References

Abrams, D. A., Nicol, T., Zecker, S., & Kraus, N. (2008). Right-hemisphere auditory cortex is dominant for coding syllable patterns in speech. *Journal of Neuroscience, 28*, 3958–3965.

Ackermann, H. (2008). Cerebellar contributions to speech production and speech perception: psycholinguistic and neurobiological perspectives. *Trends in Neurosciences, 31*, 265–272.

Ackermann, H., & Hertrich, I. (1994). Speech rate and rhythm in cerebellar dysarthria: an acoustic analysis of syllabic timing. *Folia Phoniatrica et Logopaedica, 46*, 70–78.

Ackermann, H., & Riecker, A. (2004). The contribution of the insula to motor aspects of speech production: a review and a hypothesis. *Brain and Language, 89*, 320–328.

Ackermann, H., & Riecker, A. (2010). The contribution(s) of the insula to speech production: a review of the clinical and functional imaging literature. *Brain Structure & Function, 214*, 419–433.

Ackermann, H., Vogel, M., Petersen, D., & Poremba, M. (1992). Speech deficits in ischaemic cerebellar lesions. *Journal of Neurology, 239*, 223–227.

Albus, J. S. (1971). A theory of cerebellar function. *Mathematical Biosciences, 10*, 25–61.

Alexander, G. E., & Crutcher, M. D. (1990). Neural representations of the target (goal) of visually guided arm movements in three motor areas of the monkey. *Journal of Neurophysiology, 64*, 164–178.

Alexander, G. E., DeLong, M. R., & Strick, P. L. (1986). Parallel organization of functionally segregated circuits linking basal ganglia and cortex. *Annual Review of Neuroscience, 9*, 357–381.

Amunts, K., & Zilles, K. (2012). Architecture and organizational principles of Broca's region. *Trends in Cognitive Sciences, 16*, 418–426.

Andersen, B. B., Korbo, L., & Pakkenberg, B. (1992). A quantitative study of the human cerebellum with unbiased stereological techniques. *Journal of Comparative Neurology, 326*, 549–560.

Ardila, A. (1999). The role of insula in language: an unsettled question. *Aphasiology, 13*, 79–87.

Augustine, J. R. (1996). Circuitry and functional aspects of the insular lobe in primates including humans. *Brain Research. Brain Research Reviews, 22*, 229–244.

Baddeley, A. D. (1992). Working memory. *Science, 255*, 556–559.

Baddeley, A. D., & Hitch, G. (1974). Working memory. In G. H. Bower (Ed.), *The psychology of learning and motivation: advances in research and theory* (Vol. 8, pp. 47–89). New York: Academic Press.

Baldo, J. V., Wilkins, D. P., Ogar, J., Willock, S., & Dronkers, N. F. (2011). Role of the precentral gyrus of the insula in complex articulation. *Cortex*, *47*, 800–807.

Barlow, S. M. (1999). *Handbook of clinical speech physiology*. San Diego: Singular.

Barris, R. W., & Schuman, H. R. (1953). Bilateral anterior cingulate gyrus lesions: syndrome of the anterior cingulate gyri. *Neurology*, *3*, 44–53.

Bartholow, R. (1874). Experimental investigations into the functions of the human brain. *American Journal of the Medical Sciences*, *134*, 305–313.

Bastian, A. J., & Thach, W. T. (2002). Structure and function of the cerebellum. In M. U. Manto & M. Pandolfo (Eds.), *The cerebellum and its disorders* (pp. 6–29). Cambridge, UK: Cambridge University Press.

Behbehani, M. M. (1995). Functional characteristics of the midbrain periaqueductal gray. *Progress in Neurobiology*, *46*, 575–605.

Benson, D. F. (1979). *Aphasia, alexia and agraphia*. New York: Churchill Livingstone.

Blakemore, S.-J., Frith, C. D., & Wolpert, D. M. (2001). The cerebellum is involved in predicting the sensory consequences of action. *Neuroreport*, *12*, 1879–1884.

Bohland, J. W., Bullock, D., & Guenther, F. H. (2010). Neural representations and mechanisms for the performance of simple speech sequences. *Journal of Cognitive Neuroscience*, *22*, 1504–1529.

Bohland, J. W., & Guenther, F. H. (2006). An fMRI investigation of syllable sequence production. *NeuroImage*, *32*, 821–841.

Bouchard, K. E., Mesfarani, N., Johnson, K., & Chang, E. F. (2013). Functional organization of human sensorimotor cortex for speech articulation. *Nature*, *495*, 327–332.

Broca, P. (1861). Perte de la parole: ramollissement chronique et destructions partielle du lobe antérior gauche du cerveau. *Bulletin de la Société d'Anthropologie*, *2*, 235–238.

Brodal, P., & Bjaalie, J. G. (1992). Organization of the pontine nuclei. *Neuroscience Research*, *13*, 83–118.

Brodmann, K. (1909). *Vergleichende lokalisationslehre der großhirnrinde in ihren prinzipien dargestellt auf grund des zellenbaues*. Liepzig: Barth.

Brooks, V. B. (1986). *The neural basis of motor control*. Oxford: Oxford University Press.

Brown, J. W., Bullock, D., & Grossberg, S. (2004). How laminar frontal cortex and basal ganglia circuits interact to control planned and reactive saccades. *Neural Networks*, *17*, 471–510.

Brown, S., Ingham, R. J., Ingham, J. C., Laird, A. R., & Fox, P. T. (2005). Stuttered and fluent speech production: an ALE meta-analysis of functional neuroimaging studies. *Human Brain Mapping*, *25*, 105–117.

Brown, T. G. (1915). Note on physiology of basal ganglia and midbrain of anthropoid ape, especially in reference to the act of laughter. *Journal of Physiology*, *49*, 195–207.

Bryan, K. L. (1988). Language prosody and the right hemisphere. *Aphasiology*, *3*, 285–299.

Callan, D. E., Tsytsarev, V., Hanakawa, T., Callan, A. M., Katsuhara, M., Fukuyama, H., et al. (2006). Song and speech: brain regions involved with perception and covert production. *NeuroImage*, *31*, 1327–1342.

Chen, S. H., & Desmond, J. E. (2005). Cerebrocerebellar networks during articulatory rehearsal and verbal working memory tasks. *NeuroImage*, *24*, 332–338.

Cheney, P. D., & Fetz, E. E. (1980). Functional classes of primate corticomotoneuronal cells and their relation to active force. *Journal of Neurophysiology*, *4*, 773–791.

Chiricozzi, F. R., Clausi, S., Molinari, M., & Leggio, M. G. (2008). Phonological short-term store impairment after cerebellar lesion: a single case study. *Neuropsychologia*, *46*, 1940–1953.

Colin, F., Ris, L., & Godaux, E. (2002). Neuroanatomy of the cerebellum. In M. U. Manto & M. Pandolfo (Eds.), *The cerebellum and its disorders* (pp. 6–29). Cambridge, UK: Cambridge University Press.

Crottaz-Herbette, S., Anagnoson, R. T., & Menon, V. (2004). Modality effects in verbal working memory: differential prefrontal and parietal responses to auditory and visual stimuli. *NeuroImage*, *21*, 340–351.

Curio, G., Neuloh, G., Numminen, J., Jousmäki, V., & Hari, R. (2000). Speaking modifies voice-evoked activity in the human auditory cortex. *Human Brain Mapping*, *9*, 183–191.

Darley, F. L., Aronson, A. E., & Brown, J. R. (1975). *Motor speech disorders*. Philadelphia: Saunders.

Dehaene, S. (2009). *Reading in the brain*. New York: Penguin.

Dejerine, J. (1914). *Sémiologie des affections du système nerveux*. Paris: Masson.

Desmond, J. E., & Fiez, J. A. (1998). Neuroimaging studies of the cerebellum: language, learning and memory. *Trends in Cognitive Sciences*, *2*, 355–362.

Desmond, J. E., Gabrieli, J. D., Wagner, A. D., Ginier, B. L., & Glover, G. H. (1997). Lobular patterns of cerebellar activation in verbal working memory and finger-tapping tasks as revealed by functional MRI. *Journal of Neuroscience*, *17*, 9675–9685.

Devinsky, O., Morrell, M. J., & Vogt, B. A. (1995). Contributions of anterior cingulate cortex to behaviour. *Brain*, *118*, 279–306.

Dronkers, N. F. (1996). A new brain region for coordinating speech articulation. *Nature*, *384*, 159–161.

Dronkers, N. F., Plaisant, O., Iba-Zizen, M. T., & Cabanis, E. A. (2007). Paul Broca's historic cases: high resolution MR imaging of the brains of Leborgne and Lelong. *Brain*, *130*, 1432–1441.

Duffy, J. R. (1995). *Motor speech disorders: substrates, differential diagnosis, and management*. St. Louis: Mosby.

Dum, R. P., & Strick, P. L. (2003). An unfolded map of the cerebellar dentate nucleus and its projections to the cerebral cortex. *Journal of Neurophysiology*, *89*, 634–639.

Düsterhöft, F., Häusler, U., & Jürgens, U. (2004). Neuronal activity in the periaqueductal gray and bordering structures during vocal communication in the squirrel monkey. *Neuroscience*, *123*, 53–60.

Esposito, A., Demeurisse, G., Alberti, B., & Fabbro, F. (1999). Complete mutism after midbrain periaqueductal gray lesion. *Neuroreport*, *10*, 681–685.

Evarts, E. V. (1968). Relation of pyramidal tract activity to force exerted during voluntary movement. *Journal of Neurophysiology*, *31*, 14–27.

Fadiga, L., Craighero, L., & D'Ausilio, A. (2009). Broca's area in language, action, and music. *Annals of the New York Academy of Sciences*, *1169*, 448–458.

Farrell, D. F., Burbank, N., Lettich, E., & Ojemann, G. A. (2007). Individual variation in human motor-sensory (Rolandic) cortex. *Journal of Clinical Neurophysiology*, *24*, 286–293.

Fiez, J. A. (1997). Phonology, semantics, and the role of the left inferior prefrontal cortex. *Human Brain Mapping*, *4*, 79–83.

Freedman, M., Alexander, M. P., & Naeser, M. A. (1984). Anatomic basis of transcortical motor aphasia. *Neurology*, *34*, 409–417.

Freud, S. (1891). *On aphasia: a critical study*. London: Imago.

Fried, I., Katz, A., McCarthy, G., Sass, K. J., Williamson, P., Spencer, S. S., et al. (1991). Functional organization of the human supplementary motor cortex studied by electrical stimulation. *Journal of Neuroscience*, *11*, 3656–3666.

Fritsch, G., & Hitzig, E. (1870). Über die elektrische Erregbarkeit des Grosshirns. *Archiv für Anatomie, Physiologie und wissenschaftliche Medicin*, *37*, 300–332.

Gelmers, H. J. (1983). Non-paralytic motor disturbances and speech disorders: the role of the supplementary motor area. *Journal of Neurology, Neurosurgery, and Psychiatry*, *46*, 1052–1054.

Georgopoulos, A. P., Kalaska, J. F., Caminiti, R., & Massey, J. T. (1982). On the relations between the direction of two-dimensional arm movements and cell discharge in primate motor cortex. *Journal of Neuroscience*, *2*, 1527–1537.

Georgopoulos, A. P., Kalaska, J. F., Crutcher, M. D., Caminiti, R., & Massey, J. T. (1984). The representation of movement direction in the motor cortex: single cell and population studies. In G. M. Edelman, W. E. Gall, & W. M. Cowan (Eds.), *Dynamic aspects of neocortical function* (pp. 501–524). New York: Wiley.

Ghosh, S. S., Tourville, J. A., & Guenther, F. H. (2008). A neuroimaging study of premotor lateralization and cerebellar involvement in the production of phonemes and syllables. *Journal of Speech, Language, and Hearing Research*, *51*, 1183–1202.

Gil Robles, S., Gatignol, P., Capelle, L., Mitchell, M.-C., & Duffau, H. (2005). The role of dominant striatum in language: a study using intraoperative electrical stimulations. *Journal of Neurology, Neurosurgery, and Psychiatry*, *76*, 940–946.

Golfinopoulos, E., Tourville, J. A., Bohland, J. W., Ghosh, S. S., Nieto-Castanon, A., & Guenther, F. H. (2011). fMRI investigation of unexpected somatosensory feedback perturbation during speech. *NeuroImage*, *55*, 1324–1338.

Groswasser, Z., Korn, C., & Groswasser-Reider, I. (1988). Mutism associated with buccofacial apraxia and bihemispheric lesions. *Brain and Language*, *34*, 157–168.

Guenther, F. H., Brumberg, J. S., Wright, E. J., Nieto-Castanon, A., Tourville, J. A., Panko, M., et al. (2009). A wireless brain-machine interface for real-time speech synthesis. *PLoS One*, *4*, e8218.

Guenther, F. H., Ghosh, S. S., & Tourville, J. A. (2006). Neural modeling and imaging of the cortical interactions underlying syllable production. *Brain and Language*, *96*, 280–301.

Guenther, F. H., Tourville, J. A., & Bohland, J. W. (2015). Speech production. In A. W. Toga (Ed.), *Brain mapping: an encyclopedic reference* (vol. 3, pp. 435–444). Oxford: Elsevier.

Habib, M., Daquin, G., Milandre, L., Royere, M. L., Rey, M., Lanteri, A., et al. (1995). Mutism and auditory agnosia due to bilateral insular damage—role of the insula in human communication. *Neuropsychologia*, *33*, 327–339.

Harris, L. J., & Almerigi, J. B. (2009). Probing the human brain with stimulating electrodes: the story of Roberts Bartholow's experiment on Mary Rafferty. *Brain and Cognition*, *70*, 92–115.

Hashimoto, Y., & Sakai, K. L. (2003). Brain activations during conscious self-monitoring of speech production with delayed auditory feedback: an fMRI study. *Human Brain Mapping*, *20*, 22–28.

Heim, S., Opitz, B., & Friederici, A. D. (2003). Distributed cortical networks for syntax processing: Broca's area as the common denominator. *Brain and Language*, *85*, 402–408.

Henson, R. N., Burgess, N., & Frith, C. D. (2000). Recoding, storage, rehearsal and grouping in verbal short-term memory: an fMRI study. *Neuropsychologia*, *38*, 426–440.

Hickok, G., Erhard, P., Kassubek, J., Helms-Tillery, A. K., Naeve-Valguth, S., Strupp, J. P., et al. (2000). A functional magnetic resonance imaging study of the role of left posterior superior temporal gyrus in speech production: implications for the explanation of conduction aphasia. *Neuroscience Letters*, *287*, 156–160.

Hillis, A. E., Work, M., Barker, P. B., Jacobs, M. A., Breese, E. L., & Maurer, K. (2004). Re-examining the brain regions crucial for orchestrating speech articulation. *Brain*, *127*, 1479–1487.

Holmes, G. (1917). The symptoms of acute cerebellar injuries due to gunshot injuries. *Brain*, *40*, 461–535.

Houde, J. F., Nagarajan, S. S., Sekihara, K., & Merzenich, M. M. (2002). Modulation of the auditory cortex during speech: an MEG study. *Journal of Cognitive Neuroscience*, *14*, 1125–1138.

Humphrey, D. R., & Tanji, J. (1991). What features of voluntary motor control are encoded in the neuronal discharge of different cortical motor areas? In D. R. Humphrey & H. J. Freund (Eds.), *Motor control: concepts and issues* (pp. 413–443). New York: Wiley.

Indefrey, P., & Levelt, W. J. M. (2004). The spatial and temporal signatures of word production components. *Cognition*, *92*, 101–144.

Ito, M. (1984). *Cerebellum and neural control*. New York: Raven.

Johansen-Berg, H., Behrens, T. E. J., Robson, M. D., Drobnjak, I., Rushworth, M. F. S., Brady, J. M., et al. (2004). Changes in connectivity profiles define functionally distinct regions in human medial frontal cortex. *Proceedings of the National Academy of Sciences of the United States of America*, *101*, 13335–13340.

Johnson, M. D., & Ojemann, G. A. (2000). The role of the thalamus in language and memory: evidence from electrophysiological studies. *Brain and Cognition*, *42*, 218–230.

Jonas, S. (1981). The supplementary motor region and speech emission. *Journal of Communication Disorders*, *14*, 349–373.

Jonas, S. (1987). The supplementary motor region and speech. In E. Perecman (Ed.), *The frontal lobes revisited* (pp. 241–250). New York: IRBN Press.

Jonides, J., Schumacher, E. H., Smith, E. E., Koeppe, R. A., Awh, E., Reuter-Lorenz, P. A., et al. (1998). The role of parietal cortex in verbal working memory. *Journal of Neuroscience*, *18*, 5026–5034.

Jürgens, U. (1984). The efferent and afferent connections of the supplementary motor area. *Brain Research*, *300*, 63–81.

Jürgens, U. (2009). The neural control of vocalization in mammals: a review. *Journal of Voice, 23,* 1–10.

Jürgens, U., & Richter, K. (1986). Glutamate-induced vocalization in the squirrel monkey. *Brain Research, 272,* 249–258.

Kalaska, J. F., Cohen, D. A., Hyde, M. L., & Prud'homme, M. (1989). A comparison of movement direction-related versus load direction-related activity in primate cortex, using a two-dimensional reaching task. *Journal of Neuroscience, 9,* 2080–2102.

Kalaska, J. F., & Crammond, D. J. (1992). Cerebral cortical mechanisms of reaching movements. *Science, 255,* 1517–1523.

Kent, R. D., Kent, J. F., Rosenbek, J. C., Vorperian, H. K., & Weismer, G. (1997). A speaking task analysis of the dysarthria in cerebellar disease. *Folia Phoniatrica et Logopaedica, 49,* 63–82.

Knecht, S., Dräger, B., Deppe, M., Bobe, L., Lohmann, H., Flöel, A., et al. (2000). Handedness and hemispheric language dominance in healthy humans. *Brain, 123,* 2512–2518.

Krainik, A., Lehéricy, S., Duffau, H., Capelle, L., Chainay, H., Cornu, P., et al. (2003). Postoperative speech disorder after medial frontal surgery: role of the supplementary motor area. *Neurology, 60,* 587–594.

Kropotov, J. D., & Etlinger, S. C. (1999). Selection of actions in the basal ganglia-thalamocortical circuits: review and model. *International Journal of Psychophysiology, 31,* 197–217.

Krubitzer, L. A., & Kaas, J. H. (1990). The organization and connections of somatosensory cortex in marmosets. *Journal of Neuroscience, 10,* 952–974.

Kruger, L., & Porter, P. (1958). A behavioral study of the functions of the Rolandic cortex in the monkey. *Journal of Comparative Neurology, 109,* 439–469.

Laplane, D., Talairach, J., Meininger, V., Bancaud, J., & Orgogozo, J. M. (1977). Clinical consequences of corticectomies involving the supplementary motor area in man. *Journal of the Neurological Sciences, 34,* 301–314.

Larson, C. (2004). Neural mechanisms of vocalization. In R. D. Kent (Ed.), *The MIT encyclopedia of communication disorders* (pp. 59–63). Cambridge, MA: MIT Press.

Llano, D. (2013). Voices below the surface: is there a role for the thalamus in language? *Brain and Language, 126,* 20–21.

Lotze, M., Seggewies, G., Erb, M., Grodd, W., & Birbaumer, N. (2000). The representation of articulation in the primary sensorimotor cortex. *Neuroreport, 11,* 2985–2989.

Luppino, G., Matelli, M., Camarda, R., & Rizzolatti, G. (1993). Corticocortical connections of area F3 (SMA-proper) and area F6 (preSMA) in the macaque monkey. *Journal of Comparative Neurology, 338,* 114–140.

Malis, L. I., Pribram, K. H., & Kruger, L. (1953). Action potentials in "motor" cortex evoked by peripheral nerve stimulation. *Journal of Neurophysiology, 16,* 161–167.

Mateer, C. (1978). Asymmetric effects of thalamic stimulation on rate of speech. *Neuropsychologia, 16,* 497–499.

Marr, D. (1969). A theory of cerebellar cortex. *Journal of Physiology, 202,* 437–470.

Matsuzaka, Y., Aizawa, H., & Tanji, J. (1992). A motor area rostral to the supplementary motor area (presupplementary motor area) in the monkey: neuronal activity during a learned motor task. *Journal of Neurophysiology, 68,* 653–662.

Mendez, M. F. (2004). Aphemia-like syndrome from a right supplementary motor area lesion. *Clinical Neurology and Neurosurgery, 106,* 337–339.

Middleton, F. A., & Strick, P. L. (2000). Basal ganglia and cerebellar loops: motor and cognitive circuits. *Brain Research. Brain Research Reviews, 31,* 236–250.

Minagar, A., & David, N. J. (1999). Bilateral infarction in the territory of the anterior cerebral arteries. *Neurology, 52,* 886–888.

Mink, J. W. (1996). The basal ganglia: focused selection and inhibition of competing motor programs. *Progress in Neurobiology, 50,* 381–425.

Mink, J. W., & Thach, W. T. (1993). Basal ganglia intrinsic circuits and their role in behavior. *Current Opinion in Neurobiology, 3*, 950–957.

Mochizuki, H., & Saito, H. (1990). Mesial frontal lobe syndromes: correlations between neurological deficits and radiological localizations. *Tohoku Journal of Experimental Medicine, 161*(Suppl), 231–239.

Morel, A., Gallay, M. N., Baechler, A., Wyss, M., & Gallay, D. S. (2013). The human insula: architectonic organization and postmortem MRI registration. *Neuroscience, 236*, 117–135.

Mottolese, C., Richard, N., Harquel, S., Szathmari, A., Sirigu, A., & Desmurget, M. (2013). Mapping motor representations in the human cerebellum. *Brain, 136*, 330–342.

Németh, G., Hegedus, K., & Molnár, L. (1988). Akinetic mutism associated with bicingular lesions: clinicopathological and functional anatomical correlates. *European Archives of Psychiatry and Neurological Sciences, 237*, 218–222.

Nicolai, J., van Putten, M. J. A. M., & Tave, D. L. J. (2001). BIPLEDs in akinetic mutism caused by bilateral anterior cerebral infarction. *Clinical Neurophysiology, 112*, 1726–1728.

Nielsen, J. M., & Jacobs, L. L. (1951). Bilateral lesions of the anterior cingulate gyri; report of a case. *Bulletin of the Los Angeles Neurological Societies, 16*, 231–234.

Nota, Y., & Honda, K. (2003). Possible role of the anterior insula in articulation. In S. Palethorpe & M. Tabain (Eds.), *Proceedings of the 6th international seminar on speech production*, Sydney, Australia (pp. 191–194).

Numminen, J., & Curio, G. (1999). Differential effects of overt, covert and replayed speech on vowel-evoked responses of the human auditory cortex. *Neuroscience Letters, 272*, 29–32.

Ogar, J., & Gorno-Tempini, M. L. (2007). The orbitofrontal cortex and the insula. In B. L. Miller & J. L. Cummings (Eds.), *The human frontal lobes: functions and disorders* (2nd ed., pp. 59–67). New York: Guilford Press.

Okada, K., & Hickok, G. (2006). Left posterior auditory-related cortices participate both in speech perception and speech production: neural overlap revealed by fMRI. *Brain and Language, 98*, 112–117.

Okada, K., Smith, K. R., Humphries, C., & Hickok, G. (2003). Word length modulates neural activity in auditory cortex during covert object naming. *Neuroreport, 14*, 2323–2326.

Oorschot, D. E. (1996). Total number of neurones in the neostriatal, pallidal, subthalamic, and substantia nigral nuclei of the rat basal ganglia: a stereological study using the Cavalieri and optical dissector methods. *Journal of Comparative Neurology, 366*, 580–599.

Pai, M. C. (1999). Supplementary motor area aphasia: a case report. *Clinical Neurology and Neurosurgery, 101*, 29–32.

Palay, S. L., & Chan-Palay, V. (1974). *Cerebellar cortex: cytology and organization*. Berlin: Springer-Verlag.

Paulesu, E., Frith, C. D., & Frackowiak, R. S. (1993). The neural correlates of the verbal component of working memory. *Nature, 362*, 245–342.

Paus, T. (2001). Primate anterior cingulate cortex: where motor control, drive, and cognition interface. *Nature Reviews. Neuroscience, 2*, 417–424.

Paus, T., Petrides, M., Evans, A. C., & Meyer, E. (1993). Role of the human anterior cingulate cortex in the control of oculomotor, manual, and speech responses: a positron emission tomography study. *Journal of Neurophysiology, 70*, 453–469.

Pell, M. D. (1999). Fundamental frequency encoding of linguistic and emotional prosody in right hemisphere-damaged speakers. *Brain and Language, 69*, 161–192.

Penfield, W., & Rasmussen, T. (1950). *The cerebral cortex of man*. New York: MacMillan.

Penfield, W., & Roberts, L. (1959). *Speech and brain mechanisms*. Princeton, NJ: Princeton University Press.

Penfield, W., & Welch, K. (1951). The supplementary motor area of the cerebral cortex: a clinical and experimental study. *Archives of Neurology and Psychiatry, 66*, 289–317.

Picard, N., & Strick, P. L. (1996). Motor areas of the medial wall: a review of their location and functional activation. *Cerebral Cortex, 6*, 342–353.

Pickett, E. R., Kuniholm, E., Protopapas, A., Friedman, J., & Lieberman, P. (1998). Selective speech motor, syntax and cognitive deficits associated with bilateral damage to the putamen and the head of the caudate nucleus: a case study. *Neuropsychologia*, *36*, 173–188.

Pickles, J. O. (1988). *An introduction to the physiology of hearing* (2nd ed.). Bingley, UK: Emerald Group.

Poeppel, D. (2003). The analysis of speech in different temporal integration windows: cerebral lateralization as "asymmetric sampling in time." *Speech Communication*, *41*, 245–255.

Poldrack, R. A., Wagner, A. D., Prull, M. W., Desmond, J. E., Glover, G. H., & Gabrieli, J. D. E. (1999). Functional specialization for semantic and phonological processing in the left inferior prefrontal cortex. *NeuroImage*, *10*, 15–35.

Prescott, T. J., González, F. M. M., Gurney, K., Humphries, M. D., & Redgrave, P. (2006). A robot model of the basal ganglia: behavior and intrinsic processing. *Neural Networks*, *19*, 31–61.

Price, C. J. (2012). A review and synthesis of the first 20 years of PET and fMRI studies of heard speech, spoken language and reading. *NeuroImage*, *62*, 816–847.

Prud'homme, M. J. L., & Kalaska, J. F. (1994). Proprioceptive activity in primate primary somatosensory cortex during active arm reaching movements. *Journal of Neurophysiology*, *72*, 2280–2301.

Pugh, K. R., Mencl, W. E., Jenner, A. R., Katz, L., Frost, S. J., Lee, J. R., et al. (2001). Neurobiological studies of reading and reading disability. *Journal of Communicable Diseases*, *34*, 479–492.

Redgrave, P., Prescott, T. J., & Gurney, K. (1999). The basal ganglia: a vertebrate solution to the selection problem? *Neuroscience*, *89*, 1009–1023.

Richter, S., Dimitrova, A., Hein-Kropp, C., Wilhelm, H., Gizewski, E., & Timmann, D. (2005). Cerebellar agenesis: II. Motor and language functions. *Neurocase*, *11*, 103–113.

Riecker, A., Ackermann, H., Wildgruber, D., Dogil, G., & Grodd, W. (2000 a). Opposite hemispheric lateralization effects during speaking and singing at motor cortex, insula and cerebellum. *Neuroreport*, *11*, 1997–2000.

Riecker, A., Ackermann, H., Wildgruber, D., Meyer, J., Dogil, G., Haider, H., et al. (2000 b). Articulatory/phonetic sequencing at the level of the anterior perisylvian cortex: a functional magnetic resonance imaging (fMRI) study. *Brain and Language*, *75*, 259–276.

Riecker, A., Kassubek, J., Groschel, K., Grodd, W., & Ackermann, H. (2006). The cerebral control of speech tempo: opposite relationship between speaking rate and BOLD signal changes at striatal and cerebellar structures. *NeuroImage*, *29*, 46–53.

Rosenbek, J. C. (2004). Neurogenic mutism. In R. D. Kent (Ed.), *The MIT encyclopedia of communication disorders* (pp. 145–147). Cambridge, MA: MIT Press.

Rubens, A. B. (1975). Aphasia with infarction in the territory of the anterior cerebral artery. *Cortex*, *11*, 239–250.

Sahin, N. T., Pinker, S., & Halgren, E. (2006). Abstract grammatical processing of nouns and verbs in Broca's area: evidence from fMRI. *Cortex*, *42*, 540–562.

Sakai, S. T., Inase, M., & Tanji, J. (2002). The relationship between MI and SMA afferents and cerebellar and pallidal efferents in the macaque monkey. *Somatosensory & Motor Research*, *19*, 39–48.

Sakurai, Y., Momose, T., Iwata, M., Sudo, Y., Ohtomo, K., & Kanazawa, I. (2001). Cortical activity associated with vocalization and reading proper. *Brain Research. Cognitive Brain Research*, *12*, 161–165.

Schaltenbrand, G. (1975). The effects on speech and language of stereotactical stimulation in the thalamus and corpus callosum. *Brain and Language*, *2*, 70–77.

Schmahmann, J. D. (1997). *The cerebellum and cognition*. San Diego: Academic Press.

Schmahmann, J. D., Doyon, J., Toga, A. W., Petrides, M., & Evans, A. C. (2000). *MRI atlas of the human cerebellum*. San Diego: Academic Press.

Schmahmann, J. D., & Pandya, D. N. (1997). The cerebrocerebellar system. *International Review of Neurobiology*, *41*, 31–60.

Schoch, B., Dimitrova, A., Gizewski, E. R., & Timmann, D. (2006). Functional localization in the human cerebellum based on voxel-wise statistical analysis: a study of 90 patients. *NeuroImage*, *30*, 36–51.

Schwartz, A. B., Kettner, R. E., & Georgopoulos, A. P. (1988). Primate motor cortex and free arm movements to visual targets in three-dimensional space: I. Relations between single cell discharge and direction of movement. *Journal of Neuroscience, 8*, 2913–2927.

Shima, K., Mushiake, H., Saito, N., & Tanji, J. (1996). Role for cells in the presupplementary motor area in updating motor plans. *Proceedings of the National Academy of Sciences of the United States of America, 93*, 8694–8698.

Shima, K., & Tanji, J. (1998 a). Both supplementary and presupplementary motor areas are crucial for the temporal organization of multiple movements. *Journal of Neurophysiology, 80*, 3247–3260.

Shima, K., & Tanji, J. (1998 b). Role for cingulate motor area cells in voluntary movement selection based on reward. *Science, 282*, 1335–1338.

Shima, K., & Tanji, J. (2000). Neuronal activity in the supplementary and presupplementary motor areas for temporal organization of multiple movements. *Journal of Neurophysiology, 84*, 2148–2160.

Shuren, J. (1993). Insula and aphasia. *Journal of Neurology, 240*, 216–218.

Silveri, M. C., Di Betta, A. M., Filippini, V., Leggio, M. G., & Molinari, M. (1998). Verbal short-term store-rehearsal system and the cerebellum: evidence from a patient with a right cerebellar lesion. *Brain, 121*, 2175–2187.

Sörös, P., Sokoloff, L. G., Bose, A., McIntosh, A. R., Graham, S. J., & Stuss, D. T. (2006). Clustered functional MRI of overt speech production. *NeuroImage, 32*, 376–387.

Takai, O., Brown, S., & Liotti, M. (2010). Representation of the speech effectors in the human motor cortex: somatotopy or overlap? *Brain and Language, 113*, 39–44.

Tanji, J. (2001). Sequential organization of multiple movements: involvement of cortical motor areas. *Annual Review of Neuroscience, 24*, 631–651.

Tanji, J., & Shima, K. (1994). Role for supplementary motor area cells in planning several movements ahead. *Nature, 371*, 413–416.

Tourville, J. A., Reilly, K. J., & Guenther, F. H. (2008). Neural mechanisms underlying auditory feedback control of speech. *NeuroImage, 39*, 1429–1443.

Türe, U., Yasargil, D. C. H., Al-Mefty, O., & Yasargil, M. G. (1999). Topographic anatomy of the insular region. *Journal of Neurosurgery, 90*, 720–733.

Turkeltaub, P., Eden, G. F., Jones, K. M., & Zeffiro, T. A. (2002). Meta-analysis of the functional neuroanatomy of single-word reading: method and validation. *NeuroImage, 16*, 765–780.

Ulu, M. O., Tanriöver, N., Ozlen, F., Sanus, G. Z., Tanriverdi, T., Ozkara, C., et al. (2008). Surgical treatment of lesions involving the supplementary motor area: clinical results of 12 patients. *Turkish Neurosurgery, 18*, 286–293.

Urban, P. P., Marx, J., Hunsche, S., Gawehn, J., Vucurevic, G., Wicht, S., et al. (2003). Cerebellar speech representation: lesion topography in dysarthria as derived from cerebellar ischemia and functional magnetic resonance imaging. *Archives of Neurology, 60*, 965–972.

Van Buren, J. M. (1963). Confusion and disturbance of speech from stimulation in the vicinity of the head of the caudate nucleus. *Journal of Neurosurgery, 20*, 148–157.

Vigneau, M., Beaucousin, V., Hervé, P. Y., Duffau, H., Crivello, F., Houdé, O., et al. (2006). Meta-analyzing left hemisphere language areas: phonology, semantics, and sentence processing. *NeuroImage, 30*, 1414–1432.

von Cramon, D., & Jürgens, U. (1983). The anterior cingulate cortex and the phonatory control in monkey and man. *Neuroscience and Biobehavioral Reviews, 7*, 423–425.

Weintraub, S., Mesulam, M.-M., & Kramer, L. (1981). Disturbances in prosody: a right-hemisphere contribution to language. *Archives of Neurology, 38*, 742–744.

Wernicke, C. (1874). *Der aphasiche symptomencomplex*. Breslau: Cohn and Weigert.

Wildgruber, D., Ackermann, H., & Grodd, W. (2001). Differential contributions of motor cortex, basal ganglia, and cerebellum to speech motor control: effects of syllable repetition rate evaluated by fMRI. *NeuroImage, 13*, 101–109.

Wildgruber, D., Ackermann, H., Klose, U., Kardatzki, B., & Grodd, W. (1996). Functional lateralization of speech production at primary motor cortex: a fMRI study. *Neuroreport, 7*, 2791–2795.

Wise, R. J. S., Greene, J., Büchel, C., & Scott, S. K. (1999). Brain regions involved in articulation. *Lancet, 353*, 1057–1061.

Woolsey, C. N., Travis, A. M., Barnard, J. W., & Ostenso, R. S. (1953). Motor representation in the postcentral gyrus after chronic ablation of precentral and supplementary motor areas. *Federation Proceedings, 12*, 160.

Zemlin, W. R. (1998). *Speech and hearing science: anatomy and physiology* (4th ed.). Needham Heights, MA: Allyn & Bacon.

Zheng, T., & Wilson, C. J. (2002). Corticostriatal combinatorics: the implications of cortico-striatal axonal arborizations. *Journal of Neurophysiology, 87*, 1007–1017.

Ziegler, W., Kilian, B., & Deger, K. (1997). The role of the left mesial frontal cortex in fluent speech: evidence from a case of left supplementary motor area hemorrhage. *Neuropsychologia, 35*, 1197–1208.

3

Overview of Speech Motor Control

This chapter introduces basic concepts regarding the characterization of speech production as a control process. An overview of the control task facing the neural system responsible for speech production is first provided. The DIVA model of speech production is then introduced; this model will serve as a theoretical framework that will guide much of the remainder of this book. The chapter closes with a brief overview of the development of speaking skills in infants, with particular attention to the developmental time course for the main computational components of the speech motor control system represented in the DIVA model.

3.1 Segmental and Prosodic Components of Speech

Speech production is often viewed as the result of two somewhat separable control processes. The first process involves the creation of acoustic signals that convey the linguistic units of speech, that is, phonemes, syllables, and words. These units are referred to as *segments*[1] in the linguistics literature, and we will use the term *segmental control* to refer to this process. The second control process is concerned with *prosody*, which refers to the rhythm, stress, and intonation patterns of speech. These patterns are referred to as *suprasegmental* because they often span many segments. For example, the intonation contour of a sentence—in particular, whether or not the pitch rises at the end—conveys whether the utterance is a statement or a question. Because linguistic information is conveyed, this is an example of *linguistic prosody*. Prosodic information can also convey information regarding affect, such as happiness or sadness; this type of prosody is referred to as *affective prosody*.

As we will see in later chapters, these two control processes share many of the same neural substrates and appear to use somewhat similar control schemes; for example, both control processes involve a combination of feedforward and sensory feedback control. However, there are also important differences between these control processes, both in terms of functional requirements and neural regions involved. For example, segmental and prosodic control processes operate on different timescales. The segmental controller must be capable of generating the finely timed laryngeal and articulatory events that make up phonemes and syllables. These events typically vary in duration from about 10–200 ms during conversational speech. Prosodic events, in contrast, range from about 200 ms to several seconds in duration. The timing demands on prosodic control are thus much less stringent, and it stands to reason that brain areas involved in processing rapidly varying sensory stimuli or articulator movements will be more heavily involved in segmental control than prosodic control. We will focus primarily on segmental control for the remainder of this chapter as well as chapters 5 through 8. In chapter 9 we will return to the topic of prosody.

3.2 The Control Task of Speech Production

The term *motor control* refers to the control of movements of an effector or articulator system, such as a robotic arm or a human vocal tract. The basic control problem is illustrated in generic form in figure 3.1.

The *plant* is the physical system being controlled. For speech production, the plant consists of the musculature, bones, cartilage, and other tissues that make up the vocal tract. The input to the plant comes from motoneurons in the brain stem and spinal cord (see section 2.2 for details), and the output of the plant consists of articulator positions and an acoustic signal. Both the input and the output of the plant consist of continuously varying signals, and (in the general case) the output at a given point in time depends not only on the current input but also on previous inputs as well as the current and previous states of the plant.

The *controller* is the system that sends the continuously varying inputs to the plant in order to achieve the desired output. For speech production, the controller consists of the subset of the nervous system involved in speech motor control. We will refer to the output of the controller, which specifies the positions and velocities of the muscles and articulators, as the *motor command*.[2]

There are a number of ways to address the control problem in figure 3.1. These different methods can be broadly divided into two classes. *Feedforward control*, also called *open-loop control*, refers to a control system that does not monitor the outputs of the plant; instead it generates a stored control program for achieving the desired output. In the case

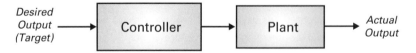

Figure 3.1
Generic motor control problem. The *controller* must send appropriate commands to the *plant* (the physical system being controlled) in order to achieve the desired output.

of speech production, a feedforward control system is one that generates the motor commands for a syllable or phoneme completely from memory, without monitoring the ongoing acoustic signal or monitoring the somatosensory state that results from these commands. *Feedback control*, also called *closed-loop control*, refers to a system that monitors the output of the plant in order to determine the command to send to the plant, for example by subtracting the actual output from the desired output to form an error signal that is then transformed into a corrective motor command. As we will see shortly, a particular control system may utilize both feedforward and feedback control mechanisms; such a system is involved in speech motor control.

Pure Feedback Control

In one of the first attempts to explain speech production using control theory concepts, Fairbanks (1954) proposed a *servosystem* (also called a *servo mechanism*, or just *servo*) model of speech production that utilized auditory feedback control. A basic auditory feedback servosystem for speech motor control is illustrated in figure 3.2. The input to the controller is a desired auditory signal, and the output of the plant is the actual auditory signal created by the vocal tract. Note that this controller does not address how the desired auditory signal, or *auditory target*, for a particular speech sound is generated—in other words, it does not address the transformation of a discrete speech segment into a corresponding acoustic signal—it only addresses how the desired auditory signal is achieved by the articulatory musculature.

In order to understand how the system in figure 3.2 works, consider a system consisting of a guitar string that you are trying to tune to a particular frequency by turning a tuning peg that tightens or loosens the string. In this system the guitar string and pegs are the plant, and you act as the controller in figure 3.2. The first step in the control process is determining the difference between the desired signal (in this case, the target tuning frequency) and the actual signal (the current frequency produced by the string). If the actual frequency is too low, an error signal is generated in your brain and you transform it into a rotation of the

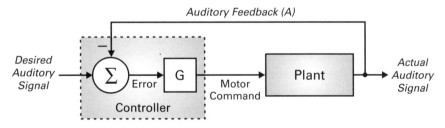

Figure 3.2
A simple *servosystem* control scheme for speech production (cf. Fairbanks, 1954). The desired auditory signal is compared to the actual auditory signal, and the difference (error) is multiplied by a gain factor (G) and transmitted to the input of the plant to serve as a corrective motor command.

peg that tightens the string, leading to a higher frequency of the actual signal. This process would continue until the string is tight enough to generate the proper frequency, at which point the error signal goes to zero and no further motor commands are issued.

The sign of the gain, labeled G in figure 3.2, determines the direction of the relationship between the error signal and the motor command. When the actual frequency is too low, the error signal is positive since it results from subtracting the actual from the desired frequency. If we assume that a positive motor command leads to a tightening movement and a negative motor command leads to a loosening movement, then the gain G would be positive in our example, thereby transforming a positive error into a tightening movement. If instead the rotational direction of the motor command was reversed such that a positive command loosened rather than tightened the string, then a negative gain would be needed.

The magnitude of the gain G determines the size of the corrective motor command for a given size of error. A high gain will lead to a system that attempts to very quickly correct for any errors. This may be desirable for a system in which there is little or no delay between the time the motor command is issued and the time its acoustic effect is available for comparison to the target by the controller. However, high gains become unacceptable in servosystems with significant feedback delays because they lead to *unstable behavior* such as oscillation. To see this, consider what would happen if the sound of the string were delayed by several seconds before being sent to your ear. When you initially hear that the string's frequency is too low, you start tightening the knob. However, the frequency that you hear will not change for several seconds because of the delay; you will still hear the frequency as too low even though you have tightened the knob, so you will continue tightening the knob during this delay period. By the time you hear the actual frequency reach the desired frequency and thus stop tightening the knob, you will have tightened it too much, and the actual frequency will *overshoot* beyond the target value. If the gain is too high, this will be followed by an overshoot in the opposite direction, and so on, resulting in unwanted oscillatory behavior. For this reason feedback control systems involving significant delays typically must use low gains, corresponding to very slow turns of the knob to correct perceived errors in our simplified system. In this case the knob will not be tightened all that much during the several seconds of delay, so there will be minimal overshoot and stable behavior can be maintained. Unfortunately, however, such a system is very sluggish and incapable of fast, stable movements.

In the feedback controller schematized in figure 3.2, the command sent to the plant is proportional to the error signal. More sophisticated feedback control schemes supplement this *proportional* command with additional commands that can, among other things, improve control in the face of delays such as those mentioned above. In *a proportional-derivative (PD) controller*, for example, the proportional command is supplemented with a command that is formed by multiplying the derivative of the error signal by a gain. The sign of this derivative indicates whether the error is increasing or decreasing. A positive derivative is indicative of increasing error; in this case the controller may need to send a larger control signal, which can be achieved by multiplying the derivative by a positive gain and adding that to the proportional command. In a *proportional-integral (PI) controller*, the proportional command is supplemented with a command formed by multiplying the integral of the error signal by a gain. If the integral of the error is large, then a lot of error has accumulated over time, which may indicate the need for a larger control signal; this can be achieved by multiplying the integral by a positive gain and adding it to the proportional command. Finally, a *PID controller* combines proportional, integral, and derivative control. It is likely that biological motor control involves proportional, integral, and derivative control components; for example, Bullock, Cisek, and Grossberg (1998) propose a neural model of how PID control of arm movements may be carried out by the cerebral cortex. We will primarily focus on the proportional component of feedback control in this book for clarity's sake, with the understanding that integral and derivative components are likely in place as well.

It has been noted that a pure auditory feedback control scheme such as that schematized in figure 3.2 cannot account for speech production by itself because of delays in the system related to the time it takes for a motor command to contract the muscle as well as the time it takes the auditory system to transduce the acoustic signal produced by the vocal tract (e.g., Neilson and Neilson, 1987). Furthermore, our ability to speak intelligibly even when we cannot hear ourselves indicates that our brains are not simply using an auditory feedback control scheme, even of the more sophisticated PID type, since all such schemes rely on detecting differences between the intended auditory signal and the actual auditory signal. Nonetheless, a simple feedback controller like the one in figure 3.2 adequately characterizes many important aspects of auditory and somatosensory feedback control in speech production, as discussed in chapters 5 and 6.

Feedback Error Learning

A number of methods have been developed to deal with problems that occur when using pure feedback control. For the purposes of this book, we will focus mainly on one of these methods, the *feedback error learning scheme* (e.g., Kawato & Gomi, 1992), which appears well suited for addressing speech acquisition and production. Figure 3.3 schematizes a feedback error learning system for speech motor control. The system combines a feedback control servosystem with a feedforward controller, and the overall motor command is the sum of a feedforward control command and a feedback control command. During initial movement attempts, the feedforward controller is not yet tuned and the motor commands are dominated by the feedback control system. The feedforward controller becomes tuned over time by using the output from the feedback controller as a *teaching signal* (dashed arrow) for updating the feedforward command for the next attempt. Eventually the feedforward command becomes accurate enough to carry out movements with little or no error, thus greatly reducing reliance on the feedback controller. This type of controller is an

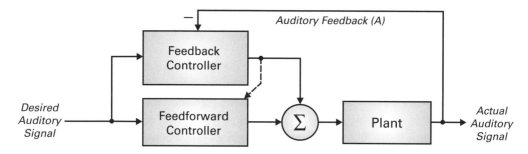

Figure 3.3
A *feedback error learning model* of speech production. In this system, the feedforward controller becomes tuned over time by using the output from the feedback controller as a teaching signal (dashed arrow) for updating the feedforward command for the next attempt.

example of *adaptive control* since the parameters of the controller change over time to correct for errors in performance. The neural controller for speech is an example of a sophisticated adaptive controller that is capable of adapting to the large changes in vocal tract morphology that occur over the course of a lifetime. In the next section we will describe the DIVA model, which is an adaptive neural network model that uses an elaborated feedback error learning scheme to account for the neural control of speech.

A note on forward models, inverse models, and internal models. Motor control researchers use several terms to refer to brain networks that encode some aspect of the relationship between motor commands and their sensory consequences. A *forward model* is a network that predicts the sensations that will arise when motor commands are executed. For example, when the muscles of the larynx are contracted to close the vocal folds while the respiratory muscles are contracting to build air pressure in the lungs, phonation occurs. The formant frequencies[3] of the resulting sound depend on the muscle commands sent to the tongue, jaw, and lip muscles. A forward model of this system would predict the resulting acoustic signal based on the motor commands that produce it. An *inverse model* encodes the opposite relationship: given a desired sensory consequence as input, it generates motor commands that achieve the sensory result. The controllers in figure 3.2 and figure 3.3 are examples of inverse models.

Unfortunately, these terms are not always used consistently in the literature. For example, some writers conflate the terms *forward model* and *feedforward command*, but the two terms mean different things. Further muddying the situation is the widely used but vague term *internal model*, which may refer to a forward model, an inverse model, or even a combination of the two. Because of these vagaries, the term *internal model* will be avoided in this book in favor of more precise terminology.

3.3 The DIVA Model

Since the early 1990s, our laboratory has developed and refined a neural network model that provides a quantitative account of the neural computations underlying speech motor control (Guenther, 1992, 1994, 1995, 2006; Guenther et al., 1998, 2006; Golfinopoulos et al., 2010; Tourville & Guenther, 2011; Guenther & Vladusich, 2012). Because a central aspect of the model concerns how the brain transforms desired movement *d*irections (in sensory space) *i*nto *v*elocities of the *a*rticulators (described further below), the model is called the *DIVA model*. The model has been refined over time via a repeated cycle of (1) generating predictions from the model in the form of articulator movements, brain activity, and/or acoustic output for a particular speech task (such as producing syllables while receiving a jaw or lip perturbation); (2) collecting experimental data from human subjects performing the same speech task; (3) comparing the model's outputs to those of the human subjects; and (4) refining the model if necessary to account for any aspects of the experimental data that mismatch model predictions.

The components of the DIVA model correspond to brain regions involved in the generation of articulator movements that produce desired speech sound "chunks," with each sound chunk having a stored *motor program* (or *motor target*) for its production. For brevity we will simply use the term *speech sound* to refer to a sound chunk with its own motor program throughout this book. These chunks can be phonemes, syllables, or words. In keeping with a number of prior proposals (e.g., Kozhevnikov & Chistovich, 1965; Compton, 1981; Levelt, 1989; MacNeilage & Davis, 1990; Levelt & Wheeldon, 1994), we hypothesize that the *syllable* is the most common sound chunk with an optimized motor program. However, we hypothesize that motor programs also exist for individual phonemes as well as very common multisyllabic utterances, for example, the names of family members or pets. The model presumes that higher-level brain regions translate an intended linguistic message into a sequence of speech sounds, and that motor sequencing circuits activate the appropriate nodes[4] of a speech sound map that acts as the highest processing level in the DIVA model. The motor sequencing circuitry for speech is addressed in chapter 8.

The DIVA model is defined at two levels. The first level describes the network of brain regions involved in speech motor control, illustrated by the DIVA model schematic provided in figure 3.5 (in a later section). The second level of description is a computer-implemented *artificial neural network* (*ANN*) consisting of equations for neural activities and synaptic weights, as introduced in chapter 1, section 1.3. For tractability, these equations currently involve only a subset of the brain regions treated in the first level of description. The neural network is implemented in computer simulations in which it controls an articulatory synthesizer (e.g., Maeda, 1990; Nieto-Castanon et al., 2005; see chapter 1, section 1.3) in order to produce simulated articulator movements and acoustic signals.

The remainder of this section provides an overview of the DIVA model, to be fleshed out in later chapters. We start by describing the control scheme implemented in computer simulations of the DIVA model neural network (i.e., the second level of definition described above). This is followed by a description of the model in terms of the brain regions that implement this control scheme (i.e., the first level of description described above), followed by an overview of learning in the model.

The DIVA Model Control Scheme

The control scheme utilized by the DIVA model is depicted in figure 3.4. This scheme is similar to the feedback error learning control scheme illustrated in figure 3.3. However, there are several key differences. The first is the inclusion of a somatosensory feedback controller in addition to the auditory feedback controller. The second concerns the input to the system. The feedback error learning system of figure 3.3 is provided with the desired auditory signal (or auditory target) that characterizes the speech sound to be produced. In the DIVA model, production of a speech sound starts with activation of a neural representation of the sound in a speech sound map (to be defined further below) by higher-level language processing areas in the brain, which in turn leads to the readout of auditory, somatosensory, and motor targets. A third key difference between the feedback error learning system of figure 3.3 and the DIVA control scheme is that learning is not limited to the feedforward controller in the DIVA model; several additional components of the model are adaptive, as described later in this chapter.

In later chapters, it will be illuminating to investigate a few of the mathematical properties of the control scheme illustrated in figure 3.4. To do this, we will need to define a number of variables where each takes the form of a vector representing information within a particular reference frame (see chapter 1, section 1.2, for a discussion of reference frames involved in speech production). Three reference frames are represented in figure 3.4: a motor reference frame, an auditory reference frame, and a somatosensory reference frame.

The *motor state* (M) is a vector whose components (or dimensions) correspond to the commanded positions of the articulators responsible for speech production, including the tongue, lips, jaw, larynx, and soft palate. To fix ideas, it is useful to think of M as the motor cortex representation of the speech articulator positions. Most of the DIVA model simulations to date have involved an articulatory synthesizer that utilizes a set of 10 articulators; in this case M is a 10-dimensional vector, with each component of M representing the position of one of the 10 articulators. Although based on actual vocal tract movements of human speakers (see Maeda, 1990, and Nieto-Castanon et al., 2005), this articulator representation is highly simplified compared to a real vocal tract. A more realistic treatment would include many more dimensions. For our current purposes the exact number of dimensions of M is not important; it suffices to note that the components of M together fully specify the shape of the vocal tract.

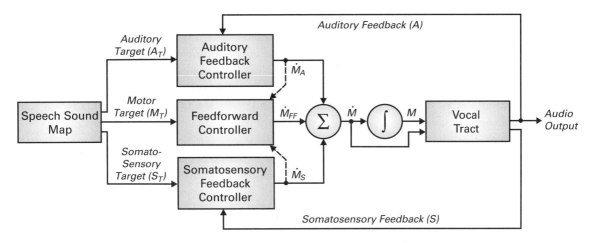

Figure 3.4
Control scheme utilized by the DIVA model for speech sound production. Production starts with activation of a neural representation of the sound in a *speech sound map*. This causes the readout of three signals: (1) a desired auditory signal (*auditory target,* characterized mathematically as a vector of auditory parameters, denoted by A_T) for the sound; (2) a desired somatosensory signal (*somatosensory target*, characterized by the vector S_T); and (3) a stored motor program (*motor target*, characterized by the vector M_T) for producing the sound. Auditory and *somatosensory feedback controllers* compare incoming auditory (A) and somatosensory (S) feedback to A_T and S_T, respectively. If there is a mismatch between the desired and actual sensory feedback signals, corrective movement commands ($\dot{M}_A$ and $\dot{M}_S$) are generated, where the over-dot represents the time derivative; in other words $\dot{M}_A$ and $\dot{M}_S$ represent motor velocities. The *feedforward controller* generates a feedforward movement command ($\dot{M}_{FF}$) by comparing the motor target to the current motor state (projection of M to the feedforward controller is not shown for clarity). The three movement commands ($\dot{M}_A$, $\dot{M}_S$, and $\dot{M}_{FF}$) are summed to produce the overall movement command $\dot{M}$, which is sent to the vocal tract along with a motor position command M formed by integrating $\dot{M}$ over time. The corrective motor commands $\dot{M}_A$ and $\dot{M}_S$ are also used to update the motor target for future productions of the syllable (dashed arrows).

The *auditory state* (A) represents the perceptually important parameters of the current auditory signal such as the fundamental frequency and formant frequencies. A can be thought of as the representation of the acoustic signal in auditory cortex. Again, the exact number of dimensions of the auditory space is not important; it suffices to note that it contains all of the perceptually relevant aspects of the acoustic signal.

The *somatosensory state* (S) represents the somatosensory parameters that are important for speech production, such as the locations and degrees of key constrictions in the vocal tract (as determined from proprioceptive signals from muscle spindles) and tactile patterns on the tongue, lips, and palate. S can be thought of as the representation of afferent somatosensory signals in somatosensory cortex. The dimensionality of S is expected to be higher than the dimensionality of M since it contains tactile information in addition to articulator position information.

With these state variables defined, we can now return to the control scheme illustrated in figure 3.4. Production of a speech sound begins with activation of a neural representation of the sound in a speech sound map. This leads to the readout of three signals that define target values for the motor, auditory, and somatosensory states.

The *auditory target* (A_T) is a vector with the same dimensionality as A. Each component of A_T specifies a desired value of an auditory parameter.[5] The *auditory feedback controller* compares A to A_T and generates *corrective movement commands* ($\dot{M}_A$) if there is a mismatch between A and A_T. The dot above the M indicates the derivative with respect to time, which is equivalent to saying that $\dot{M}_A$ specifies velocities of the speech articulators. This process is treated in detail in chapter 5.

The *somatosensory target* (S_T) is a vector with the same dimensionality as S whose components specify desired values for the somatosensory parameters represented by S. The somatosensory feedback controller compares S to S_T and generates *corrective movement commands* ($\dot{M}_S$) if there is a mismatch between S and S_T. This process is detailed in chapter 6.

The *motor target* (M_T) is a stored set of motor commands for producing the sound. The feedforward controller compares M_T to M to determine the *feedforward movement command* ($\dot{M}_{FF}$). M_T is tuned over time by incorporating corrective movements generated by the auditory and somatosensory feedback control subsystems; this is indicated by the dashed arrows in figure 3.4 and is detailed in chapter 7.

The *overall movement command* ($\dot{M}$) is formed by summing the outputs of the three controllers described above:

$$\dot{M} = \dot{M}_A + \dot{M}_S + \dot{M}_{FF}. \tag{3.1}$$

The effect of $\dot{M}$ is to change the motor state M, which is formed by integrating $\dot{M}$ over time. Both the motor velocity command $\dot{M}$ and the motor position command M are then sent to the articulators of the vocal tract in order to generate the movements for producing the sound.

Neural Correlates of the DIVA Model

Figure 3.5 illustrates the neural correlates of the DIVA model. Stereotactic coordinates associated with the model components are provided in appendix D. Each box in figure 3.5 corresponds to a set of model nodes that collectively form a *neural map* that represents a particular type of information in the model. Larger boxes represent cortical regions, and smaller boxes represent subcortical regions. Arrows correspond to excitatory axonal projections between neural maps, and lines terminating in circles represent inhibitory projections. These projections can transform neural information from one reference frame into another; such transformations are encoded in the synapses at the ends of the axonal projections.

Each model node is localized in the *Montreal Neurological Institute* (*MNI*) stereotactic space, allowing the generation of fMRI-like brain activity from model simulations. The articulator movements, acoustic output, and brain activity produced by simulations of the

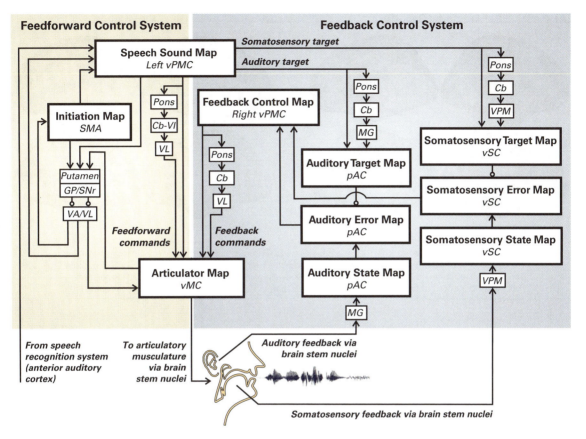

Figure 3.5
Neural correlates of the DIVA model. Cb, cerebellum; Cb-VI, cerebellum lobule VI; GP, globus pallidus; MG, medial geniculate nucleus of the thalamus; pAC, posterior auditory cortex; SMA, supplementary motor area; SNr, substantia nigra pars reticula; VA, ventral anterior nucleus of the thalamus; VL, ventral lateral nucleus of the thalamus; vMC, ventral motor cortex; VPM, ventral posterior medial nucleus of the thalamus; vPMC, ventral premotor cortex; vSC, ventral somatosensory cortex.

model can be directly compared to the results of speech experiments. For example, figure 3.6 compares brain activity measured with fMRI while human subjects read aloud single syllables (panel A) to brain activity generated by a computer simulation of the DIVA model producing similar utterances (panel B). Details regarding DIVA simulations of brain activity are provided in appendix D.

Before describing the model's components, it is helpful to address the different ways that regions of the cerebral cortex can be defined. Chapter 2, section 2.4, introduced parcellation schemes based on gross anatomical landmarks such as sulci and gyri (see appendix B) and on cytoarchitecture (in particular, Brodmann's scheme). In addition, cortical regions

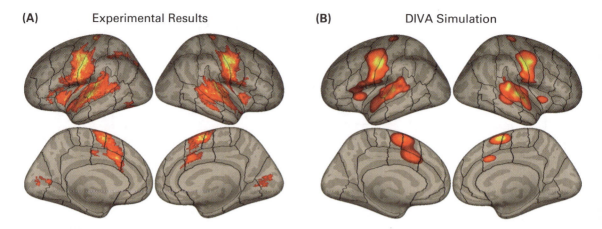

Figure 3.6
(A) Brain activity measured with fMRI while subjects read syllables aloud, contrasted with a baseline involving passive viewing of letters (see caption of figure 1.8 in chapter 1 for details) plotted on inflated cortical surfaces. Top panels show left and right lateral views; bottom panels show left and right medial views. (B) Simulated fMRI activations derived from the DIVA model's node activities during syllable production (see appendix D for details).

are often differentiated based on their functional properties (*functional parcellation*)—for example, primary auditory cortex or premotor cortex. In many cases, there is reasonably good agreement across these methods. For example, primary somatosensory cortex corresponds to BA 1, 2, and 3 and is located in the postcentral gyrus. Often, however, the correspondence between the different parcellation methods is complicated. For example, BA 4, considered to be the primary motor cortex, is typically associated with the precentral gyrus, but the more rostral portion of the ventral precentral gyrus is occupied by BA 6, which is considered premotor cortex. No gross anatomical marker divides the two regions, and BA 6 extends rostrally beyond the precentral gyrus. Beyond primary and secondary cortices, the correspondence between the three parcellation methods becomes even less straightforward. It is common practice in the neuroimaging literature to use a mixture of functional, structural, and cytoarchitectonic definitions when describing cortical areas. This practice can unfortunately result in ambiguities and inconsistencies. The problem is exacerbated when one compares regions in a neural model to brain locations from neuroimaging or cytoarchitectonic studies. To partially alleviate this problem, *cortical regions of the DIVA model will be defined in functional terms since it is their function that is directly being modeled*. These *functional maps* may span multiple cytoarchitectonic or structural regions and may vary somewhat in anatomical location across individuals. When relevant, the presumed cytoarchitectonic and/or structural locations of these functional maps will be described in the text.

As noted above, production of a speech sound in the DIVA model starts with activation of a node representing that particular sound in a *speech sound map*. Each speech sound map node is hypothesized to correspond to an ensemble of neurons primarily located in the left *ventral premotor cortex* (*vPMC*), which includes the rostral portion of the ventral precentral gyrus along with neighboring regions in the posterior inferior frontal gyrus and anterior insula. Activation of this node leads to motor commands that arrive in motor cortex via two control systems: a *feedforward control system* and a *feedback control system*. The feedback control system can be further broken into two components: an *auditory feedback control subsystem* and a *somatosensory feedback control subsystem*.

The *feedforward control system* is responsible for generating previously learned motor programs for speech sounds. This process involves two components. The first component is responsible for launching the motor program at the appropriate instant in time. This is carried out by a cortico–basal ganglia loop that involves an *initiation map* in the *supplementary motor area* (*SMA*) located on the medial wall of the frontal cortex. This loop is responsible for identifying the proper cognitive and sensorimotor context for producing the speech sound. For example, when saying the word "enter," the proper context for initiating the motor program for "ter" consists of (1) a cognitive context involving the desire to say the word "enter" and (2) a sensorimotor context signaling the impending completion of articulation for "en." We hypothesize that these contextual cues are monitored by the basal ganglia, with sensorimotor signals monitored by the putamen and cognitive signals monitored by the caudate nucleus (not shown in figure 3.5 as this component is not included in the DIVA model; see chapter 8 for a treatment of the involvement of the caudate nucleus in speech sequencing). When the appropriate context for a sound is identified, a corresponding node is activated in the initiation map via the *globus pallidus* (*GP*) and *substantia nigra pars reticula* (*SNr*) of the basal ganglia and the *ventral anterior* (*VA*) and/or *ventral lateral* (*VL*) *nuclei* of the thalamus. Activation of the initiation map node starts the readout of the learned motor program for the current speech sound. The second component of the feedforward control system comprises the motor programs themselves, which are responsible for generating feedforward commands for producing learned speech sounds. These commands are encoded by synaptic projections from the speech sound map to an *articulator map* in the *ventral primary motor cortex* (*vMC*) of the precentral gyrus bilaterally.[6] The cortico-cortical projections from left vPMC to vMC are supplemented by a cerebellar loop passing through the *pons, cerebellar cortex lobule VI* (*Cb-VI*), and the VL nucleus of the thalamus.

The *auditory feedback control subsystem* is responsible for detecting and correcting differences between the desired auditory signal for a speech sound and the current auditory feedback. According to the DIVA model, axonal projections emanate from speech sound map nodes—both directly and via a cortico-cerebellar loop involving the pons, *cerebellum*[7] (*Cb*), and *medial geniculate* (*MG*) *nucleus* of the thalamus—to an *auditory target map* in the higher-order auditory cortical areas in *posterior auditory cortex* (*pAC*), including the planum temporale and the posterior superior temporal gyrus and sulcus. These projections encode the expected auditory signal for the speech sound currently being produced. Activity in the auditory target map thus represents the auditory feedback that should arise when the speaker hears himself or herself producing the current sound. The targets consist of time-varying *regions* (or ranges) that encode the allowable variability of the acoustic signal throughout the syllable. The use of target regions rather than point targets is an important aspect of the DIVA model which provides a unified explanation for a wide range of speech production phenomena, including motor equivalence, contextual variability, anticipatory coarticulation, carryover coarticulation, and speaking rate effects (Guenther, 1995); this topic will be explored in chapter 4.

The auditory target for the current sound is compared to incoming auditory information from the auditory periphery; this information projects to cortex via MG and is represented in the model's *auditory state map*. If the current auditory feedback is outside the target region, auditory error nodes in the higher-order auditory cortical areas become active (*auditory error map* in figure 3.5). Like the auditory target map, the auditory state and error maps are hypothesized to lie in pAC. Auditory error node activities are then transformed into corrective motor commands through projections from the auditory error nodes to the *feedback control map* located primarily in right vPMC, which in turn projects to the articulator map in vMC both directly and via a loop through the pons, Cb, and VL.

The DIVA model posits a *somatosensory feedback control subsystem* operating alongside the auditory feedback control subsystem described above. The main components of the somatosensory feedback control subsystem are hypothesized to reside in *ventral somatosensory cortex* (*vSC*), including the ventral postcentral gyrus and the supramarginal gyrus. Projections from the speech sound map to the *somatosensory target map*, including cortico-cortical as well as cortico-cerebellar loop projections via the *ventral posterior medial* (*VPM*) *nucleus* of the thalamus, encode the expected somatosensory feedback (i.e., tactile and proprioceptive feedback arising from mechanoreceptors and muscle spindles in the vocal tract) during sound production. The model's *somatosensory state map* represents tactile and proprioceptive information from the speech articulators, which arrives from cranial nerve nuclei in the brain stem via VPM. Nodes in the *somatosensory error map* become active during speech if the speaker's somatosensory state deviates from the somatosensory target region for the sound being produced. The output of the somatosensory error map then propagates to the feedback control map to transform somatosensory errors into motor commands that correct those errors.

A Note Regarding Speech Motor Programs
The notion of a stored representation of well-learned movements that can be read out from memory without reliance on sensory feedback dates back at least as far as William James's classic 1890 text *The Principles of Psychology*. Since that time, the term *motor program* has often been used by the motor control community to refer to this general idea. However,

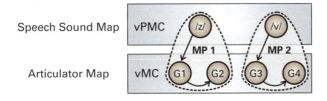

Figure 3.7
Schematic of the cortical components of motor programs for the phonemes /z/ and /v/ in the DIVA model. G, gesture; MP, motor program; vMC, ventral motor cortex; vPMC, ventral premotor cortex.

the details of what exactly constitutes a motor program vary significantly from proposal to proposal. These details include things like whether the motor program includes kinematic or dynamic information (or both) and where in the brain it is implemented. Since the concept of a speech motor program is central to much of the discussion in this book, it is important to define the term more precisely.

A speech motor program is defined herein as a set of neurons and their interconnections that, when activated, generate a learned set of articulator movements that carry out a speech sound. The focus will be on movements defined kinematically, though this is not meant as a claim that motor programs do not also include dynamic information—instead it is a simplification for the sake of tractability. Importantly, the motor program is not situated entirely in a single brain region but instead spans multiple interconnected regions.

Figure 3.7 schematizes cortical components of speech motor programs for the phonemes /z/ and /v/ within the DIVA model. The motor programs are indicated by dashed lines and are labeled MP1 and MP2. According to the model, each learned speech sound has a node representing it in the speech sound map, hypothesized to reside in left vPMC. This node projects to neurons in vMC that are responsible for carrying out the articulatory gestures (indicated as G1 and G2 for MP1 in figure 3.7) that produce the sound. Together, these nodes and their interconnections represent the cortical component of the motor programs. Although not shown in figure 3.7, the motor programs also contain projections through subcortical loops involving the basal ganglia and cerebellum, as will be detailed in subsequent chapters.

In terms of the control diagram of figure 3.4, the motor program for a sound most closely corresponds to the motor target representation for the sound, indicated by M_T. The term might also be applied to the motor commands generated by the feedforward controller, indicated by $\dot{M}_{FF}$, though the term *feedforward commands* will typically be used to refer to these commands.

Learning in the DIVA Model

Before the DIVA model can produce speech sounds, it must undergo a motor learning process analogous to that of the developing infant. For model simulations, this process is highly simplified, consisting of only two stages: a *babbling phase* and an *imitation phase*.

In the *babbling phase*, semi-random articulator movements are generated, and the model receives corresponding auditory and somatosensory feedback. The resulting combination of articulatory, auditory, and somatosensory information is used to tune inverse models that map somatosensory and auditory errors into corrective motor commands via the feedback control map in figure 3.5. The learning in this stage is not phoneme- or syllable-specific learning; the learned sensory-motor transformations will be used for all speech sounds that will be learned later.

In the *imitation phase*, the model is presented with sample speech sounds to learn, much like an infant is exposed to the sounds of his or her native language. These sounds take the form of time-varying acoustic signals corresponding to phonemes, syllables, or words spoken by a human speaker. Based on these samples, the model learns an auditory target for each sound. Learning of a sound's auditory target involves activation of a speech sound map node that will later represent the sound for production. This occurs via the *speech recognition system* when the model "hears" the sound spoken by a fluent individual, corresponding to a child's hearing a new speech sound directed at him or her by a parent, for example.[8] This in turn leads to tuning of the synapses projecting from that speech sound map node to the auditory cortex to encode the sound's auditory target.

After an auditory target for a sound has been learned, the model can attempt to produce the sound. This is carried out by activating the corresponding speech sound map node along with a node in the initiation map. On the first attempt, the model will not have a tuned motor program (i.e., the *motor target* in figure 3.4) for producing the sound in a feedforward manner, nor will it have a somatosensory target; thus it must rely heavily on the auditory feedback control subsystem to produce the sound. On each production attempt, the motor target is updated to incorporate the commands generated by the auditory feedback control subsystem on that attempt. This results in a more accurate set of feedforward commands for the next attempt. Eventually the feedforward commands by themselves are sufficient to produce the sound in normal circumstances. That is, the motor program is accurate enough that it generates very few auditory errors during production of the sound and thus does not invoke the auditory feedback control subsystem. At this point the model can fluently produce the speech sound. As the speech articulators get larger with growth, the auditory feedback control subsystem continually corrects for changes in the biomechanics of the vocal tract. These corrective commands are subsumed into the motor program, thus allowing it to stay tuned despite dramatic changes in the sizes and shapes of the speech articulators over the course of a lifetime (see Callan et al., 2000, for computer simulations verifying this property of the model).

As the model repeatedly produces a speech sound, it also learns a somatosensory target region for the sound, analogous to the auditory target region mentioned above. This target represents the expected tactile and proprioceptive sensations associated with the sound and is used in the somatosensory feedback control subsystem to detect somatosensory errors. Unlike the auditory target for a sound, the somatosensory target cannot be learned in its entirety from another speaker since crucial information about tongue shape, tactile patterns, and so forth are not available to the listener. Instead, the somatosensory target must be learned by monitoring one's own correct self-productions, a learning process that occurs later than the learning of auditory targets in the model.

The learning processes in the DIVA model represent only a subset of the developmental processes that occur in the brain of a child learning to speak. The following section provides a more thorough account of development in the speech motor system.

3.4 Speech Motor Development

The speech motor control network cannot be "hardwired" at birth. Consider, for example, that fluent speakers of a given language are no longer fluent when they attempt to produce sounds from foreign languages—in other words, our speech motor systems become specialized for the sounds and corresponding articulatory gestures of our particular language. Furthermore, once we learn to produce the sounds of our native language, we remain capable of producing them throughout our lifetime despite dramatic changes in morphology of the vocal tract that occur as we grow from children to adults; these changes affect the biomechanical and acoustic properties of the vocal tract, which in turn requires the controller to continually adjust its commands in a way that accounts for these changes.

Tuning of the speech motor system involves a number of different, though interrelated, learning processes that shape the computations performed by the speech network. This section provides a brief description of these processes, with more detailed treatments occurring in later chapters. This description is preceded by a short overview of behavioral aspects of infant babbling; it is through these behaviors that various components of the speech network become tuned.

Infant Babbling Stages

Babbling during an infant's first year of life is made up of several overlapping stages (e.g., Oller, 1980; Stark, 1980). In the first 2 months, infants pass through a *phonation stage* (Oller, 1980) wherein speech-like sounds are relatively rare. The few speech-like sounds that are seen at this stage consist largely of phonation with the mouth closed or nearly closed. This is followed by a *"goo" stage* (2 to 3 months of age) wherein infants begin to produce very crude syllable-like sequences consisting largely of velar consonant-like elements in combination with vowel-like elements. At about 4 to 6 months of age, most infants enter the *expansion stage*, characterized by the emergence of several new sound types, including bilabial and labiolingual trills ("raspberries"), squeals, growls, and a relatively small amount of *marginal babbling* consisting of vocal tract closures in combination with better-formed vowel-like utterances. These syllable-like utterances still differ significantly from adult syllables in their interarticulator timing and duration. At about 7 months of age, infants enter the *canonical babbling* stage (also called the *reduplicated babbling* stage; Stark, 1980) where for the first time syllables with adult-like timing characteristics are frequently seen. Many of the infant's utterances during this stage are reduplicated syllables such as "dadada." At the age of approximately 10 months, infants enter a stage known as *variegated* or *nonreduplicated babbling*, characterized by the use of different consonants and vowels within the same babbling sequence (e.g., "badadi"). MacNeilage and Davis (1990) have hypothesized that the variegated babbling stage is the stage during which infants begin learning to produce the various phonemes of their native language.

One conclusion that can be drawn from these findings is that many nonspeech vocalizations and articulator movements occur well before the onset of frequent speech sounds. In accordance with this view, the speech motor learning process in the DIVA model occurs in two stages: an early stage during which sensory-motor relationships between the auditory, somatosensory, and motor systems are learned, and a later stage in which individual speech sounds from the native language are learned. Although these stages are typically carried out sequentially in model simulations for convenience, in actuality the speech motor learning process is not so discrete—babbling in infants only gradually becomes less random and more like adult speech in the native language environment (e.g., de Boysson-Bardies, Sagart, & Durand, 1984; de Boysson-Bardies et al., 1989), suggesting substantial overlap between the learning of general sensory-motor transformations that characterize vocal tract movements and the learning of individual speech sounds from the native language. Infant babbles are already shaped by their linguistic environment by 8 months of age (de Boysson-Bardies, Sagart, & Durand, 1984), presumably through auditory input from speakers of the native language.[9]

The following subsections address the development of the main computational components of the speech motor system as characterized by the DIVA model (figure 3.5). Model component names are highlighted in italics in the passages that describe their development.

Development of Sensory Maps
The ability to produce the speech sounds of a language depends heavily on the ability to perceive these sounds, both in terms of their auditory properties (such as formant frequencies) and in terms of phonological units such as phonemes and syllables. Regarding the former, researchers have established that the auditory representation of an infant, corresponding to the *auditory state map* in the DIVA model (figure 3.5), already shows signs of language specificity by 6 months of age (e.g., Kuhl et al., 1992); in other words, exposure to our native language shapes the development of our auditory maps in order to optimally extract information that is important for understanding and speaking our native language. One well-known example of this is the inability of native Japanese listeners to

reliably distinguish English /r/ from /l/. These sounds differ primarily in terms of their third formant frequency. Although Japanese and American infants are equally sensitive to this distinction at around 6 to 8 months of age, by about 10 to 12 months of age the Japanese infants have lost sensitivity to the F3 difference, which does not carry phonological information in their native language (Tsushima et al., 1994; Kuhl et al., 2006). It is noteworthy that much of the shaping of our auditory representations by our native language can occur at a very early stage of development since it can be driven by the statistical nature of the acoustic signals that are processed by our auditory system without knowledge of the phonological units that make up the language (e.g., Guenther & Gjaja, 1996; Guenther et al., 1999).

The somatosensory representations involved in speech, corresponding to the *somatosensory state map* in figure 3.5, must also undergo development. Unlike auditory signals for speech, which infants receive from fluent speakers of the native language starting from birth (and probably even in utero; e.g., Graven & Brown, 2008), the somatosensory patterns associated with the sounds of a language cannot be gleaned from other speakers. Vision can provide some somatosensory information about speech—movements of a speaker's lips and jaw can be detected, for example—but infants cannot know what it "feels like" to produce most speech gestures until they have produced them themselves. Thus, development of the somatosensory maps for speech likely lags behind development of auditory maps during the very early stages of infancy, when self-generated movements of the speech articulators are limited. As the infant starts producing more and more speech-like articulatory movements in the expansion, canonical babbling, and variegated babbling stages, it is likely that the infant's somatosensory maps become increasingly sensitive to the somatosensory patterns resulting from these movements. As the developing child's articulatory gestures become more and more language specific, the somatosensory maps likely gain sensitivity to the gestures of the native language and lose sensitivity to gestures that are not used in the language, in a manner that mirrors the development of auditory maps described above. Furthermore, the somatosensory maps underlying speech must remain plastic at least into adulthood since the somatosensory pattern resulting from an articulatory gesture will change as the vocal apparatus changes size and shape as the child grows.

So far we have described development in the auditory and somatosensory systems as occurring independently. However, it is likely that the infant's brain also learns the relationships between auditory and somatosensory patterns. Clear evidence of cross-modal sensory interactions comes from a number of experiments. For example, it has been shown that somatosensory stimulation can modulate auditory perceptual responses under some conditions (Ito, Tiede, & Ostry, 2009), auditory stimulation can alter skin sensation (Ito & Ostry, 2012), and viewing speech articulation can alter auditory perception, a phenomenon referred to as the McGurk effect (McGurk & MacDonald, 1976).

Learning of Sensory-Motor Transformations

The first movements of speech-related body parts begin almost immediately after birth, when an infant uses his or her vocal folds and respiratory system to cry and their lips, jaw, and tongue to feed. These movements generate somatosensory feedback and, in some cases, auditory feedback. As mentioned above, our motor systems have the ability to predict sensory effects of movements commanded by motor cortical activity via forward models. Tuning of these predictions likely begins with these early nonspeech actions and then accelerates as the infant creates more and more speech-like utterances as he or she moves through the goo, expansion, canonical, and variegated babbling stages.

The articulatory movements generated during babbling can also be used to tune transformations in the reverse direction, that is, sensory-to-motor transformations, or inverse models. One particular form of inverse model that is crucial for motor control relates changes in sensory state to the motor actions that cause them. Suppose, for example, that a babbling child makes an articulatory movement that causes the formant frequencies to change in a particular direction, for example, increasing F1 and decreasing F2, and that the child's brain learns an association between this direction of change in formant frequency space and the articulator velocities that produced it. Later, the child attempts to produce a newly learned syllable, and the child's auditory system detects that F1 is too low and F2 is too high. Correcting this error requires an increase in F1 and a decrease in F2, which can be carried out by the learned association. Such transformations from sensory *d*irections *i*nto *v*elocities of *a*rticulators give the DIVA model its name, and they are embodied by projections from the *auditory error map* and *somatosensory error map* to the *articulator map* via the *feedback control map* in figure 3.5. It is noteworthy that, prior to the development of auditory and somatosensory targets for speech sounds, nodes in the auditory and somatosensory error maps are not yet signaling "errors" per se but instead represent changes in the auditory and somatosensory state that are occurring because of ongoing movements of the speech articulators. This allows tuning of auditory-motor and somatosensory-motor transformations well before an infant develops awareness of phonological units such as phonemes and words. As auditory and somatosensory targets are learned, the nodes in the auditory and somatosensory error maps begin to reflect *desired* changes in sensory state (corresponding to differences between sensory targets and the current sensory state) rather than *ongoing* changes in the sensory state.

Tuning of sensory-motor transformations in the speech network does not stop when babbling ceases; this tuning must continue as the child grows and the biomechanics of his or her vocal apparatus change. In fact, these transformations appear to remain plastic well into adulthood and possibly throughout the life span; this is evidenced by the lasting changes in articulation that arise when auditory or somatosensory feedback is systematically perturbed in adults (e.g., Houde & Jordan, 1998; Tremblay, Shiller, & Ostry, 2003; Villacorta, Perkell, & Guenther, 2007).

Development of Phonological Awareness and Speech Recognition

The learning processes described thus far do not require explicit knowledge of the phonemes, syllables, or words of a language. Instead, they tune transformations between the largely continuous auditory, somatosensory, and motor spaces without regard for the discrete phonological units that make up a language. These transformations form the heart of the *feedback control system* schematized in figure 3.5.

Ultimately, of course, the goal of the speech motor system is to articulate the discrete set of speech sounds used by the native language. Before a child can learn to articulate these sounds, he or she must learn to parse continuous auditory signals into discrete phonological categories such as words, syllables, and phonemes. This learning process corresponds to tuning of the *speech recognition system* in figure 3.5, and it is the focus of a tremendous amount of research in the speech perception literature. Since the current book is focused on speech production rather than perception, we will only briefly discuss the development of speech recognition capabilities here; see Swingley (2009) and Gervain and Mehler (2010) for reviews.

To get a sense of the challenge facing the infant, consider how difficult it is to identify word boundaries when listening to fluent connected speech in a foreign language. Yet infants are capable of identifying learned word forms (i.e., the sound patterns that make up words, without regard for meaning) by 8 months of age (Jusczyk & Hohne, 1997), and by 10 months of age they show sensitivity to word forms embedded in sentences (Gout, Christophe, & Morgan, 2004). These abilities appear to rely on prosodic cues such as phrase boundaries (Gout, Christophe, & Morgan, 2004; Seidl & Johnson, 2006) and stress patterns (Jusczyk, Cutler, & Redanz, 1993) as well as segmental cues such as consonant combinations (Mattys & Jusczyk, 2001). Learning of the meanings of first words starts around the end of the first year (e.g., Huttenlocher, 1974; Benedict, 1979), and by the middle of the second year this process is in full swing (Gervain & Mehler, 2010). Roughly speaking, children are initially sensitive to words and syllables, only later becoming sensitive to syllable onsets and rhymes,[10] and still later to individual phonemes within syllables or words (Anthony & Francis, 2005).

Learning of Sensory Targets and Feedforward Commands

As the infant becomes aware of discrete phonological entities, his or her brain stores information about the auditory signals that convey them. These auditory traces act as targets that the infant will try to replicate and correspond to the *auditory target map* of the DIVA model. Projections from the speech sound map to the auditory target map encode the auditory targets for sounds represented in the speech sound map so these targets can be activated later during production. As described in chapter 4, these targets take the form of *time-varying regions* in the auditory reference frame.

Imitation plays a large role in infant cognitive development, and infants have been shown to imitate facial expressions and tongue protrusions within hours of being born

(Meltzoff & Moore, 1983, 1989). As infants develop phonological awareness, they begin to imitate the sounds of speech they hear around them. These initial productions allow the infant's brain to learn *feedforward commands* for producing these sounds. According to the DIVA model, these feedforward commands are stored in synaptic projections from the speech sound map to the primary motor cortical areas, both directly and via a cortico-cerebellar loop. The auditory feedback control subsystem plays a crucial role in the development of these feedforward commands. As the infant successfully produces a speech sound, his or her brain develops a *somatosensory target map* that represents the somatic sensations that arise when properly producing the sound. As with auditory targets, the somatosensory targets take the form of time-varying regions, as described in chapter 4.

Table 3.1 provides estimates of the time courses of the developmental processes described in this section. Although the majority of development occurs during the first few years of life, our motor systems continue these learning processes through adulthood in order to stay tuned in the face of changes to morphology of the vocal apparatus that occur over time and to cope with the learning of new speech sounds, for example, from a second language.

3.5 Summary

The characterization of speech production as a control process began in the 1950s, and the subsequent decades have seen an increasing sophistication of computational models of speech. Perhaps the simplest characterization of speech motor control, that of a pure auditory feedback control process, has been shown to provide only a partial characterization of the speech production mechanism. A combination of feedforward and feedback controllers for speech has been utilized in more recent models.

The DIVA model of speech production characterizes cortical and subcortical interactions underlying the production of speech sounds. The model is specified both mathematically and neurally, and it provides a unified account for a wide range of acoustic, kinematic, and neuroimaging data. The model posits two interacting systems for the neural control of speech production: a feedback control system and a feedforward control system. The feedback control system is further broken into an auditory feedback control subsystem and a somatosensory feedback control subsystem. The feedforward control system is thought to involve cortico-cortical projections from premotor to motor cortex, as well as contributions from the cerebellum. The auditory feedback control subsystem is thought to involve projections from premotor cortex to higher-order auditory cortex; these projections encode auditory targets for speech sounds. Another set of projections from higher-order auditory cortex to motor cortex transforms auditory errors into corrective motor commands. The somatosensory feedback control subsystem is thought to involve projections from premotor cortex to higher-order somatosensory cortex that encode somatosensory targets for

Table 3.1
Time courses for development of the main computational components of the speech motor system

Neural System	Age/Development Stage								
	0–1 Mo./ Phonation	2–3 Mos./ "Goo"	4–6 Mos./ Expansion	7–10 Mos./ Canonical	11–12 Mos./ Variegated	1–2 Years/ Words	3–5 Years/ Sentences	6–18 Years/ School	> 18 Years/ Mature
Auditory state and error maps	Low	Low	Medium	High	High	High	High	Medium	Low
Somatosensory state and error maps	Low	Low	Medium	High	High	High	High	Medium	Low
Auditory-motor transformation	Low	Low	Medium	High	High	High	High	Medium	Low
Somatosensory-motor transformation	Low	Low	Medium	High	High	High	High	Medium	Low
Somatosensory-auditory transformation	Low	Low	Medium	High	High	High	High	Medium	Low
Speech sound map			Low	Medium	High	High	High	Medium	Low
Auditory target map			Low	Medium	High	High	High	Medium	Low
Feedforward commands			Low	Medium	High	High	High	Medium	Low
Somatosensory target map			Low	Low	Medium	High	High	Medium	Low

Note. Low, Medium, and High refer to the amount of learning occurring in the neural system during the time period indicated. These estimates are approximate, and considerable variation exists across individuals.

speech sounds, as well as projections from somatosensory error nodes to motor cortex that encode corrective motor commands.

Speech motor development involves a number of learning processes occurring in a quasi-parallel fashion. Auditory maps develop in a way that highlights important acoustic distinctions in a language and de-emphasizes unimportant distinctions. Analogously, somatosensory maps become sensitive to the tactile and proprioceptive feedback patterns that occur when producing sounds from the native language. Self-generated movements of the vocal tract allow the learning of sensory-motor relationships, including forward models that translate motor commands into their expected auditory and somatosensory outcomes, and inverse models that map desired changes in the auditory and somatosensory state to movements that can achieve these changes. Auditory targets for speech sound "chunks" such as phonemes, syllables, and words are formed by monitoring the productions of fluent speakers of the native language, and feedforward commands (or motor programs) are tuned as a child attempts to produce these sound chunks. This tuning process relies heavily on the auditory feedback control subsystem, which detects production errors and generates corrective motor commands. Continued practice leads to the creation of somatosensory targets for speech sounds. The somatosensory feedback control subsystem compares these targets to incoming somatosensory feedback to detect and correct production errors in parallel with the auditory feedback control subsystem.

Notes

1. Some researchers consider the term *segment* to refer specifically to the phoneme. Here we use a more general definition that includes syllables and words in addition to phonemes.

2. The term *motor* is used here to refer to a representation closely related to the musculature even if the motor dimensions do not equate in a one-to-one fashion with muscles (see related discussion of reference frames in chapter 1, section 1.2). For example, in most of the DIVA model simulations described herein, we utilize a simplified articulator model that involves only 10 articulatory dimensions, far fewer than the number of muscles involved in speech, but we nonetheless refer to the output of the model as the motor command.

3. Recall from chapter 1, section 1.2, that formant frequencies are the peaks of the acoustic spectrum and are related to the overall shape of the vocal tract. They are important acoustic cues for differentiating phonemes.

4. Recall from "Computational Modeling" in chapter 1, section 1.3, that the modeled neurons in an artificial neural network will be referred to as *nodes* herein to distinguish them from real neurons in the brain, and that each node is thought to correspond to a group of neurons in the brain rather than a single neuron.

5. This characterization will be modified in chapter 4 to accommodate the use of target *regions* that specify a range of acceptable values for each auditory parameter rather than specifying just a single target value. The same will be done for somatosensory targets.

6. There is also a dorsal portion of the motor cortex that appears to be involved in speech breathing, as described in chapter 2, section 2.4. However, since the large bulk of the speech articulator motor representations lie in vMC, we will use vMC to describe the location of speech-related motor representations.

7. Current data are insufficient to determine which lobules of the cerebellum are involved in the representation of auditory and somatosensory targets.

8. Note that it is not necessary for the child to recognize the *meaning* of the sound at this stage; it only needs to be recognized as a distinct new sound that is not in the child's verbal repertoire.

9. Auditory information is typically supplemented by visual input of the visible articulators, but the role of vision is very likely subservient to auditory information when the latter is available. Blind infants, for example,

follow a nearly normal time course of speech development (Dodd, 1983; Landau & Gleitman, 1985), in stark contrast to the substantial speech development problems that occur with deafness.

10. In English, the syllable *onset* consists of any consonant or consonant cluster that occurs at the beginning of the syllable, and the *rhyme* (sometimes spelled *rime*) consists of the rest of the syllable.

References

Anthony, J. L., & Francis, D. J. (2005). Development of phonological awareness. *Current Directions in Psychological Science, 14*, 255–259.

Benedict, H. (1979). Early lexical development: comprehension and production. *Journal of Child Language, 6*, 183–200.

Bullock, D., Cisek, P., & Grossberg, S. (1998). Cortical networks for control of voluntary arm movements under variable force conditions. *Cerebral Cortex, 8*, 48–62.

Callan, D. E., Kent, R. D., Guenther, F. H., & Vorperian, H. K. (2000). An auditory-feedback-based neural network model of speech production that is robust to developmental changes in the size and shape of the articulatory system. *Journal of Speech, Language, and Hearing Research, 43*, 721–736.

Compton, A. (1981). Syllables and segments in speech production. *Linguistics, 19*, 663–716.

de Boysson-Bardies, B., Hallé, P., Sagart, L., & Durand, C. (1989). A cross linguistic investigation of vowel formants in babbling. *Journal of Child Language, 16*, 1–17.

de Boysson-Bardies, B., Sagart, L., & Durand, C. (1984). Discernible differences in the babbling of infants according to target language. *Journal of Child Language, 11*, 1–15.

Dodd, B. (1983). The visual and auditory modalities in phonological acquisition. In A. E. Mills (Ed.), *Language acquisition in the blind child: normal and deficient* (pp. 57–61). London: Croom Helm.

Fairbanks, G. (1954). Systematic research in experimental phonetics: I. A theory of the speech mechanism as a servosystem. *Journal of Speech and Hearing Disorders, 19*, 133–139.

Gervain, J., & Mehler, J. (2010). Speech perception and language acquisition in the first year of life. *Annual Review of Psychology, 61*, 191–218.

Golfinopoulos, E., Tourville, J. A., & Guenther, F. H. (2010). The integration of large-scale neural network modeling and functional brain imaging in speech motor control. *NeuroImage, 52*, 862–874.

Gout, A., Christophe, A., & Morgan, J. L. (2004). Phonological phrase boundaries constrain lexical access: II. Infant data. *Journal of Memory and Language, 51*, 548–567.

Graven, S. N., & Brown, J. V. (2008). Auditory development in the fetus and infant. *Newborn and Infant Nursing Reviews; NAINR, 8*, 187–193.

Guenther, F. H. (1992). *Neural models of adaptive sensory-motor control for flexible reaching and speaking*. PhD dissertation, Boston University.

Guenther, F. H. (1994). A neural network model of speech acquisition and motor equivalent speech production. *Biological Cybernetics, 72*, 43–53.

Guenther, F. H. (1995). Speech sound acquisition, coarticulation, and rate effects in a neural network model of speech production. *Psychological Review, 102*, 594–621.

Guenther, F. H. (2006). Cortical interactions underlying the production of speech sounds. *Journal of Communication Disorders, 39*, 350–365.

Guenther, F. H., Ghosh, S. S., & Tourville, J. A. (2006). Neural modeling and imaging of the cortical interactions underlying syllable production. *Brain and Language, 96*, 280–301.

Guenther, F. H., & Gjaja, M. N. (1996). The perceptual magnet effect as an emergent property of neural map formation. *Journal of the Acoustical Society of America, 100*, 1111–1121.

Guenther, F. H., Hampson, M., & Johnson, D. (1998). A theoretical investigation of reference frames for the planning of speech movements. *Psychological Review, 105*, 611–633.

Guenther, F. H., Husain, F. T., Cohen, M. A., & Shinn-Cunningham, B. G. (1999). Effects of categorization and discrimination training on auditory perceptual space. *Journal of the Acoustical Society of America, 106*, 2900–2912.

Guenther, F. H., & Vladusich, T. (2012). A neural theory of speech acquisition and production. *Journal of Neurolinguistics, 25*, 408–422.

Houde, J. F., & Jordan, M. I. (1998). Sensorimotor adaptation in speech production. *Science, 279*, 1213–1216.

Huttenlocher, J. (1974). The origins of language comprehension. In R. L. Solso (Ed.), *Theories in cognitive psychology: the Loyola Symposium* (pp. 331–368). Potomac, MD: Erlbaum.

Ito, T., & Ostry, D. J. (2012). Speech sounds alter facial skin sensation. *Journal of Neurophysiology, 107*, 442–447.

Ito, T., Tiede, M., & Ostry, D. J. (2009). Somatosensory function in speech perception. *Proceedings of the National Academy of Sciences of the United States of America, 106*, 1245–1248.

James, W. (1890). *The principles of psychology*. New York: Holt.

Jusczyk, P. W., Cutler, A., & Redanz, N. J. (1993). Infants' sensitivity to the predominant stress patterns of English words. *Child Development, 64*, 675–687.

Jusczyk, P. W., & Hohne, E. A. (1997). Infants' memory for spoken words. *Science, 277*, 1984–1986.

Kawato, M., & Gomi, H. (1992). A computational model of four regions of the cerebellum based on feedback-error learning. *Biological Cybernetics, 68*, 95–103.

Kozhevnikov, V. A., & Chistovich, L. A. (1965). *Speech: articulation and perception*. Washington, DC: Joint Publication Research Service.

Kuhl, P. K., Stevens, E., Hayashi, A., Deguchi, T., Kiritani, S., & Iverson, P. (2006). Infants show a facilitation effect for native language phonetic perception between 6 and 12 months. *Developmental Science, 9*, F13–F21.

Kuhl, P. K., Willams, K. A., Lacerda, F., Stevens, K. N., & Lindblom, B. (1992). Linguistic experience alters phonetic perception in infants by 6 months of age. *Science, 255*, 606–608.

Landau, B., & Gleitman, L. R. (1985). *Language and experience: evidence from the blind child*. Cambridge, MA: Harvard University Press.

Levelt, W. (1989). *Speaking: from intention to articulation*. Cambridge, MA: MIT Press.

Levelt, W. J., & Wheeldon, L. (1994). Do speakers have access to a mental syllabary? *Cognition, 50*, 239–269.

MacNeilage, P. F., & Davis, B. L. (1990). Acquisition of speech production: frames, then content. In M. Jeannerod (Ed.), *Attention and performance XIII: Motor representation and control* (pp. 453–476). Hillsdale, NJ: Erlbaum.

Maeda, S. (1990). Compensatory articulation during speech: evidence from the analysis and synthesis of vocal tract shapes using an articulatory model. In W. J. Hardcastle & A. Marchal (Eds.), *Speech production and speech modeling* (pp. 131–149). Boston: Kluwer Academic.

Mattys, S. L., & Jusczyk, P. W. (2001). Do infants segment words or recurring contiguous patterns? *Journal of Experimental Psychology: Human Perception and Performance, 27*, 644–655.

McGurk, H., & MacDonald, J. (1976). Hearing lips and seeing voices. *Nature, 264*, 746–748.

Meltzoff, A. N., & Moore, M. K. (1983). Newborn infants imitate adult facial gestures. *Child Development, 54*, 702–709.

Meltzoff, A. N., & Moore, M. K. (1989). Imitation in newborn infants: exploring the range of gestures imitated and the underlying mechanisms. *Developmental Psychology, 25*, 954–962.

Neilson, M. D., & Neilson, P. D. (1987). Speech motor control and stuttering: a computational model of adaptive sensory-motor processing. *Speech Communication, 6*, 325–333.

Nieto-Castanon, A., Guenther, F. H., Perkell, J. S., & Curtin, H. (2005). A modeling investigation of articulatory variability and acoustic stability during American English /r/ production. *Journal of the Acoustical Society of America, 117*, 3196–3212.

Oller, D. K. (1980). The emergence of the sounds of speech in infancy. In G. H. Yeni-Komshian, J. F. Kavanagh, & C. A. Ferguson (Eds.), *Child phonology: Vol. 1. Production* (pp. 93–112). New York: Academic Press.

Seidl, A., & Johnson, E. K. (2006). Infant word segmentation revisited: edge alignment facilitates target extraction. *Developmental Science, 9*, 565–573.

Stark, R. E. (1980). Stages of speech development in the first year of life. In G. H. Yeni-Komshian, J. F. Kavanagh, & C. A. Ferguson (Eds.), *Child phonology: Vol. 1. Production* (pp. 93–112). New York: Academic Press.

Swingley, D. (2009). Contributions of infant word learning to language development. *Philosophical Transactions of the Royal Society of London. Series B, Biological Sciences, 364*, 3617–3632.

Tourville, J. A., & Guenther, F. H. (2011). The DIVA model: A neural theory of speech acquisition and production. *Language and Cognitive Processes, 26*, 952–981.

Tremblay, S., Shiller, D. M., & Ostry, D. J. (2003). Somatosensory basis of speech production. *Nature, 423*, 866–869.

Tsushima, T., Takizawa, O., Sasaki, M., Shiraki, S., Nishi, K., Kohno, M., et al. (1994). Discrimination of English /r-l/ and /w-y/ by Japanese infants at 6–12 months: language-specific developmental changes in speech perception abilities. *1994 International Conference on Spoken Language Processing* (pp. 1695–1698), Yokohama, Japan.

Villacorta, V. M., Perkell, J. S., & Guenther, F. H. (2007). Sensorimotor adaptation to feedback perturbations of vowel acoustics and its relation to perception. *Journal of the Acoustical Society of America, 122*, 2306–2319.

4

The Targets of Speech

In order to adequately characterize the neural control system for speech, it is first necessary to determine what variables are controlled; that is, what are the targets that the speech production mechanism is attempting to reach? It is tempting to conclude that the target for an utterance consists solely of a desired acoustic/auditory signal; that is, production of a phoneme, syllable, or word involves an attempt by the controller to produce a desired acoustic trajectory. However, this represents a gross simplification of the actual neural control problem for speech. For example, it has been demonstrated that, in addition to using auditory targets, speakers will automatically adapt to a somatosensory perturbation (e.g., an outward tug on the jaw) even if the perturbation has no measurable acoustic effect (Tremblay, Shiller, & Ostry, 2003), indicating the existence of a somatosensory target.

This section provides a more complete discussion of the nature of the targets of the speech production mechanism, concluding that the target for a phoneme or syllable consists of temporally varying regions in both auditory and somatosensory spaces. Because this topic is crucially intertwined with issues in speech motor development in the infant and young child, much of the following treatment will be motivated by considering the tasks facing a developing infant when learning to speak the ambient language. Along the way a number of key behavioral findings will be presented concerning kinematic and neurophysiological aspects of speech acquisition and production.

4.1 Definition of a Speech Target

In the speech motor control literature, the term *target* (or *goal*) is typically used to refer to the input to the control system for speech (see figure 3.1). In other words, what exactly is it that the speech production mechanism really cares about? It certainly cares about the acoustic signal since this is the primary conveyer of linguistic information from the speaker's brain to the listener's brain. Does it also care about the details of the muscle commands used to produce speech, or are these details left unattended as long as the acoustic signal is achieved? The simple auditory feedback controller schematized in figure 3.2 does not care about the details of the muscle commands, but some models of speech production do (e.g., MacNeilage, 1970; Payan & Perrier, 1997).

To better understand the issue, it is helpful to first consider the learning of a nonspeech motor skill that involves 3-D spatial goals since most of us are more accustomed to thinking about 3-D space than auditory space. When playing a game of darts, the player's primary target is a particular pie-slice-shaped region of the dartboard. That is, the target is defined as a simple shape in a 3-D reference frame. Amateur dart players who are new to the game concentrate primarily on achieving this target when throwing, and they (at least occasionally) succeed in doing so despite a large amount of variability in the way they move their hands, arms, and body to throw the darts on different attempts. We can at this point say that, although the inexperienced player knows the primary *spatial target*, the player does not yet have a reliable *somatosensory target* for the dart throw. In other words, the player does not yet know what a proper throw "feels like."

With some time and practice, however, the novice dart thrower's motor system will learn at least a coarse somatosensory target to accompany the spatial target. This is because the player's wildest throws (those in which he or she tried to throw too hard or with too much lower body action, etc.) will usually be unsuccessful, and the player's motor system can learn not to repeat these movements. In this way a somatosensory target which contains reasonably successful movements can be "sculpted" from the space of all possible movements.

In darts, the highly accomplished player uses a very stereotyped set of body movements to achieve the spatial goals on the dart board. Likewise, the vast majority of professional golfers use very stereotyped, fundamental body movements when swinging, compared to the high degree of motor variability produced by amateur golfers (Hogan, 1957), and the fundamentals of a professional billiard stroke involve maintaining particular relationships between segments of the arm and body (Capelle, 1995). In fact, the list of sporting skills of this type, whose primary goal is spatial but whose mastery requires learning of certain fundamental somatosensory relationships that the unpolished amateur does not maintain, is vast, including archery, baseball, basketball, bowling, and so forth.

We take the view here that, unlike the sports mentioned above which involve primary goals defined in 3-D space, the *primary goal of speech production is the generation of an acoustic signal*[1] *that will successfully convey linguistic units (phonemes, syllables, and words) to a listener*. We will use the term *auditory* to refer to acoustic signals as they are represented in the brain. Thus, in our view each speech sound has its own *auditory target* analogous to the targeted pie slice on the dart board for the dart player. We further posit that, as speech learners become more and more skilled with speech over the first few years of life, they develop *somatosensory targets* that accompany the auditory targets. They also learn *feedforward commands* that largely replace auditory and somatosensory feedback control under normal speaking conditions. The initial learning and constant tuning of these feedforward commands (as necessitated, for example, by the substantial changes in the sizes and shapes of speech articulators that take place over the first few years of life; Callan et al., 2000; Vorperian et al., 2005) depend on the auditory and somatosensory feedback control subsystems. This theoretical viewpoint, embodied by the DIVA model, will be detailed, along with supporting evidence, throughout the remainder of the current chapter as well as in chapters 5 through 7.

The view that the primary target of speech production is auditory may seem intuitively appealing, but it is not the only theoretical viewpoint. In particular, theorists associated with Haskins Laboratories have posited that it is the *articulatory gesture*, rather than the auditory signal, that is the primary target of speech production (e.g., Browman & Goldstein, 1990a,b; Fowler, 1980; Saltzman & Munhall, 1989) as well as speech perception (e.g., Liberman et al., 1967; Liberman & Mattingly, 1985). These gestures involve achieving particular constriction locations and degrees along the vocal tract, typically using parts of the tongue and/or lips to form the constriction. Because the locations and degrees of the constrictions in the vocal tract play a dominant role in generating the acoustic signal, the auditory and gestural views of speech production are often difficult to differentiate from each other experimentally, and the two views share more in common than they differ. As we will detail in the next few chapters, the DIVA model provides a means for reconciling the two views into a unified model of speech production that utilizes targets in both auditory and somatosensory spaces. Nevertheless, an important aspect of the DIVA model concerns the primary nature of the auditory target compared to the somatosensory target during development. By this, we mean that the auditory target for a speech sound typically plays the largest role in shaping the feedforward motor commands (or gestures) for producing the sound and also plays a role in shaping the somatosensory target for the sound, though it is likely that the two target types are to some degree refined in parallel over time. By referring to the auditory target as "primary" in this way, we do not mean to imply that the auditory target takes precedence in the ongoing control of speech; in our theoretical view, both auditory and somatosensory targets, along with feedforward motor commands, play major roles in the control of speech in the mature system.

The next two sections detail experimental evidence that sheds more light on the nature of speech targets and provides support for this theoretical view.

4.2 Roles of Different Sensory Systems in Speech

In this section we review the literature regarding the roles of the three types of sensory information involved in speech communication—auditory, somatosensory, and visual information—in speech development and mature speech communication.

It has long been known that each phoneme of a spoken language involves a distinct acoustic pattern; that is, all speech sounds in a language can be distinguished on the basis of the auditory signal alone, at least in ideal listening conditions. The same is not true of visual information; studies of lip reading (e.g., Woodward & Barber, 1960; Fisher, 1968;

Walden et al., 1977; see Mills, 1983, for a review) indicate that the number of visually distinctive phonemes in a language, or *visemes* (Erber, 1974), is smaller than the full number of phonemes in the language. For example, bilabial consonants such as /p/, /b/, and /m/ are not visually distinctive from each other since the features of voice onset time and nasalization that distinguish these sounds are not visible. Thus visual information alone is not enough to convey all speech sounds, unlike auditory information.

Nonetheless, it is evident that both auditory and visual information play a role in the normal development of speech. The deleterious effects of auditory deprivation on speech are clear. For example, studies of deaf infants indicate that babbling of well-formed syllables occurs much later, if at all, in these individuals (Oller, Eilers, Bull, & Carney, 1985; Stoel-Gammon & Otomo, 1986; Kent et al., 1987; Oller & Eilers, 1988), and many congenitally deaf individuals show phonological deficits into puberty (Sterne & Goswami, 2000) and beyond. Less obvious, but still evident, are delays in the development of speech in blind children (Landau and Gleitman, 1985). Importantly, however, blind children typically acquire a normal phonological system, and by the age of 3 show little if any phonological deficit relative to sighted children (Dodd, 1983; Landau & Gleitman, 1985), providing compelling evidence for the primacy of the auditory signal, as compared to visual information, in speech communication.

Based on the above evidence, it is tempting to conclude that visual information of speech gestures might play no role in adult speech mechanisms. However, there is clear evidence to the contrary in the case of speech perception (see Massaro, 1998, for a review). For example, the effectiveness of lip reading implies that articulated speech can be largely perceived through vision. Furthermore, it has been demonstrated that viewing a speaking face can significantly enhance speech perceptual abilities when the audio signal is noisy (e.g., Sumby & Pollack, 1954). In a study by McGurk and MacDonald (1976), an audio track of a speaker producing one syllable was dubbed onto a video of the speaker producing a different syllable. When subjects viewed (and heard) the resulting video, in many instances the visual signal strongly biased auditory perception away from the audio signal. For example, viewing a speaker say "gah" while listening to an audio "bah" consistently results in the auditory percept "dah" for most people, even when they are aware of the deception, a phenomenon often referred to as the *McGurk effect*. Thus, although visual information plays a secondary role to auditory information for speech development and speech production, it can have a significant effect on speech perception.

One final note regarding speech development in individuals with sensory impairment involves very rare cases of individuals with a severe somatosensory deficit involving the oral articulators but with normal hearing and preserved motor output pathways. Mac-Neilage, Rootes, and Chase (1967) describe the results of several studies involving such a patient. While the patient's speech perception was nearly normal, she was unable to produce intelligible speech. On the surface such a finding might argue for the primary importance of somatosensory targets in speech production. However, loss of somatosensory feedback would not only disrupt the ability to achieve somatosensory targets in the somatosensory feedback control subsystem, but it would also impair learning of feedforward commands since the appropriate feedforward commands are dependent on the current state of the vocal tract articulators, which is largely sensed by tactile and proprioceptive mechanisms. The inability to produce intelligible speech with a severe somatosensory deficit thus likely arises in large part from the inability to learn the intricate patterning of muscle activity required for fluent speech, as concluded by the authors of this study.

The evidence reviewed in this section indicates that deficits in auditory or somatosensory processing are far more debilitating to speech than loss of visual input, and that normal speech production is typically achieved with only minor delays in the complete absence of vision. It is thus unlikely that visual targets of speech gestures play a crucial role in speech production, although visual information, when available, does play a minor role in the development of speech and can have a strong effect on speech perception in the adult system. The evidence described here does not provide a clear answer regarding whether auditory or somatosensory targets play a larger role in the shaping of speech movements. The next section provides evidence that the auditory target plays a larger role than the somatosensory target in the development of motor commands utilized to produce speech sounds.

4.3 Motor Equivalence in Speech

The term *motor equivalence* is used to describe the ability to carry out the same task using different motor means. For example, people are capable of producing written letters with very similar shapes on a chalkboard using their wrist and fingers, or alternatively with their shoulder and elbow (Merton, 1972), their nondominant arm (Raibert, 1977; Wright, 1990), and even using pens attached to their feet or held in their teeth (Raibert, 1977). Motor equivalence is seen in a wide variety of human behaviors, including not only handwriting but also reaching (e.g., Cruse, Brüwer, & Dean, 1993) and speaking (e.g., Abbs & Gracco, 1984; Lindblom, Lubker, & Gay, 1979; Perkell et al., 1993; Savariaux, Perrier, & Orliaguet, 1995; Guenther et al., 1999; Savariaux et al., 1999; Nieto-Castanon et al., 2005), and in a wide variety of nonhuman species including turtles (Stein, Mortin, & Robertson, 1986) and frogs (Berkinblit, Gelfand, & Feldman, 1986). The ubiquity of motor equivalence is no doubt the evolutionary result of its utility: animals capable of using different motor means to carry out a task under different environmental conditions have a tremendous advantage over those that cannot.

An example of motor equivalent behavior in speech is the ability to use redundant motoric degrees of freedom to compensate for externally imposed constraints on, or perturbations to, the articulators while producing speech. For example, people normally use jaw movements during speech, but they can also successfully produce phonemes with a bite block clenched in their teeth by adjusting lip and tongue movements to compensate for the fixed jaw. Compensation occurs immediately and automatically, without requiring practice with the bite block and even before auditory feedback from the first attempt is available (Lindblom, Lubker, & Gay, 1979), though a smaller additional increment in performance is gained with some practice with an articulatory perturbation (McFarland & Baum, 1995; Baum, McFarland, & Diab, 1996). Computer simulations have shown that both gestural (Saltzman & Munhall, 1989; Guenther, 1994) and auditory target models (Guenther, 1995; Guenther, Hampson, & Johnson, 1998) can account for bite block compensation as well as compensations for unexpected perturbations to the lips or jaw (e.g., Abbs & Gracco, 1984; Kelso et al., 1984).

The phenomenon of motor equivalence provides an opportunity to investigate the reference frames used to plan speech movements. As alluded to above, the auditory signal is closely related to the locations and degrees of constrictions in the vocal tract, and therefore gestural and acoustic models make similar predictions for most speech sounds and tasks (see, e.g., Zandipour, 2007, who demonstrates statistically indistinguishable trajectories for vowel-to-vowel movements using auditory and motor target models to control movements of a biomechanically realistic vocal tract model). However, in some cases the two views make differential predictions. For example, insertion of a lip tube that prevents subjects from reaching the small bilabial constriction normally used for French /u/ should, according to the auditory target view, cause subjects to compensate by changing the locations of constrictions in the vocal tract, along with other auditory maneuvers such as changing voice pitch. Such behavior was observed experimentally by Savariaux et al. (1995, 1999), and computer simulations of an auditory target model have been shown to account for these data (Guenther, Hampson, & Johnson, 1998). Gestural models, on the other hand, provide no means for compensation for the lip tube since they do not provide a means for using different constrictions or voicing adjustments to compensate for the perturbed lip constriction. Such compensatory trade-offs between constrictions have also been identified in normal (unperturbed) productions (e.g., Perkell et al., 1993) and are often referred to as *articulatory trading relations*.

The American English phoneme /r/ provides a striking example of the primacy of an auditory target over articulatory or constriction targets. This phoneme demonstrates a particularly large amount of articulatory variability in its production in different phonetic contexts and across speakers (Delattre & Freeman, 1968; Hagiwara, 1994; Boyce & Espy-Wilson, 1997; Ong & Stone, 1998; Westbury, Hashi, & Lindstrom, 1998; Guenther et al., 1999). In many speakers, productions of /r/ across many phonetic contexts fall into a bimodal distribution of articulations; that is, the speaker appears to use two different articulatory configurations for /r/, with some phonetic contexts involving one configuration and others involving a second (Guenther et al., 1999). These different configurations can often be roughly characterized as *bunched* and *retroflexed* as in figure 4.1, though the configurations for a particular speaker will typically differ somewhat from the canonical examples in the figure.

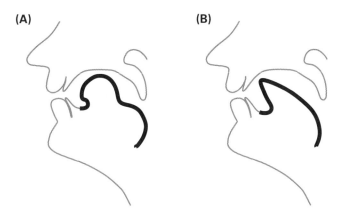

Figure 4.1
(A) Bunched and (B) retroflexed tongue configurations (bold lines) for the American English phoneme /r/ (cf. Delattre & Freeman, 1968).

The DIVA model provides a simple and natural explanation for these bimodal distributions. Two aspects of the model are key to this account: (1) the use of an auditory target as the primary target for speech early in development and (2) the use of a mapping from directions in auditory space to changes in articulator velocities, that is, the *d*irections *i*nto *v*elocities of *a*rticulators mapping that gives the model its name. The following paragraphs detail this account.

An important concept in this account is that of a *region in sensory space*. It is easy to imagine a region in Cartesian 3-D space; for example, a baseball takes up a sphere-shaped region in this space. To imagine a region in auditory space, one can, for example, replace the x dimension of Cartesian space with the first formant frequency, the y dimension with the second formant frequency, and the z dimension with the third formant frequency. A particular production of a vowel can be characterized by its first three formant frequencies, which together define a point in this auditory space (more specifically, in *formant space*). If we plotted three different productions of a vowel and connected them together with lines, we would define a triangle in auditory space; plotting more and more examples of the vowel would lead to a more complex 3-D shape, and if enough examples are added, this shape could be said to characterize the distribution of that vowel in formant space. Although we will typically use 2-D representations of sensory spaces in the figures of this book for visualization purposes, it is important to note that a given sensory space as represented in the brain may have many dimensions, not just two or three.

Panel A of figure 4.2 shows a simple *convex region*[2] in formant space that approximates the ranges of F1 and F2 for the phoneme /r/. Panel B shows the corresponding region in two dimensions of articulator space. This figure was produced by fixing five of the seven articulatory degrees of freedom in the Maeda articulatory synthesizer (Maeda, 1990) and

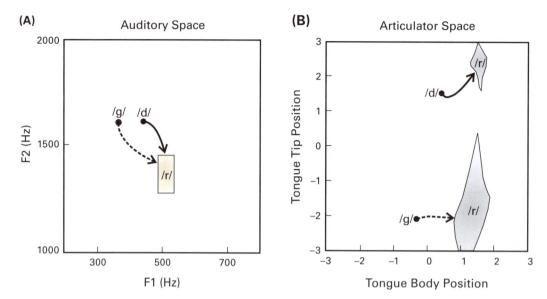

Figure 4.2
(A) A simple convex region target for /r/ (beige rectangle) in an auditory space defined by the first two formant frequencies (F1 and F2). (B) The portion of an articulator space defined by tongue tip position and tongue body position that corresponds to the /r/ target in (A), which consists of two distinct subregions (blue). Arrows indicate DIVA model trajectories when producing /r/ starting from a /d/ configuration (solid lines) and from a /g/ configuration (dashed lines). Despite moving to the same auditory target for /r/, different subregions of articulator space are reached for the different initial consonants. The top subregion in articulator space corresponds to a retroflexed tongue configuration that occurs when /r/ follows /d/, and the bottom subregion corresponds to a bunched configuration that occurs when /r/ follows /g/.

varying the remaining two, which are related to tongue shape, through their entire ranges to determine which articulatory configurations result in formants in the ranges specified in panel A. Note that the simple convex region in auditory space corresponds to a complex, nonconvex shape with two distinct subregions in articulatory space. The top subregion corresponds to a high tongue tip position as in the retroflexed /r/ shown in panel B of figure 4.1, and the bottom subregion corresponds to a bunched tongue configuration as in panel A of figure 4.1.

The production of /r/ from two different starting positions (corresponding to two different phonetic contexts) is schematized by the arrows in figure 4.2. The solid arrows represent the trajectory produced by the DIVA model in auditory space (panel A) and articulatory space (panel B) when /d/ precedes /r/ as in the word "drag," and the dashed arrows represent the trajectory produced when /g/ precedes /r/ as in "grab." Much like human speakers (e.g., Boyce & Espy-Wilson, 1997; Guenther et al., 1999), the model ends up in a retroflexed tongue configuration for /r/ when it is preceded by /d/ and a bunched configuration when /r/ follows /g/. In both human speakers and the model, the movements for /r/ are *efficient* in the sense that they require relatively little articulator movement to achieve the desired acoustic target for /r/ (Guenther, Hampson, & Johnson, 1998; Guenther et al., 1999). When /r/ is preceded by /g/, the tongue is already in a bunched configuration for the /g/ and thus relatively little movement is needed to achieve the bunched /r/ configuration. When /r/ is preceded by /d/, in contrast, the retroflexed configuration is closer to the /d/ configuration and thus requires less articulatory movement than the bunched /r/ configuration.

The qualitative account described thus far has been supplemented with a number of quantitative analyses that strongly support the DIVA model's account of articulatory variability for /r/. Guenther et al. (1999) demonstrated that articulatory trade-offs across /r/ productions in different phonetic contexts are organized in a way that maintains a relatively constant third formant frequency, a key acoustic cue for /r/, across the different contexts. Nieto-Castanon et al. (2005) extended this work by demonstrating that the articulatory variability seen across /r/ productions is constrained such that variation across contexts was limited to those movement directions in articulator space that had minimal effects on F3 rather than movements that had minimal effects on vocal tract constrictions. In other words, movement variations across contexts that would have affected F3 were avoided by subjects regardless of whether these variations resulted in different constriction locations or degrees of the primary tongue constriction for /r/. Nieto-Castanon et al. (2005) further demonstrated that even the articulatory variations across repetitions of /r/ in the *same* phonetic context are organized so as to minimize variability in F3 rather than variability in the location and/or degree of constrictions in the vocal tract. Finally, Nieto-Castanon et al. (2005) demonstrated that, when the DIVA model is given a vocal tract model crafted to match the anatomy and acoustics of a particular experimental subject, the model's movement trajectories for /r/ in different contexts closely mimic those of the particular speaker being modeled (see figure 4.3).

Together, these results indicate the fundamental importance of the auditory target in the planning and production of speech movements. As described earlier in this chapter, we take the view that, as a developing speaker practices production of a newly learned sound, the speaker's brain forms a somatosensory target for the sound that supplements the auditory target. In the next section we address the nature of the auditory and somatosensory targets for speech in more detail, in particular the idea that the targets take the form of *regions* rather than points in the auditory and somatosensory spaces.

4.4 Target Regions

The inventory of sounds used in speech varies considerably across languages (Ladefoged & Maddieson, 1996). Furthermore, the amount of allowable variability across phonetic contexts in a given auditory dimension (such as the first formant frequency) or somatosensory dimension (such as size of the lip opening, or *lip aperture*) can vary significantly from

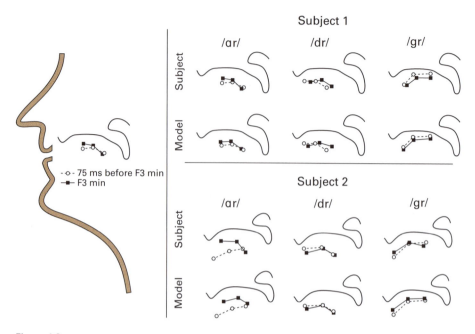

Figure 4.3
Comparison of human subject /r/ productions measured using electromagnetic midsagittal articulometry to simulations of the DIVA model (Nieto-Castanon et al., 2005). The left panel illustrates the orientation of the experimental data (along with outlines of the hard palate and velum) relative to the head. The solid lines represent the tongue shape at the acoustic "center" of /r/, characterized by the minimum in the trajectory of F3. Dashed lines represent the tongue shape 75 ms before the F3 minimum. Thus, the articulatory gesture for /r/ in a given panel can be estimated by imagining the movement between the dashed and solid tongue shapes (e.g., an upward movement of the tongue tip in the left panel). The rows marked *Subject* in the right portion of the figure indicate the /r/ articulations of two experimental subjects producing /r/ in three different phonetic contexts: /ɑr/, /dr/, and /gr/. The rows labeled *Model* indicate movements made by the DIVA model when controlling speaker-specific vocal tract models tuned to represent the anatomy and acoustics of the subjects' vocal tracts. The model's movements closely mimic those of the modeled speakers; the correlation between the modeled and experimental tongue gestures was r = +0.86 and r = +0.93 for Subjects 1 and 2, respectively.

one sound to another across languages as well as within the same language. For example, the third formant frequency must be relatively strictly controlled when producing the phonemes /r/ and /l/ in English, but the Japanese /r/ sound can be produced with a wide range of F3 variability that would spill into the English /l/ category (see figure 4.4). For many other English speech sounds, F3 is a relatively minor perceptual cue that can vary substantially. Similarly, lip aperture can vary significantly for most English vowels but not for the bilabial stop consonants /p/ and /b/ as these sounds require complete closure of the lips.

Formation of Target Regions
These considerations highlight a difficult problem for the developing infant: not only must the infant learn which sounds are used in his or her language, but he or she must also learn

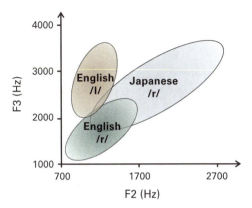

Figure 4.4
Schematic illustrating differences in the location and extent of American English /r/ and /l/ and Japanese /r/ sound categories in auditory space (specifically, in the F2/F3 plane). The two languages differ significantly in the location and extent of the /r/ sound category, with the Japanese /r/ overlapping both English /r/ and /l/. After Lotto, Sato, and Diehl (2004).

which auditory and somatosensory dimensions must be strictly controlled and which can vary for each sound in the native language. For example, how does the infant learn that, whereas lip aperture must be strictly controlled for bilabial stops, it can vary over a relatively large range for many other speech sounds, including vowels and velar, alveolar, and dental consonants? How does a Japanese speaker's nervous system know that F3 of /r/ can vary widely while the nervous system of an English speaker knows to control F3 more strictly when producing /r/ so that /l/ is not produced instead?

Guenther (1995) details a *convex region theory* of the targets of speech that addresses these questions and provides a unified account of a wide range of speech production phenomena. Here we will focus on the development of target regions and their relationship to acoustic and articulatory variability, coarticulation, speaking rate, and speech clarity.

According to our theoretical framework, infants learn a convex region target in auditory space for each speech sound from examples presented by speakers of the native language. This process is schematized for a single speech sound in figure 4.5. The first time an infant hears a sample of the sound, the target region for the sound is stored in the infant's brain as a small region around the sound sample (panel A). With each subsequent sample, the region is expanded to encompass the new sample (panel B). After many samples are presented to the infant, the target region reflects the acceptable range of variation for the sound in the native language (panel C).

Once an infant learns to produce a speech sound (initially under auditory feedback control, as described in chapters 3 and 5), the infant can learn a target region in somatosensory space for that sound by monitoring his or her own correct self-productions. Figure 4.6 schematizes the somatosensory target regions for two speech sounds, /ɪ/ (as in "bit")

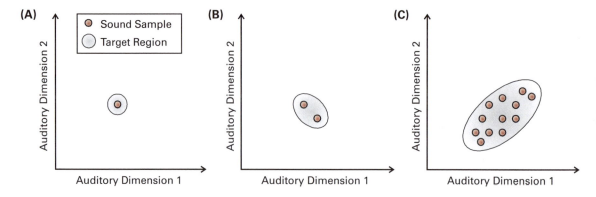

Figure 4.5
Learning of an auditory convex region target for a speech sound from samples provided by speakers of the native language. (A) When the infant hears a new sound for the first time, a target region is formed around that sample. (B) When a second sample of the sound is heard, the region expands to include both samples. (C) After many samples of the sound have been heard, the target region encodes the entire allowable variability for the sound in the native language.

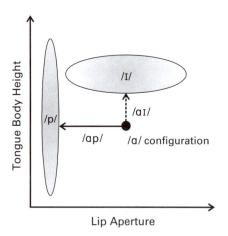

Figure 4.6
Schematized somatosensory target regions for the phonemes /ɪ/ and /p/. Also shown are idealized movement trajectories in the DIVA model to each of the targets from a configuration corresponding to the phoneme /ɑ/. The solid arrow shows the movement trajectory in somatosensory space for the utterance /ɑp/, and the dashed arrow shows the movement trajectory for /ɑɪ/. The model commands movements only along somatosensory dimensions along which the current configuration is outside the target region for the sound. Therefore movement of the lips is commanded for producing the /p/ in /ɑp/ but no movement of the tongue body is commanded, whereas movement of the tongue body but not the lips is commanded for the /ɪ/ in /ɑɪ/.

and /p/, along two dimensions in somatosensory space (corresponding to lip aperture and tongue body height) after a number of correct self-productions of each sound. Variation along the tongue body dimension has little effect on /p/ production as long as the tongue body height is not near its maximum (i.e., it is not up against the palate) whereas variation along the lip aperture dimension must be severely restricted in order to correctly produce /p/ (i.e., the lips must be completely closed). As a result, the target region for /p/ is very wide along the tongue body height dimension and narrow along the lip aperture dimension. The situation for /ɪ/ is almost the opposite; lip aperture makes little difference as long as the lips are not completely closed, but tongue body height must be within a relatively narrow range.

Now we can address the questions posed above. The convex region for /p/ does not vary over the dimension of lip aperture but varies widely over the dimension of tongue body height; this is because all bilabial stops that the infant has produced have the same lip aperture, but tongue body height has varied. In other words, the infant has learned that bilabial aperture is the important somatosensory invariant for producing the bilabial stop /p/. In contrast, the infant has learned that this dimension is not very important for /ɪ/, as indicated by the wide range of lip aperture in the target for /ɪ/. Finally, since convex region learning relies on the variability of speech sounds in the native language, the shapes of the resulting convex regions will vary from language to language.

Also shown in figure 4.6 are schematized movements from a vocal tract configuration corresponding to /ɑ/ (as in "fɑther") to each of the two targets. According to the DIVA model, movements are commanded only along a sensory dimension when the corresponding sensory feedback is outside the target region along that dimension. When producing /p/ after /ɑ/ (solid arrow in figure 4.6), there is no need to move the tongue body since it is already within the acceptable range of tongue body height for /p/. When producing /ɪ/ after /ɑ/ (dashed line), there is no need to move the lips up or down since lip aperture is already within the acceptable range for /ɪ/. In this way the model avoids unnecessary movements, leading to efficient movements from one speech sound to the next. In this sense, the convex region theory constitutes a generalization of Keating's (1990) *window model* of coarticulation, which states that the target for an articulator is specified by a "window" of acceptable values, to high-dimensional auditory and somatosensory spaces (see Guenther, 1995, for a detailed treatment).

Figure 4.6 also illustrates another interesting property of the model: it can learn to "ignore" totally unimportant auditory or somatosensory dimensions by allowing variability throughout the entire range of such dimensions. This can account for why a particular auditory or somatosensory dimension may be important for creating speech sound distinctions in one language but not another: in the former case certain speech sound targets specify a small allowable range for this dimension while in the latter case the entire possible range for this dimension is included in all of the sound targets, causing this dimension to be effectively ignored by the neural controller for speech. It should be noted, however, that in some cases the "ignoring" of auditory dimensions occurs because of development in the perceptual system rather than the motor system. For example, it is well-known that monolingual Japanese listeners have great difficulty detecting variations in F3 that distinguish English /r/ from /l/ (Miyawaki et al., 1975), and this inability to detect F3 variations is almost certainly largely (if not completely) responsible for their inability to reliably produce these distinctions without significant speech training. Indeed, one of the most effective forms of training Japanese speakers to produce these sounds properly involves training them to distinguish /r/ and /l/ perceptually (e.g., Bradlow et al., 1997, 1999). Examples of Japanese speakers who produce better /r/-/l/ contrasts than they can perceive have been noted (Goto, 1971; Sheldon & Strange, 1982); according to the DIVA model these individuals have managed to learn somatosensory and motor targets for the sounds despite not having accurate auditory targets, perhaps as the result of articulatory training (cf. Catford & Pisoni, 1970; Flege, 1991).

Articulatory Variability
The existence of target ranges along auditory and somatosensory dimensions in DIVA, rather than strict target positions, implies that variability will be seen in these dimensions during speech. This is because no movements are commanded for positions anywhere within the target range, so entering the range at different positions during different production trials (due, for example, to contextual or biomechanical influences) will lead to different places of articulation. Furthermore, because the size of the target range along an auditory or somatosensory dimension reflects the amount that the vocal tract is allowed to vary along that dimension while still adequately producing the same phoneme, more variation is expected along acoustically less important dimensions. An example of this phenomenon in speech comes from studies of place of articulation for velar stops. English speakers (and listeners) do not differentiate between velar and palatal stop consonants; as a result, wide anteroposterior variability is seen in the place of constriction for the stop consonants /k/ and /g/ in different vowel contexts (e.g., Daniloff, Schuckers, & Feth, 1980; Kent & Minifie, 1977). Kent and Minifie point out that if the target position for /k/ or /g/ is very concrete and fully specified, then the variation cannot be explained by a target position model. Furthermore, if the target positions are only loosely specified, the possibility exists for too much variation that can destroy phonemic identity. Since large anteroposterior variation is seen in /k/ and /g/ but little or no variation is allowable in the vertical position of the tongue body (i.e., the tongue body must contact the palate), it appears that neither a fully specified nor loosely specified target position will suffice. Instead, it appears that tongue body target ranges are defined separately for anteroposterior position and vertical position, with a large target range for the former and a much smaller range for the latter.

For consonants, it is clear that humans must strictly control the place of articulation along the somatosensory dimension corresponding to the constriction degree. For vowels, however, it is unlikely that any somatosensory dimension need be so strictly controlled (Lindblom, 1963). Still, the convex region theory implies that more variability will be seen for vowels along acoustically less important dimensions. The hypothesis of more articulatory variability along acoustically less important dimensions for the vowels /i/ and /a/ was tested in studies by Perkell and Nelson (1982, 1985). These reports showed more variability in tongue position along a direction parallel to the vocal tract midline than for the acoustically more important tongue position along a direction perpendicular to the vocal tract midline, supporting this hypothesis, and simulations of the DIVA model have verified this property in the model (Guenther, 1995). A similar property was found for articulatory variability during American English /r/ production by Nieto-Castanon et al. (2005).

A final example of articulatory variability comes from studies of velum position during vowel production. Production of American English vowels in different consonant contexts results in large, but not complete, variability in velum position during the vowel (Kent, Carney, & Severeid, 1974). For example, if a vowel is produced between two non-nasal consonants as in the word "dad," the velum remains completely closed throughout the utterance. When a vowel is produced between a nasal and a no-nnasal consonant as in the word "dan," the velum smoothly transitions from closed to open during the vowel. From these observations it might appear that no target velum position is specified for vowels. However, Kent, Carney and Severeid (1974) report that for a vowel between two nasal consonants, a slight but incomplete raising of the velum occurs during the vowel, followed by a lowering of the velum for the final nasal consonant. As Keating (1990) points out, these data provide a compelling case for a target range from maximally closed to largely, but not completely, open, rather than for any canonical target position.

Coarticulation

It is well-known that speech movements display *coarticulation*, which is defined as the influence of one phoneme on the sound and/or articulator configuration of neighboring phonemes. The convex region theory provides a useful framework for understanding how and why coarticulation arises.

There are two main types of coarticulation. The first is *carryover coarticulation*, also known as *perseveratory* or *left-to-right coarticulation*, which involves cases when the vocal tract configuration for one phoneme influences the configuration for a later phoneme. Some instances of carryover coarticulation may result from mechanical or inertial effects involved in moving the articulators from one sound's target to the next rather than from explicit movement planning (Baum & Waldstein, 1991; Daniloff, Schuckers, & Feth, 1980; Flege, 1988; Gay, 1977; Recasens, 1987, 1989). However, the mechano-inertial explanation is insufficient since large carryover effects are seen at low speeds and may spread over two or three segments, indicating a deliberate process for producing these effects (Daniloff & Hammarberg, 1973). This implies that many aspects of carryover coarticulation are explicitly planned by the speech motor control system.

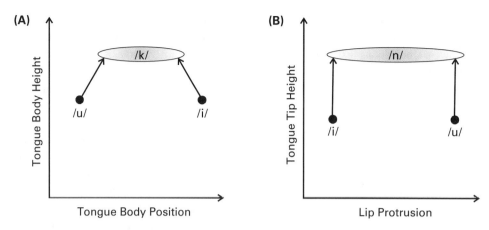

Figure 4.7
Convex region theory account of carryover coarticulation. (A) Schematized representation of the movements from /u/ to /k/ (as in "Luke") and from /i/ to /k/ (as in "leak") along the somatosensory dimensions of tongue body height and tongue body position. The movements are made to the closest points on the target region for /k/, resulting in a relatively posterior tongue body position for the /k/ in "Luke" compared to the /k/ in "leak" (cf. Daniloff et al., 1980; Kent & Minifie, 1977). (B) Schematized representation of the movements from /i/ to /n/ (as in "keen") and from /u/ to /n/ (as in "spoon"). In this case, the target region for /n/ spans the entire range of lip protrusion so no movements are made along that dimension, resulting in /n/ configurations that have the same amount of lip protrusion as the preceding vowel.

Figure 4.7 illustrates how the convex region theory accounts for carryover coarticulation effects. Panel A concerns the observation that, in American English, productions of the phoneme /k/ vary widely in anteroposterior location of the tongue body constriction in the vocal tract depending on phonetic context (e.g., Daniloff, Schuckers, & Feth, 1980; Kent & Minifie, 1977). For example, when /k/ is produced after the front vowel /i/ in the word "leak," the tongue body constriction for /k/ is anterior relative to the tongue position for /k/ after the back vowel /u/ as in "Luke." This behavior will naturally arise in the DIVA model using a convex region target for /k/ since the movement from the vowel to the /k/ target will be to the closest part of the target region. For the front vowel /i/, this will be the anterior portion of the target region whereas the movement from the back vowel /u/ will go to the posterior portion of the target region.

Panel B of figure 4.7 illustrates a case of carryover coarticulation where one aspect of a phoneme's articulatory configuration is maintained for one or more following phonemes. For example, lip protrusion for the /u/ in "spoon" is maintained through the /n/ (Daniloff & Moll, 1968). In the DIVA model, this occurs automatically when the position of the vocal tract for the preceding sound (the /u/ in this case) along the somatosensory dimension in question lies within the convex region of the target for the following sound (the /n/) along the same dimension. This case of carryover coarticulation is the result of a general tendency not to move an articulator unless it needs to be moved. In the /n/ of "spoon," the protruded lips do not need to be retracted since they already fall within the target range for the following /n/. In keeping with this account, Wood (1991) noted that instances of perseveratory coarticulation in his dynamic x-ray articulatory data "all seem to be examples of the ... tendency for individual articulators to be left idle until required again" (p. 290).

The second type of coarticulation, *anticipatory coarticulation* (also *called right-to-left coarticulation*), occurs when the vocal tract configuration for a future phoneme influences the configuration for a prior phoneme. Possible neural mechanisms underlying anticipatory coarticulation will be discussed in chapter 7. For the current purposes, it suffices to note that most theoretical accounts of anticipatory coarticulation rely on the traditional assumption that the target for each phoneme involves only a subset of the articulators, with the positions of the remaining articulators completely free to begin movements for future phonemes. These traditional "all-or-nothing" targets comprise a special case of convex region targets in which the target ranges of articulators are binarized to either the entire possible range or one particular position. For example, the velum is assumed to be a free articulator during vowels in American English, thus allowing anticipatory opening of the velum during a vowel when the following phoneme is a nasal consonant. In many cases, however, the "unused" articulators actually have some constraints on their positions. For example, we saw in the previous section that some degree of velar closing occurs for American English vowels that are produced between two nasal consonants, implying that the velum position is not completely unrestricted for vowels. In the convex region theory, movements can occur even for articulators that are not traditionally considered part of the upcoming phoneme's target, as in the case of velum height for American English vowels. The generalized form of speech targets in the convex region theory can account for a number of additional observations that contradict theories of coarticulation that involve all-or-nothing targets (see Guenther, 1995, for details).

Speech Rate and Clarity

Humans are capable of varying their speaking rate over a wide range, and speakers may change their speaking rate multiple times in a normal conversation (Miller, Grosjean, & Lomanto, 1984). Changes in speaking rate often occur automatically when a speaker is faced with different speaking situations. Lindblom (1983, 1990) postulated that the speech motor system employs a strategy of *economy of effort*, whereby speakers minimize the amount of movement effort used to speak with the constraint of maintaining intelligibility for the listener. For example, when speaking in good listening conditions with a native speaker of the language, speech movements will be reduced in magnitude compared to speaking in poor listening conditions or with a non-native speaker. Perkell et al. (2002) verified this property by measuring articulator movements under conditions requiring clear speech (as if speaking over a noisy phone line) and normal speech, noting increased movement extents and durations for the clear speech condition compared to the normal speech condition.

The convex region theory accounts for these findings as follows. When speakers are forced to speak clearly, the target regions for speech sounds shrink, leading to more precise articulations. This is illustrated in panel A of figure 4.8, which shows two dimensions (tongue body height and lip aperture) of the convex region target for the vowel /ɪ/ under clear (darker blue shade) and fast (lighter blue) speaking conditions. For a given initial vocal tract configuration (represented by the black dot in the figure), the movement distance needed to reach the target for the clear condition (D_C) is longer than the movement needed in the fast condition (D_F). Such a tendency for vowel articulations to show decreased displacements at faster speech rates is commonplace in speech; it is termed *vowel reduction* (e.g., Lindblom, 1963, 1983) and accords well with Lindblom's theory of economy of effort as well as the convex region theory.

It has been noted that consonants do not reduce as much as vowels at faster speech rates (e.g., Gay et al., 1974). The model's account of this finding is illustrated in panel B of figure 4.8. Compared to vowels, consonants require more strict control along important somatosensory dimensions to ensure either full vocal tract closure (for stops) or near-complete closure (for fricatives). Because the target region is already very narrow along these important somatosensory dimensions (lip aperture for /p/ in the figure), shrinking the

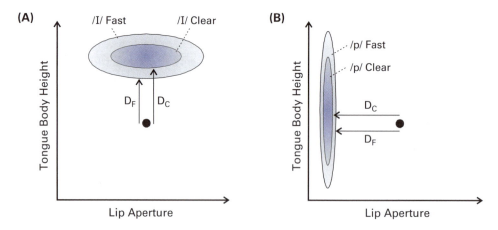

Figure 4.8
Convex region account of differences in speech movements for clear versus fast speech. (A) Movements for the vowel /ɪ/ from an initial vocal tract configuration indicated by the black dot. The target for /ɪ/ during clear speech (darker shade) is shrunk relative to the target during fast speech (lighter shade). Movement distance from the initial configuration for clear speech (D_C) is thus longer compared to fast speech (D_F). (B) Movements for the consonant /p/ from the initial vocal tract configuration indicated by the black dot. Since the target range along the key dimension of lip aperture is already very narrow, shrinking the target region for clear speech results in very little change in the distance of the required movement compared to fast speech.

target has very little effect on the amount of movement needed to reach the target under clear (D$_C$) versus fast (D$_F$) speech conditions.

Time-Varying Target Regions

Thus far we have described the targets for sounds as static (as opposed to time-varying) regions, and we have focused on the phoneme as the speech sound unit. Fowler (1980) convincingly argues that static targets cannot account for many aspects of phoneme production. For example, many phonemes have time clearly built into their motor programs. Stop consonants with the same place of articulation, such as /b/ and /p/ in English, differ from each other primarily in *voice onset time* (*VOT*), the time from opening of the vocal tract constriction to the beginning of voicing, with voiced stops such as /b/ having very short VOTs (around 0 ms or even prevoiced, i.e., vocal fold vibration begins *before* the stop closure is released), whereas voiceless stops such as /p/ typically have VOTs in the 20–40 ms range in spontaneous speech. Furthermore, sounds such as /w/ as in "*w*ay" and /j/ as in "*y*ellow," known as *glides*, are characterized by dynamic rather than static formant patterns. These observations highlight the fact that the targets for these phonemes must be time-varying rather than static.

The need for time-varying targets becomes even more obvious when one considers syllables to be speech sound units (in the sense of having their own stored motor programs) as in the DIVA model. For this reason, the model utilizes *time-varying target regions* for each speech sound. An example of this is presented in figure 4.9 for the auditory target of the syllable "bah." The target consists of target ranges for each formant frequency that vary with time. If the model successfully produces an auditory trajectory that falls within this

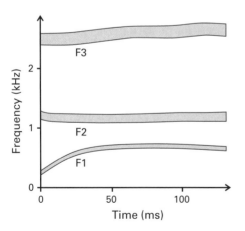

Figure 4.9
Time-varying auditory target region for the first three formants (F1, F2, and F3) of the syllable "bah" in the DIVA model.

target region at all points in time, then the sound "bah" has been properly produced. The somatosensory targets in the model vary with time in a similar fashion.

Talker and Time Normalization

An important issue not treated in this discussion, nor in the remainder of this book, concerns how the infant's brain transforms adult speech, which consists of lower pitch and formant frequencies than the infant vocal tract is capable of producing, into a *talker-normalized* auditory representation that allows the infant to match his or her own vocalizations to those of adults. Such a representation is assumed in the account of auditory target learning and auditory feedback control proposed in this book. Although proposals for talker-normalized auditory representations have been made in the literature (e.g., Syrdal & Gopal, 1986; Miller, 1989) and such a representation has been used in simulations of the DIVA model (e.g., Guenther, Hampson, & Johnson, 1998), the neural instantiation of these representations remains an important area for future research. Additionally, the sound examples presented to a developing infant are produced at different speech rates, highlighting the need for *time normalization* in addition to talker normalization when learning auditory targets for speech sounds. It is simply presumed in the DIVA model that a neural solution to this problem exists (that is, a "non-neural" algorithmic time normalization process is used in DIVA model simulations), leaving the neural processes underlying this time normalization as another important area for future research.

4.5 The Neural Bases of Speech Targets

The Mirror Neuron System

When infants learn sensory targets for speech sounds, these targets become associated with motor acts—specifically, the vocal tract movements for producing the sounds once the speech motor system is "tuned up." According to the DIVA model, this linkage occurs at a *speech sound map* located in *ventral premotor cortex* (*vPMC*), primarily in the left hemisphere. A key property of speech sound map nodes in the DIVA model is that they are active both when *perceiving* the sound as produced by another talker and when *producing* the sound (Guenther, 1992, 1994, 1995). During perception they are involved in learning sensory targets for speech sounds, and during production they are responsible for activating the motor programs for producing these sounds.

Researchers investigating the grasping system in monkeys were the first to identify neurons of this type, that is, neurons that are active both when a monkey perceives a particular act and when it produces that act itself (di Pellegrino et al., 1992; Rizzolatti et al., 1996). Since their discovery, these cells, termed *mirror neurons*, have garnered a tremendous amount of attention in the motor control and neuroscience literatures (see Rizzolatti & Craighero, 2004, and Iacoboni & Dapretto, 2006, for reviews). In our theoretical view, a mirror neuron system devoted to speech plays a crucial role in the learning of sensory targets for speech production. This is in keeping with the general perception of mirror neurons as crucial components in imitation-based learning (e.g., Rizzolatti & Arbib, 1998; Rizzolatti & Craighero, 2004; Iacoboni & Dapretto, 2006) and language (Rizzolatti & Arbib, 1998; Arbib, 2005).

Mirror neurons for grasping were first found in monkey *area F5* (di Pellegrino et al., 1992), a ventral premotor area that is often described as the homologue to *Brodmann area (BA) 44* (e.g., Binkofski & Buccino, 2004; Rizzolatti & Arbib, 1998) in the *inferior frontal gyrus pars opercularis (IFo)*. In the left hemisphere, this region forms part of Broca's area, along with *BA 45* in the *inferior frontal gyrus pars triangularis (IFt)*. Together, IFo and IFt are often referred to as the *posterior inferior frontal gyrus (pIFG)*.

Mirror neurons in monkey area F5 that are sensitive to mouth movements were later identified (Ferrari et al., 2003), as were audiovisual mirror neurons that respond when the monkey either hears or sees a particular act being performed (Kohler et al., 2002; Keysers et al., 2003). Furthermore, neurons with mirror neuron properties have been identified in the *rostral inferior parietal lobule (rIPL*; Gallese et al., 2002). These parietal neurons are hypothesized to receive input from purely sensory visual nodes in the *posterior superior temporal sulcus (pSTS*; Rizzolatti & Craighero, 2004).

It is widely believed that humans possess a similar mirror neuron system to monkeys, as schematized in panel A of figure 4.10 (adapted from Iacoboni & Dapretto, 2006), although no single neuron recording data are available to verify this at present. Instead, the results of neuroimaging and transcranial magnetic stimulation studies have been used to provide support for the existence of mirror neurons in the human brain. Panel B of figure 4.10 details three possible sites of speech-related mirror neurons in the inferior frontal cortex, derived from the results of the studies described below. These sites include the *dorsal portion of the inferior frontal gyrus pars opercularis (dIFo)*, the *ventral portion of the inferior frontal gyrus pars opercularis (vIFo)*, and a site located in *rostral precentral gyrus (rPrCG)* near the posterior-most extent of the inferior frontal sulcus. This latter site is likely located in *BA 6* whereas the other two sites are in BA 44.

Neuroimaging studies of the mirror system for hand movements in humans indicate a bilateral mirror neuron representation in pIFG (Iacoboni et al., 1999; Iacoboni & Dapretto, 2006; Aziz-Zadeh et al., 2006). The meta-analysis of seven fMRI studies performed by Molnar-Szakacs et al. (2004) further localized this representation to dIFo. In contrast, vIFo was active during imitation of hand movements but not during action observation without imitation (as opposed to mirror neurons, which are active during both observation and action). vIFo was also inactive when a similar motor task was performed in the absence of imitation, suggesting a specialized role in imitation. The Molnar-Szakacs meta-analysis also identified mirror-like properties near rPrCG in figure 4.10B. Although mirror neuron studies of visually presented hand movements typically identify bilateral activation, Aziz-Zadeh et al. (2004) note that listening to a motor act (as opposed to viewing it) appears to invoke only left-hemisphere motor areas, suggesting that mirror neurons in the right

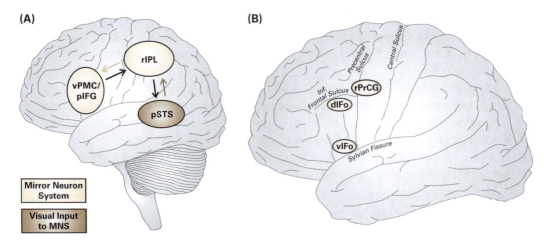

Figure 4.10
(A) Schematic of the mirror neuron system for grasping in humans (beige) and its visual input pathways (brown). Black arrows indicate efference copies of motor commands. After Iacoboni and Dapretto (2006). (B) Detail of potential mirror neuron sites in the ventral premotor cortex and posterior inferior frontal gyrus; see the text for details. dIFo, dorsal inferior frontal gyrus pars opercularis; Inf., inferior; MNS, mirror neuron system; pIFG, posterior inferior frontal gyrus; pSTS, posterior superior temporal sulcus; rIPL, rostral inferior parietal lobule; rPrCG, rostral precentral gyrus; vIFo, ventral inferior frontal gyrus pars opercularis; vPMC, ventral premotor cortex.

hemisphere may be purely visual whereas left-hemisphere mirror neurons may be audiovisual or multimodal (Aziz-Zadeh et al., 2006; Iacoboni & Dapretto, 2006).

The mirror neuron system for speech has also been investigated in a number of studies. The fMRI studies of Wilson and colleagues indicate that listening to speech bilaterally activates rPrCG (Wilson et al., 2004; Wilson & Iacoboni, 2006). These authors report that Broca's area is not strongly active during passive listening to meaningless speech. However, other neuroimaging studies of auditory speech perception have found activity in pIFG. For example, although Callan et al. (2003) noted frontal activity primarily in left rPrCG when Japanese subjects identified English /b/ and /g/ sounds, a more difficult task involving identification of English /r/ and /l/ sounds produced activity in left dIFo, vIFo, and rPrCG. After receiving training on the English /r/-/l/ contrast, greater activity was found in all three regions bilaterally. A later study (Callan et al., 2004) found that English speakers performing the /r/-/l/ identification task had left hemisphere activity in rPrCG, dIFo, and vIFo, with right hemisphere activity mainly found in rPrCG.

Transcranial magnetic stimulation (*TMS*) has also provided evidence for mirror neurons during speech listening. Fadiga et al. (2002) utilized a *motor evoked potential* technique in which a TMS pulse was applied to left motor cortex at a near-threshold level (i.e., a level just below that which results in a measurable muscle activation) while subjects listened to speech stimuli. Tongue muscles were simultaneously monitored for evoked potentials using electromyography. Because the TMS pulse was applied at a subthreshold level, the presence of motor evoked potentials during stimulus processing was taken to indicate additional activation in motor cortex due to the stimulus. The authors noted that listening to speech increases the excitability of tongue muscles, with the biggest effect occurring for a speech sound requiring precise tongue control, thus suggesting that auditorily presented speech activates the motor areas that would normally be used to produce that speech, presumably via the mirror neuron system.

Viewing a speaking face in the absence of auditory information also appears to invoke the mirror neuron system. Although early studies of visual speech processing concentrated on activation within auditory cortex (e.g., Calvert et al., 1997), later studies identified activation in the inferior frontal cortex. For example, Campbell et al. (2001) note activity in the left inferior frontal gyrus and premotor cortex when contrasting viewing a speaking face to viewing a face making nonspeech gurning movements. Buccino et al. (2004) also note a strong, left-lateralized response in the left dIFo, vIFo, and rPrCG when subjects viewed a speaking face contrasted with viewing a still image of the face. Nishitani and Hari (2002) used magnetoencephalography to investigate the time course of activation in the cortex when viewing still pictures of verbalizing lip forms. Activity started in the visual cortex approximately 115 ms after stimulus onset, then moved in 20- to 70-ms steps to the superior temporal region (in/near the superior temporal sulcus; cf. figure 4.10), the inferior parietal lobule, the inferior frontal gyrus, and finally to the primary motor cortex approximately 310 ms after picture presentation. Viewing nonverbal lip forms did not result in activity in the inferior frontal gyrus or motor cortex, suggesting specialization for speech processing in this circuit.

Taken together, the results summarized in this section suggest that the human brain contains mirror neurons in inferior frontal cortex, likely residing in portions of both BA 6 and 44, with a left-hemisphere bias for mirror neurons encoding speech perceived via auditory or audiovisual means. In the remainder of this book, the term *vPMC* will be used to describe the location of speech-related mirror neurons, keeping in mind that this definition of vPMC includes portions of both the posterior inferior frontal gyrus and the rostral precentral gyrus.

Mirror Neurons and Speech Targets

According to the DIVA model, speech mirror neurons (corresponding to the model's *speech sound map*) are used to learn sensory targets for speech sounds. These targets are encoded in synaptic projections from the mirror neurons in left vPMC to higher-order auditory and somatosensory cortical areas, both directly and via the cortico-cerebellar loop, as schematized in figure 4.11. When an infant hears an adult speaker producing a new speech sound, a speech sound map node that comes to represent that sound is activated, and synaptic projections from this node to speech-related auditory cortical areas in the posterior superior temporal gyrus (BA 22) are tuned to encode the auditory cortical

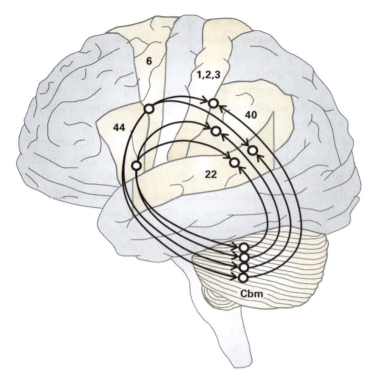

Figure 4.11
Projections from the speech-related premotor areas are hypothesized to encode expected sensory consequences, or sensory *targets*, corresponding to the articulator movements currently being commanded by the premotor cortex. Numbers indicate Brodmann's areas (BA). Projections link speech sound map nodes in premotor areas (BA 6, 44) to higher-order auditory (BA 22) and somatosensory (BA 1, 2, 3, 40) cortical areas, via both direct cortico-cortical pathways and indirectly via the pons, cerebellum, and thalamus (pathways simplified for clarity). These projections are hypothesized to represent the expected auditory and somatic sensation patterns corresponding to the speech sound currently being produced. Cbm, cerebellum.

activity associated with the sound. Projections similar to these are sometimes referred to as *forward models* in the motor control literature. If the infant is also viewing the adult speaker, somatosensory information available through vision is stored in synaptic projections from the speech sound map in inferior frontal cortex to the speech-related somatosensory cortical areas in the inferior parietal cortex (BA 1, 2, 3, and 40), though these somatosensory targets will necessarily be incomplete since the infant cannot see the tongue movements used to produce the speech sound. The somatosensory target becomes refined as the infant monitors his or her own correct self-productions of the sound. This refinement process continues into adulthood, adjusting the somatosensory targets to account for the dramatic changes in the sizes and shapes of vocal tract articulators that occur because of growth. During production of the sound, the auditory and somatosensory targets are used for auditory and somatosensory feedback control, respectively, as detailed in the next two chapters.

4.6 Summary

Characterization of speech production as a motor control process requires identification of the controlled variables. This chapter has provided evidence that the targets for speech sounds consist of time-varying regions in auditory and somatosensory spaces. Evidence for somatosensory targets comes from studies indicating compensation for articulatory perturbations that do not affect the acoustic signal. Compensation is also seen for auditory perturbations that do not affect the somatosensory state, indicative of the use of auditory targets.

Evidence of a primary role for auditory targets in shaping motor output comes from studies showing large articulatory variability when producing a speech sound in different phonetic contexts. This variability is not random, instead taking the form of articulatory trading relations that maintain acoustic stability for the sound across contexts. Furthermore, developing infants are provided with complete auditory targets for speech sounds from adult speakers, but they cannot directly perceive somatosensory information about tongue position that is crucial for the production of most sounds. Therefore, it is proposed that the auditory target for a speech sound is primarily responsible for shaping the motor program for that sound, and somatosensory targets are learned at a later stage of development as the speech production mechanism monitors its own correct productions to determine the somatosensory correlates for each sound.

To explain how infants learn language-specific limits on variability in different auditory and somatosensory dimensions, the *convex region theory* of speech targets posits that the targets take the form of regions, rather than points, in auditory and somatosensory spaces. Infants learn the limits of acceptable variability by monitoring the distribution of speech sounds in the ambient language along each auditory dimension. Similarly, somatosensory target regions are learned by monitoring the amount of somatosensory variability during correct self-productions. In addition to accounting for important aspects of speech development, the convex region theory provides a simple and intuitive explanation for a number of experimental findings concerning speech movements, including acoustic and articulatory variability, motor equivalence, coarticulation, and speaking rate and clarity adjustments.

Auditory targets for speech sounds are believed to be effectively "stored" in synaptic projections from mirror neurons in a speech sound map in left ventral premotor cortex (including portions of the rostral precentral gyrus and adjoining posterior inferior frontal gyrus) to higher-order auditory cortical areas in the posterior superior temporal gyrus. These projections likely involve direct cortico-cortical connections as well as projections that pass through subcortical structures, most notably the pons, cerebellum, and thalamus.

Somatosensory targets for speech sounds are thought to be encoded in projections from the speech sound map to higher-order somatosensory areas in the postcentral gyrus and supramarginal gyrus, again likely involving both direct cortico-cortical projections and indirect projections through the cerebellum and other subcortical structures.

Notes

1. In this way speech is more closely related to musical skill development than to the visuomotor skill development required for most sports. This topic is addressed in detail by Patel (2008).

2. In mathematics, a *convex region* is a multidimensional region that has the following property: for any two points in the region, all points on a line segment connecting these two points are also in the region. Such a region is easier to learn from a sparse number of samples than a nonconvex region; for this reason, the DIVA model utilizes convex auditory target regions for speech sounds.

References

Abbs, J. H., & Gracco, V. L. (1984). Control of complex motor gestures: orofacial muscle responses to load perturbations of lip during speech. *Journal of Neurophysiology, 51,* 705–723.

Arbib, M. A. (2005). From monkey-like action recognition to human language: an evolutionary framework for neurolinguistics. *Behavioral and Brain Sciences, 28,* 105–124.

Aziz-Zadeh, L., Iacoboni, M., Zaidel, E., Wilson, S., & Mazziotta, J. (2004). Left hemisphere motor facilitation in response to manual action sounds. *European Journal of Neuroscience, 19,* 2609–2612.

Aziz-Zadeh, L., Koski, L., Zaidel, E., Mazziotta, J., & Iacoboni, M. (2006). Lateralization of the human mirror neuron system. *Journal of Neuroscience, 26,* 2964–2970.

Baum, S. R., McFarland, D. H., & Diab, M. (1996). Compensation to articulatory perturbation: perceptual data. *Journal of the Acoustical Society of America, 99,* 3791–3794.

Baum, S. R., & Waldstein, R. S. (1991). Perseveratory coarticulation in the speech of profoundly hearing-impaired and normally hearing children. *Journal of Speech, Language, and Hearing Research: JSLHR, 34,* 1286–1292.

Berkinblit, M. B., Gelfand, I. M., & Feldman, A. G. (1986). A model of the aiming phase of the wiping reflex. In S. Grillner, P. S. G. Stein, D. G. Stuart, H. Forssberg, & R. M. Herman (Eds.), *Neurobiology of vertebrate locomotion* (pp. 217–227). London: Macmillan.

Binkofski, F., & Buccino, G. (2004). Motor functions of the Broca's region. *Brain and Language, 89,* 362–369.

Boyce, S., & Espy-Wilson, C. Y. (1997). Coarticulatory stability in American English /r/. *Journal of the Acoustical Society of America, 101,* 3741–3753.

Bradlow, A. R., Akahane-Yamada, R., Pisoni, D. B., & Tohkura, Y. (1999). Training Japanese listeners to identify English /r/ and /l/: long-term retention of learning in perception and production. *Perception & Psychophysics, 61,* 977–985.

Bradlow, A. R., Pisoni, D. B., Akahane-Yamada, R., & Tohkura, Y. (1997). Training Japanese listeners to identify English /r/ and /l/: IV. Some effects of perceptual learning on speech production. *Journal of the Acoustical Society of America, 101,* 2299–2310.

Browman, C., & Goldstein, L. (1990 a). Tiers in articulatory phonology, with some implications for casual speech. In J. Kingston & M. E. Beckman (Eds.), *Papers in laboratory phonology: Vol. 1. Between the grammar and physics of speech* (pp. 341–376). Cambridge, UK: Cambridge University Press.

Browman, C., & Goldstein, L. (1990 b). Gestural specification using dynamically-defined articulatory structures. *Journal of Phonetics, 18,* 299–320.

Buccino, G., Lui, F., Canessa, N., Patteri, I., Lagravinese, G., Benuzzi, F., et al. (2004). Neural circuits involved in the recognition of actions performed by nonconspecifics: an fMRI study. *Journal of Cognitive Neuroscience, 16,* 114–126.

Callan, D. E., Jones, J. A., Callan, A. M., & Akahane-Yamada, R. (2004). Phonetic perceptual identification by native- and second-language speakers differentially activates brain regions involved with acoustic phonetic processing and those involved with articulatory-auditory/orosensory internal models. *NeuroImage, 22,* 1182–1194.

Callan, D. E., Kent, R. D., Guenther, F. H., & Vorperian, H. K. (2000). An auditory-feedback-based neural network model of speech production that is robust to developmental changes in the size and shape of the articulatory system. *Journal of Speech, Language, and Hearing Research, 43,* 721–736.

Callan, D. E., Tajima, K., Callan, A. M., Kubo, R., Masaki, S., & Akahane-Yamada, R. (2003). Learning-induced neural plasticity associated with improved identification performance after training of a difficult second-language phonetic contrast. *NeuroImage, 19,* 113–124.

Calvert, G. A., Bullmore, E. T., Brammer, M. J., Campbell, R., Williams, S. C., McGuire, P. K., et al. (1997). Activation of auditory cortex during silent lipreading. *Science, 276,* 593–596.

Campbell, R., MacSweeney, M., Surguladze, S., Calvert, G., McGuire, P., Suckling, J., et al. (2001). Cortical substrates for the perception of face actions: an fMRI study of the specificity of activation for seen speech and for meaningless lower-face acts (gurning). *Brain Research. Cognitive Brain Research, 12,* 233–243.

Capelle, P. (1995). *Play your best pool: secrets to winning 8-ball and 9-ball.* Midway City, CA: Billiards Press.

Catford, J. C., & Pisoni, D. (1970). Auditory vs. articulatory training in exotic sounds. *Modern Language Journal, 54,* 477–481.

Cruse, H., Brüwer, M., & Dean, J. (1993). Control of three- and four-joint arm movement: strategies for a manipulator with redundant degrees of freedom. *Journal of Motor Behavior, 25,* 131–139.

Daniloff, R., & Hammarberg, R. E. (1973). On defining coarticulation. *Journal of Phonetics, 1,* 239–248.

Daniloff, R., & Moll, K. (1968). Coarticulation of lip rounding. *Journal of Speech and Hearing Research, 11,* 707–721.

Daniloff, R., Schuckers, G., & Feth, L. (1980). *The physiology of speech and hearing: an introduction.* Englewood Cliffs, NJ: Prentice-Hall.

Delattre, P., & Freeman, D. C. (1968). A dialect study of American r's by x-ray motion picture. *Linguistics, 44,* 29–68.

di Pellegrino, G., Fadiga, L., Fogassi, L., Gallese, V., & Rizzolatti, G. (1992). Understanding motor events: a neurophysiological study. *Experimental Brain Research, 91,* 176–180.

Dodd, B. (1983). The visual and auditory modalities in phonological acquisition. In A. E. Mills (Ed.), *Language acquisition in the blind child: normal and deficient* (pp. 57–61). London: Croom Helm.

Erber, N. P. (1974). Visual perception of speech by deaf children: recent developments and continuing needs. *Journal of Speech and Hearing Disorders, 39,* 178–185.

Fadiga, L., Craighero, L., Buccino, G., & Rizzolatti, G. (2002). Speech listening specifically modulates the excitability of tongue muscles: a TMS study. *European Journal of Neuroscience, 15,* 399–402.

Ferrari, P. F., Gallese, V., Rizzolatti, G., & Fogassi, L. (2003). Mirror neurons responding to the observation of ingestive and communicative mouth actions in the monkey ventral premotor cortex. *European Journal of Neuroscience, 17,* 1703–1714.

Fisher, C. G. (1968). Confusions among visually perceived consonants. *Journal of Speech and Hearing Research, 11,* 796–804.

Flege, J. E. (1988). Anticipatory and carry-over nasal coarticulation in the speech of children and adults. *Journal of Speech and Hearing Research, 31,* 525–536.

Flege, J. E. (1991). Perception and production: the relevance of phonetic input to L2 phonological learning. In J. Hueber & C. Ferguson (Eds.), *Crosscurrents in second language acquisition and linguistic theories* (pp. 249–289). Amsterdam: John Benjamins.

Fowler, C. A. (1980). Coarticulation and theories of extrinsic timing. *Journal of Phonetics, 8,* 113–133.

Gallese, V., Fogassi, L., Fadiga, V., & Rizzolatti, G. (2002). Action representation and the inferior parietal lobe. In W. Prinz & B. Hommel (Eds.), *Attention and performance XIX: Common mechanisms in perception and action* (pp. 247–266). Oxford: Oxford University Press.

Gay, T. (1977). Articulatory movements in VCV sequences. *Journal of the Acoustical Society of America*, *62*, 183–193.

Gay, T., Ushijima, T., Hirose, H., & Cooper, F. S. (1974). Effects of speaking rate on labial consonant-vowel articulation. *Journal of Phonetics*, *2*, 47–63.

Goto, H. (1971). Auditory perception by normal Japanese adults of the sounds "l" and "r." *Neuropsychologia*, *9*, 317–323.

Guenther, F. H. (1992). *Neural models of adaptive sensory-motor control for flexible reaching and speaking.* PhD dissertation, Boston University.

Guenther, F. H. (1994). A neural network model of speech acquisition and motor equivalent speech production. *Biological Cybernetics*, *72*, 43–53.

Guenther, F. H. (1995). A modeling framework for speech motor development and kinematic articulator control. In K. Elenius & P. Branderud (Eds.), *Proceedings of the XIIIth international congress of phonetic sciences*, Stockholm, Sweden, 13–19 August, 1995 (Vol. 2, pp. 92–99). Stockholm: KTH and Stockholm University.

Guenther, F. H., Espy-Wilson, C. Y., Boyce, S. E., Matthies, M. L., Zandipour, M., & Perkell, J. S. (1999). Articulatory tradeoffs reduce acoustic variability during American English /r/ production. *Journal of the Acoustical Society of America*, *105*, 2854–2865.

Guenther, F. H., Hampson, M., & Johnson, D. (1998). A theoretical investigation of reference frames for the planning of speech movements. *Psychological Review*, *105*, 611–633.

Hagiwara, R. (1994). Three types of American /r/. *UCLA Working Papers in Phonetics*, 88, 63–90.

Hogan, B. (1957). *Five lessons: the modern fundamentals of golf.* New York: Simon & Schuster.

Iacoboni, M., & Dapretto, M. (2006). The mirror neuron system and the consequences of its dysfunction. *Nature Reviews. Neuroscience*, *7*, 942–951.

Iacoboni, M., Woods, R. P., Brass, M., Bekkering, H., Mazziotta, J. C., & Rizzolatti, G. (1999). Cortical mechanisms of human imitation. *Science*, *286*, 2526–2528.

Keating, P. A. (1990). The window model of coarticulation: articulatory evidence. In J. Kingston & M. E. Beckman (Eds.), *Papers in laboratory phonology: Vol. 1. Between the grammar and physics of speech* (pp. 451–470). Cambridge, UK: Cambridge University Press.

Kelso, J. A., Tuller, B., Vatikiotis-Bateson, E., & Fowler, C. A. (1984). Functionally specific articulatory cooperation following jaw perturbations during speech: evidence for coordinative structures. *Journal of Experimental Psychology: Human Perception and Performance*, *10*, 812–832.

Kent, R. D., Carney, P., & Severeid, L. (1974). Velar movement and timing: evaluation of a model for binary control. *Journal of Speech and Hearing Research*, *17*, 470–488.

Kent, R. D., & Minifie, F. D. (1977). Coarticulation in recent speech production models. *Journal of Phonetics*, *5*, 115–133.

Kent, R. D., Osberger, M. J., Netsell, R., & Hustedde, C. G. (1987). Phonetic development in identical twins differing in auditory function. *Journal of Speech and Hearing Disorders*, *52*, 64–75.

Keysers, C., Kohler, E., Umiltà, M. A., Nanetti, L., Fogassi, L., & Gallese, V. (2003). Audiovisual mirror neurons and action recognition. *Experimental Brain Research*, *153*, 628–636.

Kohler, E., Keysers, C., Umiltà, M. A., Fogassi, L., Gallese, V., & Rizzolatti, G. (2002). Hearing sounds, understanding actions: action representation in mirror neurons. *Science*, *297*, 846–848.

Ladefoged, P., & Maddieson, I. (1996). *The sounds of the world's languages.* Oxford: Blackwell.

Landau, B., & Gleitman, L. R. (1985). *Language and experience: evidence from the blind child.* Cambridge, MA: Harvard University Press.

Liberman, A. M., Cooper, F. S., Shankweiler, D. P., & Studdert-Kennedy, M. (1967). Perception of the speech code. *Psychological Review*, *74*, 431–461.

Liberman, A. M., & Mattingly, I. G. (1985). The motor theory of speech revisited. *Cognition*, *21*, 1–36.

Lindblom, B. (1963). Spectrographic study of vowel reduction. *Journal of the Acoustical Society of America*, *35*, 1773–1781.

Lindblom, B. (1983). Economy of speech gestures. In P. F. MacNeilage (Ed.), *The production of speech* (pp. 217–245). New York: Springer-Verlag.

Lindblom, B. (1990). Explaining phonetic variation: a sketch of the H&H theory. In W. J. Hardcastle & A. Marchal (Eds.), *Speech production and speech modeling* (pp. 403–440). Dordrecht: Kluwer.

Lindblom, B., Lubker, J., & Gay, T. (1979). Formant frequencies of some fixed-mandible vowels and a model of speech motor programming by predictive simulation. *Journal of Phonetics, 7*, 147–161.

Lotto, A. J., Sato, M., & Diehl, R. L. (2004). Mapping the task for the second language learner: the case of Japanese acquisition of /r/ and /l/. In J. Slifka, S. Manuel, & M. Matthies (Eds.), *From sound to sense: 50+ years of discoveries in speech communication.* Electronic conference proceedings.

MacNeilage, P. F. (1970). Motor control of serial ordering in speech. *Psychological Review, 77*, 182–196.

MacNeilage, P. F., Rootes, T. P., & Chase, R. A. (1967). Speech production and perception in a patient with severe impairment of somesthetic perception and motor control. *Journal of Speech and Hearing Research, 10*, 449–467.

Maeda, S. (1990). Compensatory articulation during speech: evidence from the analysis and synthesis of vocal tract shapes using an articulatory model. In W. J. Hardcastle & A. Marchal (Eds.), *Speech production and speech modeling* (pp. 131–149). Boston: Kluwer Academic.

Massaro, D. W. (1998). *Perceiving talking faces: from speech perception to a behavioral principle.* Cambridge, MA: MIT Press.

McFarland, D. H., & Baum, S. R. (1995). Incomplete compensation to articulatory perturbation. *Journal of the Acoustical Society of America, 97*, 1865–1873.

McGurk, H., & MacDonald, J. (1976). Hearing lips and seeing voices. *Nature, 264*, 746–748.

Merton, P. A. (1972). How we control the contraction of our muscles. *Scientific American, 226*, 30–37.

Miller, J. D. (1989). Auditory-perceptual interpretation of the vowel. *Journal of the Acoustical Society of America, 85*, 2114–2134.

Miller, J. L., Grosjean, F., & Lomanto, C. (1984). Articulation rate and its variability in spontaneous speech: a reanalysis and some implications. *Phonetica, 41*, 215–225.

Mills, A. E. (1983). *Language acquisition in the blind child: normal and deficient.* London: Croom Helm.

Miyawaki, K., Strange, W., Verbrugge, R., Liberman, A. M., Jenkins, J. J., & Fujimura, O. (1975). An effect of linguistic experience: the discrimination of [r] and [l] by native speakers of Japanese and English. *Perception & Psychophysics, 18*, 331–340.

Molnar-Szakacs, I., Iacoboni, M., Koski, L., & Mazziotta, J. (2004). Functional segregation within pars opercularis of the inferior frontal gyrus: evidence from fMRI studies of imitation and action observation. *Cerebral Cortex, 15*, 986–994.

Nieto-Castanon, A., Guenther, F. H., Perkell, J. S., & Curtin, H. (2005). A modeling investigation of articulatory variability and acoustic stability during American English /r/ production. *Journal of the Acoustical Society of America, 117*, 3196–3212.

Nishitani, N., & Hari, R. (2002). Viewing lip forms: cortical dynamics. *Neuron, 36*, 1211–1220.

Oller, D. K., & Eilers, R. E. (1988). The role of audition in infant babbling. *Child Development, 59*, 441–449.

Oller, D. K., Eilers, R. E., Bull, D. H., & Carney, A. E. (1985). Prespeech vocalizations of a deaf infant: a comparison with normal metaphonological development. *Journal of Speech and Hearing Research, 28*, 47–63.

Ong, D., & Stone, M. (1998). Three dimensional vocal tract shapes in [r] and [l]: a study of MRI, ultrasound, electropalatography, and acoustics. *Phonoscope, 1*, 1–14.

Patel, A. D. (2008). *Music, language, and the brain.* New York: Oxford University Press.

Payan, Y., & Perrier, P. (1997). Synthesis of V–V sequences with a 2D biomechanical tongue model controlled by the equilibrium point hypothesis. *Speech Communication, 22*, 185–205.

Perkell, J. S., Matthies, M. L., Svirsky, M. A., & Jordan, M. I. (1993). Trading relations between tongue-body raising and lip rounding in production of the vowel /u/: a pilot "motor equivalence" study. *Journal of the Acoustical Society of America, 93*, 2948–2961.

Perkell, J. S., & Nelson, W. L. (1982). Articulatory targets in speech motor control: a study of vowel production. In S. Grillner, A. Persson, B. Lindblom, & J. Lubker (Eds.), *Speech motor control* (pp. 187–204). New York: Pergamon.

Perkell, J. S., & Nelson, W. L. (1985). Variability in production of the vowels /i/ and /a/. *Journal of the Acoustical Society of America, 77,* 1889–1895.

Perkell, J. S., Zandipour, M., Matthies, M. L., & Lane, H. (2002). Economy of effort in different speaking conditions: I. A preliminary study of intersubject differences and modeling issues. *Journal of the Acoustical Society of America, 112,* 1627–1641.

Raibert, M. H. (1977). *Motor control and learning by the state space model.* Technical Report AI-M-351, Massachusetts Institute of Technology.

Recasens, D. (1987). An acoustic analysis of V-to-C and V-to-V coarticulatory effect in Caralan and Spanish VCV sequences. *Journal of Phonetics, 15,* 299–312.

Recasens, D. (1989). Long range coarticulation effect for tongue dorsum contact in VCVCV sequence. *Speech Communication, 8,* 293–307.

Rizzolatti, G., & Arbib, M. A. (1998). Language within our grasp. *Trends in Neurosciences, 21,* 188–194.

Rizzolatti, G., & Craighero, L. (2004). The mirror-neuron system. *Annual Review of Neuroscience, 27,* 169–192.

Rizzolatti, G., Fadiga, L., Gallese, V., & Fogassi, L. (1996). Premotor cortex and the recognition of motor actions. *Brain Research. Cognitive Brain Research, 3,* 131–141.

Saltzman, E. L., & Munhall, K. G. (1989). A dynamical approach to gestural patterning in speech production. *Ecological Psychology, 1,* 333–382.

Savariaux, C., Perrier, P., & Orliaguet, J. P. (1995). Compensation strategies for the perturbation of the rounded vowel [u] using a lip tube: a study of the control space in speech production. *Journal of the Acoustical Society of America, 98,* 2428–2442.

Savariaux, C., Perrier, P., Orliaguet, J. P., & Schwartz, J. L. (1999). Compensation strategies for the perturbation of French [u] using a lip tube: II. Perceptual analysis. *Journal of the Acoustical Society of America, 106,* 381–393.

Sheldon, A., & Strange, W. (1982). The acquisition of /r/ and /l/ by Japanese learners of English: evidence that speech production can precede speech perception. *Applied Psycholinguistics, 3,* 243–261.

Stein, P. S. G., Mortin, L. I., & Robertson, G. A. (1986). The forms of a task and their blends. In S. Grillner, P. S. G. Stein, D. G. Stuart, H. Forssberg, & R. M. Herman (Eds.), *Neurobiology of vertebrate locomotion* (pp. 201–216). London: Macmillan.

Sterne, A., & Goswami, U. (2000). Phonological awareness of syllables, rhymes, and phonemes in deaf children. *Journal of Child Psychology and Psychiatry and Allied Disciplines, 41,* 609–625.

Stoel-Gammon, C., & Otomo, K. (1986). Babbling development of hearing-impaired and normally hearing subjects. *Journal of Speech and Hearing Disorders, 51,* 33–41.

Sumby, W., & Pollack, I. (1954). Visual contribution to speech intelligibility in noise. *Journal of the Acoustical Society of America, 26,* 212–215.

Syrdal, A. K., & Gopal, H. S. (1986). A perceptual model of vowel recognition based on the auditory representation of American English vowels. *Journal of the Acoustical Society of America, 79,* 1086–1100.

Tremblay, S., Shiller, D. M., & Ostry, D. J. (2003). Somatosensory basis of speech production. *Nature, 423,* 866–869.

Vorperian, H. K., Kent, R. D., Lindstrom, M. J., Kalina, C. M., Gentry, L. R., & Yandell, B. S. (2005). Development of vocal tract length during early childhood: a magnetic resonance imaging study. *Journal of the Acoustical Society of America, 117,* 338–350.

Walden, B. E., Prosek, R. A., Montgomery, A. A., Scherr, C. K., & Jones, C. J. (1977). Effects of training on the visual recognition of consonants. *Journal of Speech and Hearing Research, 20,* 130–145.

Westbury, J. R., Hashi, M., & Lindstrom, M. J. (1998). Differences among speakers in lingual articulation of American English /r/. *Speech Communication, 26,* 203–226.

Wilson, S. M., & Iacoboni, M. (2006). Neural responses to non-native phonemes varying in producibility: evidence for the sensorimotor nature of speech perception. *NeuroImage*, *33*, 316–325.

Wilson, S. M., Saygin, A. P., Sereno, M. I., & Iacoboni, M. (2004). Listening to speech activates motor areas involved in speech production. *Nature Neuroscience*, *7*, 701–702.

Wood, S. A. J. (1991). X-ray data on the temporal coordination of speech gestures. *Journal of Phonetics*, *19*, 281–292.

Woodward, M. F., & Barber, C. G. (1960). Phoneme perception in lipreading. *Journal of Speech and Hearing Research*, *3*, 212–222.

Wright, C. E. (1990). Generalized motor programs: reexamining claims of effector independence in writing. In M. Jeannerod (Ed.), *Attention and performance XIII: Motor representation and control* (pp. 294–320). Hillsdale, NJ: Erlbaum.

Zandipour, M. (2007). *A modeling investigation of vowel-to-vowel movement planning in acoustic and muscle spaces*. PhD dissertation, Boston University.

5

Auditory Feedback Control

The use of auditory feedback to tune motor programs for the speech articulators has long been recognized as crucial for achieving fluent speech. The auditory feedback control subsystem plays its largest role as a child first learns to produce the words of his or her language, starting around the end of the first year of life. Although the importance of auditory feedback control starts to decrease after the bulk of the word inventory of the native language has been learned, it continues to operate through adulthood. This chapter addresses the neural circuitry underlying these processes.

5.1 The Auditory Feedback Control Subsystem

Figure 5.1 highlights the auditory feedback control subsystem within the overall control system for speech introduced in chapter 3, section 3.3. Recall that A is a vector whose components correspond to the current auditory state, A_T is a vector specifying the current auditory target, M is a vector specifying the current motor positional command that determines the overall shape of the vocal tract, $\dot{M}$ is a vector representing the overall movement[1] command to the speech articulators, and $\dot{M}_A$ is a vector representing the corrective motor command issued by the auditory feedback controller. Production of a speech sound begins when the *speech sound map* node for that sound is activated, leading to the readout from memory of the time-varying auditory target for the sound. This target is sent to the *auditory feedback controller*, where it is compared to incoming auditory feedback (or *sidetone*) from the ongoing speech. If there is a discrepancy between the auditory target and auditory feedback, corrective motor commands are generated; these commands sum with commands from the feedforward and somatosensory feedback controllers to form the overall movement command to the vocal tract musculature.

Figure 5.2 provides an expanded view of the auditory feedback controller (dashed box) as implemented in the DIVA model. In this system, an auditory error signal ΔA is calculated by comparing the auditory target for the current sound (A_T) to incoming feedback regarding the auditory state (A). This comparison is represented by the circle labeled Σ in figure 5.2. An auditory error signal ΔA is generated if any of the components of A fall outside the

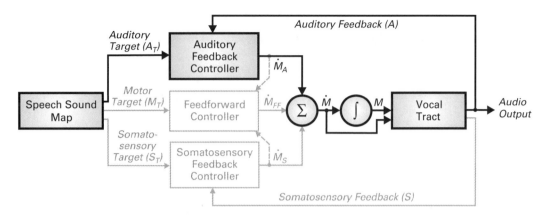

Figure 5.1
The auditory feedback control subsystem highlighted within the overall control system for speech production introduced in chapter 3. M, motor positional command; $\dot{M}$, overall movement command; $\dot{M}_A$, corrective movement command issued by the auditory feedback controller.

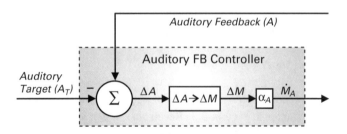

Figure 5.2
Expanded view of the auditory feedback controller implemented in the DIVA model. α_A, auditory gain factor; ΔA, auditory error; ΔM, motor error; FB, feedback; $\dot{M}_A$, corrective movement command issued by the auditory feedback controller.

auditory target region specified by A_T. This error signal represents the change, in auditory coordinates, needed to move the speech signal to the target region. If F1 is too low, for example, then ΔA represents an increase in F1.

Ultimately, the auditory feedback controller must generate commands to the vocal tract musculature. This means ΔA must be transformed into motor coordinates. This transformation is represented by the box labeled $\Delta A \rightarrow \Delta M$ in figure 5.2, where ΔM specifies the motor error, that is, the changes in the positions of the speech articulators needed to achieve the auditory signal change specified by ΔA. The transformation $\Delta A \rightarrow \Delta M$ constitutes a form of *inverse model* (see chapter 3, section 3.2). We will use the term *directional mapping* for this type of transformation since it maps desired movement directions in sensory space into movement directions in motor space. According to the DIVA model, the brain learns this directional mapping during babbling by associating changes in the auditory signal with movements of the speech articulators and storing this information in a matrix of synaptic weights, Z_{AM}. In the final processing stage shown in figure 5.2, ΔM is scaled by a gain factor α_A to create the corrective movement command $\dot{M}_A$. The processing performed by the auditory feedback controller is summarized by the following equation:[2]

$$\dot{M}_A = \alpha_A Z_{AM}[A - A_T]. \tag{5.1}$$

The DIVA model derives its name from this mapping from *d*irections in auditory space *i*nto *v*elocities of *a*rticulators. We have shown elsewhere how controllers that utilize this type of directional mapping[3] have a number of desirable properties that can account for a wide range of observations regarding speech movements (e.g., Guenther & Micci Barreca, 1997; Guenther, Hampson, & Johnson, 1998; Micci Barreca & Guenther, 2001). One key property of this type of controller is that it produces highly efficient movements in the following sense: it can generate the smallest motor velocity signal (out of an infinite number of possible motor velocity signals) that achieves the desired change in auditory signal (see footnote 3 for details). This minimizes the overall amount of movement required to reach auditory targets, which in turn minimizes the amount of energy required to speak.

The next section reviews behavioral data that further support this view of auditory feedback control of speech and provide additional details concerning the auditory feedback controller. This is followed by a treatment of the neural substrates of the auditory feedback control subsystem.

5.2 Behavioral Studies of Auditory Feedback Control in Speech

Effects of Deafness and Hearing Restoration

Even the earliest human cultures had clear evidence that the ability to speak depends on the ability to hear since infants who are born deaf rarely learn to speak fluently. In one of the earliest recorded references, Plato's *Cratylus* cites the signing of deaf individuals as an alternative means of communication for those who cannot speak. Congenital deafness also changes the nature of infant babbling, including delaying its onset. The restoration of hearing through hearing aids or cochlear implants at an early age counteracts this tendency (Oller, 2000; Colletti et al., 2005), and as a result of these technologies many individuals who are born deaf (*congenital hearing loss*) or become deaf before learning to speak (*prelingual deafness*) now learn to speak fluently. When the onset of deafness occurs in adulthood (*postlingual deafness*), speech typically remains intelligible for years, even in the absence of hearing aids or cochlear implants. However, the quality of this speech is degraded, with higher than normal F0 (or pitch), more variability in F0, and decreased vowel and consonant contrasts (Leder et al., 1987; Waldstein, 1990; Lane & Webster, 1991).

Perkell, Lane, and colleagues performed a number of studies investigating the speech of postlingually deaf adults who received cochlear implants. Generally speaking, these studies indicate improvements in speech intelligibility within 6 to 12 months (Gould et al., 2001) although testing 1 month after implantation showed reduced sound contrasts compared to pre-implant speech (Lane et al., 2007b), likely because the neural signals induced by the implant do not match the signals that were received from the cochlea prior to deafness. In other words, the implant recipients needed time to retrain their auditory feedback control subsystems to handle the novel sound input provided by the cochlear implant. In an investigation of the effects of rapid changes in hearing status produced by turning the cochlear implant off or on, Perkell et al. (2007) noted two different auditory parameter types: *postural parameters* such as sound intensity, F0, and duration that change relatively rapidly with a change in hearing status, and *segmental parameters* such as formant frequencies that typically change much more slowly with a change in hearing status. This finding suggests that postural parameters, which play large roles in speech prosody, are more directly influenced by the auditory feedback control subsystem while segmental parameters are more influenced by feedforward control mechanisms that change only gradually when hearing status changes.

Although speech production generally improves after implantation, some aspects of speech perception (Lane et al., 2007a) and production (Matthies et al., 2008) are still impaired in adult cochlear implant recipients relative to neurologically normal control participants even a year or more after implantation. Studies of cochlear implantation of children indicate that age of implantation is a significant factor in the amount of speech improvement obtained from the implant. For example, Habib et al. (2010) note that children who receive cochlear implants before 2 years of age show highly intelligible speech when tested several years after implantation, but this was true of only some children who were implanted in their third year. Leigh et al. (2013) note that children implanted by 12 months of age demonstrated better language development than those implanted between 13 and 24 months of age. In a comparison of children implanted by 6 months of age to those implanted between 6 months and 2 years of age, Colletti, Mandalà, and Colletti (2012) noted that age of implantation was a significant factor in speech perception, speech production, and language outcomes, further supporting the notion that the younger an individual is when implanted, the better the long-term outcome. Together these findings indicate that plasticity in the speech network is at its highest within the first year of life and declines gradually thereafter.

Auditory Feedback Loudness Manipulations
Although the studies described thus far imply a key role for auditory feedback in speech development, they do not speak to its role during ongoing speech in the fluent speaker. Lombard (1911) introduced the notion that the auditory environment affects our ongoing speech, noting that speakers increase the loudness (intensity) of their speech in the presence of background noise or when the speaker's sidetone is reduced. This phenomenon is often referred to as the *Lombard reflex* or *Lombard effect*. The speaker is usually not aware

of this change in intensity; that is, it is an automatic, rather than voluntary, response, as clearly evidenced when trying to converse with someone wearing earphones or headphones. A literature review by Lane and Tranel (1971) found that speakers typically compensate by increasing or decreasing their volume by about half the amount of the change in noise or sidetone level, an observation that supports the authors' hypothesis that the speaker's goal is to maintain an acceptable signal-to-noise ratio for the speech signal. It is noteworthy, however, that speakers make few speech errors even in the presence of loud masking noise (Ringel & Steer, 1963; Schliesser & Coleman, 1968; Gammon et al., 1971), indicating that auditory feedback control provides a relatively small contribution to segmental aspects of speech in fluent adults.

Using real-time perturbations of voice loudness that were unexpected and short in duration, Bauer et al. (2006) identified several additional properties of the Lombard reflex. The average latency of the compensatory responses was 157 ms, and although the average response magnitude increased for larger perturbations (6-dB vs. 1-dB perturbations), the gain of these responses (i.e., the response magnitude divided by the perturbation magnitude) was larger for small perturbations.

Delayed Auditory Feedback

The invention of the microphone and speaker led to another discovery regarding the strong effect auditory feedback can have on ongoing speech: *delayed auditory feedback* (*DAF*) of our own voice during speech can severely disrupt our ability to maintain fluency (e.g., Lee, 1950; Yates, 1963). This can happen to a talker whose voice is projecting from a distant loudspeaker—for example, in a sports stadium—since the sound is delayed in reaching the talker's ear because of the distance it must travel. Little to no disruption occurs for delays below 50 ms (Stuart et al., 2002), but if the delay is approximately 200 ms, speakers exhibit a range of dysfluent behaviors, including slowing of speech, part-word prolongations and repetitions, raising of loudness and pitch, and even complete cessation of speech (Lee, 1950; Stuart et al., 2002; Corey & Cuddapah, 2008). It is likely that the disruption caused by DAF is related to the mismatch between the speaker's auditory expectations (auditory targets) and the delayed feedback. This view is supported by the observation that DAF affects the vocalizations of children in the 28- to 35-month age range but not in the 21- to 26-month range (Yeni-Komshian et al., 1968) if one assumes that younger children have less well-formed auditory expectations for their own vocalizations.

Spectrally Altered Speech

In recent years researchers have used more subtle manipulations of auditory feedback to investigate its role in speech production. Of particular note are studies that manipulate auditory feedback in real time, that is, with delays of 20 ms or less, which are not noticeable to the talker. These experiments involve transduction of the sound output of the talker by a microphone that routes the signal to a digital signal processing system, which then manipulates the sound signal in some way before sending it back to the talker over a set of headphones or earphones. Most auditory perturbation studies involve manipulations of fundamental frequency (*pitch perturbation experiments*) or one or more formant frequencies (*formant perturbation experiments*). Perturbations to the acoustic signal can be either *unexpected* (occurring only on a small percentage of randomly chosen production trials) or *sustained* (applied over many consecutive trials). In a mature talker, few if any auditory errors will occur under normal auditory feedback conditions, and therefore the auditory feedback control subsystem will not significantly impact the overall motor commands to the articulators. In contrast, an unexpected perturbation of the talker's pitch or formant frequencies will generate an auditory error signal (ΔA in figure 5.2) that is transformed into a corrective motor command ($\dot{M}_A$) by the auditory feedback controller. Unexpected auditory perturbations thus highlight the workings of the auditory feedback controller. If the auditory perturbation is sustained, the commands generated by the feedforward controller in figure 5.1 will be modified to compensate for the perturbation and thereby decrease the auditory error signal. This modification process, often referred to as *sensorimotor adaptation*, is guided by the output of the auditory feedback controller (indicated by dashed arrow in figure 5.1). Because we are interested in auditory feedback control in the current chapter, the focus here will be on studies involving unexpected perturbations of auditory feedback. Studies involving sustained auditory perturbations are discussed in chapter 7, which describes the feedforward control system.

Pitch Perturbation Experiments Elman (1981) was the first to demonstrate compensatory responses to real-time auditory perturbations, employing a device called the Lexicon Varispeech to perturb the pitch of a talker's voice by electronically shifting all frequencies of the speech signal up by approximately 10%. Speakers responded to the perturbation by decreasing the pitch of their speech under the perturbation. This compensatory response, which we will refer to as the *pitch shift response*, occurred even for talkers who did not consciously notice the pitch shift.

Since the late 1990s, a number of carefully controlled studies by Charles Larson and colleagues have identified numerous properties of the pitch shift response and, by extension, the auditory feedback controller for speech. Burnett et al. (1998) showed that the pitch shift response was not sensitive to the intensity of auditory feedback, and the magnitude of the response was on average smaller than the perturbation. For perturbations ranging from 25 to 300 cents (where 1 cent is 1/1,200 of an octave), the average magnitude of the response was approximately 30 cents and did not differ significantly for different perturbation magnitudes. The authors also noted that not all trials showed a response, and not all responses were in the opposite direction as the shift (*compensatory responses*); a minority of the responses was in the same direction as the shift (*following responses*). However, other studies have found very few or no following responses. For example, Larson et al. (2000) and Jones and Munhall (2002) both used pitch shifts whose onset was gradual (as opposed to the abrupt changes used in the studies mentioned above) and found almost no following responses. Together, these findings suggest that gradual onset of deviations from the desired auditory trajectory, which are more like naturally occurring deviations during speech than are abrupt onset perturbations, may be more efficiently processed by the auditory feedback control subsystem.

By simultaneously applying pitch and loudness perturbations during sustained vowel productions, Larson et al. (2007) demonstrated that the pitch response and loudness response are largely independent; for example, the average response to an upward pitch shift was in the downward direction regardless of whether loudness was being perturbed upward or downward.

Hain et al. (2000) showed that perturbations lasting 500 ms or more resulted in two responses: a relatively automatic response about 150 ms after perturbation onset and a later voluntary response about 350 ms after perturbation onset. The first, more automatic response is most likely generated by the auditory feedback control subsystem. The voluntary response may involve feedforward control mechanisms invoked to follow the instructions provided to the subjects (which were to oppose, follow, or ignore the shift).

Chen et al. (2007) compared the early, automatic component of the pitch shift response in running speech to the response during sustained vowels (which had been used in most prior studies). They found larger responses during running speech than during isolated vowels. Furthermore, unlike the response to isolated vowel perturbations, the responses to running speech perturbations scaled with stimulus magnitude. Together, these observations suggest that the auditory feedback controller may respond more robustly during running speech than during isolated steady state vowel production.

Some pitch perturbation studies have noted asymmetry in the pitch shift response. For example, Chen et al. (2007) noted a larger response for downward shifts occurring during running speech. The shift in this study occurred just prior to a rise in pitch that was necessary to convey that the speech stimulus formed a question ("You know Nina?"). A downward shift opposes the upcoming pitch rise; the increased response to downward versus upward shifts thus suggests that the auditory feedback controller is more sensitive to perceived deviations in pitch that negatively impact linguistic intent. Further support for this conjecture comes from studies of pitch shifts during Mandarin bitonal disyllables (Xu et al., 2004). Mandarin is a *tonal language* in which the pitch contour can change the meaning of a word. Xu et al. (2004) found that shifts in the direction opposite of an intersyllabic tonal transition resulted in shorter latencies and larger compensations than did shifts in the same direction (Xu et al., 2004).

Formant Perturbation Experiments Although pitch shift experiments have shed significant light on auditory feedback control mechanisms, in nearly all European languages pitch is a suprasegmental parameter that does not significantly affect which phonemes are perceived. Formant frequencies, in contrast, provide crucial information about phoneme identity. Formant frequency perturbations thus get at the neural circuitry involved in controlling the amazingly rapid movements of the tongue and other supralaryngeal articulators that are responsible for the bulk of the linguistic content of speech. The first experiment to investigate the effects of shifting formant frequencies in real time was performed by Houde and Jordan (1998); this study used a sustained perturbation and will thus be discussed in chapter 7. Subsequent studies have used similar techniques to investigate the response to unexpected formant perturbations.

Purcell and Munhall (2006) applied unexpected perturbations of F1 during prolonged isolated productions of the vowel /ɛ/ (as in "bet"). The perturbations were either upward, resulting in auditory feedback that sounded like /æ/ (as in "bat") when producing /ɛ/, or downward, resulting in auditory feedback that sounded like /ɪ/ (as in "bit"). As with pitch shifts, subjects responded to the formant shifts by adjusting their formants in the opposite direction of the shift, and the compensatory responses were on average smaller than the applied shift (approximately 11% and 16% of the shifts in the /ɪ/ and /æ/ directions, respectively).

The formant shifts applied by Purcell and Munhall (2006) were applied gradually (over 500 ms) rather than abruptly. Tourville et al. (2008) utilized abrupt upward or downward shifts of F1 by 30% while subjects produced one-syllable words with the vowel /ɛ/. The average compensation magnitude was about 13% for both upward and downward shifts, which is very similar to the values measured in response to a gradual shift by Purcell and Munhall (2006). The average latencies of the compensatory responses in the up and down conditions were 107 ms and 165 ms, respectively; these values are similar to values found for abrupt pitch shifts as described above. Tourville et al. (2008) performed simulations of the DIVA model producing the same speech stimuli under perturbed and unperturbed conditions that verified the model's ability to quantitatively account for these compensatory responses.

Niziolek and Guenther (2013) investigated whether the amount of compensation to a formant shift was dependent on whether the shift was likely to interfere with linguistic intent. For each participant, two shifts of F1 and F2 were applied. The shifts were the same size in frequency space, but one shift direction was chosen to cause a change in the perceived phoneme, whereas the other shift direction was *subphonemic*—although it was in the direction of another phoneme, the shift was not large enough to change the perceived identity of the phoneme. The authors found that shifts that were likely to push the auditory feedback across a phoneme category boundary resulted in faster (256 ms compared to 400 ms) and larger (25% compared to 3%) compensations to shifts that did not push auditory feedback across a boundary. Similarly, the sustained perturbation study of Mitsuya et al. (2011) found that perturbations toward another vowel resulted in more adaptation than perturbations that moved the speech signal away from any other vowels. These studies provide further evidence that the auditory feedback controller is biased toward correcting changes that would alter the linguistic content of the speech signal. Although Niziolek and Guenther (2013) found larger responses to perturbations that crossed a vowel category boundary, it is important to note that compensations (and corresponding brain activations) were still found for perturbations that did not cross a category boundary. In other words, the auditory feedback controller for segmental aspects of speech is responsive to subphonemic variations in the speech signal, although to a lesser degree than for variations that cross a phoneme boundary.

Conclusions from Unexpected Spectral Perturbation Studies Several common themes arise from the pitch and formant perturbation studies described above.

First, both pitch and formant perturbations result in a relatively rapid, subconscious compensatory response that starts as early as 100 to 150 ms after an abrupt perturbation onset. This delay, which is similar to the delay found by Bauer et al. (2006) for perturbations of sound intensity, is short enough to permit online correction within the duration of a typical syllable (Hillenbrand et al., 2001; Ferguson & Kewley-Port, 2002). These latencies are sufficiently long to allow for a cortically mediated compensatory response (see Guenther, Ghosh, & Tourville, 2006, for discussion) and are much longer than brain stem–mediated auditory perioral reflex responses (McClean & Sapir, 1981), suggesting cortical involvement in the auditory feedback controller's response to deviations from the desired auditory signal during ongoing speech.

Second, both perturbation types appear to result in larger responses if the perturbation is in a direction that might affect linguistic content, though responses are still found for perturbations that are too small to alter the linguistic message. It is well-known that our auditory perceptual spaces are warped in a way that emphasizes small auditory differences for sounds near phoneme category boundaries compared to sounds near phoneme category centers, a phenomenon first discovered for consonants and termed *categorical perception* (e.g., Liberman et al., 1961) and later demonstrated for vowels in a more subtle form, termed the *perceptual magnet effect* (Kuhl, 1991). This warping may account for the response magnitude differences noted in formant perturbation studies. A similar warping of perceptual space for lexical pitch has been found for native speakers of a tonal language (Mandarin) but not for native English speakers (Xu et al., 2006; Liu, 2013). In contrast, there does not appear to be significantly heightened discriminability for pitch near suprasegmental category boundaries, such as the boundary between rising and flat pitch patterns that signify questions versus statements (Liu & Rodriguez, 2012) or the boundary between normal and emphasized versions of a word (Ladd & Morton, 1997). These findings indicate that response biases in the control of pitch, and possibly other suprasegmental cues, in nontonal languages are not simply due to perceptual differences and likely involve additional control mechanisms beyond those illustrated in figure 5.2. Liu and Rodriguez (2012) also failed to find heightened discriminability near the question/statement pitch contour boundary for Mandarin speakers, suggesting a differentiation between control mechanisms for lexical and suprasegmental pitch in tonal languages.

A third common finding in pitch and formant perturbation studies, as well as the intensity perturbation study of Bauer et al. (2006), is that compensation is incomplete, especially for larger perturbations. The formant perturbation studies described above found compensation ranging from 3% for subphonemic perturbations to 25% for perturbations that pushed auditory feedback across a category boundary (Niziolek & Guenther, 2013).

There are at least two possible reasons why compensation is incomplete. The first explanation involves a competition of sorts between the auditory and somatosensory feedback control subsystems. Spectral perturbations invoke errors in the auditory feedback control subsystem, but not in the somatosensory feedback control subsystem. Compensatory responses to the auditory perturbation will cause somatosensory feedback to deviate from normal, resulting in perceived somatosensory errors and corrective commands that counter the compensatory responses to auditory feedback (Villacorta et al., 2007; see also Katseff et al., 2012). In keeping with this account, Larson et al. (2008) noted larger compensatory responses to to pitch perturbations when the vocal folds were anesthetized. Furthermore, the mismatch between auditory feedback and somatosensory feedback resulting from an auditory perturbation may cause the talker's brain to determine that auditory feedback is currently unreliable, resulting in a "down-weighting" of auditory feedback compared to somatosensory feedback. This is equivalent to decreasing the gain parameter α_A in the control system shown in figure 5.2, thereby reducing the proportion of compensation to the auditory perturbation.

The second explanation for incomplete compensation to unexpected perturbations is that long lags in a feedback control system can lead to instabilities, as discussed in section 3.2 of chapter 3. As we have seen, the auditory feedback controller has a delay of about 100 to 150 ms. This delay represents the approximate time it takes for (1) activity in motor cortex to effect a sound-inducing movement via brain stem nuclei; (2) the resulting sound to be transduced into neural signals by the cochlea; (3) neural information from the cochlea to reach auditory cortex, where it is compared to an auditory expectation; and (4) any resulting auditory error to be transferred back to motor cortex for a corrective action (see Guenther, Ghosh, & Tourville, 2006, for further discussion). With the exception of a few stressed vowels and diphthongs, phonemes typically last less than 150 ms in spontaneous speech (Greenberg et al., 2003), and some last as little as 50 ms. This means that corrective movements commanded by the auditory feedback control subsystem often do not "kick in" during the current phoneme. If the gain of the feedback system is high, the (relatively large) corrective commands will continue well past the end of the phoneme, causing movement overshoots and potential oscillations—in other words, the system would be unstable. This explanation may also account, at least partially, for why the proportion of compensation is smaller for large perturbations; in order to maintain stability, smaller gains are needed to prevent these large error signals from generating large corrective commands that would interfere with future phonemes.

5.3 Neural Circuitry Underlying Auditory Feedback Control

The neural circuitry responsible for auditory feedback control according to the DIVA model is shown in figure 5.3. Stereotactic coordinates associated with the model components in figure 5.3 are provided in appendix D.

As discussed in section 2.2, auditory information from the cochlea makes its way to auditory cortex via brain stem nuclei (including the cochlear nucleus, superior olive, and inferior colliculus) and the *medial geniculate (MG) nucleus* of the thalamus. This

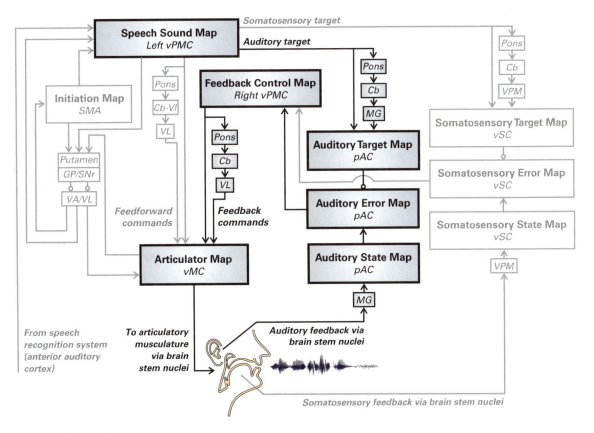

Figure 5.3
Neural components of the auditory feedback control subsystem according to the DIVA model. Cb, cerebellum; MG, medial geniculate nucleus of the thalamus; pAC, posterior auditory cortex; VL, ventral lateral nucleus of the thalamus; vMC, ventral motor cortex; vPMC, ventral premotor cortex.

information is represented in the model's *auditory state map*, which contains a talker-normalized representation related to the formant frequencies of the speech signal. This highly processed auditory information is hypothesized to reside in the higher-order *posterior auditory cortex* (*pAC*) in the *posterior superior temporal gyrus* (*pSTG*), including the planum temporale and dorsal bank of the posterior superior temporal sulcus.[4] When the *speech sound map* node for a sound (i.e., its mirror neuron) is activated in left *ventral premotor cortex* (*vPMC*), which lies in the inferior frontal cortex, axonal projections from this node to pAC transmit the sound's auditory target to the model's *auditory target map*. This target takes the form of a multidimensional, time-varying region in auditory space (see figure 4.9 in chapter 4 for an example). Functional connections between inferior frontal cortex and pSTG, such as those proposed here, have been demonstrated in humans using cortico-cortical evoked potentials prior to epilepsy surgery (Matsumoto et al., 2004), and relevant functional and structural connectivity maps derived from from magnetic resonance imaging are provided in figures C.1, C.19, C.20, C.21, C.23, C.24, and C.28 in appendix C.

According to the DIVA model, the functional connections that carry auditory target information from vPMC to pAC are mediated by both cortico-cortical axonal projections and projections through the cortico-cerebellar loop described in chapter 2, section 2.3. The cortico-cortical projections are likely contained in one or both of two fiber pathways: a dorsal pathway within the *superior longitudinal fasciculus*, in particular the *arcuate fasciculus*, which courses from the posterior inferior frontal gyrus to the posterior superior temporal gyrus and inferior parietal cortex (Kelly et al., 2010), and a ventral pathway through the *extreme capsule fasciculus*, which connects the inferior frontal gyrus to the superior and middle temporal gyri (Kelly et al., 2010), including terminations in the anterior insula (Weiller et al., 2011). The dorsal stream has been more commonly associated with sensorimotor integration whereas the ventral stream is more commonly associated with meaning (Hickok and Poeppel, 2004; Weiller et al., 2011), suggesting that the arcuate fasciculus is the more likely cortico-cortical pathway for auditory targets for the speech production mechanism. The pathways carrying these targets may include a stop in the inferior parietal cortex before reaching pAC (Catani, Jones, & ffytche, 2005; Kelly et al., 2010).

The cortico-cerebellar loop component of the auditory target projections in figure 5.3 is motivated by two considerations. The first consideration is that the auditory signals produced during speech involve fine timing distinctions; for example, the difference in voice onset time between voiced and voiceless stop consonants is on the order of 40 ms. As described in earlier chapters, the cerebellum is ideally suited to learning these fine distinctions. Furthermore, the auditory cortex response to speech is delayed by tens of milliseconds relative to the motor cortex signals generating vocalization because of the time it takes for (1) motor commands to activate movements in the muscles and (2) auditory feedback processing to occur in the ascending auditory pathway. The cerebellum is widely considered to be an important locus for timing, making it a likely candidate for predicting the delayed and rapidly changing auditory signals arising from speech articulation.

Second, numerous studies have indicated that the cerebellum is involved in predicting sensory consequences of ongoing movements, though the bulk of these studies investigated somatosensory rather than auditory consequences of movements (e.g., Blakemore, Wolpert, & Frith, 1999, 2000; Blakemore, Frith, & Wolpert, 2001). Studies investigating a role for the cerebellum in auditory prediction are scarce. However, Knolle et al. (2012) studied suppression of the auditory N100 event-related potential in neurologically normal individuals and individuals with cerebellum damage. Suppression of the N100 was found in neurologically normal individuals during self-generated movements (finger taps that created tone sounds) compared to hearing the tones without the finger taps, as reported in prior studies (Schäfer & Marcus, 1973; Hazemann, Audin, & Lille, 1975; McCarthy & Donchin, 1976; Martikainen, Kaneko, & Hari, 2005; Baess, Jacobsen, & Schröger, 2008). In contrast, patients with cerebellar damage did not show significant attenuation of the N100 response to self-generated sounds, indicating a role for the cerebellum in prediction of auditory consequences of self-generated sounds. The locations of the cerebellar lesions of the participants in Knolle et al. (2012) were highly variable, making it impossible to determine which portion of the cerebellum was responsible for auditory prediction.

According to the DIVA model, the role of auditory target projections from vPMC is to essentially cancel out incoming auditory information; that is, they have an inhibitory effect on *auditory error map* nodes that compare incoming auditory information to the auditory target. However, long-range inputs to a cortical region from another cortical region or from the thalamus are generally excitatory[5] (e.g., Amaral & Strick, 2013). Thus, the projections from the model's speech sound map to auditory cortex excite nodes in an *auditory target map* before inhibiting nodes in the auditory error map. If the incoming auditory signal is within the auditory target region, this inhibition cancels the excitatory effects of the incoming auditory signal. If the incoming auditory signal is outside the target region, the inhibitory influence will not completely cancel the excitatory input from the auditory periphery, resulting in activation of auditory error map nodes.

As noted earlier, projections such as these, which predict the sensory state resulting from a movement, are often described as representing a *forward model* of the movement (e.g., Miall & Wolpert, 1996; Kawato, 1999; Desmurget & Grafton, 2000; Davidson & Wolpert, 2005). The use of forward models in motor control has been postulated for nearly a century (Helmholtz, 1924). Early work focused on how stored sensory predictions of motor output could be used to maintain a stable sensory image during self-generated movements (Sperry, 1950; von Holst & Mittelstaedt, 1950). von Holst and Mittelstaedt (1950) proposed the *principle of reafference*, which stated that a copy of the expected sensory consequences of a motor command, termed an *efference copy*, is effectively subtracted from the realized sensory consequences. The efference copy, according to von Holst and Mittelstaedt, provided a means to distinguish the effects of *reafferent* sensory activation (i.e., activation resulting from self-generated motion rather than external stimulation) from those due to *exafferent* activation (i.e., activation resulting from external stimulation).

Numerous lines of evidence support the hypothesis that the expected auditory consequences of articulation and resulting auditory feedback are compared in posterior superior temporal cortex. At the single-neuron level, Eliades and Wang (2005) note that both excitatory and inhibitory influences occur in auditory cortex during vocalization in monkeys, consistent with the DIVA model's inclusion of auditory target and state maps that receive excitatory input and an auditory error map that receives inhibitory input. Evidence of inhibition in auditory cortical areas in the superior temporal gyrus during one's own speech (compared to listening to speech while not speaking) comes from several different sources, including signals measured by *magnetoencephalography* (*MEG*; Numminen & Curio, 1999; Numminen, Salmelin, & Hari, 1999; Curio et al., 2000; Houde et al., 2002; Franken, Hagoort, & Acheson, 2015), *positron emission tomography* (*PET*; Paus et al., 1996; Wise et al., 1999) *functional magnetic resonance imaging* (*fMRI*; Christoffels, Formisano, & Schiller, 2007), and single- and multi-unit recordings during open brain surgery (Creutzfeldt, Ojemann, & Lettich, 1989a,b).

In the DIVA model, auditory error map nodes become active when auditory feedback falls outside the current auditory target region. In support of this, the MEG study of Heinks-Maldonado, Nagarajan, and Houde (2006) reported a reduced degree of speech-induced suppression (in other words, an increased auditory response) when auditory feedback does not match the speaker's expectation because of a pitch shift. Similar results have been reported using PET and fMRI when auditory feedback of ongoing speech is pitch shifted (McGuire et al., 1996; Fu et al., 2006; Toyomura et al., 2007; Parkinson et al., 2012), delayed (Hashimoto & Sakai, 2003; Takaso et al., 2010), corrupted by masking noise (Christoffels et al., 2007), or transformed to sound like the speech of another person (Fu et al., 2006). The MEG study of Niziolek, Nagarajan, and Houde (2013) used token-to-token variability to show that less prototypical productions were less suppressed, again supporting the idea of an auditory error map that becomes active when auditory feedback mismatches the current auditory target.

Detected auditory errors must be transformed into a motor coordinate frame in order to generate corrective movements; this transformation is the directional mapping described in section 5.1. According to the DIVA model, this transformation is carried out by projections from the auditory error map to a *feedback control map* in right vPMC and then on to the articulator map in *ventral motor cortex* (*vMC*). At this time it remains an open question whether the directional mapping is implemented in the pAC to right vPMC projections, the right vPMC to vMC projections, or both of these projections (with right vPMC acting as an intermediate stage between auditory and motor reference frames). In addition to direct cortico-cortical projections from right vPMC to bilateral vMC, the DIVA model includes indirect projections via a cerebellar loop. Numerous studies suggest that sensory error representations in cerebellum influence corrective movements (e.g., Diedrichsen et al., 2005; Grafton et al., 2008; Penhune & Doyon, 2005), and the model accordingly posits that the cerebellum contributes to the feedback motor command, most likely via the *ventral lateral (VL) nucleus* of the thalamus, which provides the majority of cerebellar input to motor cortex (see section 2.3). At present it is not entirely clear which portions of the cerebellar cortex are involved in generating feedback-based corrective commands. Evidence from two sensory perturbation studies during speech suggest that inferior cerebellar cortex, in/near lobule VIII, is involved in auditory (Tourville, Reilly, & Guenther, 2008) and somatosensory (Golfinopoulos et al., 2011) feedback control. As noted in section 2.3, however, this region appears to be involved in a number of tasks, including speech sequencing (Bohland & Guenther, 2006) and verbal working memory (Desmond et al., 1997; Chen and Desmond, 2005). Furthermore, nonspeech studies have implicated superior cerebellar cortex in sensory feedback control (e.g., Seidler, Noll, & Thiers, 2004; Diedrichsen et al., 2005; Donchin et al., 2012) including lobule VI, which is heavily involved in speech production (see section 2.3).

The DIVA model's account of auditory feedback control was directly tested in two fMRI studies involving real-time perturbation of formant frequencies during speech (Tourville et al., 2008; Niziolek & Guenther, 2013). In both studies, speakers produced monosyllabic utterances under normal feedback conditions and, in a randomly distributed subset of trials, under perturbed feedback conditions when F1 and/or F2 were shifted in real time during the vowel. The formant shifts have the effect of "warping" the perceived vowel toward another vowel in the vowel space, a perturbation expected to result in auditory error signals as well as compensatory articulator movements according to the DIVA model. Acoustic analysis of the speech produced under perturbed conditions verified that subjects compensated (partially) for the perturbations, as described in the previous section. These adjustments occurred subconsciously as the speakers were generally unaware that their speech had been modified. The results of the *perturbed speech–unperturbed speech* contrast from a pooled analysis of fMRI data from the 25 subjects in these two studies are shown in panel A of figure 5.4. Bilateral activity is seen in pSTG, including the planum temporale on the supratemporal plane as well as the dorsal bank of the precentral sulcus, supporting the DIVA model prediction of auditory error maps in these areas (see model simulation results in panel B of figure 5.4). Parkinson et al. (2012) report bilateral posterior superior temporal gyrus activity during pitch-perturbed speech that largely overlaps with the activity in figure 5.4A, a topic we will return to in chapter 9, which discuss the neural bases of prosody. Together, these results suggest that auditory feedback control of both pitch and formant frequencies involves bilateral auditory error maps in the posterior superior temporal gyrus. A left-hemisphere peak of activity is found at the border between the parietal operculum and planum temporale, a region that Hickok and colleagues have implicated as an auditory-motor interface for speech (Buchsbaum, Hickok, & Humphries, 2001; Hickok et al., 2003);

they refer to this region as *area Spt* (for Sylvian parietal-temporal). The results of Tourville et al. (2008) and Niziolek and Guenther (2013) suggest that the auditory-motor interface is not limited to Spt in the left hemisphere but instead involves a significantly larger portion of the posterior superior temporal cortex in both hemispheres.

The *perturbed speech–unperturbed speech* contrast in figure 5.4A also reveals activity in right *inferior frontal gyrus pars opercularis* (*IFo*) and *pars triangularis* (*IFt*). We interpret this activity as a feedback control map involved in transforming auditory errors into corrective motor commands. Tourville et al. (2008) used structural equation modeling to show increased functional connectivity between bilateral posterior superior temporal gyrus and right inferior frontal cortex during perturbed speech, supporting this interpretation.

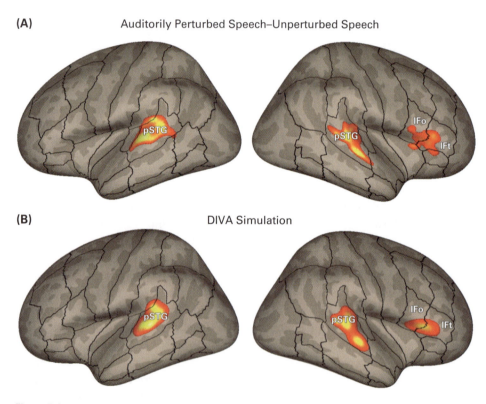

Figure 5.4
(A) Blood-oxygen-level-dependent response in the *perturbed speech–unperturbed speech* contrast from a pooled analysis of the Tourville et al. (2008) and Niziolek and Guenther (2013) formant perturbation studies plotted on inflated left-hemisphere and right-hemisphere cortical surfaces. Black lines indicate anatomical region boundaries; see appendix B for region-of-interest definitions. Functional data were first smoothed on the cortical surface; individual cortical vertices were then thresholded at p < 0.005 uncorrected before a cluster-level analysis thresholded at a family-wise error rate of 0.05. (B) Activity generated by a DIVA model simulation of the auditory perturbation experiment; see appendix D for details. IFo, inferior frontal gyrus pars opercularis; IFt, inferior frontal gyrus pars triangularis; pSTG, posterior superior temporal gyrus.

Based on their fMRI results during pitch-perturbed and unperturbed speech, Toyomura et al. (2007) similarly conclude that feedback control of pitch is right-lateralized in the cerebral cortex. Several nonspeech studies have shown right inferior frontal cortex activation in nonspeech tasks where sensory input dictates alteration of a motor response, for example, successful response inhibition or rapid switching of an ongoing task (see Aron et al., 2004, for review). Stevens et al. (2000) found right inferior frontal cortex activity when subjects silently counted rare ("oddball") sensory stimuli. Together, these findings suggest that right inferior frontal cortex is involved when sensory inputs directly influence motor processing, including when auditory feedback of one's own speech indicates the need for corrective motor responses.

The right-lateralized inferior frontal activity during auditory feedback control (figure 5.4A) contrasts with left-lateralized activity in approximately the same areas during unperturbed speech (see figure 2.14 in chapter 2). These findings suggest the following dichotomy: left-hemisphere inferior frontal regions are primarily responsible for feedforward control (which dominates during unperturbed adult speech) whereas right-hemisphere inferior frontal regions are primarily responsible for feedback control. This dichotomy is consistent with lesion data indicating that acquired apraxia of speech generally arises from left-hemisphere frontal lesions (discussed in further detail in chapter 10) whereas right-hemisphere frontal regions leave speech largely intact (with the possible exception of prosodic aspects of speech; see chapter 9). Once the feedforward commands for speech are learned, feedback control becomes far less important since few speech errors are made. Thus, damage to the right-hemisphere regions responsible for feedback control will be less disruptive to speech output than left-hemisphere lesions involving the feedforward control system.

Despite the fact that motor compensation was evident in acoustic analyses that compared perturbed and unperturbed speech, little or no primary motor cortex activity was found in the *perturbed speech–unperturbed speech* contrasts of Tourville et al. (2008), Niziolek and Guenther (2013), or the pooled analysis of these studies (figure 5.4A). There are at least two possible explanations for this finding. First, compensatory motor responses typically involve increasing the activity of one set of muscles while decreasing the activity of antagonist muscles. Decreased activity of antagonist muscles likely involves decreased motor cortical activity; this activity decrease may essentially cancel out activity increases needed for agonist muscles, resulting in little or no overall change in motor cortical activity during perturbed speech. Alternatively, brain regions other than primary motor cortex may be responsible for the muscle activity changes underlying compensation to a perturbation. The region of right inferior frontal cortex activity in figure 5.4 may be a candidate, though the existence of direct projections from this area to brain stem motor nuclei has not been demonstrated. Tourville et al. (2008) note activity in right inferior cerebellar lobule VIII in the *perturbed speech–unperturbed speech* contrast, making this region another candidate for projections to the brain stem motor nuclei. Cerebellar lobule VIII has been associated with motor error correction by Diedrichsen et al. (2005), who noted activity in this region when sensory error was induced during a reaching task, and Golfinopoulos et al. (2011) noted lobule VIII activity in response to unexpected blocking of the jaw during speech. While further study is required, these results collectively suggest a role for lobule VIII in the monitoring and/or adjustment of movements when sensory feedback indicates movement error, via projections either to cortex or to the brain stem.

5.4 Summary

Auditory feedback control of speech involves (1) comparing incoming auditory information from ongoing speech to an auditory target for the speech sound currently being produced and (2) generating corrective movement commands for any discrepancy between the auditory target and the incoming sound signal. According to the DIVA model, auditory errors, which can be interpreted as desired movement directions in auditory space, are translated into articulator movements that will correct the auditory errors. This transformation constitutes a form of inverse model. A desirable property of this transformation is that it can generate highly efficient articulatory movement commands for reaching auditory targets, which in turn minimizes the amount of energy required to speak.

Behavioral observations provide a number of important insights into auditory feedback control of speech. The severely impaired speech abilities of congenitally deaf individuals highlight the importance of the auditory feedback controller for the development of normal speaking skills, a topic that will be treated in further detail in chapter 7. Studies of the effects of cochlear implantation of postlingually deafened adults indicate that, although speech may degrade slightly in the first weeks after implantation because of the novelty of the neural signals generated by the implant, improvements in speech production are typically evident within a year. Studies of cochlear implantation of infants and children indicate that an earlier age of implantation typically yields more complete recovery of speaking skills.

Abrupt changes in hearing status created by turning a cochlear implant off and on reveal two different types of auditory parameters. *Postural parameters* such as sound intensity, F0, and duration (which all play roles in speech prosody) change relatively rapidly with a change in hearing status. *Segmental parameters* such as formant frequencies change much more slowly with a change in hearing status. This suggests that postural parameters are more heavily dependent on auditory feedback control mechanisms than segmental parameters, which are primarily under feedforward control.

Experimental manipulations of auditory feedback during speech provide further information into the workings of the auditory feedback controller. Altering the pitch or formant frequencies of auditory feedback during speech leads to rapid, subconscious compensatory responses as early as 100 to 150 ms after the onset of the perturbation. A similar delay is seen in response to a sudden decrease in the amplitude of the auditory feedback signal, which results in an increase in speech amplitude. This phenomenon is closely related to the

Lombard effect, in which speakers increase loudness in response to masking noise. Compensatory responses to auditory perturbations are generally incomplete, indicative of a relatively low gain in the auditory feedback controller. This property is likely necessary to maintain the integrity of the speech signal since feedback controllers with large gains and substantial delays in the feedback loop exhibit undesirable behaviors such as overshoots and oscillations.

Neuroimaging and lesion studies provide information regarding the neural substrates underlying auditory feedback control of speech. The DIVA model posits that *mirror neurons* in premotor cortical areas of the left inferior frontal cortex form a *speech sound map* that is central to the learning and readout of auditory targets for speech sounds. When an infant is presented with a new speech sound, a speech sound map node representing this sound becomes active, and synaptic projections from this node to higher-order auditory cortical areas in the posterior superior temporal gyrus (both directly and via the cerebellum) are adjusted to encode the auditory properties of the speech sound. In other words, these synapses encode the auditory target for the sound. When the infant attempts to produce the sound, the corresponding speech sound map node is activated, which effectively reads out the auditory target to the posterior superior temporal gyrus, where it is compared to incoming auditory information. An *auditory error map* in posterior superior temporal gyrus becomes active when a mismatch is detected between the auditory target and incoming auditory feedback. These auditory errors are transmitted via projections to a *feedback control map* in right ventral premotor cortex, where they are transformed into corrective movement commands that project to bilateral ventral motor cortex.

Notes

1. We use the term *movement* rather than *motor* here to distinguish $\dot{M}$, which indicates the commanded *change* in position, from the positional motor command M.

2. This equation defines ΔA as $A - A_T$, rather than $A_T - A$, because there is evidence that auditory targets for speech sounds are represented by inhibitory signals in the auditory cortical areas. The direction of this subtraction does not change performance of the controller since the sign change can be subsumed into the matrix Z_{AM} or the gain parameter α_A. Furthermore, comparison of the auditory target to the auditory state is represented as a simple subtraction here, though this simplification would strictly hold only for point targets rather than target regions. For target regions, $\Delta A = [A - A_{Tmax}]^+ - [A_{Tmin} - A]^+$, where A_{Tmax} and A_{Tmin} are vectors defining the maximum and minimum values of the target range along each auditory dimension and $[x]^+$ is a rectification function such that $[x]^+ = 0$ for $x \leq 0$ and $[x]^+ = x$ for $x > 0$. Guenther (1995) describes a simple neural implementation that accounts for both point and region targets.

3. Mathematically, the transformation defined by equation 5.1 is closely related to the *inverse of the Jacobian matrix* relating the motor and auditory spaces. When there are more motor dimensions than auditory dimensions (as is usually assumed for speech, which involves redundant articulatory degrees of freedom), there are an infinite number of possible motor movements that can achieve a particulator change in the auditory signal. Correspondingly, there are an infinite number of possible inverses (referred to as *generalized inverses*, or *pseudoinverses*) of the Jacobian matrix. Controllers that utilize a particular generalized inverse called the *Moore-Penrose pseudoinverse* (Moore, 1920; Penrose, 1955) have the desirable property of using the smallest motor velocity signal that can be used to achieve the desired auditory change. The synaptic weights Z_{AM} in the DIVA model approximate the Moore-Penrose pseudoinverse.

4. Recall from chapter 3 that the locations of cortical components of the DIVA model are specified in functional terms (such as *auditory cortex*) rather than in anatomical terms (such as *posterior superior temporal gyrus*)

since (1) it is the region's function that is being directly modeled and (2) the functional maps may span multiple anatomical structures and vary somewhat in anatomical location across individuals.

5. Although it is widely assumed that long-range cortico-cortical projections are excitatory, there is evidence for a relatively small number of inhibitory long-range connections in cortex (e.g., Tamamaki & Tomioka, 2010). To date the function of these projections is poorly understood; however, their relatively small number make them unlikely candidates for representing the auditory or somatosensory targets for speech production.

References

Amaral, D. G., & Strick, P. L. (2013). The organization of the central nervous system. In E. R. Kandel, J. H. Schwartz, T. M. Jessell, S. A. Siegelbaum, & A. J. Hudspeth (Eds.), *Principles of neural science* (5th ed., pp. 337–355). New York: McGraw-Hill Medical.

Aron, A. R., Robbins, T. W., & Poldrack, R. A. (2004). Inhibition and the right inferior frontal cortex. *Trends in Cognitive Sciences*, *8*, 170–177.

Baess, P., Jacobsen, T., & Schröger, E. (2008). Suppression of the auditory N1 event-related potential component with unpredictable self-initiated tones: evidence for internal forward models with dynamic stimulation. *International Journal of Psychophysiology*, *70*, 137–143.

Bauer, J. J., Mittal, J., Larson, C. R., & Hain, T. C. (2006). Vocal responses to unanticipated perturbations in voice loudness feedback: an automatic mechanism for stabilizing voice amplitude. *Journal of the Acoustical Society of America*, *119*, 2363–2371.

Blakemore, S.-J., Frith, C. D., & Wolpert, D. M. (2001). The cerebellum is involved in predicting the sensory consequences of action. *Neuroreport*, *12*, 1879–1884.

Blakemore, S. J., Wolpert, D. M., & Frith, C. D. (1999). The cerebellum contributes to somatosensory cortical activity during self-produced tactile stimulation. *NeuroImage*, *10*, 448–459.

Blakemore, S.-J., Wolpert, D., & Frith, C. (2000). Why can't you tickle yourself? *Neuroreport*, *11*, R11–R16.

Bohland, J. W., & Guenther, F. H. (2006). An fMRI investigation of syllable sequence production. *NeuroImage*, *32*, 821–841.

Buchsbaum, B. R., Hickok, G., & Humphries, C. (2001). Role of left posterior superior temporal gyrus in phonological processing for speech perception and production. *Cognitive Science*, *25*, 663–678.

Burnett, T. A., Freedland, M. B., Larson, C. R., & Hain, T. C. (1998). Voice F0 responses to manipulations in pitch feedback. *Journal of the Acoustical Society of America*, *103*, 3153–3161.

Chen, S. H., & Desmond, J. E. (2005). Cerebrocerebellar networks during articulatory rehearsal and verbal working memory tasks. *NeuroImage*, *24*, 332–338.

Chen, S. H., Liu, H., Xu, Y., & Larson, C. R. (2007). Voice $F0$ responses to pitch-shifted voice feedback during English speech. *Journal of the Acoustical Society of America*, *121*, 1157–1163.

Christoffels, I. K., Formisano, E., & Schiller, N. O. (2007). Neural correlates of verbal feedback processing: an fMRI study employing overt speech. *Human Brain Mapping*, *28*, 868–879.

Colletti, V., Carner, M., Miorelli, V., Guida, M., Colletti, L., & Fiorino, F. G. (2005). Cochlear implantation at under 12 months: report on 10 patients. *Laryngoscope*, *115*, 445–449.

Colletti, L., Mandalà, M., & Colletti, V. (2012). Cochlear implants in children younger than 6 months. *Otolaryngology—Head and Neck Surgery*, *147*, 139–146.

Corey, D. M., & Cuddapah, V. A. (2008). Delayed auditory feedback effects during reading and conversation tasks: gender differences in fluent adults. *Journal of Fluency Disorders*, *33*, 291–305.

Creutzfeldt, O., Ojemann, G., & Lettich, E. (1989 a). Neuronal activity in the human lateral temporal lobe: I. Responses to speech. *Experimental Brain Research*, *77*, 451–475.

Creutzfeldt, O., Ojemann, G., & Lettich, E. (1989 b). Neuronal activity in the human lateral temporal lobe: II. Responses to the subjects own voice. *Experimental Brain Research*, *77*, 476–489.

Curio, G., Neuloh, G., Numminen, J., Jousmaki, V., & Hari, R. (2000). Speaking modifies voice-evoked activity in the human auditory cortex. *Human Brain Mapping*, *9*, 183–191.

Davidson, P. R., & Wolpert, D. M. (2005). Widespread access to predictive models in the motor system: a short review. *Journal of Neural Engineering*, *2*, S313–S319.

Desmond, J. E., Gabrieli, J. D., Wagner, A. D., Ginier, B. L., & Glover, G. H. (1997). Lobular patterns of cerebellar activation in verbal working memory and finger-tapping tasks as revealed by functional MRI. *Journal of Neuroscience, 17,* 9675–9685.

Desmurget, M., & Grafton, S. (2000). Forward modeling allows feedback control for fast reaching movements. *Trends in Cognitive Sciences, 4,* 423–431.

Diedrichsen, J., Hashambhoy, Y., Rane, T., & Shadmehr, R. (2005). Neural correlates of reach errors. *Journal of Neuroscience, 25,* 9919–9931.

Donchin, O., Rabe, K., Diedrichsen, J., Lally, N., Schoch, B., Gizewski, E. R., et al. (2012). Cerebellar regions involved in adaptation to force field and visuomotor perturbation. *Journal of Neurophysiology, 107,* 134–147.

Eliades, S. J., & Wang, X. (2005). Dynamics of auditory-vocal interaction in monkey auditory cortex. *Cerebral Cortex, 15,* 1510–1523.

Elman, J. L. (1981). Effects of frequency-shifted feedback on the pitch of vocal productions. *Journal of the Acoustical Society of America, 70,* 45–50.

Ferguson, S. H., & Kewley-Port, D. (2002). Vowel intelligibility in clear and conversational speech for normal-hearing and hearing-impaired listeners. *Journal of the Acoustical Society of America, 112,* 259–271.

Franken, M. K., Hagoort, P., & Acheson, D. J. (2015). Modulations of the auditory M100 in an imitation task. *Brain and Language, 142,* 18–23.

Fu, C. H., Vythelingum, G. N., Brammer, M. J., Williams, S. C., Amaro, E., Jr., Andrew, C. M., et al. (2006). An fMRI study of verbal self-monitoring: neural correlates of auditory verbal feedback. *Cerebral Cortex, 16,* 969–977.

Gammon, S. A., Smith, P. J., Daniloff, R. G., & Kim, C. W. (1971). Articulation and stress/juncture production under oral anesthetization and masking. *Journal of Speech and Hearing Research, 14,* 271–282.

Golfinopoulos, E., Tourville, J. A., Bohland, J. W., Ghosh, S. S., Nieto-Castanon, A., & Guenther, F. H. (2011). fMRI investigation of unexpected somatosensory feedback perturbation during speech. *NeuroImage, 55,* 1324–1338.

Gould, J., Lane, H., Vick, J., Perkell, J. S., Matthies, M. L., & Zandipour, M. (2001). Changes in speech intelligibility of postlingually deaf adults after cochlear implantation. *Ear and Hearing, 22,* 453–460.

Grafton, S. T., Schmitt, P., Van Horn, J., & Deidrichsen, J. (2008). Neural substrates of visuomotor learning based on improved feedback control and prediction. *NeuroImage, 39,* 1383–1395.

Greenberg, S., Carvey, H., Hitchcock, L., & Chang, S. (2003). Temporal properties of spontaneous speech—a syllable-centric perspective. *Journal of Phonetics, 31,* 465–485.

Guenther, F. H. (1995). Speech sound acquisition, coarticulation, and rate effects in a neural network model of speech production. *Psychological Review, 102,* 594–621.

Guenther, F. H., Ghosh, S. S., & Tourville, J. A. (2006). Neural modeling and imaging of the cortical interactions underlying syllable production. *Brain and Language, 96,* 280–301.

Guenther, F. H., Hampson, M., & Johnson, D. (1998). A theoretical investigation of reference frames for the planning of speech movements. *Psychological Review, 105,* 611–633.

Guenther, F. H., & Micci Barreca, D. (1997). Neural models for flexible control of redundant systems. In P. Morasso & V. Sanguineti (Eds.), *Self-organization, computational maps, and motor control* (pp. 383–421). Amsterdam: Elsevier-North Holland.

Habib, M. G., Waltzman, S. B., Tajudeen, B., & Svirsky, M. A. (2010). Speech production intelligibility of early implanted pediatric cochlear implant users. *International Journal of Pediatric Otorhinolaryngology, 74,* 855–859.

Hain, T. C., Burnett, T. A., Kiran, S., Larson, C. R., Singh, S., & Kenney, M. K. (2000). Instructing subjects to make a voluntary response reveals the presence of two components to the audio-vocal reflex. *Experimental Brain Research, 130,* 133–141.

Hashimoto, Y., & Sakai, K. L. (2003). Brain activations during conscious self-monitoring of speech production with delayed auditory feedback: an fMRI study. *Human Brain Mapping, 20,* 22–28.

Hazemann, P., Audin, G., & Lille, F. (1975). Effect of voluntary self-paced movements upon auditory and somatosensory evoked potentials in man. *Electroencephalography and Clinical Neurophysiology, 39,* 247–254.

Heinks-Maldonado, T. H., Nagarajan, S. S., & Houde, J. F. (2006). Magnetoencephalographic evidence for a precise forward model in speech production. *Neuroreport, 17*, 1375–1379.

Helmholtz, H. (1924). *Treatise on physiological optics* (J. P. C. Southall, Trans.). New York: Optical Society of America.

Hickok, G., Buchsbaum, B., Humphries, C., & Muftuler, T. (2003). Auditory-motor interaction revealed by fMRI: speech, music, and working memory in area Spt. *Journal of Cognitive Neuroscience, 15*, 673–682.

Hickok, G., & Poeppel, D. (2004). Dorsal and ventral streams: a framework for understanding aspects of the functional anatomy of language. *Cognition, 92*, 67–99.

Hillenbrand, J. M., Clark, M. J., & Nearey, T. M. (2001). Effects of consonant environment on vowel formant patterns. *Journal of the Acoustical Society of America, 109*, 748–763.

Houde, J. F., & Jordan, M. I. (1998). Sensorimotor adaptation in speech production. *Science, 279*, 1213–1216.

Houde, J. F., Nagarajan, S. S., Sekihara, K., & Merzenich, M. M. (2002). Modulation of the auditory cortex during speech: an MEG study. *Journal of Cognitive Neuroscience, 14*, 1125–1138.

Jones, J. A., & Munhall, K. G. (2002). The role of auditory feedback during phonation: studies of Mandarin tone production. *Journal of Phonetics, 30*, 303–320.

Katseff, S., Houde, J., & Johnson, K. (2012). Partial compensation for altered auditory feedback: a tradeoff with somatosensory feedback? *Language and Speech, 55*, 295–308.

Kawato, M. (1999). Internal models for motor control and trajectory planning. *Current Opinion in Neurobiology, 9*, 718–727.

Kelly, C., Uddin, L. Q., Shehzad, Z., Margulies, D., Castellanos, F. X., Milham, M. P., et al. (2010). Broca's region: linking human brain functional connectivity data and non-human primate tracing anatomy studies. *European Journal of Neuroscience, 32*, 383–398.

Knolle, F., Schröger, E., Baess, P., & Kotz, S. A. (2012). The cerebellum generates motor-to-auditory predictions: ERP lesion evidence. *Journal of Cognitive Neuroscience, 24*, 698–706.

Kuhl, P. K. (1991). Human adults and human infants show a "perceptual magnet effect" for the prototypes of speech categories, monkeys do not. *Perception & Psychophysics, 50*, 93–107.

Ladd, R. D., & Morton, R. (1997). The perception of intonational emphasis: continuous or categorical? *Journal of Phonetics, 25*, 313–342.

Lane, H., Denny, M., Guenther, F. H., Hanson, H. M., Marrone, N., Matthies, M. L., et al. (2007 a). On the structure of phoneme categories in listeners with cochlear implants. *Journal of Speech, Language, and Hearing Research, 50*, 2–14.

Lane, H., Matthies, M. L., Guenther, F. H., Denny, M., Perkell, J. S., Stockmann, E., et al. (2007 b). Effects of short- and long-term changes in auditory feedback on vowel and sibilant contrasts. *Journal of Speech, Language, and Hearing Research, 50*(4), 913–927.

Lane, H., & Tranel, B. (1971). The Lombard sign and the role of hearing in speech. *Journal of Speech, Language, and Hearing Research, 14*, 677–709.

Lane, H., & Webster, J. W. (1991). Speech deterioration in postlingually deafened adults. *Journal of the Acoustical Society of America, 89*, 859–866.

Larson, C. R., Altman, K. W., Liu, H., & Hain, T. C. (2008). Interactions between auditory and somatosensory feedback for voice F0 control. *Experimental Brain Research, 187*, 613–621.

Larson, C. R., Burnett, T. A., Kiran, S., & Hain, T. C. (2000). Effects of pitch-shift velocity on voice F0 responses. *Journal of the Acoustical Society of America, 107*, 559–564.

Larson, C. R., Sun, J., & Hain, T. C. (2007). Effects of simultaneous perturbations of voice pirch and loudness feedback on voice F0 and amplitude control. *Journal of the Acoustical Society of America, 121*, 2862–2872.

Leder, S. B., Spitzer, J. B., & Kircher, J. C. (1987). Speaking fundamental frequency of postlingually profoundly deaf adult men. *Annals of Otology, Rhinology, and Laryngology, 96*, 322–324.

Lee, B. (1950). Some effects of side-tone delay. *Journal of the Acoustical Society of America, 22*, 639–640.

Leigh, J., Dettman, S., Dowell, R., & Briggs, R. (2013). Communication development in children who receive a cochlear implant by 12 months of age. *Otology & Neurotology, 34*, 443–450.

Liberman, A. M., Harris, K. S., Kinney, J. A., & Lane, H. (1961). The discrimination of relative onset time of the components of certain speech and nonspeech patterns. *Journal of Experimental Psychology*, *61*, 379–388.

Liu, C. (2013). Just noticeable difference of tone pitch contour change for English- and Chinese-native listeners. *Journal of the Acoustical Society of America*, *134*, 3011–3020.

Liu, C., & Rodriguez, A. (2012). Categorical perception of intonation contrasts: effects of listeners' language background. *Journal of the Acoustical Society of America*, *131*, EL427–EL433.

Lombard, E. (1911). Le signe de l'elevation de la voix. *Annales des Maladies de l'Oreille du Larynx*, *37*, 101–119.

Martikainen, M. H., Kaneko, K., & Hari, R. (2005). Suppressed responses to self-triggered sounds in the human auditory cortex. *Cerebral Cortex*, *15*, 299–302.

Matsumoto, R., Nair, D. R., LaPresto, E., Najm, I., Bingaman, W., Shibasaki, H., et al. (2004). Functional connectivity in the human language system: a cortico-cortical evoked potential study. *Brain*, *127*, 2316–2330.

Matthies, M. L., Guenther, F. H., Denny, M., Perkell, J. S., Burton, E., Vick, J., et al. (2008). Perception and production of /r/ allophones improve with hearing from a cochlear implant. *Journal of the Acoustical Society of America*, *124*, 3191–3202.

McCarthy, G., & Donchin, E. (1976). The effects of temporal and event uncertainty in determining the waveforms of the auditory event related potential (ERP). *Psychophysiology*, *13*, 581–590.

McClean, M. D., & Sapir, S. (1981). Some effects of auditory stimulation on perioral motor unit discharge and their implications for speech production. *Journal of the Acoustical Society of America*, *69*, 1452–1457.

McGuire, P. K., Silbersweig, D. A., & Frith, C. D. (1996). Functional neuroanatomy of verbal self-monitoring. *Brain*, *119*, 907–917.

Miall, R. C., & Wolpert, D. M. (1996). Forward models for physiological motor control. *Neural Networks*, *9*, 1265–1279.

Micci Barreca, D., & Guenther, F. H. (2001). A modeling study of potential sources of curvature in human reaching movements. *Journal of Motor Behavior*, *33*, 387–400.

Mitsuya, T., MacDonald, E. N., Purcell, D. W., & Munhall, K. G. (2011). A cross-language study of compensation in response to real-time formant perturbation. *Journal of the Acoustical Society of America*, *130*, 2978–2986.

Moore, E. H. (1920). On the reciprocal of the general algebraic matrix. *Bulletin of the American Mathematical Society*, *26*, 394–395.

Niziolek, C., & Guenther, F. H. (2013). Vowel category boundaries enhance cortical and behavioral responses to speech feedback alterations. *Journal of Neuroscience*, *33*, 12090–12098.

Niziolek, C., Nagarajan, S. S., & Houde, J. F. (2013). What does motor efference copy represent? Evidence from speech production. *Journal of Neuroscience*, *33*, 16110–16116.

Numminen, J., & Curio, G. (1999). Differential effects of overt, covert and replayed speech on vowel-evoked responses of the human auditory cortex. *Neuroscience Letters*, *272*, 29–32.

Numminen, J., Salmelin, R., & Hari, R. (1999). Subject's own speech reduces reactivity of the human auditory cortex. *Neuroscience Letters*, *265*, 119–122.

Oller, K. (2000). *The emergence of the speech capacity*. Mahwah, NJ: Erlbaum.

Parkinson, A. L., Flagmeier, S. G., Manes, J. L., Larson, C. R., Rogers, B., & Robin, D. A. (2012). Understanding the neural mechanisms involved in sensory control of voice production. *NeuroImage*, *61*, 314–322.

Paus, T., Perry, D. W., Zatorre, R. J., Worsley, K. J., & Evans, A. C. (1996). Modulation of cerebral blood flow in the human auditory cortex during speech: role of motor-to-sensory discharges. *European Journal of Neuroscience*, *8*, 2236–2246.

Penhune, V. B., & Doyon, J. (2005). Cerebellum and M1 interaction during early learning of timed motor sequences. *NeuroImage*, *26*, 801–812.

Penrose, R. (1955). A generalized inverse for matrices. *Proceedings of the Cambridge Philosophical Society*, *51*, 406–413.

Perkell, J. S., Lane, H., Denny, M., Matthies, M. L., Tiede, M., Zandipour, M., et al. (2007). Time course of speech changes in response to unanticipated short-term changes in hearing state. *Journal of the Acoustical Society of America, 121*, 2296–2311.

Purcell, D. W., & Munhall, K. G. (2006). Compensation following real-time manipulation of formants in isolated vowels. *Journal of the Acoustical Society of America, 119*, 2288–2297.

Ringel, R. L., & Steer, M. D. (1963). Some effect of tactile and auditory alterations on speech output. *Journal of Speech and Hearing Research, 6*, 369–378.

Schäfer, E. W., & Marcus, M. M. (1973). Self-stimulation alters human sensory brain responses. *Science, 181*, 175–177.

Schliesser, H. F., & Coleman, R. O. (1968). Effectiveness of certain procedures for alteration of auditory and oral tactile sensation for speech. *Perceptual and Motor Skills, 26*, 275–281.

Seidler, R. D., Noll, D. C., & Thiers, G. (2004). Feedforward and feedback processes in motor control. *NeuroImage, 22*, 1775–1783.

Sperry, R. W. (1950). Neural basis of the spontaneous optokinetic response produced by visual inversion. *Journal of Comparative and Physiological Psychology, 43*, 482–489.

Stevens, A. A., Skudlarski, P., Gatenby, J. C., & Gore, J. C. (2000). Event-related fMRI of auditory and visual oddball tasks. *Magnetic Resonance Imaging, 18*, 495–502.

Stuart, A., Kalinowski, J., Rastatter, M. P., & Lynch, K. (2002). Effect of delayed auditory feedback on normal speakers at two speech rates. *Journal of the Acoustical Society of America, 111*, 2237–2241.

Takaso, H., Eisner, F., Wise, R. J. S., & Scott, S. K. (2010). The effect of delayed auditory feedback on activity in the temporal lobe while speaking: a positron emission tomography study. *Journal of Speech, Language, and Hearing Research, 53*, 226–236.

Tamamaki, N., & Tomioka, R. (2010). Long-range GABAergic connections distributed throughout the neocortex and their possible function. *Frontiers in Neuroscience, 4*, 202.

Tourville, J. A., Reilly, K. J., & Guenther, F. H. (2008). Neural mechanisms underlying auditory feedback control of speech. *NeuroImage, 39*, 1429–1443.

Toyomura, A., Koyama, S., Miyamoto, T., Terao, A., Omori, T., Murohashi, H., et al. (2007). Neural correlates of auditory feedback control in human. *Neuroscience, 146*, 499–503.

Villacorta, V., Perkell, J. S., & Guenther, F. H. (2007). Sensorimotor adaptation to perturbations of vowel acoustics and its relation to perception. *Journal of the Acoustical Society of America, 122*, 2306–2319.

von Holst, E., & Mittelstaedt, H. (1950). The reafference principle: interaction between the central nervous system and the periphery. In *Selected papers of Erich von Holst, volume 1: the behavioural physiology of animals and man* (pp. 139–173; translated from German by R. Martin). London: Methuen.

Waldstein, R. S. (1990). Effects of postlingual deafness on speech production: implications for the role of auditory feedback. *Journal of the Acoustical Society of America, 88*, 2099–2114.

Weiller, C., Bormann, T., Saur, D., Musso, M., & Rijntjes, M. (2011). How the ventral pathway got lost—and what its recovery might mean. *Brain and Language, 118*, 29–39.

Wise, R. J., Greene, J., Buchel, C., & Scott, S. K. (1999). Brain regions involved in articulation. *Lancet, 353*, 1057–1061.

Xu, Y., Gandour, J. T., & Francis, A. L. (2006). Effects of language experience and stimulus complexity on the categorical perception of pitch direction. *Journal of the Acoustical Society of America, 120*, 1063–1074.

Xu, Y., Larson, C. R., Bauer, J. J., & Hain, T. C. (2004). Compensation for pitch-shifted auditory feedback during the production of Mandarin tone sequences. *Journal of the Acoustical Society of America, 116*, 1168–1178.

Yates, A. J. (1963). Delayed auditory feedback. *Psychological Bulletin, 60*, 213–232.

Yeni-Komshian, G., Chase, P. A., & Mobley, R. L. (1968). The development of auditory feedback monitoring: II. Delayed auditory feedback studies on the speech of children between two and three years of age. *Journal of Speech and Hearing Research, 11*, 307–315.

# 6

Somatosensory Feedback Control

The cerebral cortex receives abundant somatosensory information from the vocal tract during speech production. *Tactile* information arises from mechanoreceptors in the skin surface, which transmit signals to the central nervous system when the skin is touched. For example, when producing the alveolar fricative consonant /s/, the tongue tip touches the alveolar ridge to form a constriction of the vocal tract whose area must be carefully controlled. Mechanoreceptors in the tongue surface and palate provide detailed information concerning the location and degree of this constriction. *Proprioception*, the sense of body position and motion, arises from several sources in addition to the skin mechanoreceptors, most notably from muscle spindles that provide information about the lengths and shortening velocities of the vocal tract musculature.

To get a sense of the amount of somatosensory information available to the brain during speech, consider the most complex speech articulator, the tongue. Collectively, the muscles of the tongue are innervated by approximately 1,000 muscle spindles (Kubota, Negishi, & Masegi, 1975). Though reliable numbers for the density of tactile mechanoreceptors in the tongue are scarce, comparison with the fingers (which, along with the tongue, constitute the most sensitive body parts to tactile stimulation) suggests a density of a few thousand mechanoreceptors per square centimeter, innervated by a few hundred nerve fibers (Gardner, Martin, & Jessell, 2000). The surface of the tongue is on the order 100 cm^2 (Naumova et al., 2013), resulting in an estimate of a few hundred thousand mechanoreceptors feeding tens of thousands of afferent nerve fibers.

Somatosensory information is central to the DIVA model both for specifying the targets of speech and for activating appropriate articulator movements to reach these targets. A number of investigators have hypothesized speech targets within a somatosensory frame. Based on the results of bite block experiments showing automatic compensation even on the first glottal pulse (i.e., in the absence of acoustic feedback), Lindblom, Lubker, and Gay (1979) hypothesized that "the target of a vowel segment is coded neurophysiologically in terms of its area function by means of corresponding [somato]sensory information" (p. 157). Similarly, Perkell (1980) posited that acoustic goals are transformed into corresponding somatosensory goals during the production process. The task dynamic model of

Saltzman and Munhall (1989) hypothesizes a vocal tract reference frame existing between the levels of acoustic goals and motor realization. Because these tract variables characterize key constrictions in the vocal tract, they could be interpreted as another example of sound targets in a somatosensory reference frame.

6.1 The Somatosensory Feedback Control Subsystem

Figure 6.1 highlights the somatosensory feedback control subsystem within the overall control system for speech introduced in section 3.3. Recall from section 3.3 that S is a vector whose components correspond to the current somatosensory state, S_T is a vector specifying the current somatosensory target, M is a vector specifying the current motor positional command, $\dot{M}$ is a vector representing the overall movement command to the speech articulators, and $\dot{M}_S$ is a vector representing the corrective motor command issued by the somatosensory feedback controller. Production of a speech sound begins when the *speech sound map* node for that sound is activated, leading to the readout from memory of the sound's time-varying somatosensory target, that is, the tactile and proprioceptive feedback expected to occur when the syllable is properly produced. This target is sent to the *somatosensory feedback controller*, where it is compared to incoming somatosensory feedback from the speech articulators. If there is a discrepancy between the somatosensory target and the somatosensory feedback, corrective motor commands are generated; these commands sum with commands from the feedforward and auditory feedback controllers to form the overall motor command to the vocal tract musculature.

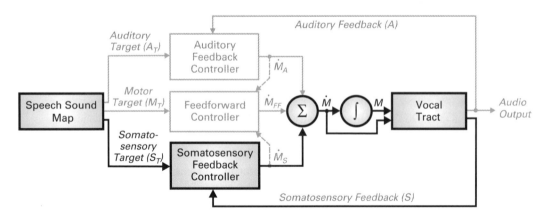

Figure 6.1
The somatosensory feedback control subsystem highlighted within the overall control system for speech production introduced in chapter 3. M, motor positional command; $\dot{M}$, overall movement command; $\dot{M}_S$, corrective movement command issued by the somatosensory feedback controller.

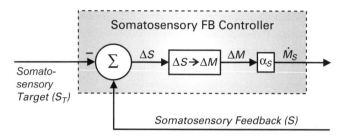

Figure 6.2
Expanded view of the somatosensory feedback controller implemented in the DIVA model. α_S, somatosensory gain factor; ΔM, motor error; ΔS, somatosensory error; FB, feedback; $\dot{M}_S$, corrective movement command issued by the somatosensory feedback controller.

Figure 6.2 provides an expanded view of the somatosensory feedback controller (dashed box) as implemented in the DIVA model. If somatosensory feedback falls outside the range specified by the somatosensory target, a somatosensory error signal ΔS is generated. This error signal represents the change, within the somatosensory reference frame, needed to get the speech signal within the target region. The somatosensory error signal is then transformed into an error signal in motor coordinates, represented by the box labeled $\Delta S \rightarrow \Delta M$ in figure 6.2; this transformation constitutes a form of inverse model (see chapter 3, section 3.2). According to the DIVA model, the brain learns this directional mapping during babbling by associating changes in the somatosensory state with movements of the speech articulators and storing this information in a matrix of synaptic weights, Z_{SM}. ΔM is then scaled by a gain factor α_S to create the corrective movement command $\dot{M}_S$, where the subscript S indicates that this is the somatosensory feedback controller's contribution to the overall movement command. The processing performed by the somatosensory feedback controller is summarized by the following equation:[1]

$$\dot{M}_S = \alpha_S Z_{SM}[S - S_T]. \qquad (6.1)$$

As was the case for the auditory feedback controller, somatosensory feedback controllers that utilize a directional mapping of this type have a number of desirable movement properties, including the use of very efficient movement commands for reaching somatosensory targets (see chapter 5, section 5.1, for further details).

The remainder of this chapter reviews behavioral data that provide further information regarding somatosensory feedback control in speech, followed by a treatment of the neural substrates of the somatosensory feedback control subsystem.

6.2 Behavioral Studies of Somatosensory Feedback Control in Speech

Although the role of somatosensory feedback in speech may be less obvious than the role of auditory feedback, a number of experimental studies have clearly demonstrated that somatosensory feedback is used in the control of speech movements. These studies fall into two main classes: those that involve the removal of somatosensory information and those that involve perturbing the articulators during speech.

Deprivation of Somatosensory Information
Unlike deafness, the selective and profound loss of somatic sensation is very rare. The few available reports indicate substantial deficits in speech articulation when the somatic sensation of the oral cavity is profoundly disturbed (MacNeilage, Rootes, & Chase, 1967; Cole & Paillard, 1998), though in one case of sudden loss of somatic sensation in adulthood deficits were only apparent when the speech articulators were impeded by a bite block (Hoole, 1987). These cases are of limited utility for characterizing somatosensory feedback control, however, since it is typically not possible to rule out damage to the feedforward control system in addition to the loss of somatic sensation. Additional evidence for a role for somatosensory feedback control in speech comes from studies in which topical or local anesthesia is applied to the articulators (e.g., Ringel & Steer, 1963; Gammon et al., 1971; Schliesser & Coleman, 1968; Hutchinson & Putnam, 1974; Hardcastle, 1975; Putnam & Ringel, 1976). For the most part, these studies report nearly normal speech production if auditory feedback is available, though some articulation errors and loss of movement precision are found, particularly for fricative consonants that require precise control of the articulators (Gammon et al., 1971; Hardcastle, 1975; Putnam & Ringel, 1976). Ringel and Steer (1963) note that local anesthesia, in the form of nerve blocks, resulted in more articulation errors than topical anesthesia. Hutchinson and Putnam (1974) report that intraoral air pressure is higher than normal under local anesthesia, and Hardcastle (1975) reported articulation overshoots with topical anesthetic applied to the oral cavity. These adjustments may reflect attempts to provide increased sensory information regarding the reaching of articulatory targets. If both auditory and somatosensory feedback are removed, the number of articulation errors increases (Ringel & Steer, 1963; Gammon et al., 1971).

Disruption of somatosensory feedback does not appear to disrupt temporal sequencing of speech sounds. Gammon et al. (1971) found no effects on stress pattern or juncture[2] even when anesthesia was combined with masking noise.

Collectively these studies indicate that segmental aspects of speech production depend significantly on somatosensory feedback control, but suprasegmental aspects such as temporal sequencing and stress do not.

Perturbations to the Speech Articulators
Anyone who has spoken with a pipe or pen clenched between his or her teeth has experienced the ability of the speech motor control system to deal with constraints on articulatory movement. This phenomenon was investigated experimentally by Lindblom, Lubker, and Gay (1979), who had subjects produce vowels while holding a bite block between their

teeth. The bite block prevented jaw movement and held the jaw at an unnaturally large or small opening for each vowel, thus requiring compensation in other speech articulators. Despite the bite block, subjects were able to produce approximately normal formant frequencies for all of the vowels immediately after bite block insertion.

The somatosensory feedback control subsystem's influence on articulator movements can be studied in more detail by applying brief, unexpected perturbations to the speech articulators. Unexpected perturbations avoid the potential confound of motor learning in the feedforward control system that occurs with sustained perturbations such as false palates or bite blocks. In an early study of this sort, Folkins and Abbs (1975) applied an unexpected downward load on the jaw during closing movements for bilabial stops in speech utterances. Compensatory responses were found in the upper and lower lips. These compensations were complete in the sense of achieving the goal of lip closure; this contrasts with the typically incomplete compensations to auditory perturbations described in section 5.2.

The response latency of the somatosensory feedback control subsystem also appears to be shorter than that of the auditory feedback control subsystem. Abbs and Gracco (1984) applied unexpected loads to the lower lip during lip closing gestures for bilabial stop consonants in the utterances "aba" and "sapapple." Compensatory responses were seen in independently controlled muscles of the upper and lower lips, with response latencies in the 22- to 75-ms range. These compare to latencies of approximately 100 to 150 ms for unexpected perturbations of auditory feedback. Abbs and Gracco (1984) note that the response latencies to lip perturbations were longer than expected for brain stem responses, suggesting that the response is cortically mediated. A later study by Ito, Kimura, and Gomi (2005) showed that transcranial magnetic stimulation over speech motor cortex facilitates the compensatory response, verifying that cortical processing plays a role in compensation to somatosensory perturbations.

The Folkins and Abbs (1975) and Abbs and Gracco (1984) findings also highlight the fact that compensation for a perturbation to a speech articulator is not simply reflexive resistance to the load in the perturbed articulator; responses are found in synergistic articulators as well (a form of *motor equivalence*, which was treated in chapter 4, section 4.3). This implies that somatosensory targets for speech are defined at a higher level than the individual articulators. These higher-level somatosensory goals are more directly related to the acoustic signal, which depends much more on the locations and degrees of constrictions in the vocal tract than on individual articulator positions. Researchers from Haskins Laboratories introduced the term *tract variables* to differentiate these higher-level goals from the positions of individual articulators (Saltzman & Munhall, 1989; Browman & Goldstein, 1990). The key tract variable for bilabial consonants is the size of the opening between the lips, or *lip aperture*. The bilabial stop consonant /b/ can be successfully produced with a wide range of lower lip, upper lip, and jaw positions, as long as the summed effect of these articulator positions is a lip aperture of zero, corresponding to complete closure. Additional tract variables define the location and degree of constrictions of the tongue tip, tongue dorsum, velum, and glottis.

One might imagine that subjects are using auditory feedback to compensate for the articulator perturbations described thus far; that is, the subject hears that his or her speech is not quite right because of the perturbation and so uses auditory feedback control to correct it. Several experimental findings contradict this viewpoint. Lindblom, Lubker, and Gay (1979) report that compensation for a bite block occurred before the subject uttered the first glottal pulse for the vowel; that is, the compensation had occurred before any auditory feedback was available. Nasir and Ostry (2008) noted compensation for even very small somatosensory perturbations in cochlear implant recipients with their processors turned off, thus essentially eliminating auditory feedback. Tremblay, Shiller, and Ostry (2003) applied a velocity-dependent jaw perturbation during the jaw opening phase of the utterance /siæt/ that had no measurable acoustic effect. Nonetheless, participants compensated for the jaw perturbation by adjusting jaw movements in the opposite direction to the perturbation, indicating the existence of a somatosensory target. Nasir and Ostry (2006) extended this finding to additional utterances, involving both vowels and consonants. Together, these studies verify that corrections for somatosensory perturbations are carried out at least in part by somatosensory feedback control.

Compensatory responses to articulator perturbations during speech appear to occur primarily in articulators that are necessary for the ongoing speech gesture. For example, Abbs and Gracco (1983) noted upper and lower lip compensation for a downward lower lip perturbation during the bilabial consonant in /bɑ/, but no upper lip compensatory movements when the same perturbation was applied during the /f/ in /ɑfɑ/, presumably since upper lip position is not important for producing /f/. Kelso et al. (1984) applied downward perturbations of the jaw during consonant closures for the final consonant in two utterances, /bæb/ and /bæz/. The authors report compensatory adjustments that were specific to the phoneme being produced: lip compensation was found for the bilabial stop /b/ whereas tongue compensation was found for the alveolar consonant /z/.

The DIVA model accounts for the articulatory specificity of compensatory articulator movements through the use of sensory target *regions*, rather than point targets, as described in chapter 4. Consider the consonants /b/ and /z/ from the Kelso et al. (1984) study. For /b/, the DIVA model learns a somatosensory target that specifies complete lip closure (since lip closure is mandatory for the bilabial consonant /b/) but allows a wide range of variation in tongue position, as schematized in figure 6.3. When the jaw is perturbed downward during /b/, the lips are pushed outside of the target range of lip aperture for /b/, as represented by the red arrow in panel A of figure 6.3. However, the tongue position change induced by the jaw perturbation does not push the tongue tip constriction outside of its target range for /b/ since a wide range of tongue tip constrictions are acceptable. As a result, a compensatory response (green arrow) is needed only for lip aperture. As shown panel B of figure 6.3, the situation is reversed for /z/: the model's target region for /z/ includes a wide range of lip

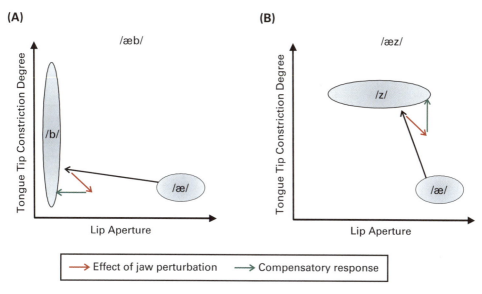

Figure 6.3
DIVA model account of Kelso et al. (1984) jaw perturbation results. (A) The articulatory movement from /æ/ to /b/ in the utterance /bæb/ (black arrow). The red arrow represents the effect of a downward perturbation to the jaw. The somatosensory target region for /b/ can be reached with a compensatory decrease in lip aperture (green arrow); no compensation is needed for tongue tip constriction degree. (B) The movement from /æ/ to /z/ in /bæz/. In this case compensation for the perturbation requires a compensatory increase in tongue tip constriction degree with no lip aperture compensation necessary.

positions since /z/ production does not rely on lip aperture, but the target includes a very small range of tongue tip constrictions since precise control of this constriction is crucial for /z/. Thus a jaw perturbation will not push the lips outside their target range but will push the tongue tip constriction outside its target range, leading to a compensatory response in tongue tip constriction but not lip aperture. Guenther, Ghosh, and Tourville (2006) report computer simulations of the DIVA model that verify its ability to account for the Kelso et al. (1984) findings, as well those of the Abbs and Gracco (1984) lip perturbation experiment.

6.3 Neural Circuitry Underlying Somatosensory Feedback Control

The neural circuitry responsible for somatosensory feedback control according to the DIVA model is shown in figure 6.4. Stereotactic coordinates associated with the model components in figure 6.4 are provided in appendix D.

As discussed in chapter 2, section 2.2, tactile and proprioceptive information from mechanoreceptors in the speech articulators projects to somatosensory cortex via cranial nerve nuclei (especially the trigeminal sensory nucleus) and the *ventral posterior medial*

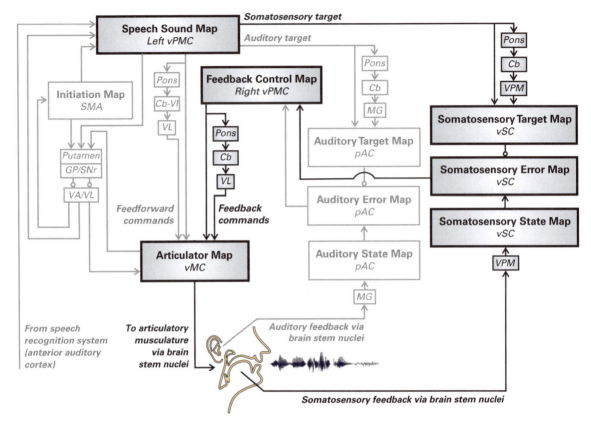

Figure 6.4
Neural components of the somatosensory feedback control subsystem according to the DIVA model. Cb, cerebellum; VL, ventral lateral nucleus of the thalamus; vMC, ventral motor cortex; VPM, ventral posterior medial nucleus of the thalamus; vPMC, ventral premotor cortex; vSC, ventral somatosensory cortex.

(*VPM*) *nucleus* of the thalamus. This information is represented in the model's *somatosensory state map*, which is hypothesized to reside in *ventral somatosensory cortex* (*vSC*), including the ventral postcentral gyrus and supramarginal gyrus. We further hypothesize that it is higher-order somatosensory cortical areas, rather than primary areas, that are most involved in speech since the key somatosensory variables for speech are more closely related to the locations and degrees of vocal tract constrictions than to the positions of individual articulators, as discussed in the previous section. Since each constriction involves multiple articulators, the constriction representation most likely resides in higher-order somatosensory cortex.

In the fully developed system,[3] when the *speech sound map* node for a sound is activated in left *ventral premotor cortex* (*vPMC*), axonal projections from this node transmit the sound's somatosensory target to the model's *somatosensory target map*, hypothesized to reside in vSC. As described in chapter 4, the somatosensory targets take the form of time-varying multidimensional regions in somatosensory space, and they are effectively stored in synaptic weights in the axonal projections between the speech sound map and the somatosensory target map. These projections constitute a forward model that complements the auditory forward model described in chapter 5. As with the auditory target projections, we hypothesize that both cortico-cortical projections and projections through a cerebellar loop are involved. The superior longitudinal fasciculus has been shown to project from vPMC to vSC (e.g., Catani, Jones, & ffytche, 2005; Kelly et al., 2010) and is likely the key white matter tract for cortico-cortical somatosensory target projections. (See figures C.1, C.19, C.20, C.21, C.23, C.24, and C.28 in appendix C for relevant functional and structural connectivity maps derived from from magnetic resonance imaging.)

The inclusion of a cerebellar component of the somatosensory target projections is motivated by the considerations discussed for auditory target projections in chapter 5, section 5.3, as well as data indicating cerebellar involvement in sensory prediction. For example, Blakemore, Wolpert, & Frith (1999) noted that tactile stimulation of the hand generated by one's own movements produced decreased activity in anterior superior cerebellum compared to when tactile stimulation was externally generated. The authors conclude that the cerebellum is involved in predicting the sensory consequences of movements (see also Blakemore, Wolpert, & Frith, 2000; Blakemore, Frith, & Wolpert, 2001). Current data are insufficient to determine which lobules of the cerebellum are involved in somatosensory prediction for speech.

The current somatosensory state is compared to the somatosensory target at the *somatosensory error map*. Nodes in this map become active during speech if somatosensory feedback from the vocal tract deviates from the somatosensory target region for the sound being produced. The net effect of projections from the speech sound map to the somatosensory error map is inhibitory, effectively canceling out somatosensory feedback that is expected to arise during production of the current sound. The hypothesized inhibition of somatosensory cortex is supported by numerous studies reporting reduced sensory cortical responses immediately prior to and during voluntary movement. For instance, recordings from monkey somatosensory cortex during peripheral stimulation are depressed prior to and during voluntary limb movements (Chapman et al., 1988). Somatosensory response suppression during self-generated movements[4] has been identified in humans using fMRI (Blakemore, Wolpert, & Frith, 1999) and MEG (Wasaka et al., 2003; Wasaka et al., 2007). It has also been shown that sensory attenuation during a self-generated action is predictive of the timing of the action and the expected sensory results of that action (Blakemore, Frith, & Wolpert, 1999; Bays, Wolpert, & Flanagan, 2005; Bays, Flanagan, & Wolpert, 2006; Voss et al., 2006). The predictive nature of the attenuation suggests that the brain has learned a forward model of the sensory consequences of motor commands. The modulation of somatosensory responses during speech appears to occur in the cerebral cortex rather than at lower levels of the nervous system. McClean et al. (1990) performed presurgical microelectrode recordings in the VPM during speech and during mechanical stimulation of the lips and tongue. Responses of VPM neurons during speech were similar to responses elicited by mechanical stimulation, suggesting that speech-induced suppression does not occur in somatosensory thalamus or in the brain stem nuclei that project to it.

We hypothesize that somatosensory error information is transmitted to a *feedback control map* in right vPMC, where it is transformed into corrective movement commands that are transmitted to the *articulator map* in *ventral motor cortex* (*vMC*) bilaterally. In addition to direct projections from right vPMC to bilateral vMC, we hypothesize indirect projections via a loop through the *cerebellum* (*Cb*) and *ventral lateral* (*VL*) *nucleus* of the thalamus based on studies indicating that the cerebellum influences corrective movements (e.g., Diedrichsen et al., 2005; Grafton et al., 2008; Penhune & Doyon, 2005). As noted in chapter 5, section 5.3, it is not clear at present which portion of the cerebellar cortex is involved in these projections, though there is evidence implicating lobule VIII (Diedrichsen et al., 2005; Tourville, Reilly, & Guenther, 2008; Golfinopoulos et al., 2011) and/or lobule VI (Seidler, Noll, & Thiers, 2004; Diedrichsen et al., 2005; Donchin et al., 2012) in sensory feedback control.

The DIVA model's account of somatosensory feedback control of speech was directly tested in an fMRI study performed by our laboratory (Golfinopoulos et al., 2011). This study utilized a custom device to rapidly and unexpectedly block jaw movement during production of vowel-consonant-vowel utterances that started with the vowel /ɑ/ and ended with the vowel /i/ (e.g., /ɑbi/, /ɑgi/). In one of seven productions (randomly dispersed), a computer-controlled pneumatic device rapidly inflated a small, stiff balloon lying between the molars to a diameter of 1 to 1.5 cm during the first vowel of the utterance; this has the effect of blocking upward jaw movement for the start of the second syllable. As described in section 6.2, prior studies indicate that subjects quickly compensate for such perturbations by increasing the movement of the tongue and other unperturbed articulators to reach vocal tract constriction targets.

The main results of this study are illustrated in panel A of figure 6.5. The model predicts increased activation in somatosensory error cells in the ventral postcentral gyrus and supramarginal gyrus when contrasting jaw-perturbed speech with unperturbed speech (see the model simulations in panel B of figure 6.5). This hypothesis was upheld in the experimental results. The model also predicts increased activity in the feedback control map located in right vPMC, including the rostral precentral gyrus and posterior inferior frontal gyrus. This hypothesis was also upheld, and effective connectivity analyses indicated an increase in the influence of the anterior supramarginal gyrus on right vPMC during perturbed speech compared to nonperturbed speech, in keeping with the model. The model's prediction of cerebellar involvement in correcting somatosensory errors was also upheld, with activity noted in lobule VIII bilaterally (not shown).

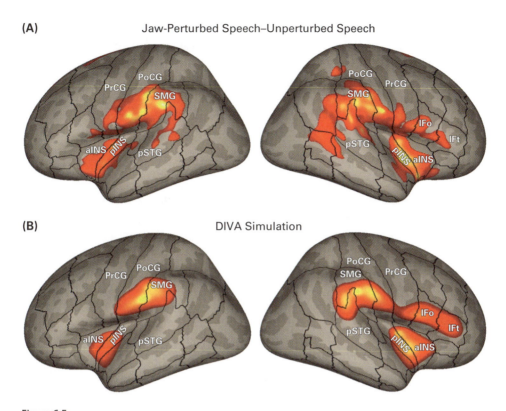

Figure 6.5
(A) Regions of significant activation in the *perturbed speech–unperturbed speech* contrast of an fMRI experiment investigating the effects of unexpected jaw perturbation during bisyllable production (Golfinopoulos et al., 2011) shown on inflated left and right lateral surfaces of the cerebral cortex. Functional data were first smoothed on the cortical surface; individual cortical vertices were then thresholded at $p < 0.001$ uncorrected before a cluster-level analysis thresholded at a family-wise error rate of 0.05. (B) Activity generated by a DIVA model simulation of the jaw perturbation experiment; see appendix D for details. aINS, anterior insula; IFo, inferior frontal gyrus pars opercularis; IFt, inferior frontal gyrus pars triangularis; pINS, posterior insula; PoCG, postcentral gyrus; PrCG, precentral gyrus; pSTG, posterior superior temporal gyrus; SMG, supramarginal gyrus.

The *perturbed speech–unperturbed speech* contrast also identified activity in the insula bilaterally, particularly around the central insular sulcus. Interpreted within the DIVA model framework, this activity may correspond to somatosensory error detection, generation of corrective motor commands, or both. However, as discussed in chapter 4, section 2.4, the insula has been implicated in a very wide range of behaviors and cognitive states, precluding the drawing of strong conclusions regarding insula activity in this task. For example, insula activity may simply reflect increased awareness of one's own movements during the perturbed trials (cf. Karnath & Baier, 2010).

Although we have treated the auditory and somatosensory representations as distinct entities in this discussion, there are likely combined somato-auditory representations, and somato-auditory error maps, that involve relatively highly processed combinations of speech-related somatosensory and auditory information. Thus we expect a continuum of sensory error map representations in the superior temporal gyrus, Sylvian fissure, and supramarginal gyrus, rather than entirely distinct auditory and somatosensory error maps as described thus far.

6.4 Summary

Speech involves a somatosensory feedback control process that acts in parallel with the auditory feedback controller described in chapter 5. This process consists of comparing afferent somatosensory information to a learned somatosensory target and generating corrective motor commands if somatosensory feedback is not within the target region for the sound. According to the DIVA model, the transformation between somatosensory errors and motor corrective commands (which constitutes a form of inverse model) takes the form of a directional mapping that results in very efficient movement commands for reaching somatosensory targets.

Deprivation of somatosensory feedback through the application of anesthesia has only a small effect on intelligibility of speech, with errors primarily occurring for sounds that require precise articulation such as fricative consonants. Removing auditory feedback in addition to somatosensory feedback results in an increased number of errors, though still most sounds are produced intelligibly. The timing of speech gestures does not appear to degrade even with combined auditory and somatosensory deprivation.

Studies involving mechanical perturbations to the articulators during speech have produced a more detailed picture of the somatosensory feedback controller for speech. The compensatory response for a somatosensory perturbation begins approximately 25 to 75 ms after the perturbation is applied. This is substantially shorter than the response delay for auditory perturbations, which is approximately 100 to 150 ms. Perturbing one articulator such as the jaw can lead to compensation in additional articulators such as the lips and tongue, indicating that the somatosensory targets for speech sounds include higher-level somatosensory goals such as constriction locations and degrees that are affected by multiple articulators. Compensatory responses to articulator perturbations are typically reduced or absent in articulators that are not heavily involved in a sound's production, indicating that somatosensory targets encode ranges of acceptable articulator positions rather than specifying precise targets for all articulator positions or somatosensory dimensions.

We propose that projections from a *speech sound map* in left ventral premotor cortex (including portions of the rostral precentral gyrus and posterior inferior frontal gyrus) carry somatosensory expectations to a *somatosensory target map* in higher-order somatosensory cortical areas in ventral somatosensory cortex (including portions of the ventral postcentral gyrus and supramarginal gyrus). These projections include direct cortico-cortical connections between left ventral premotor cortex and bilateral ventral somatosensory cortex as well as indirect connections between these regions via a cerebellar side loop. Ventral somatosensory cortex also contains a *somatosensory state map* that represents highly processed somatosensory feedback from mechanoreceptors in the speech articulators. Nodes in a *somatosensory error map*, also in ventral somatosensory cortex, represent any discrepancies between afferent somatosensory information and the somatosensory target. These somatosensory errors are transformed into corrective motor commands via projections to a feedback control map in right ventral premotor cortex (including portions of the rostral precentral gyrus and posterior inferior frontal gyrus) and then on to bilateral ventral motor cortex both directly and via a cerebellar loop.

Notes

1. See footnote 2 in chapter 5 for caveats regarding this characterization of the calculation of somatosensory errors as a simple subtraction (which holds only for point targets) as well as a formulation that accommodates target regions.

2. *Juncture* describes the relationship between neighboring phonemes; e.g., "might rain" and "my train" involve the same phonemes but different juncture patterns.

3. Recall from chapter 3, section 3.4, that the somatosensory feedback control subsystem is hypothesized to provide relatively little contribution to speech motor control early in development since a child must learn to properly produce a speech sound before he or she can build a somatosensory target for the sound.

4. Although none of these studies involved speech production, in chapter 5, section 5.3, we cite several studies that provide evidence for an analogous suppression of auditory cortex during speech.

References

Abbs, J. H., & Gracco, V. L. (1983). Sensorimotor actions in the control of multi-movement speech gestures. *Trends in Neurosciences, 6*, 391–395.

Abbs, J. H., & Gracco, V. L. (1984). Control of complex motor gestures: orofacial muscle responses to load perturbations of lip during speech. *Journal of Neurophysiology, 51*, 705–723.

Bays, P. M., Flanagan, J. R., & Wolpert, D. M. (2006). Attenuation of self-generated tactile sensations is predictive, not postdictive. *PLoS Biology, 4*, e28.

Bays, P. M., Wolpert, D. M., & Flanagan, J. R. (2005). Perception of the consequences of self-action is temporally tuned and event driven. *Current Biology, 15*, 1125–1128.

Blakemore, S.-J., Frith, C. D., & Wolpert, D. M. (1999). Spatio-temporal prediction modulates the perception of self-produced stimuli. *Journal of Cognitive Neuroscience, 11*, 551–559.

Blakemore, S.-J., Frith, C. D., & Wolpert, D. M. (2001). The cerebellum is involved in predicting the sensory consequences of action. *Neuroreport, 12*, 1879–1884.

Blakemore, S. J., Wolpert, D. M., & Frith, C. D. (1999). The cerebellum contributes to somatosensory cortical activity during self-produced tactile stimulation. *NeuroImage, 10*, 448–459.

Blakemore, S.-J., Wolpert, D., & Frith, C. (2000). Why can't you tickle yourself? *Neuroreport, 11*, R11–R16.

Browman, C., & Goldstein, L. (1990). Gestural specification using dynamically-defined articulatory structures. *Journal of Phonetics, 18*, 299–320.

Catani, M., Jones, D. K., & ffytche, D. H. (2005). Perisylvian language networks of the human brain. *Annals of Neurology, 57*, 8–16.

Chapman, C. E., Jiang, W., & Lamarre, Y. (1988). Modulation of lemniscal input during conditioned arm movements in the monkey. *Experimental Brain Research*, *72*, 316–334.

Cole, J., & Paillard, J. (1998). Living without touch and peripheral information about body position and movement: studies with deafferented subjects. In J. Bermúdez (Ed.), *The body and the self* (pp. 245–266). Cambridge, MA: MIT Press.

Diedrichsen, J., Hashambhoy, Y., Rane, T., & Shadmehr, R. (2005). Neural correlates of reach errors. *Journal of Neuroscience*, *25*, 9919–9931.

Donchin, O., Rabe, K., Diedrichsen, J., Lally, N., Schoch, B., Gizewski, E. R., et al. (2012). Cerebellar regions involved in adaptation to force field and visuomotor perturbation. *Journal of Neurophysiology*, *107*, 134–147.

Folkins, J., & Abbs, J. (1975). Lip and jaw motor control during speech: responses to resistive loading of the jaw. *Journal of Speech and Hearing Research*, *18*, 207–220.

Gammon, S. A., Smith, P. J., Daniloff, R. G., & Kim, C. W. (1971). Articulation and stress/juncture production under oral anesthetization and masking. *Journal of Speech and Hearing Research*, *14*, 271–282.

Gardner, E. P., Martin, J. H., & Jessell, T. M. (2000). The bodily senses. In E. R. Kandel, J. H. Schwartz, & T. M. Jessell (Eds.), *Principles of neural science* (pp. 430–450). New York: McGraw-Hill.

Golfinopoulos, E., Tourville, J. A., Bohland, J. W., Ghosh, S. S., Nieto-Castanon, A., & Guenther, F. H. (2011). fMRI investigation of unexpected somatosensory feedback perturbation during speech. *NeuroImage*, *55*, 1324–1338.

Grafton, S. T., Schmitt, P., Van Horn, J., & Deidrichsen, J. (2008). Neural substrates of visuomotor learning based on improved feedback control and prediction. *NeuroImage*, *39*, 1383–1395.

Guenther, F. H., Ghosh, S. S., & Tourville, J. A. (2006). Neural modeling and imaging of the cortical interactions underlying syllable production. *Brain and Language*, *96*, 280–301.

Hardcastle, W. J. (1975). Some aspects of speech production under controlled conditions of oral anaesthesia and auditory masking. *Journal of Phonetics*, *3*, 197–214.

Hoole, P. (1987). Bite-block speech in the absence of oral sensibility. *Proceedings of the 11th International Congress on Phonetic Sciences*, *4*, 16–19.

Hutchinson, J. M., & Putnam, A. H. (1974). Aerodynamic aspect of sensory deprived speech. *Journal of the Acoustical Society of America*, *56*, 1612–1617.

Ito, T., Kimura, T., & Gomi, H. (2005). The motor cortex is involved in reflexive compensatory adjustment of speech articulation. *Neuroreport*, *16*, 1791–1794.

Karnath, H. O., & Baier, B. (2010). Right insula for our sense of limb ownership and self-awareness of actions. *Brain Structure & Function*, *214*, 411–417.

Kelly, C., Uddin, L. Q., Shehzad, Z., Margulies, D., Castellanos, F. X., Milham, M. P., et al. (2010). Broca's region: linking human brain functional connectivity data and non-human primate tracing anatomy studies. *European Journal of Neuroscience*, *32*, 383–398.

Kelso, J. A., Tuller, B., Vatikiotis-Bateson, E., & Fowler, C. A. (1984). Functionally specific articulatory cooperation following jaw perturbations during speech: evidence for coordinative structures. *Journal of Experimental Psychology: Human Perception and Performance*, *10*, 812–832.

Kubota, K., Negishi, T., & Masegi, T. (1975). Topological distribution of muscle spindles in the human tongue and its significance in proprioception. *Bulletin of Tokyo Medical and Dental University*, *22*, 235–242.

Lindblom, B., Lubker, J., & Gay, T. (1979). Formant frequencies of some fixed-mandible vowels and a model of speech motor programming by predictive simulation. *Journal of Phonetics*, *7*, 147–161.

MacNeilage, P. P., Rootes, T. P., & Chase, R. A. (1967). Speech production and perception in a patient with severe impairment of somesthetic perception and motor control. *Journal of Speech and Hearing Research*, *10*, 449–467.

McClean, M. D., Dostrovsky, J. O., Lee, L., & Tasker, R. R. (1990). Somatosensory neurons in human thalamus respond to speech-induced orofacial movements. *Brain Research*, *513*, 343–347.

Nasir, S. M., & Ostry, D. J. (2006). Somatosensory precision in speech production. *Current Biology*, *16*, 1918–1923.

Nasir, S. M., & Ostry, D. J. (2008). Speech motor learning in profoundly deaf adults. *Nature Neuroscience, 11*, 1217–1222.

Naumova, E. A., Dierkes, T., Sprang, J., & Arnold, W. H. (2013). The oral mucosal surface and blood vessels. *Head & Face Medicine, 9*, 8.

Penhune, V. B., & Doyon, J. (2005). Cerebellum and M1 interaction during early learning of timed motor sequences. *NeuroImage, 26*, 801–812.

Perkell, J. S. (1980). Phonetic features and the physiology of speech production. In B. Butterworth (Ed.), *Language production: Vol 1: Speech and talk* (pp. 337–372). New York: Academic Press.

Putnam, A. H., & Ringel, R. L. (1976). A cineradiographic study of articulation in two talkers with temporarily induced oral sensory deprivation. *Journal of Speech and Hearing Research, 19*, 247–266.

Ringel, R. L., & Steer, M. D. (1963). Some effect of tactile and auditory alterations on speech output. *Journal of Speech and Hearing Research, 6*, 369–378.

Saltzman, E. L., & Munhall, K. G. (1989). A dynamical approach to gestural patterning in speech production. *Ecological Psychology, 1*, 333–382.

Schliesser, H. F., & Coleman, R. O. (1968). Effectiveness of certain procedures for alteration of auditory and oral tactile sensation for speech. *Perceptual and Motor Skills, 26*, 275–281.

Seidler, R. D., Noll, D. C., & Thiers, G. (2004). Feedforward and feedback processes in motor control. *NeuroImage, 22*, 1775–1783.

Tourville, J. A., Reilly, K. J., & Guenther, F. H. (2008). Neural mechanisms underlying auditory feedback control of speech. *NeuroImage, 39*, 1429–1443.

Tremblay, S., Shiller, D. M., & Ostry, D. J. (2003). Somatosensory basis of speech production. *Nature, 423*, 866–869.

Voss, M., Ingram, J. N., Haggard, P., & Wolpert, D. M. (2006). Sensorimotor attenuation by central motor command signals in the absence of movement. *Nature Neuroscience, 9*, 26–27.

Wasaka, T., Hoshiyama, M., Nakata, H., Nishihira, Y., & Kakigi, R. (2003). Gating of somatosensory evoked magnetic fields during the preparatory period of self-initiated finger movement. *NeuroImage, 20*, 1830–1838.

Wasaka, T., Kida, T., Nakata, H., Akatsuka, K., & Kakigi, R. (2007). Characteristics of sensori-motor interaction in the primary and secondary somatosensory cortices in humans: a magnetoencephalography study. *Neuroscience, 149*, 446–456.

7

Feedforward Control

The average human is capable of remarkably fast and precise movements of the speech articulators. Many researchers have noted that the rate of conversational speech is too fast to rely entirely on feedback control mechanisms given the time lags inherent in neural processing of sensory inputs (e.g., Lashley, 1951; Lenneberg, 1967; Neilson & Neilson, 1987). Using our example from chapter 1, a speaker produces the word "dilapidated" in 1 second or less, implying an average movement time of less than 100 ms for each of the 11 phonemes in this word. In chapter 5 we noted that the auditory feedback control subsystem has an inherent delay of approximately 100–150 ms. This means that the entire pattern of muscle activities for producing a phoneme often needs to be generated *before* any auditory feedback from the commanded articulatory movements is available, eliminating the possibility of pure auditory feedback control. As Lashley (1951) points out, this fact implies the existence of "some central nervous mechanism which fires with predetermined intensity and duration or activates different muscles in predetermined order" (p. 516). In our terminology, this mechanism is the feedforward control system.

7.1 The Feedforward Control System

Figure 7.1 highlights the feedforward control system within the overall control system for speech introduced in chapter 3, section 3.3. Production of a sound begins when the *speech sound map* node for that sound is activated, leading to the readout from memory of a learned set of articulatory gestures for producing the syllable, which we term the *motor target* (M_T). The *feedforward controller* compares this motor target to the current motor state[1] (M) to generate feedforward movement commands ($\dot{M}_{FF}$) that sum with commands from the auditory and somatosensory feedback controllers to form the overall movement command to the vocal tract.

The feedforward motor commands for speech sounds cannot be entirely innate; they must be tuned during infancy and childhood to allow generation of acoustically appropriate phonemes and syllables from the child's native language using the child's particular vocal tract morphology. In the DIVA model, the motor target is tuned through teaching signals

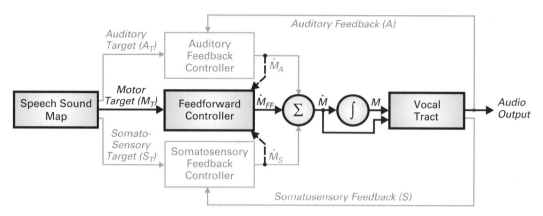

Figure 7.1
The feedforward control system highlighted within the overall control system for speech production introduced in chapter 3. M, motor positional command; $\dot{M}$, overall movement command; $\dot{M}_A$, corrective movement command issued by the auditory feedback controller; $\dot{M}_{FF}$, feedforward movement command; $\dot{M}_S$, corrective movement command issued by the somatosensory feedback controller.

(indicated by dashed lines in figure 7.1) that arise from the auditory and somatosensory feedback controllers. When the model attempts to produce a new speech sound, the contribution of the feedforward controller to the overall motor command will be imprecise since the motor target has not yet been tuned. Therefore, during the first few productions, the model will rely heavily on feedback control (see also Kawato & Gomi, 1992), in particular, auditory feedback control, since the system will not yet have a somatosensory target for the sound. The motor target and associated feedforward commands get better and better over time, all but eliminating the need for feedback-based control except when external constraints are applied to the articulators (e.g., a bite block) or auditory feedback is artificially perturbed. As the speech articulators get larger with growth, the auditory feedback control subsystem provides corrective commands that are eventually subsumed into the motor target, thus allowing the feedforward control system to stay tuned over the course of a lifetime.

A number of different proposals have been put forth regarding the nature of the feedforward controller for speech. Figure 7.2 illustrates perhaps the simplest and most straightforward feedforward control scheme[2] (Guenther, Ghosh, & Tourville, 2006). According to this scheme, the brain stores the motor commands it has learned for successfully producing a speech sound[3] as a motor target, M_T. The motor target specifies the time course of articulator positions that has been learned for producing the sound and can be thought of as the sound's *motor program* or *gestural score* (cf. Browman & Goldstein, 1990). The current motor state, M, is subtracted from M_T to obtain the change in motor state, ΔM, required to achieve the motor target. ΔM is then multiplied by a constant gain factor, α_{FF}, and then scaled by a *GO signal* (G, cf. Bullock & Grossberg, 1988) that arises from the

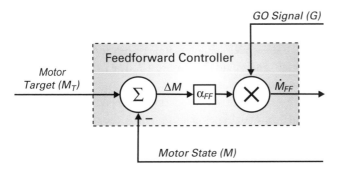

Figure 7.2
Expanded view of the feedforward control scheme implemented in the DIVA model. α_{FF}, feedforward gain factor; ΔM, motor error; $\dot{M}_{FF}$, feedforward movement command.

cortico–basal ganglia motor loop as described later in this chapter. The GO signal controls speaking rate and is typically constrained to range between 0 (no movement) and 1 (maximal movement speed). This control process is summarized by the following equation:[4]

$$\dot{M}_{FF} = \alpha_{FF} G [M_T - M]. \tag{7.1}$$

Thus far we have largely ignored processing delays inherent in feedback control for the sake of simplicity. However, these delays become important when considering how the motor target is updated. Recall that corrective motor commands generated by the auditory and somatosensory feedback controllers ($\dot{M}_A$ and $\dot{M}_S$, respectively) occur only *after* a sensory error has been detected, and furthermore these corrective responses are delayed due to the time it takes the nervous system to detect sensory errors and translate them into corrective motor commands. To avoid the same error in the future, the *corrective commands must occur earlier in time in future productions* by amounts equal to the auditory and somatosensory feedback controller delays,[5] which we will refer to as τ_A and τ_S, respectively. The following equation describes the desired update to the motor target (ΔM_T) given corrective motor commands $\dot{M}_A$ and $\dot{M}_S$, with the motor commands now explicitly treated as functions of time (t):

$$\Delta M_T(t) = \lambda_A \dot{M}_A(t + \tau_A) + \lambda_S \dot{M}_S(t + \tau_S), \tag{7.2}$$

where λ_A and λ_S are learning rate parameters that determine how much the motor target is updated on each production attempt based on the corrective commands generated by the auditory and somatosensory feedback control subsystems, respectively.[6] This equation states that the change in the motor target for the next production of the current sound is proportional to the corrective commands generated by the auditory and somatosensory feedback systems shifted back in time by τ_A and τ_S, respectively. In computer simulations of the DIVA model, this time alignment is done algorithmically; we discuss possible neural mechanisms that may underlie the time alignment process later in this chapter.

The results of simulations of the DIVA model using this type of feedforward controller to learn a new utterance ("good doggie") are provided in figure 7.3. The top plot shows the spectrogram of the acoustic signal created by a human speaker saying "good doggie." The model uses this signal as the auditory target for its production attempts. The remaining plots show the 1st, 3rd, 5th, 7th, and 9th model attempts to produce the utterance. In the

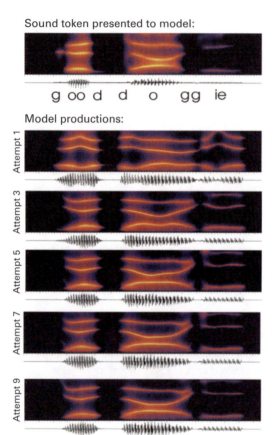

Figure 7.3
Spectrograms showing the first three formants of the utterance "good doggie" as produced by an adult male speaker (top panel) and by the DIVA model (bottom panels). The y-axis represents frequency, and the x-axis represents time. Hot colors represent frequencies with high energy in the acoustic signal, that is, formant frequencies. The model first learns an acoustic target for the utterance based on the sample presented to it (top panel). Then the model attempts to produce the sound, at first heavily under feedback control (attempt 1), then with progressively improved feedforward commands (attempts 3, 5, 7, and 9). By the 9th attempt the feedforward control signals are accurate enough for the model to reproduce the formant trajectories from the sample utterance with no auditory errors produced.

first attempt, overshoots of the formant frequencies can be seen; these overshoots, related to the auditory feedback control delay described above, are indicative of a heavy reliance on auditory feedback control mechanisms rather than feedforward control. With each trial, however, the feedforward control system subsumes corrective commands generated by the auditory feedback controller (shifted back in time by τ_A) to compensate for the auditory error signals that arose during that trial. This results in an improved feedforward command for the next trial. After a few iterations, the model's speech (bottom panels) closely follows the formant trajectories of the original target (top panel), indicating that an accurate motor target has been learned and the model is relying primarily on feedforward control.

Through this learning process, the motor target and feedforward movement commands for a speech sound are primarily shaped by the auditory target for the sound since the model learns to mimic the auditory trajectory specified by the auditory target. In other words, *planning of speech movements is carried out primarily in auditory space* (see chapter 4 for further details and supporting evidence). Only after many correct productions does the model learn an accurate somatosensory target for the sound. This does not imply that somatosensory information plays *no* role in learning of the feedforward commands for a speech sound. Viewing a fluent speaker's face can provide somatosensory information for certain sounds such as bilabials, whose articulation is clearly visible. Furthermore, past experience producing similar sounds can lead to partial somatosensory targets for a new sound. Though incomplete and possibly imprecise, these partial somatosensory targets may also contribute to shaping the motor target and feedforward commands for a new sound.

7.2 Behavioral Studies of Feedforward Control Mechanisms

Sensory Deprivation

A number of studies in the 1960s and 1970s investigated the feedforward control system by combining masking noise, which incapacitates the auditory feedback control subsystem, with local anesthesia, which incapacitates the somatosensory feedback system. The most salient result from these studies is that speech remains intelligible with relatively minor degradation. For example, Schliesser and Coleman (1968) noted that speech under masking noise and oral anesthesia exhibited a degree of degradation that was less than that of an individual with a clinically moderate speech problem. Ringel and Steer (1963) investigated the effects of two types of anesthesia, topical anesthesia and nerve block anesthesia, with and without masking noise. Nerve block anesthesia produced greater articulation error scores than topical anesthesia, and adding masking noise to either form of anesthesia resulted in further degradation of speech output. Different aspects of speech production are differentially impaired by sensory deprivation. Gammon et al. (1971) noticed no increase in vowel misarticulations under oral anesthesia and masking noise, but a 20% increase in misarticulation of consonants, with the most severe degradation occurring for fricative and affricates, presumably because these consonants rely more heavily on somatosensory feedback to attain the precise constriction sizes needed for frication. Notably, stress and juncture were not impaired by disruption of auditory and/or somatosensory feedback, suggesting that these aspects of speech production are controlled almost entirely in a feedforward manner.

Sensorimotor Adaptation

As discussed in chapter 5, applying a sustained perturbation to sensory feedback during speech—for example a consistent formant frequency shift or force applied to the jaw—leads to a sustained modification of the motor commands used to produce speech sounds, a process referred to as *sensorimotor adaptation*. In the DIVA model, sustained perturbation of auditory or somatosensory feedback will lead to reorganization of the feedforward commands for speech sounds. This occurs because the model's feedforward control system constantly monitors the corrective commands generated by the feedback control system, gradually incorporating repeatedly occurring corrections into the feedforward command. If the feedback perturbation is then removed, the adaptation will persist for a time; that is, the nonperturbed productions will still include the compensatory adaptation until the feedforward commands are retuned under normal feedback conditions. However, since the perturbation is no longer in place, these responses lead to sensory errors that are in a direction opposite to that of the errors created by the perturbation. For this reason they are often called *negative aftereffects*. In the following paragraphs we review studies of sustained somatosensory and auditory perturbation to gain insight into the process of tuning feedforward commands for speech sounds.

Adaptation to Somatosensory Perturbations As noted in chapter 6, speakers produce immediate compensatory articulations in response to a bite block (Lindblom, Lubker, & Gay, 1979). However, these compensations are not always complete, and additional compensation can occur after practice with a bite block. For example, McFarland and Baum (1995) found that a 15-minute conversation with a bite block led to improved vowel productions compared to before the conversation, which we interpret as the result of adaptation of feedforward commands for vowels with extended exposure. Notably, consonants did not show this effect in the McFarland and Baum (1995) study, suggesting that updates to the feedforward commands for consonants may occur more slowly. This hypothesis was supported by an experiment involving artificial palates that effectively change the distance between the tongue and the hard palate (Aasland, Baum, & McFarland, 2006). This study found that 1 hour of practice with the artificial palate led to improvements in fricative production. Furthermore, negative aftereffects were found when the palate was removed, indicating that the feedforward motor programs for fricatives were modified over time to accommodate the artificial palate.

David Ostry and colleagues have utilized sophisticated robotic perturbations to the jaw to demonstrate that perturbation of somatosensory feedback alone (i.e., in the absence of any acoustic change) is sufficient to induce adaptation in the motor commands for both vowels and consonants. Tremblay, Shiller, and Ostry (2003) performed an experiment in which jaw motion during syllable production (specifically, during a high-vowel to low-vowel transition) was modified by application of a force to the jaw that did not measurably affect the acoustics of the syllable productions. Subjects compensated for the jaw perturbation, indicating that they were using somatosensory targets like those in the DIVA model. Negative aftereffects were also found, indicating that adaptation had occurred. No compensation or adaptation occurred when the same jaw perturbation was applied to non-speech jaw movements. Nasir and Ostry (2006) extend this finding to consonant gestures.[7] Tremblay, Houle, and Ostry (2008) investigated the degree to which adaptation generalizes to utterances that were not produced during the jaw perturbation phase of the experiment. Adaptation in this study was highly specific to the utterance that was perturbed; even utterances with similar jaw movements to the perturbed utterance did not show adaptation. Furthermore, adaptation to a voiced version of the utterance did not transfer to a silent version and vice versa. Notably, the subjects trained on only a single utterance. The authors posit that generalization would likely be found if multiple utterances had been used during training, analogous with findings from arm movement perturbation experiments (e.g., Gandolfo, Mussa-Ivaldi, & Bizzi, 1996; Ghahramani & Wolpert, 1997; Malfait, Gribble, & Ostry, 2005; Mattar & Ostry, 2007). However, it is noteworthy that arm movement experiments involving training on a single movement (analogous to Tremblay, Houle, & Ostry, 2008) typically showed a small degree of generalization to movement patterns that were similar to, but not exactly the same as, the training movements, in contrast with the complete lack of generalization for speech movements noted by Tremblay, Houle, and Ostry (2008).

Adaptation to the jaw perturbations applied by Ostry and colleagues likely depended primarily on proprioception from muscle spindles rather than on tactile inputs from cutaneous mechanoreceptors. Ito and Ostry (2010) extended these results by demonstrating adaptation to a perturbation applied to the facial skin, which has plentiful cutaneous mechanoreceptors but essentially no muscle spindles.

One outstanding question regarding the modification of feedforward commands based on somatosensory errors concerns whether adaptation depends on the generation of real-time motor corrections by the feedback control system or whether the detection of sensory errors alone (i.e., in the absence of motor corrections) is sufficient to induce changes in the feedforward commands. Although data from speech studies on this topic are lacking, Tseng et al. (2007) found that adaptation of arm movements does not depend on actually generating corrective movements—sensory prediction errors are sufficient to induce adaptation.

Adaptation to Auditory Perturbations The first study to investigate the effects of sustained perturbation of formant frequencies during speech was performed by Houde and Jordan (1998), who modified the auditory feedback of talkers by shifting the first two formant frequencies of whispered monosyllabic utterances containing the vowel /ɛ/ in real time. Talkers adapted to the perturbation by changing the formants of their speech in the direction opposite the shift. Productions of the same utterances with auditory feedback masked by noise also showed the compensation, indicating adaptation of feedforward commands for these utterances. Similar results were found in the formant perturbation experiment of Purcell and Munhall (2006), who additionally demonstrated negative aftereffects that decayed gradually after normal auditory feedback conditions were restored.

Adaptation has also been noted for perturbations of auditory feedback involving parameters other than formant frequencies. Talkers will adapt to sustained perturbations of pitch, again with negative aftereffects when pitch feedback is returned to normal (Jones & Munhall, 2000). Jones and Munhall (2002) noted that adaptation to pitch perturbation also occurs in speakers of a tonal language (Mandarin). Shiller et al. (2009) shifted the centroid of the frequency spectrum while speakers produced words beginning with /s/, a shift that moves the sound /s/ toward the sound /ʃ/ (as in "s*h*e"). Analogous to the formant shift experiments described above, talkers adapted to the shift by adjusting their /s/ productions in the direction opposite the shift. This adjustment persisted after the perturbation was removed, resulting in a negative aftereffect. Interestingly, the auditory perceptual representation of the category boundary between /s/ and /ʃ/ also changed, indicating that adaptation is not limited to the motor domain but also involves changes in sensory representations.

Talkers will adapt to perturbations of auditory feedback regardless of whether they are conscious of the perturbation. In most of the studies mentioned above, talkers were unaware of the auditory perturbations because they were ramped up gradually. Keough, Hawco, and Jones (2013) investigated adaptation when talkers were explicitly informed that the pitch of their voice would be manipulated. Adaptation occurred regardless of whether the participants were told to compensate for the perturbation or to ignore it, indicating that sensorimotor adaptation to auditory perturbations is automatic and not under conscious control.

Interpreted within the DIVA model framework, the sensorimotor adaptation results described above indicate that the feedforward controller for speech continuously monitors auditory feedback and is modified when that feedback does not meet expectations, for example, because of a pitch or formant perturbation. Furthermore, the model predicts that speakers with relatively good auditory acuity (i.e., those who can detect small differences in auditory signals) will more readily detect the F1 changes that occur because of perturbation. Villacorta, Perkell and Guenther (2007) directly tested the DIVA model's account of sensorimotor adaptation to auditory perturbations in an experiment involving sustained perturbations of F1 of monosyllabic words involving the vowel /ɛ/. The auditory acuity of the experimental subjects for speech sounds similar to those presented in the experiment was first measured. Then the subjects performed a sensorimotor adaptation experiment that involved four phases during which subjects repeated a list of nine monosyllabic training words. Each repetition of the list will be referred to as an *epoch*. A *baseline phase* consisted of 15 epochs with normal auditory feedback. This was followed by a *ramp phase* of 5 epochs, over which a perturbation of F1 was gradually increased to a full shift of 30% of the original F1. This perturbation was maintained during a *training phase* of 25 epochs. Auditory feedback was then returned to normal for 20 epochs during the *posttest phase*. A measure of *adaptive response* (*AR*) was calculated as the percent change in F1 (compared to the average F1 during the baseline phase) in the direction opposite the perturbation.

The results of this study are provided in figure 7.4. AR (expressed in decimal form) is indicated by solid lines with standard error bars, and dashed vertical lines demarcate the four experimental phases. Adaptation was found in the first epoch of the ramp phase, suggesting that subjects adapt to even small shifts[8] in F1 and that a small number of perturbed productions (nine in this case) are sufficient to produce measurable adaptation. The amount of adaptation increased during the ramp and test phases before leveling off near the end of the training phase. As expected, a negative aftereffect was found as F1 only gradually returned to baseline values in the posttest phase. Furthermore, the amount a subject adapted to the perturbation was positively correlated with the subject's auditory acuity, though this

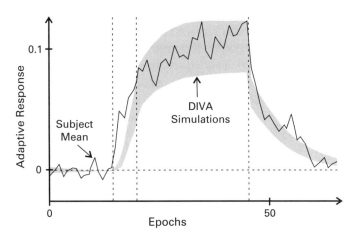

Figure 7.4
Adaptive response (AR) to systematic perturbation of F1 during a sensorimotor adaptation experiment (solid line) compared to DIVA model simulations of the same experiment (shaded area). The shaded region represents the 95% confidence interval derived from DIVA model simulations. The vertical dashed lines indicate the transitions between the baseline, ramp, training, and posttest phases over the course of the experiment (approximately 100 minutes' total duration). The horizontal dashed line indicates the baseline F1 value. Adapted from Villacorta (2006).

correlation was rather weak as will be discussed further below. Villacorta, Perkell, and Guenther (2007) also performed computer simulations of the DIVA model that provide a quantitative account of the experimental results. A version of the model was created for each experimental subject, with the auditory acuity of the subject determining the size of the model's auditory target regions for the experimental stimuli. The results of these simulations are indicated by the blue shaded area in figure 7.4, which represents the 95% confidence interval of the productions of the different versions of the model (see Villacorta, 2006, for details). With the exception of one epoch in the ramp phase, the model's productions did not differ significantly from the experimental results.

A common finding in auditory sensorimotor adaptation experiments is that the amount of adaptation levels off well before full compensation for the perturbation is attained. For example, in Villacorta, Perkell, and Guenther (2007) adaptation leveled off at about 35% to 50% of the applied F1 perturbation. This can be seen in figure 7.4, where the average AR at the end of the training phase was only about 12% for a 30% shift of F1. Similarly, adaptation to formant perturbations in Houde and Jordan (1998) and Purcell and Munhall (2006) corrected for only about 30% to 50% of the applied perturbation. This property is also inherent in the DIVA model, as illustrated by the blue shaded region in figure 7.4. The model's account of incomplete adaptation to auditory perturbations is as follows. Recall that both the auditory and somatosensory feedback control subsystems contribute to adaptation of feedforward commands. When auditory feedback is perturbed during production of a speech sound, the perceived auditory signal is pushed outside the auditory target for the sound, but initially somatosensory feedback remains within the somatosensory target region. However, as the feedforward command for the sound adapts to the auditory perturbation, somatosensory feedback during production starts to mismatch the somatosensory target for the sound. This causes the somatosensory feedback control subsystem to kick in, leading to corrective motor commands that counteract the corrective commands generated by the auditory feedback control subsystem. At some point the corrective commands from the auditory and somatosensory feedback control subsystems cancel each other out and adaptation stops prior to full compensation to the auditory perturbation.

Relatedly, Lametti, Nasir, and Ostry (2012) performed an experiment involving two types of perturbation applied separately and in combination: an acoustic perturbation of F1 and a jaw perturbation that did not affect F1. For a given subject, the amount of compensation to both perturbations combined was about the same as for each perturbation individually. Interestingly, there was a negative correlation across subjects between the amount of compensation for the auditory perturbation and for the somatosensory perturbation. That is, the subjects who adapted most for the jaw perturbation tended to show little or no compensation for the F1 perturbation, and vice versa. These results indicate that (1) auditory and somatosensory feedback control mechanisms for speech appear to operate somewhat independently (as in the DIVA model) and (2) adult subjects vary in the degree to which they adapt feedforward commands in response to auditory versus somatosensory perturbations. This last finding may reflect differences in auditory and somatosensory acuity across subjects. However, studies that have looked for correlations between sensory acuity and amount of adaptation have tended to show either weakly significant correlations (Villacorta et al., 2007) or no significant correlations (Feng et al., 2011). In another study that combined somatosensory and auditory feedback perturbations, Feng et al. (2011) found that speakers do not adapt to a jaw perturbation if they receive a simultaneous auditory perturbation in the opposite direction, leading the authors to conclude that the auditory modality likely plays a dominant role in speech sensorimotor adaptation.

A final point of interest regarding auditory perturbation experiments is the degree to which adaptation generalizes to sounds/words not produced under perturbed conditions. Houde and Jordan (1998) noted generalization to words with the same vowel as the training utterances but in a different consonant context as well as words with the same consonants as a training word but with different vowels. This generalization amounted to approximately 30% to 70% of the adaptation in the training words. Villacorta, Perkell, and Guenther (2007) noted a similar amount of adaptation generalization to utterances that were similar to those in the training set. Jones and Munhall (2005) noted that adaptation to a pitch perturbation in Mandarin speakers generalized to an utterance with the same phonemes but a different tonal profile, though this adaptation extinguished more quickly than for the training utterance when feedback was returned to normal. Together, these findings indicate partial generalization of adaptation to sounds/words that are similar to the training utterances.

Motor Equivalence

In chapter 4, section 4.3, we introduced the concept of *motor equivalence*, which is the ability to carry out the same task using different motor means. In chapter 5, section 5.1, and chapter 6, section 6.1, we noted how the auditory and somatosensory feedback controllers can explain a number of motor equivalence phenomena that occur when perturbations are applied to the auditory signal or speech articulators. In the DIVA model this property arises through the use of a directional mapping between errors in sensory space and corrective movement velocities, as described in note 3 of chapter 5 and detailed elsewhere (Guenther, 1995; Guenther, Hampson, & Johnson, 1998; Guenther, Ghosh, & Tourville, 2006). However, motor equivalence is not limited to cases of feedback control—it is also inherent in feedforward commands. An example of this was provided in chapter 4, section 4.3, where we noted that the articulator movements used to produce the phoneme /r/ are highly dependent on the phonetic context in which /r/ appears, even when no perturbations are applied.[9]

The DIVA model's account of this phenomenon is as follows. We posit that every common syllable of a language has an optimized motor program in the mature brain. For example, the feedforward commands for producing /grɑ/ and /drɑ/ are stored as distinct motor programs. The tuning of feedforward commands for syllables depends very heavily on auditory feedback control. Specifically, for a new syllable, an auditory target is first learned, and then the auditory feedback control subsystem is used to generate corrective commands to achieve that auditory target. These corrective commands are used to update the feedforward commands, and eventually the feedforward commands become sufficient by themselves to produce the sound. In essence, the feedforward control system is "taught" how to produce a speech sound by the auditory feedback control subsystem. In chapter 4, section 4.3, we described how the auditory feedback control subsystem will generate different motor commands for /r/ in different phonetic contexts. Because the tongue shapes for /g/ and /d/ differ dramatically, the most efficient movements to get to the auditory target for /r/ from those configurations is very different (see figure 4.3 in chapter 4). Different articulatory movements for /r/ will thus get encoded in the motor programs for /grɑ/ and /drɑ/. Through this process, the speech motor system learns highly efficient feedforward commands for producing commonly occurring syllables. In many cases this leads to *coarticulation*, which was introduced in chapter 4 and is described in the next subsection.

Coarticulation
Carryover coarticulation refers to cases where the vocal tract configuration for one phoneme influences the configuration for a later phoneme. The /r/ example just described is a case of carryover coarticulation since the tongue configurations used to produce /r/ in different contexts are related to the tongue position for the phoneme that preceded the /r/. Chapter 4, section 4.4, illustrated how the DIVA model accounts for carryover coarticulation since the point in the target region that is used to produce a sound depends on the vocal tract configuration for the preceding sound; specifically, the model moves from the current vocal tract configuration to the closest point on the target region.

Anticipatory coarticulation refers to cases where one phoneme influences the configuration of a preceding phoneme. At least some instances of anticipatory coarticulation (and carryover coarticulation, for that matter) can be explained because the auditory targets a developing speaker learns from fluent speakers already have anticipatory coarticulation built into them. For example, when fluent speakers of American English pronounce "coo," lip rounding begins during the /k/, whereas when they pronounce "key," lip spreading begins during the /k/. Although the phoneme /k/ is correctly perceived by listeners in both cases, there are acoustic differences between the two /k/ productions (i.e., they are *allophones*) because of the differences in lip configuration. These acoustic differences will be encoded in the auditory targets for "coo" and "key," and achieving these auditory targets during production will therefore entail anticipatory lip rounding. This type of anticipatory coarticulation occurs naturally in the DIVA model.

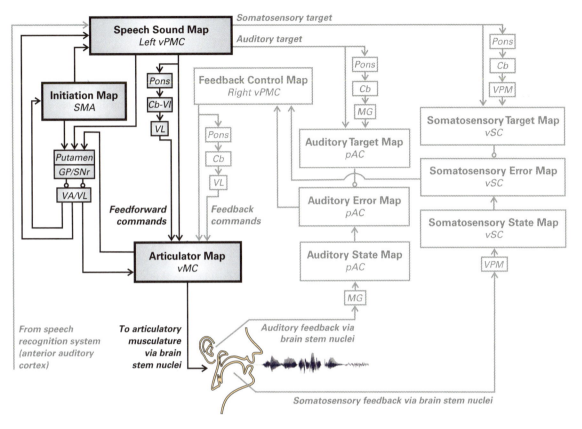

Figure 7.5
Neural components of the feedforward control system according to the DIVA model. Cb-VI, cerebellum lobule VI; GP, globus pallidus; SMA, supplementary motor area; SNr, substantia nigra pars reticulata; VA, ventral anterior nucleus of the thalamus; VL, ventral lateral nucleus of the thalamus; vMC, ventral motor cortex; vPMC, ventral premotor cortex.

Although this explanation provides a plausible account of many examples of anticipatory coarticulation, it begs the question of how anticipatory coarticulation came to exist in the first place. One possibility is that anticipatory coarticulation arises from a general tendency of the speech motor system to produce the movements within a movement sequence (e.g., the articulatory gestures within a syllable) closer together in time with each iteration until they start to interfere with successful production of the goal(s) of the movement sequence. Such a property would also account for the "sliding back in time" of motor commands generated by the feedback control system when tuning feedforward commands mentioned earlier in this chapter. Below we identify possible neural mechanisms for this property.

7.3 Neural Circuits Underlying Feedforward Control of Speech

The neural substrates underlying feedforward control in the DIVA model are illustrated in figure 7.5. The model's feedforward control system consists of two interacting subsystems: an *initiation circuit*, responsible for selecting and initiating the motor program for the sound to be produced, and an *articulation circuit*, responsible for generating the time-varying motor commands for producing the sound. The initiation circuit heavily involves the *supplementary motor area* (*SMA*) and the basal ganglia, particularly the putamen and *globus pallidus* (*GP*). The articulation circuit involves the *ventral motor cortex* (*vMC*) and cerebellum, particularly *lobule VI* (*Cb-VI*). Both circuits involve the left *ventral premotor cortex* (*vPMC*) as well as the *ventral lateral* (*VL*) *nucleus* of the thalamus. These circuits are addressed in the following subsections. Stereotactic coordinates associated with the model components in figure 7.5 are provided in appendix D.

Initiation Circuit

In the DIVA model, the initiation circuit is responsible for (1) selecting the proper motor program to produce next and (2) initiating the motor program (which is generated by the articulation circuit) at the proper instant in time. We propose that the speech initiation circuit corresponds to the cortico–basal ganglia motor loop introduced in chapter 2, section 2.3, where it was noted that the architecture of this loop—in particular the large amount of "funneling" of information from a wide expanse of cerebral cortex into a small number of basal ganglia output channels—is suitable for selecting one output from a set of competing alternatives but poorly suited for generating precise motor commands to the speech articulators. In other words, the basal ganglia motor loop is more likely involved in initiating motor programs that are stored elsewhere in the brain, specifically in the articulation circuit discussed in the following subsection.

At the heart of the initiation circuit is an *initiation map*, hypothesized to reside in SMA. As noted in chapter 2, damage to SMA is associated with transient total mutism, followed by a transcortical motor aphasia involving a decline in propositional (self-initiated) speech with largely spared nonpropositional speech (overlearned and automatic speech such as counting or repeating words) and only rare distorted articulations. In keeping with these findings, we posit that initiation of speech sound production requires the activation of nodes in the initiation map. Furthermore, initiation map nodes do not themselves encode the time course of articulator movements used to produce the sounds—instead they activate motor and premotor cortex nodes in the articulation system that generates the motor commands for the sound being produced. This is carried out via the GO signal (see figure 7.2 and equation 7.1) in the DIVA controller.

Activation of SMA initiation map nodes can happen in two different ways. First, higher-level circuitry involved in the sequencing of speech sounds, located in the preSMA, can activate initiation map nodes through cortico-cortical connections. We consider this pathway to be involved in speech sound sequencing, which is outside the purview of the DIVA model and is instead treated by the GODIVA model, which will be covered in chapter 8. These connections are thus omitted from figure 7.5.

The second way an initiation map cell can be activated is via the basal ganglia motor loop. An expanded schematic of this loop is presented in figure 7.6. The *putamen* (*Pu*), which is part of the striatum of the basal ganglia, receives excitatory inputs from neurons in several regions of the cerebral cortex, most notably the primary motor cortex (BA4), primary somatosensory cortex (BA 1, 2, and 3), and premotor areas SMA (medial BA 6)

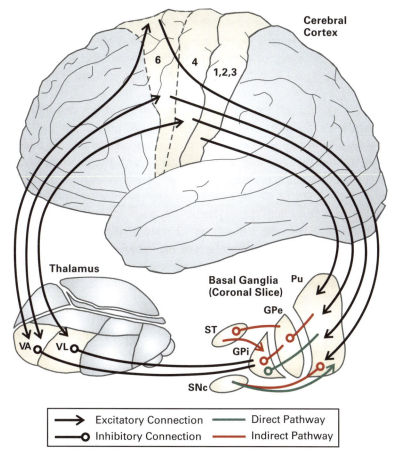

Figure 7.6
Simplified schematic of the cortico–basal ganglia–thalamo–cortical motor circuit (cf. Alexander and Crutcher, 1990; Smith et al., 1998). Excitatory projections from cerebral cortex to ST and from thalamus to Pu are omitted for clarity. Double-headed arrows represent bidirectional projections. GPe, external segment of the globus pallidus; GPi, internal segment of the globus pallidus; Pu, putamen; SNc, substantia nigra pars compacta; ST, subthalamic nucleus; VA, ventral anterior thalamic nucleus; VL, ventral lateral thalamic nucleus.

and vPMC (lateral BA 6). It has been noted that the convergence of a wide range of sensorimotor and cognitive information in the striatum, including Pu, makes this structure well-suited for monitoring cortex for the proper cognitive and sensorimotor context for launching a goal-directed action (Mink, 1996). The output neurons of the basal ganglia are located in two closely related structures: the *internal segment of the globus pallidus* (*GPi*) and the *substantia nigra pars reticulata* (*SNr*; not shown in figure 7.6). Both GPi (Manes et al., 2014) and SNr (Inchul et al., 2005) have been implicated in speech/orofacial movements.

Pu projects to GPi-SNr via two pathways: a *direct pathway* (indicated in green in figure 7.6) and an *indirect pathway* (indicated in red in figure 7.6). The direct pathway involves inhibitory axonal projections from Pu to GPi-SNr. The indirect pathway involves inhibitory projections from PU to neurons in the *external segment of the globus pallidus* (*GPe*), which in turn sends inhibitory projections to *subthalamic nucleus* (*ST*) and to GPi-SNr. The indirect pathway also includes diffuse excitatory projections from ST to GPi-SNr.

Neural activity in GPi-SNr reflects the combined effects of the direct and indirect pathways. GPi-SNr then sends inhibitory projections to the VL and *ventral anterior* (*VA*) nuclei of the thalamus, which in turn send excitatory projections back to primary motor cortex (primarily from VL) and premotor cortex (primarily from VA). Because it involves two inhibitory projections in sequence, *the main effect of the direct pathway on cortex is excitatory*, whereas *the main effect of the indirect pathway on cortex is inhibitory* since it involves three inhibitory projections in sequence. Because the indirect pathway projections from ST to GPi-SNr are more diffuse than the more localized direct pathway projections from Pu to GPi-SNr, the inhibitory effect of the indirect pathway on cortex is likely relatively broad, having an inhibitory effect on many competing motor programs, compared to the focalized excitatory effect of the direct pathway on the chosen motor program.

The direct and indirect pathways of the basal ganglia are differentially modulated by dopaminergic inputs from the *substantia nigra pars compacta* (*SNc*) to the striatum, including Pu (see figure 7.6). Dopamine has an excitatory effect on Pu neurons in the direct pathway and an inhibitory effect on neurons in the indirect pathway. Both of these actions have an excitatory effect on cerebral cortex and movement via the cortico–basal ganglia motor loop. Dopaminergic signals from SNc to striatum are thought to play an important role in motor learning, perhaps acting as a *teaching signal* that strengthens successful actions within the current cognitive and sensorimotor context. This hypothesis is supported by the study of Shan et al. (2014), who demonstrated that learning goal-directed actions induces an increase in synaptic excitability in direct pathway striatal neurons and a decrease in synaptic excitability in indirect pathway striatal neurons in mice.

Additional pathways not illustrated in figure 7.6 also exist within the cortico–basal ganglia motor loop. ST receives fast excitatory projections from motor areas of cerebral cortex which, along with excitatory projections from ST to GPi, form a *hyperdirect pathway* that may be involved in inhibiting large areas of cortex related to potentially relevant motor programs before the correct motor program is activated via the direct pathway (Mink, 1996; Nambu, Tokuno, & Takada, 2002). Also, projections from thalamic nuclei VA and VL to Pu provide a means for output signals from both basal ganglia and cerebellum to contribute contextual information to Pu that supplements contextual information from cerebral cortex. This would, for example, allow precisely timed cerebellar outputs to serve as indicators of the correct sensorimotor context for subsequent movements in a movement sequence.

Mink (1996) proposed an influential model of the basic function of the cortico–basal ganglia motor loop. According to this model, the initiation of a movement or movement sequence starts with activity in premotor and motor cortical areas, including SMA, vPMC, and *motor cortex* (*MC*). These regions send projections to ST, which produces a rapid and broad inhibition of movements related to the current cognitive and sensorimotor context via the hyperdirect pathway. This is followed by focused excitation of the proper movement for the current context via the direct pathway. Further focusing of the motor output, including inhibition of competing motor programs, occurs via the indirect pathway. In the words of Mink (1996, p. 414), "The net result of basal ganglia activity during voluntary movement is the braking of competing motor patterns and focused release of the brake for the selected voluntary movement pattern generators." For speech, the competing motor patterns may correspond to other phonemes/articulatory gestures represented in MC and/or other syllables represented in vPMC.

In addition to a role in enabling the proper action and suppressing competing actions, the cortico–basal ganglia motor loop also appears to play a role in the sequencing of motor actions (Marsden, 1987; Brotchie, Iansek, & Horne, 1991). Given that most movement-related basal ganglia neurons begin firing shortly after onset of the first movement in a sequence, it is unlikely that the cortico–basal ganglia motor loop is responsible for initiating this first movement (Mink, 1996). Instead, the basal ganglia may be responsible for initiating subsequent movements in a well-learned motor sequence. Within the context of syllable production, this suggests that cortical mechanisms may be responsible for initiating the first articulatory gesture in the syllable whereas the basal ganglia initiate subsequent gestures at the correct instant in time, that is, when the precise sensorimotor context for initiating the gesture is detected in Pu. In keeping with this view, Brotchie, Iansek, and Horne (1991) noted that many basal ganglia neurons fire just before the end of each component movement of a motor sequence, in effect acting as "completion signals" indicating the end of the current movement so the next movement can initiated. The role of basal ganglia in motor sequencing will be addressed in more detail in chapter 8, and in chapter 10 we will discuss malfunction of the cortico–basal ganglia motor loop as a potential underlying cause of several speech disorders, including hypokinetic and hyperkinetic dysarthria and stuttering.

Articulation Circuit

According to the DIVA model, the readout of feedforward motor commands for speech sounds begins with activation of speech sound map neurons in left ventral premotor cortex. The fact that apraxia of speech, which involves an inability to generate appropriate motor programs for speech sounds, occurs with damage to the language-dominant cortical hemisphere but not the nondominant hemisphere indicates that, at some level, motor programs for speech are left lateralized. Neuroimaging studies of short speech utterances (see figure 2.14 in chapter 2) provide evidence that this level is vPMC, including the rostral precentral gyrus, posterior portions of the inferior frontal gyrus, and anterior portions of the insula. These areas show left-lateralized activity during normal (unperturbed) speech, suggesting their involvement in feedforward control, which dominates during unperturbed speech in the fully developed system. Furthermore, Peeva et al. (2010) provide evidence for a syllable-level representation in left vPMC, as would be expected for a region representing the motor programs for speech (which are predominantly syllabic),[10] but no syllable representations in the right hemisphere.

The model proposes that projections from the speech sound map to the *articulator map* in bilateral primary motor cortex constitute the feedforward motor commands for a speech sound (see figure 7.5). Using magnetoencephalography during single-word production, Salmelin et al. (2000) identified activity in left inferior frontal cortex and premotor cortex prior to bilateral motor cortical activation, in keeping with this view. From motor cortex, the motor commands are sent to the brain stem motor nuclei via the corticobulbar tract. As discussed in chapter 2, each hemisphere of motor cortex sends projections to brain stem motor nuclei bilaterally, though with a contralateral bias for some articulators. The primary motor and premotor cortices are well-known to be strongly interconnected (e.g., Passingham, 1993; also see cortico-cortical functional and structural connectivity maps provided in appendix C, section C.1). These connections include direct cortico-cortical projections as well as projections via the cerebellum. The cerebellum has long been associated with both motor learning and learning of precise timing between motor events. In accord with this view, damage to the superior paravermal region of the cerebellar cortex, particularly in/near lobule VI, results in ataxic dysarthria, a motor speech disorder characterized by slurred, poorly coordinated speech (e.g., Ackermann et al., 1992). This topic will be addressed further in chapter 10.

The central components of the cortico-cerebellar loop involved in generating feedforward commands for speech according to the DIVA model are schematized in figure 7.7. Projections from vPMC to the *pons* represent the speech sound being produced. Additional projections to the pons from auditory, somatosensory, and motor/premotor cortical areas (not shown) provide important contextual information concerning the ongoing articulation. As described in chapter 2, mossy fibers arising from the pons distribute this information over a large number of *granule cells* in the cerebellar cortex in a *sparse coding* scheme. Cerebellar *Golgi cells* (not shown), which receive input from granule

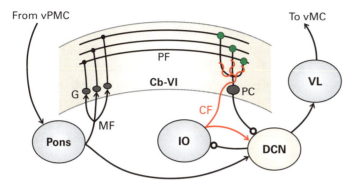

Figure 7.7
Schematic of the main components of the cerebellar loop involved in speech production. Green circles represent parallel fiber–Purkinje cell synapses thought to be central to motor learning. Noncerebellar structures (blue) are contralateral to the cerebellar structures (beige). Inhibitory interneurons are omitted for clarity. Cb-VI, cerebellum lobule VI; CF, climbing fiber; DCN, deep cerebellar nuclei; G, granule cells; IO, inferior olive; MF, mossy fibers; PC, Purkinje cell; PF, parallel fibers; VL, ventral lateral nucleus of the thalamus; vMC, ventral motor cortex; vPMC, ventral premotor cortex.

cells and mossy fibers, may contribute to the sparseness of this code through inhibition of granule cells. The granule cells give rise to *parallel fibers*, which course along the cerebellar cortex (particularly lobule VI for speech), making contact with a large number of Purkinje cells that are oriented perpendicularly to the parallel fibers. This geometry allows each Purkinje cell to receive input from a large number of parallel fibers, in effect providing the Purkinje cell with a wealth of information regarding the current sensorimotor context. Each Purkinje cell becomes sensitive to a particular sensorimotor context that is appropriate for the articulatory musculature associated with the cell; for example, Purkinje cells whose outputs affect voicing will become sensitive to sensorimotor contexts that call for voicing. The outputs of the Purkinje cells project through inhibitory synapses (indicated by circles) to the *deep cerebellar nuclei*[11] (*DCN*), which in turn project to the motor cortex via the VL nucleus of the thalamus.

Motor learning in the cerebellum is thought to heavily involve changes in the strengths of the parallel fiber–Purkinje cell synapses. Two different learning processes have been clearly identified at these synapses. The first, called *long-term potentiation* (*LTP*), occurs when a parallel fiber and Purkinje cell are active at the same time. This causes an increase in the synaptic weight; that is, the excitatory effect of the parallel fiber on the Purkinje cell is enhanced. The second, called *long-term depression* (*LTD*), involves a decrease in synaptic efficacy and is heavily influenced by inputs to the cerebellum from the *inferior olivary nucleus*, also called the *inferior olive* (*IO*). IO is the largest nucleus of the brain stem, and it has been associated with error signals in motor learning and timing (e.g., DeZeeuw et al., 1998; Rhodes & Bullock, 2002). Ablation of IO results in ataxia-like symptoms similar to those induced by ablation of the contralateral cerebellum (to which it projects). IO neurons send projections called *climbing fibers* to Purkinje cells in the cerebellar cortex as well as to the DCN. Each Purkinje cell receives input from only one climbing fiber, while each climbing fiber can project to several Purkinje cells. The synaptic connection between a climbing fiber and a Purkinje cell is one of the strongest connections in the nervous system, with a single action potential from a climbing fiber resulting in a *complex spike* consisting of a burst of action potentials from the Purkinje cell as well as a decrease in the strength of synapses from parallel fibers that are active during the complex spike. This LTD process in effect "punishes" the Purkinje cell for firing in the current sensorimotor context, making it less likely to do so in the future.

To understand how LTD at the parallel fiber–Purkinje cell synapse can contribute to learning of precisely timed motor commands, consider the scenario schematized in figure 7.8, which involves learning of the intergestural interval between two gestures, G1 and G2, that occur sequentially within a speech sound. To fix ideas, consider G1 to be the lip closure gesture for the stop consonant /p/ in the syllable /pa/ and G2 to be the onset of voicing for the /a/. The schematic represents a simplified version of the dynamics of the *recurrent slide and latch model* of cerebellar learning (Rhodes & Bullock, 2002). Shown are six consecutive production attempts, with the first attempt indicated by the top G1-G2 pair. The shaded area in the figure denotes the acceptable range of the intergestural interval (denoted by

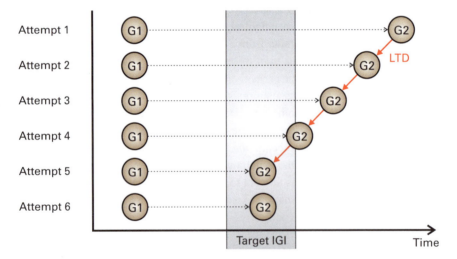

Figure 7.8
Schematic of hypothesized effect of long-term depression (LTD) in parallel fiber–Purkinje cell synapses on the intergestural timing (cf. Rhodes & Bullock, 2002) between two speech gestures (G1 and G2). After each of the first four production attempts, climbing fiber error signals are produced because the intergestural interval (IGI) is outside the target range (shaded area). These error signals result in LTD, causing the second gesture (G2) to occur slightly earlier in time (see the text for details). On the fifth attempt, G2 occurs within the target IGI range. Thus no error signal or LTD occurs, and the IGI remains "latched" at the proper value for subsequent attempts.

dashed arrows), which in this case corresponds to the acceptable range of voice onset time for the unvoiced consonant /p/. This target interval has been learned from listening to the productions of fluent speakers of the language. Furthermore, assume that the intergestural interval for a production attempt is determined by the delay between the arrival of the sensorimotor context associated with G1 on the parallel fibers and the *pause in firing* of a Purkinje cell affiliated with G2. This pause in firing releases DCN neurons corresponding to G2 from inhibition, allowing them to become active and thereby initiate G2 through projections to motor cortex. Finally, assume that the climbing fiber associated with this Purkinje cell becomes active whenever a production attempt falls above the acceptable intergestural interval range, thereby providing an error signal that induces LTD in the parallel fiber–Purkinje cell synapses.

In the first four production attempts (the top four G1-G2 pairs in figure 7.8), the intergestural interval is longer than the target range for this interval. Each time, LTD (indicated by the red arrow) occurs in the synapses between parallel fibers representing the current sensorimotor context and the Purkinje cell associated with G2. This has the effect of making the Purkinje cell pause its firing slightly earlier relative to G1 on the subsequent attempt, thereby releasing G2 slightly earlier. On the fifth production attempt, G2 occurs within the target intergestural interval, so no error signal or LTD is generated. As a result, the intergestural interval remains "latched" at this value for subsequent production attempts.

The process characterized in figure 7.8 has the overall effect of *sliding subsequent gestures in a sequence earlier in time* until they are within an acceptable intergestural interval. Section 7.1 described how such a process was necessary for adjusting feedforward commands based on corrective motor commands generated by the auditory and somatosensory feedback control subsystems. The process can also account for anticipatory coarticulation since gestures from future phonemes can slide forward in time as long as they do not interfere with successful production of the current phoneme. This, in turn, leads to the prediction that coarticulation should be evident in motor cortical cell activities since the cerebellar output projects back to motor cortex. This prediction was recently verified in an electrocorticography study of speech production (Bouchard & Chang, 2014). There is also abundant evidence that cerebellar outputs play a causal role in generating anticipatory components of motor cortex activations. For example, anticipatory, but not other, movement components disappear when the cerebellum's dentate nucleus is temporarily disabled (Vilis & Hore, 1980).

Another way to view the cerebellar contribution to motor learning arises from considering that the net effect of the cortico-cerebellar loop is inhibitory on motor output. From this view we can see that LTP in the parallel fiber–Purkinje cell synapses will result in a decrease in overall muscle activity and corresponding force, while LTD induced by climbing fiber error signals would increase force. In other words, the net effect of the LTP/LTD dynamic is to slowly and continuously decrease the force used for movements until such decreases induce an error signal, at which time force is increased again to maintain successful completion of the movement goal. This process can account for the common observation that the muscle force used to perform a motor task decreases with practice, leading to more and more efficient movements over time.

In chapter 1, section 1.1, we introduced the difference between *inverse kinematics* (involving generation of desired joint angles or muscle lengths to achieve a spatial or auditory goal) and *inverse dynamics* (involving generation of appropriate muscle forces or motoneuron activation levels to achieve target joint angles or muscle lengths given articulator mass, external loads, inertia, gravity, and other force-related variables). Although this book focuses primarily on inverse kinematics, learning of the feedforward motor commands for speech also includes inverse dynamics. To date, the neuroimaging literature on cerebellar contributions to learning of inverse kinematics and dynamics has primarily focused on studies of reaching. These studies indicate that kinematic and dynamic learning processes can occur largely independently (Krakauer et al., 1999), and although they both involve the cerebellar cortex, including lobules V and VI (Diedrichsen et al., 2005; Donchin et al., 2012), they may involve somewhat distinct subregions within the superior cerebellar cortex (Donchin et al., 2012). More posterior regions of the cerebellar cortex, including lobules VII and VIII, may also be involved in these learning processes, but the findings for these regions are not as consistent.

7.4 Summary

Speech is carried out through remarkably fast and precise articulations, a feat which cannot be achieved without feedforward control mechanisms that do not rely on relatively slow sensory feedback channels. The feedforward controller must learn appropriate motor commands, or *motor programs*, for producing sounds of the native language. According to the DIVA model, this occurs by incorporating motor commands generated by the auditory and somatosensory feedback control subsystems to correct for perceived errors on the current production attempt into the feedforward motor commands for the next production attempt. To avoid the same errors in the future, the corrective commands must occur earlier in time in future productions; in other words, they must be "slid earlier in time" by the feedforward control mechanism. Although both the auditory and somatosensory feedback control subsystems contribute to the learning of feedforward commands, in the early stages of development the auditory feedback control subsystem is the primary influence. This is because auditory targets can be learned simply by listening to fluent speakers whereas somatosensory targets can be learned only through correct self-productions.

A number of behavioral experimental paradigms have been used to identify properties of the feedforward controller for speech. Studies that combine masking noise with local anesthesia indicate that speech remains largely intelligible with only minor degradation, indicating that feedforward control mechanisms by themselves are accurate enough to produce intelligible speech in the fully developed brain. The errors that do occur are primarily on fricative and affricate consonants, which rely more heavily on sensory feedback for correct production than vowels and stop consonants because of the requirement of precise control of vocal tract constrictions to produce frication.

Sensorimotor adaptation studies indicate that sustained perturbations applied to either auditory or somatosensory feedback result in adaptation of the feedforward commands to counteract the perturbation. These studies verify that both auditory and somatosensory feedback controllers are involved in updating feedforward motor commands for speech in the fully developed brain. Adaptation to auditory feedback perturbations is typically incomplete, likely because adaptation to the auditory perturbation leads to somatosensory feedback that mismatches the somatosensory target, causing the auditory and somatosensory feedback controllers to generate commands that cancel each other out. After removal of a sustained perturbation, negative aftereffects are seen in the first few speech productions, indicating that the feedforward commands have been adjusted in response to the perturbations. These aftereffects decay exponentially as the speaker re-adapts to normal feedback conditions. The learning curves exhibited by subjects in sensorimotor adaptation studies can be accurately accounted for by a learning process that involves updating feedforward commands based on the motor commands generated by auditory and somatosensory feedback controllers, as in the DIVA model.

Speech production studies have identified two ubiquitous aspects of speech articulations that contribute substantially to speed and efficiency: *motor equivalence* and *coarticulation*. Motor equivalence refers to the use of different articulations to produce the same sound, for example, in different phonetic contexts. The DIVA model accounts for this phenomenon because feedforward commands are tuned by the auditory feedback controller early in development, and the same auditory target is reached by different articulations when starting from different initial configurations of the vocal tract.

Coarticulation refers to cases where articulations for neighboring sounds overlap. In *carryover coarticulation*, aspects of the vocal tract configuration for one phoneme continue into the next phoneme. The DIVA model accounts for carryover coarticulation because the point in the target region that is used to produce a sound depends on the vocal tract configuration for the preceding sound; the model moves from the current vocal tract configuration to the closest point on the target region. In *anticipatory coarticulation*, aspects of the vocal tract configuration for a future phoneme begin during a preceding phoneme. Since the sound examples provided to a developing child by fluent speakers are already coarticulated, learning to replicate these sound targets will lead to similar anticipatory coarticulation. Another possible source of anticipatory coarticulation may be a general tendency to slide subsequent gestures in a sequence earlier and earlier in time (as described above for updating feedforward commands) until they start to interfere with preceding phonemes.

The neural mechanisms underlying feedforward control can be broken down into two subsystems: an *initiation circuit* and an *articulation circuit*. The initiation circuit is responsible for initiating motor programs for speech sounds but not for generating the detailed motor commands. According to the DIVA model, an *initiation map* in supplementary motor area is responsible for initiating motor programs. Neurons in this map can be activated via two routes: via cortico-cortical projections from the pre-supplementary motor area, a higher-level region involved in speech sequencing, or via the cortico–basal ganglia motor loop. Early in learning, the pre-supplementary motor area is heavily involved in initiating the individual articulatory gestures making up a speech sound, but with practice the cortico–basal ganglia loop takes over this process, making it more "automatic" and thereby freeing up pre-supplementary motor area and other sequencing circuitry in the cortex for higher-level aspects of sound sequencing as discussed further in the next chapter. The loop through the basal ganglia involves two competing pathways: a *direct pathway* responsible for activating the appropriate motor program, and an *indirect pathway* that suppresses competing motor programs. Dopamine inputs from the substantia nigra pars compacta are important for learning in the basal ganglia, acting as *teaching signals* that strengthen successful actions within the current cognitive and sensorimotor context. This is done by increasing the strengths of synapses within the direct pathway and decreasing the strengths of synapses in the indirect pathway.

The articulation circuit is responsible for generating the precisely timed muscle activations that make up the motor programs for speech sounds. According to the DIVA model, the articulation circuit involves projections from a *speech sound map* in left ventral premotor cortex to an *articulator map* in bilateral primary motor cortex, including both cortico-cortical projections as well as projections through the pons, cerebellum (particularly paravermal lobule VI), and the ventral lateral nucleus of the thalamus. This cortico-cerebellar loop is ideally suited to learning precisely timed motor commands through long-term depression and long-term potentiation of synapses between cerebellar parallel fibers and Purkinje cells. Learning in these synapses may also be responsible for sliding subsequent gestures earlier in time as needed for tuning feedforward commands based on corrective commands from the sensory feedback controller. This same process also may be responsible for anticipatory coarticulation by sliding gestures earlier and earlier in time until they start to interfere with preceding phonemes.

Notes

1. The projection of motor state information to the feedforward controller is omitted from figure 7.1 for clarity. This motor state information can come from motor cortical cells as well as somatosensory information regarding muscle state.

2. One alternative scheme uses a forward model to predict the sensory outcome of ongoing motor commands, then compares the predicted sensory state to the sensory target to generate "corrective" motor commands

(e.g., Guenther et al., 1998; Houde & Nagarajan, 2011; Hickok, 2012), a process sometimes termed *state feedback control*. This proposal essentially equates feedforward control to sensory feedback control with actual sensory feedback replaced by predicted sensory feedback. Predicted sensory feedback does not suffer from the substantial delays associated with actual sensory feedback, thereby avoiding the instabilities inherent in feedback control with long delays described in section 3.2 (see also Desmurget & Grafton, 2000). Early versions of the DIVA model utilized such a scheme for feedforward control (Guenther, 1994, 1995; Guenther, Hampson, & Johnson, 1998) while in later versions it is replaced with the simpler scheme of figure 7.2 in order to simplify the model. To our knowledge, current data do not definitively favor one of these accounts over the other, and it is quite possible that both types of feedforward control are involved in speech production.

3. Recall from earlier chapters that the syllable is probably the most common sound "chunk" to have its own optimized motor program, though we hypothesize that motor programs also exist for individual phonemes as well as very common multisyllabic utterances.

4. This formulation, which holds only for point targets, can be extended to accommodate motor target regions as described in footnote 2 of chapter 5.

5. In chapter 5 we estimated τ_A to be approximately 100 to 150 ms, and in chapter 6 we estimated τ_S to be approximately 25 to 75 ms.

6. The values of λ_A and λ_S are limited to a range between 0 (corresponding to no learning) and 1 (corresponding to adding the entire corrective command to the motor target for the next iteration).

7. Although Nasir and Ostry (2006) found adaptation to the jaw perturbation, they did not find negative aftereffects, in contrast to the results of Tremblay, Shiller, and Ostry (2003) and Tremblay, Houle, and Ostry (2008). The reason for this difference is unclear, but it may result from the use of different strategies for compensation depending on the context of the jaw perturbation. Nasir and Ostry (2006) attribute their results to impedance control—i.e., a change in the *stiffness* of the jaw likely due to increased co-contraction of jaw muscles during the perturbation. The results of Tremblay, Shiller, and Ostry (2003) and Tremblay, Houle, and Ostry (2008) are more consistent with a change in the planned motor trajectory (as in the DIVA model) rather than a simple increase in stiffness.

8. Purcell and Munhall (2006) note that adaptation for F1 perturbations typically will not occur for shifts smaller than 60 Hz.

9. Because this study involved fluent adults, we can assume that very little sensory error occurs during unperturbed productions of /r/ and thus the sensory feedback control systems contribute very little to the generation of articulator movements.

10. See the discussion in chapter 3, section 3.3, regarding the size of optimized motor programs for speech.

11. Of the DCN, the dentate nucleus (which makes up 90% of the DCN in humans) is most strongly associated with speech; see related discussion in chapter 2, section 2.3.

References

Aasland, W. A., Baum, S. R., & McFarland, D. H. (2006). Electropalatographic, acoustic, and perceptual data on adaptation to a palatal perturbation. *Journal of the Acoustical Society of America*, *119*, 2372–2381.

Ackermann, H., Vogel, M., Petersen, D., & Poremba, M. (1992). Speech deficits in ischaemic cerebellar lesions. *Journal of Neurology*, *239*, 223–227.

Alexander, G. E., & Crutcher, M. D. (1990). Neural representations of the target (goal) of visually guided arm movements in three motor areas of the monkey. *Journal of Neurophysiology*, *64*, 164–178.

Bouchard, K. E., & Chang, E. F. (2014). Control of spoken vowel acoustics and the influence of phonetic context in human speech sensorimotor cortex. *Journal of Neuroscience*, *34*, 12662–12672.

Brotchie, P., Iansek, R., & Horne, M. K. (1991). Motor function of the monkey globus pallidus: II. Cognitive aspects of movement and phasic neuronal activity. *Brain*, *114*, 1685–1702.

Browman, C. P., & Goldstein, L. (1990). Gestural specification using dynamically-defined articulatory structures. *Journal of Phonetics*, *18*, 299–320.

Bullock, D., & Grossberg, S. (1988). Neural dynamics of planned arm movements: emergent invariants and speed-accuracy properties during trajectory formation. *Psychological Review*, *95*, 49–90.

Desmurget, M., & Grafton, S. (2000). Forward modeling allows feedback control for fast reaching movements. *Trends in Cognitive Sciences, 4*, 423–431.

DeZeeuw, C. I., Hoogenraad, C. C., Koekkoek, S. K. E., Ruigrok, T. J. H., Galjart, N., & Simpson, J. I. (1998). Microcircuitry and function of the inferior olive. *Trends in Neurosciences, 21*, 391–400.

Diedrichsen, J., Hashambhoy, Y., Rane, T., & Shadmehr, R. (2005). Neural correlates of reach errors. *Journal of Neuroscience, 25*, 9919–9931.

Donchin, O., Rabe, K., Diedrichsen, J., Lally, N., Schoch, B., Gizewski, E. R., et al. (2012). Cerebellar regions involved in adaptation to force field and visuomotor perturbation. *Journal of Neurophysiology, 107*, 134–147.

Feng, Y., Gracco, V. L., & Max, L. (2011). Integration of auditory and somatosensory error signals in the neural control of speech movements. *Journal of Neurophysiology, 106*, 667–679.

Gandolfo, F., Mussa-Ivaldi, F. A., & Bizzi, E. (1996). Motor learning by field approximation. *Proceedings of the National Academy of Sciences of the United States of America, 93*, 3843–3846.

Gammon, S. A., Smith, P. J., Daniloff, R. G., & Kim, C. W. (1971). Articulation and stress/juncture production under oral anesthetization and masking. *Journal of Speech and Hearing Research, 14*, 271–282.

Ghahramani, Z., & Wolpert, M. (1997). Modular decomposition in visuomotor learning. *Nature, 386*, 392–395.

Guenther, F. H. (1994). A neural network model of speech acquisition and motor equivalent speech production. *Biological Cybernetics, 72*, 43–53.

Guenther, F. H. (1995). Speech sound acquisition, coarticulation, and rate effects in a neural network model of speech production. *Psychological Review, 102*, 594–621.

Guenther, F. H., Ghosh, S. S., & Tourville, J. A. (2006). Neural modeling and imaging of the cortical interactions underlying syllable production. *Brain and Language, 96*, 280–301.

Guenther, F. H., Hampson, M., & Johnson, D. (1998). A theoretical investigation of reference frames for the planning of speech movements. *Psychological Review, 105*, 611–633.

Hickok, G. (2012). Computational neuroanatomy of speech production. *Nature Reviews. Neuroscience, 13*, 135–145.

Houde, J. F., & Jordan, M. I. (1998). Sensorimotor adaptation in speech production. *Science, 279*, 1213–1216.

Houde, J. F., & Nagarajan, S. S. (2011). Speech production as state feedback control. *Frontiers in Human Neuroscience, 5*, 82.

Inchul, P., Amano, N., Satoda, T., Murata, T., Kawagishi, S., Yoshino, K., et al. (2005). Control of oro-facio-lingual movements by the substantia nigra pars reticulata: high-frequency electrical microstimulation and GABA microinjection findings in rats. *Neuroscience, 134*, 677–689.

Ito, T., & Ostry, D. J. (2010). Somatosensory contribution to motor learning due to facial skin deformation. *Journal of Neurophysiology, 104*, 1230–1238.

Jones, J. A., & Munhall, K. G. (2000). Perceptual calibration of F0 production: evidence from feedback perturbation. *Journal of the Acoustical Society of America, 108*, 1246–1251.

Jones, J. A., & Munhall, K. G. (2002). The role of auditory feedback during phonation: studies of Mandarin tone production. *Journal of Phonetics, 30*, 303–320.

Jones, J. A., & Munhall, K. G. (2005). Remapping auditory-motor representations in voice production. *Current Biology, 15*, 1768–1772.

Kawato, M., & Gomi, H. (1992). A computational model of four regions of the cerebellum based on feedback-error learning. *Biological Cybernetics, 68*, 95–103.

Keough, D., Hawco, C., & Jones, J. A. (2013). Auditory-motor adaptation to frequency-altered auditory feedback occurs when participants ignore feedback. *BMC Neuroscience, 9*, 14–25.

Krakauer, J. W., Ghilardi, M. F., & Ghez, C. (1999). Independent learning of internal models for kinematic and dynamic control of reaching. *Nature Neuroscience, 2*, 1026–1031.

Lametti, D. R., Nasir, S. M., & Ostry, D. J. (2012). Sensory preference in speech production revealed by simultaneous alteration of auditory and somatosensory feedback. *Journal of Neuroscience, 32*, 9351–9358.

Lashley, K. S. (1951). The problem of serial order in behavior. In L. Jeffress (Ed.), *Cerebral mechanisms in behavior* (pp. 112–136). New York: Wiley.

Lenneberg, E. H. (1967). *Biological foundations of language*. New York: Wiley.

Lindblom, B., Lubker, J., & Gay, T. (1979). Formant frequencies of some fixed-mandible vowels and a model of speech motor programming by predictive simulation. *Journal of Phonetics*, *7*, 147–161.

Malfait, N., Gribble, P. L., & Ostry, D. J. (2005). Generalization of motor learning based on multiple field exposures and local adaptation. *Journal of Neurophysiology*, *93*, 3327–3338.

Manes, J. L., Parkinson, A. L., Larson, C. R., Greenlee, J. D., Eickhoff, S. B., Corcos, D. M., et al. (2014). Connectivity of the subthalamic nucleus and globus pallidus pars interna to regions within the speech network: a meta-analytic connectivity study. *Human Brain Mapping*, *35*, 3499–3519.

Marsden, C. D. (1987). What do the basal ganglia tell premotor cortical areas? *Ciba Foundation Symposium*, *132*, 282–300.

Mattar, A. A. G., & Ostry, D. J. (2007). Modifiability of generalization in dynamics learning. *Journal of Neurophysiology*, *98*, 3321–3329.

McFarland, D. H., & Baum, S. R. (1995). Incomplete compensation to articulatory perturbation. *Journal of the Acoustical Society of America*, *97*, 1865–1873.

Mink, J. W. (1996). The basal ganglia: focused selection and inhibition of competing motor programs. *Progress in Neurobiology*, *50*, 381–425.

Nambu, A., Tokuno, H., & Takada, M. (2002). Functional significance of the cortico-subthalamo-pallidal "hyperdirect" pathway. *Neuroscience Research*, *433*, 111–117.

Nasir, S. M., & Ostry, D. J. (2006). Somatosensory precision in speech production. *Current Biology*, *16*, 1918–1923.

Neilson, M. D., & Neilson, P. D. (1987). Speech motor control and stuttering: a computational model of adaptive sensory-motor processing. *Speech Communication*, *6*, 325–333.

Passingham, R. E. (1993). *The frontal lobes and voluntary action*. Oxford: Oxford University Press.

Peeva, M. G., Guenther, F. H., Tourville, J. A., Nieto-Castanon, A., Anton, J. L., Nazarian, B., et al. (2010). Distinct representations of phonemes, syllables, and supra-syllabic sequences in the speech production network. *NeuroImage*, *50*, 626–638.

Purcell, D. W., & Munhall, K. G. (2006). Adaptive control of vowel formant frequency: evidence from real-time formant manipulation. *Journal of the Acoustical Society of America*, *120*, 966–977.

Rhodes, B. J., & Bullock, D. (2002). A scalable model of cerebellar adaptive timing and sequencing: the recurrent slide and latch (RSL) model. *Applied Intelligence*, *17*, 35–48.

Ringel, R. L., & Steer, M. D. (1963). Some effect of tactile and auditory alterations on speech output. *Journal of Speech and Hearing Research*, *6*, 369–378.

Salmelin, R., Schnitzler, A., Schmitz, F., & Freund, H. J. (2000). Single word reading in developmental stutterers and fluent speakers. *Brain*, *124*, 1184–1202.

Schliesser, H. F., & Coleman, R. O. (1968). Effectiveness of certain procedures for alteration of auditory and oral tactile sensation for speech. *Perceptual and Motor Skills*, *26*, 275–281.

Shan, Q., Ge, M., Christie, M. J., & Balleine, B. W. (2014). The acquisition of goal-directed actions generates opposing plasticity in direct and indirect pathways in dorsomedial striatum. *Journal of Neuroscience*, *34*, 9196–9201.

Shiller, D. M., Sato, M., Gracco, V. L., & Baum, S. R. (2009). Perceptual recalibration of speech sounds following speech motor learning. *Journal of the Acoustical Society of America*, *125*, 1103–1113.

Smith, Y., Bevan, M. D., Shink, E., & Bolam, J. P. (1998). Microcircuitry of the direct and indirect pathways of the basal ganglia. *Neuroscience*, *86*, 353–387.

Tremblay, S., Houle, G., & Ostry, D. J. (2008). Specificity of speech motor learning. *Journal of Neuroscience*, *28*, 2426–2434.

Tremblay, S., Shiller, D. M., & Ostry, D. J. (2003). Somatosensory basis of speech production. *Nature*, *423*, 866–869.

Tseng, Y. W., Diedrichsen, J., Krakauer, J. W., Shadmehr, R., & Bastian, A. J. (2007). Sensory prediction errors drive cerebellum-dependent adaptation of reaching. *Journal of Neurophysiology, 98*, 54–62.

Vilis, T., & Hore, J. (1980). Central neural mechanisms contributing to cerebellar tremor produced by perturbations. *Journal of Neurophysiology, 43*, 279–291.

Villacorta, V. M. (2006). *Sensorimotor adaptation to perturbations of vowel acoustics and its relation to perception*. PhD dissertation, Massachusetts Institute of Technology.

Villacorta, V. M., Perkell, J. S., & Guenther, F. H. (2007). Sensorimotor adaptation to feedback perturbations of vowel acoustics and its relation to perception. *Journal of the Acoustical Society of America, 122*, 2306–2319.

8

Sequencing of Speech Sounds

Thus far we have concentrated on the neural mechanisms responsible for learning and executing individual speech motor programs for phonological units such as phonemes and syllables, which are the domain of the DIVA model. These mechanisms represent only part of the speech production process. When we express our thoughts through speech, our brains construct grammatically structured *phrases* consisting of one or more words in a particular order. Each word, in turn, is constructed of smaller phonological units that must be produced in a particular order. During conversational speech, talkers can typically produce up to six to nine syllables (20–30 phonemes) per second, which is faster than any other form of discrete motor behavior (Kent, 2000). These observations indicate that the brain possesses the following capabilities, which we will refer to collectively as *speech sound sequencing*: (1) the ability to temporarily store phonological items within phrases in a manner that preserves their serial order and (2) the ability to sequentially activate the appropriate motor programs for producing these units at the correct instants in time. This chapter investigates the neural mechanisms responsible for these competencies. Relevant models of serial behavior, phonological structure, and working memory are first described, followed by a treatment of the neural circuitry responsible for speech sound sequencing.

8.1 Models of Serial Behavior

In a seminal paper, American psychologist Karl Lashley (1951) framed the problem of serial order in behavior, asking how the brain organizes and executes smooth, temporally integrated behaviors such as speech. Perhaps the simplest neural construct that can represent a serial ordering of individual items is an *associative chain*, as illustrated in panel A of figure 8.1. Here, the items are the phonemes that make up the word "peel," and their serial ordering is "hardwired" by directed connections between the items, represented by arrows in the figure. In neural terms, each phoneme has its own neural representation (indicated by a brown circle), and the serial order of the phonemes is represented by directed axonal projections between the representations. As Lashley notes, the primary problem with a simple associative chain representation of serial order is that each new serial ordering of the elements (e.g., the word "leap") requires new copies of those elements because of the hardwired nature of the serial order between items. Such a mechanism might work fine for a relatively small number of item sequences, but our language production mechanism is capable of generating an astronomical number of different combinations of phonemes within a phrase. Associative chain models also provide no basis for novel sequence performance, and they have difficulty simulating cognitive error data since there is no means to recover and correctly produce the remaining items after an incorrect link has been chosen (Henson et al., 1996).

Lashley (1951) proposed that serial behavior might instead be performed based on an underlying *parallel* planning representation. His proposal for the "priming of expressive units," or parallel, simultaneous activation of the items in a behavioral sequence prior to execution, is schematized in panel B of figure 8.1. Each unit has an *activity level* associated with it (indicated by the bar above the unit), and these activity levels encode the serial order of the sequence. For the word "peel," the /p/ node has the highest activity, followed by the /i/ node and then the /l/ node. There is an overwhelming amount of behavioral evidence for representing future components of an utterance in parallel with ongoing components as in such a representation (see Shattuck-Hufnagel, 2015, for a review), including the fact that substitution errors (e.g., saying "heft lemisphere" instead of "left hemisphere") are far more likely to involve elements that are upcoming in the utterance (such as the "h" from "hemisphere") than random elements not involved in the utterance. Such a parallel representation of current and future items is further supported by studies of linguistic performance errors (e.g., MacKay, 1970; Fromkin, 1971; Gordon & Meyer, 1987), reaction time experiments (e.g., Klapp, 2003), and demonstrations of anticipatory and carryover coarticulation (e.g., Ohman, 1966; Hardcastle & Hewlett, 1999).

In the parallel representation model in figure 8.1B, the same nodes can represent different sequences involving the same three units simply by varying the activity of the nodes, as illustrated for "peel," "leap" and "plea." Grossberg (1978a, 1978b) extended this idea by mathematically formulating a neural network model that can account for the formation and readout of serially ordered units in a parallel representation. Such a construction, often referred to as a *competitive queuing* (*CQ*) model, has been used in a number of subsequent models of speech, language, and motor control processes (e.g., Houghton, 1990; Dell, Burger, & Svec, 1997; Bullock & Rhodes, 2003; Bohland, Bullock, & Guenther, 2010) because it can account for a wide range of data on sequencing performance and error patterns. Figure 8.1C schematizes a CQ model primed to produce the words "peel" (left panel), "leap" (center), and "plea" (right). The first layer in the CQ model is the *plan layer*, which contains a parallel representation of the phonemes in the word to be produced. Inhibitory projections between the phoneme unit nodes (indicated by lines ending with filled circles) are crucial for maintaining the relative activities of the nodes over time. The activity pattern across these nodes can be

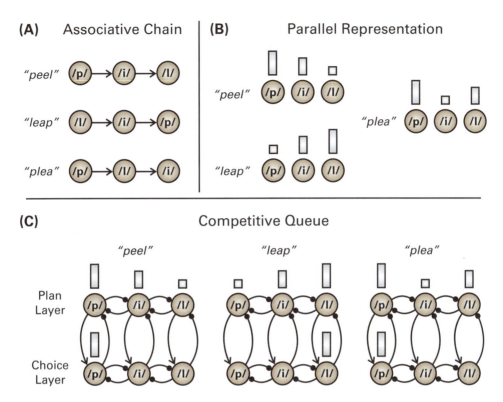

Figure 8.1
Models of neural representations of serial order. (A) The *associative chain* model, in which the serial order is hardwired by the connections between the units in the sequence. (B) A *parallel representation* wherein the activity level of the units (indicated by the bars above them) encode the serial ordering. (C) A *competitive queuing* model utilizing a parallel representation in the *plan layer* and a winner-take-all representation in the *choice layer*. Filled circles indicate inhibitory projections.

thought of as a *short-term memory* or *working memory* representation of the corresponding word. (We will expand on these concepts in a later section.) The second layer of nodes in the CQ model is the *choice layer*. Nodes in this layer receive inputs from the plan layer nodes, and they compete with each other via inhibitory connections that are stronger than those in the plan layer. Because of this strong internode inhibition, only one of the nodes in the choice layer becomes active—the one with the largest input from the plan layer—and it inhibits the other nodes such that their activity goes to zero. The unit corresponding to the chosen node (e.g., the /p/ node for "peel") is then produced. The plan and choice layer nodes for this unit are then extinguished (not shown in figure 8.1C), which causes the second-most-active node in the plan layer to become the most active node, leading to its choice layer node's becoming active, leading to readout of the corresponding unit, and so forth. CQ models have received substantial support from direct neurophysiological recordings in monkeys (Averbeck et al., 2003) and from chronometric analyses of seriation errors (Farrell & Lewandowsky, 2004). The CQ architecture thus appears to capture key aspects of the neural mechanisms underlying speech sequence planning and execution.

8.2   Frame-Content Models of Phonological Representation

In chapter 3, section 3.1, we introduced the notion that speech can be broken into two somewhat separate control processes: *segmental control* (e.g., the generation of the phonemes, syllables, and words that convey linguistic information) and *suprasegmental* or *prosodic control* of the rhythm, stress, and intonation patterns of speech. Modern theories of phonological encoding and speech planning commonly propose some form of factorization of the global structure (often associated with prosody) and the phonological content of an utterance (e.g., Lashley, 1951; Fromkin, 1971; Shattuck-Hufnagel, 1983, 1992, 2015; Stemberger, 1984; Dell, 1988; Levelt, 1989; MacNeilage, 1998; Roelofs & Meyer, 1998). The process of merging these two aspects of speech just prior to articulation is sometimes referred to as *phonological encoding* (e.g., Roelofs & Meyer, 1998), though the meaning of this term varies substantially across authors in different domains of speech and language research.

The topic of prosodic structure is a complex one that will be addressed in more detail in chapter 9. For the current purposes, it suffices to note that several models of speech planning posit syllable-sized or word-sized prosodic units called *structural frames* (also called *prosodic frames*, *syllabic frames*, or simply *frames*) along with phoneme-sized *content* elements, as illustrated in panel A of figure 8.2. The basic idea is that, at some level or levels of the speech planning process, structural frames are represented largely independently of their phonological content. For example, the word "stoop" is represented in memory by an abstract syllabic frame consisting of *slots* that can be filled with different phonemic items (in this case a CCVC frame, where C denotes a consonant slot and V denotes a vowel slot). Also in memory are phonemic content elements for the current and upcoming words in the utterance; these are represented by brown circles. Producing the word "stoop" involves filling the structural frame with the appropriate content elements, as indicated by arrows in figure 8.2A.

The frame/content distinction is motivated in large part by the pattern of errors observed in spontaneously occurring slips of the tongue. MacKay (1970), in his study of *spoonerisms*, or phoneme exchange errors (e.g., saying "heft lemisphere" instead of the intended "left hemisphere"), noted the prominence of the syllable position constraint, in which exchanges are greatly biased to occur between phonemes occupying the same positional slot in different planned syllables. In other words, exchanges very rarely occur across the dashed line boundaries separating content elements associated with different frame elements in figure 8.2 but instead occur between content elements associated with the same

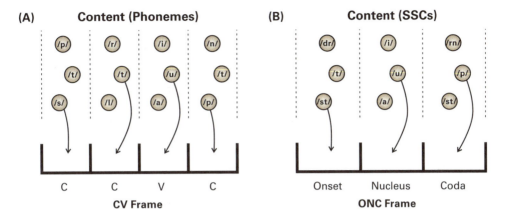

Figure 8.2
Two possible syllabic frame structures along with corresponding phonological content items. (A) Consonant-vowel (CV) frame structure for a CCVC syllable. Arrows indicate phonemic content elements for constructing the word "stoop." (B) Onset-nucleus-coda (ONC) frame structure and corresponding subsyllabic constituent (SSC) content elements. Arrows indicate the SSC content elements for constructing the word "stoop."

slot in different syllables. This constraint appears to be the strongest pattern observed in speech errors. Shattuck-Hufnagel (1979), for example, found that 207 of 211 exchange errors involved transpositions to and from similar syllabic positions. More recently, Vousden et al. (2000) found that approximately 90% of consonant movement errors followed this constraint. Treiman and Danis (1988) also noted that during nonword repetition, most errors are phonemic substitutions that preserve syllable structure. Such exchanges also follow a transposition distance constraint (MacKay, 1970), in that phonemes are more likely to exchange between neighboring rather than distant syllables. Beyond speech error data, priming studies have demonstrated effects in speech production based purely on CV structure (while controlling for content) at the syllable and word level (Meijer, 1996; Sevald, Dell, & Cole, 1995).

The CV frame is only one type of syllable frame that has been proposed in the phonological literature. An alternative type of syllabic frame is shown in panel B of figure 8.2. In this frame, the syllable is broken into three *subsyllabic constituents*[1] (*SSCs*): an *onset* (one or more consecutive consonants at the beginning of the syllable), a *nucleus* (a vowel, diphthong, or sonorant consonant like /r/ that forms the "core" of the syllable), and a *coda* (one or more consecutive consonants at the end of the syllable). Such a breakdown will be termed an *ONC frame* herein. In this case, the phonological content elements are not phonemes per se but instead are SSCs that include consonant clusters as well as individual phonemes. For example, the word "stoop" consists of three SSCs, as illustrated in figure 8.2B: the onset consonant cluster /st/, the nucleus vowel /u/, and the coda consonant /p/. A number of studies have found experimental support for consonant clusters as cohesive "motor units." For example, Fromkin (1971) noted that initial consonant clusters are often (though not always) substituted, deleted, or added to an utterance as a single unit in speech errors, and the articulatory kinematics study by Loevenbruck et al. (1999) found evidence that consonant clusters act as units of speech motor programming. These findings suggest that SCCs consisting of consonant clusters may act as phonological content elements for speech production, in addition to individual phonemes.

In another characterization of a structural frame, the syllable consists of only two SSCs, an *onset* and a *rime* (or *rhyme*), termed an *OR frame* herein. A rime is essentially a nucleus and coda combined.[2] In addition to speech error patterns, data from word games (Fowler et al., 1993) and short-term memory errors (Treiman & Danis, 1988) provide strong support for the rime as a speech unit at some level of the production process.

Although the exact nature of the structural frames and phonological content elements used at various stages of speech and language processing is still under considerable debate (and likely varies by language), a preponderance of evidence supports the general assertion that syllabic structure and phonological content are processed somewhat independently by the brain during planning of speech sequences. This distinction is at the heart of the GODIVA model of speech sound sequencing, as described later in this chapter.

8.3 The Baddeley and Hitch Model of Working Memory

Since the nineteenth century, memory processes in the brain have been broadly separated into two types: *short-term memory*, which lasts at most for a few seconds, and *long-term memory*, which can last for hours, days, years, or even decades. In the latter half of the twentieth century, the concept of short-term memory was largely supplanted by the concept of a multicomponent *working memory* system, which involves both short-term storage and manipulation of information needed to perform a cognitive task. The most widely accepted model of working memory has been developed by British psychologist Alan Baddeley and colleagues, originating with the model proposed by Baddeley and Hitch (1974; see also Baddeley, 1986) schematized in figure 8.3. According to this model, the working memory system consists of three main components: the *central executive*, the *phonological loop*, and the *visuospatial sketch pad*.

The central executive is the attentional control system, responsible for directing attention to relevant information and coordinating information from the two "slave systems," the phonological loop and visuospatial sketch pad.[3] The central executive is most often associated with lateral prefrontal cortex, though it may be better thought of as a conceptual description than as a unitary module located entirely within a single brain region (Baddeley, 1998).

The visuospatial sketch pad stores spatial and visual information, effectively constituting an "inner eye." This system has limited capacity and a short memory duration (on the order of seconds). For longer storage of visually presented information, items must

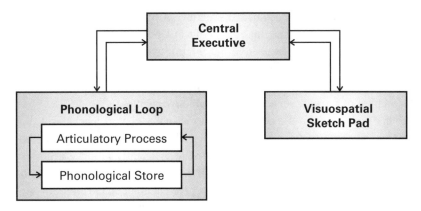

Figure 8.3
Schematic of the Baddeley and Hitch (1974) multicomponent model of working memory. The model consists of a *central executive* that coordinates between two "slave systems": a *phonological loop* and a *visuospatial sketch pad*.

be translated into phonological form (words) and transferred to the phonological loop. The visuospatial sketch pad most likely involves visual areas of the occipital and parietal lobes.

The phonological loop is responsible for maintaining verbal information in working memory. This loop contains two main components, the *articulatory process* and the *phonological store*. The articulatory process is the "inner voice," which is responsible for rehearsing the items in working memory. Each rehearsal activates the phonological store, which acts as an "inner ear" that represents sound structure. (Imagine rehearsal of a phone number by repeatedly saying it in your head, then hearing yourself say it, then silently repeating what you heard yourself say, etc.) Thus, although the phonological store by itself can only represent information for at most a few seconds, the rehearsal process allows for indefinite storage within the phonological loop, at least until the phonological loop is needed for another task. As we will address further below, the articulatory process has been associated with the left inferior frontal gyrus and sulcus and ventral premotor cortex (e.g., Bohland, Bullock, & Guenther, 2010; Herman et al., 2013) while the phonological store has been associated with the inferior parietal cortex including the supramarginal gyrus (Jonides et al., 1998) and the posterior superior temporal gyrus (Herman et al., 2013).

8.4 Phonological Working Memory in Speech Production

The language mechanism is capable of generating a tremendous number of possible linguistic sequences. For example, consider the set of five-syllable sentences such as "Please take out the trash." Such a sequence is short enough to be stored in its entirety in working memory. If we conservatively estimate the number of syllables familiar to an adult speaker to be on the order of 1,000,[4] then there are on the order of 1 quadrillion (10^{15}) possible five-syllable sequences that might need to be stored in working memory. Although this may be an overestimate as it assumes that any syllable can occur in any location independently of any other syllable, the larger point is clear: our brains are capable of generating a staggering number of possible sequences of syllables/phonemes.

Now consider the problem from the perspective of the motor control mechanism. For the current purposes, assume that the basic units of speech motor programming are phonemic gestures, and that the motor commands needed to efficiently produce a gesture depend on phonetic context. In order to produce efficient movements (i.e., articulations that use as little muscle force as possible to complete a given utterance), the motor system learns *motor programs* consisting of optimized feedforward motor commands for producing commonly occurring phoneme subsequences. Strong evidence exists for such optimized motor programs, especially at the syllabic level; they are thus often referred to collectively as a *mental syllabary* (Levelt & Wheeldon, 1994; Levelt et al., 1999; Cholin et al., 2006).

Operating between the language system and the motor control system lies working memory. Fluent speech requires formulation and temporary storage of linguistic passages that contain multiple syllables to ensure that the next syllable is available when the motor control system finishes the current syllable. This is a form of working memory, which we will call *phonological working memory* with the caveat that we are specifically concerned with phonological working memory processes involved in speech production, which may differ somewhat from those involved in speech perception.

It is well-known that working memory has a limited capacity. One crucial determinant of the capacity is the nature of the individual items, or *units,* being stored. For phonological working memory, the units might be individual phonemes, or they might be short strings of phonemes (often referred to as *chunks* in the working memory literature) such as syllables or words. For example, the sentence "I'm going to the pub" could be represented as 5 word-sized chunks, 6 syllable-sized chunks, or 13 phoneme-sized chunks. Larger chunks may seem advantageous since fewer of them will need to be stored in working memory to represent a given linguistic sequence. However, there is a cost to using larger chunks that arises from the very large number of such chunks that the system must be capable of representing. To see this, note that in CQ-style models of working memory, each possible item is associated with a different *node* corresponding roughly to a population of neurons in the brain. If the individual items represented in working memory are phonemes, then for English approximately 44 nodes are needed since the language utilizes approximately 44 different phonemes.[5] The number of necessary nodes (and, by analogy, cortical neurons) goes up drastically if the items stored are multiphoneme chunks; a syllable representation would require thousands of nodes, and a word representation would require tens of thousands.

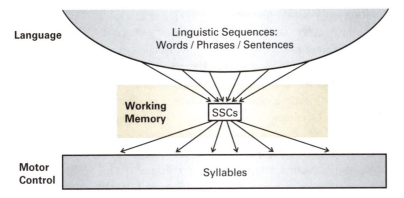

Figure 8.4
Schematized view of the process of translating a linguistic message into syllabic motor programs via a working memory system. An essentially unlimited number of linguistic messages can be represented in working memory as different sequences of a relatively small set of *subsyllabic constituents* (*SSCs*), such as phonemes and/or consonant clusters. This working memory can then access a large number of optimized syllabic motor programs.

These considerations lead to the conclusion that there is likely an optimal chunk size for phonological working memory that best satisfies the trade-off between number of items that must be stored simultaneously and amount of neural real estate required to represent the full set of possible items. Based on behavioral findings (particularly from the speech error literature described in section 8.2) as well as our own neuroimaging studies, *we propose that the phonological working memory component of the language production system utilizes an SSC representation* that contains on the order of 100 distinct units.[6] Within this view, SSCs act as a compact representation for interfacing between a nearly limitless linguistic inventory and thousands of optimized syllabic motor programs, as schematized in figure 8.4. At present the exact nature of the SSCs utilized in phonological working memory—in particular whether the individual units are phonemes or some other type of SSC such as onsets, nuclei, codas, and/or rimes—is not entirely clear, and furthermore it is likely to differ across languages and even across individuals who speak the same language (e.g., as a function of developmental stage). In acknowledgment of these uncertainties, the term SSC is used herein to refer in a general sense to units that are smaller than an entire syllable but not smaller than an individual phoneme, thereby accommodating any of the SSC representations discussed in section 8.2.

8.5 The Neural Bases of Speech Sound Sequencing

The neural mechanisms underlying the planning of multisyllabic utterances are less well-characterized than those underlying speech motor control at the single-syllable level. Nonetheless, a number of important insights can be gained from the neurological and neuroimaging literatures, as detailed in the following subsections.

Speech Sound Sequencing and Working Memory

It is not surprising that the neural circuitry involved in planning the production of upcoming speech sounds overlaps substantially with brain regions underlying working memory. This is illustrated in figure 8.5, which compares the results of an activity likelihood estimate meta-analysis of 113 working memory neuroimaging studies (Rottschy et al., 2012) to the results of an fMRI study comparing the production of complex versus simple sound sequences (Bohland & Guenther, 2006). The blue spheres indicate left-hemisphere activation foci for the *core working memory network*, common to both verbal and nonverbal working memory tasks, identified by the Rottschy et al. (2012) meta-analysis. This network includes two activation foci in the *inferior frontal sulcus*[7] (*IFS*) and individual foci in the *supramarginal gyrus* (*SMG*), *anterior insula* (*aINS*), and the *pre-supplementary motor area* (*preSMA*). The Rottschy et al. (2012) meta-analysis identified only one location that was preferentially activated in verbal working memory tasks when contrasted with nonverbal working memory tasks; this locus is indicated by a brown sphere in figure 8.5 and lies in left IFS. Its preferential role in verbal rather than nonverbal working memory implicates this region as part of the phonological loop of Baddeley's model, in particular the articulatory process component that acts as an "inner voice" during verbal working memory tasks (Paulesu, Frith, & Frackowiak, 1993; Awh et al., 1996).

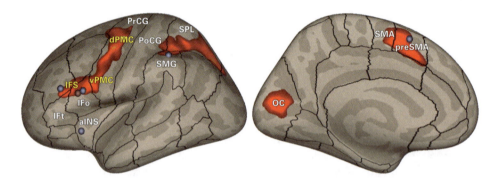

Figure 8.5
Comparison of left-hemisphere activation foci for the core working memory network identified in the Rottschy et al. (2012) activation likelihood estimate meta-analysis (blue spheres) with fMRI results showing brain areas that are more active when producing speech sequences involving three unique syllables compared to simpler sequences consisting of the same syllable repeated three times (hot colormap). The brown sphere corresponds to the single activation focus from the Rottschy et al. meta-analysis that was preferentially active in verbal versus nonverbal working memory tasks. Speech sequence complexity fMRI results are from an analysis of data from Bohland and Guenther (2006) in which individual cortical surface vertices were thresholded at $p < 0.005$ uncorrected before a cluster-level analysis thresholded at a family-wise error rate of 0.05. White labels indicate regions of interest as defined in appendix B. Yellow labels indicate additional regions discussed in the main text. aINS, anterior insula; dPMC, dorsal premotor cortex; IFo, inferior frontal gyrus pars opercularis; IFt, inferior frontal gyrus pars triangularis; IFS, inferior frontal sulcus; OC, occipital cortex; PoCG, postcentral gyrus; PrCG, precentral gyrus; preSMA, pre-supplementary motor area; SMA, supplementary motor area; SMG, supramarginal gyrus; SPL, superior parietal lobule; vPMC, ventral premotor cortex.

Several researchers have suggested that the phonological store component of the phonological loop resides in parietal cortex (Paulesu, Frith, & Frackowiak, 1993; Awh et al., 1996; Jonides et al., 1998), near the SMG focus from Rottschy et al. (2012), perhaps with a left-hemisphere bias.

The hot colormap in figure 8.5 indicates regions of increased brain activity when contrasting complex syllable sequences (containing three unique syllables) to simple syllable sequences (containing the same syllable repeated three times) from the Bohland and Guenther (2006) speech production fMRI study. Notably, higher sequence complexity was accompanied by increased activity in several regions of the core working memory network, including IFS, SMG, and preSMA. Furthermore, using a more sensitive statistical analysis of the same data set, Bohland and Guenther (2006) identified activity in the region of the left aINS focus of the core working memory network. These results indicate a high degree of overlap between the network for speech sound sequencing and the core working memory network.

In addition to the core working memory network, increased sequence complexity was also accompanied by additional activity in two premotor areas associated with the generation of feedforward commands for speech sounds in chapter 7, namely, left *ventral premotor cortex* (*vPMC*) and *supplementary motor area* (*SMA*). Increased sequence complexity was also associated with higher activity in *occipital cortex* (*OC*) and medial portions of the *superior parietal lobule* (*SPL*); activity in OC and SPL is likely related to the increased visual processing requirements for reading orthographic stimuli for complex sequences compared to simple sequences. This explanation also accounts for dorsal (*dPMC*) activity near the junction of the precentral sulcus and superior frontal sulcus since the frontal eye fields (which play a central role in eye movement control) are located in this region (Pierrot-Deseilligny et al., 2004).

The following subsection describes the GODIVA neurocomputational model of speech sound sequencing, which elaborates on the computations performed by preSMA, SMA, IFS, and vPMC during the generation of speech sequences.

The GODIVA Model of Speech Sound Sequencing

The DIVA model accounts for production of individual speech motor programs, each corresponding to a different speech sound, with the typical unit of motor programming in the mature system being the syllable. The *gradient order DIVA* (*GODIVA*) model (Bohland, Bullock, & Guenther, 2010) extends DIVA to account for the neural computations underlying multisyllabic planning, timing, and coordination as evidenced from clinical and neuroimaging studies. The model focuses on several brain regions known to be involved in working memory and motor sequencing, including the lateral prefrontal cortex (specifically, the *posterior inferior frontal sulcus*, *pIFS*), vPMC, SMA, preSMA, and basal ganglia.

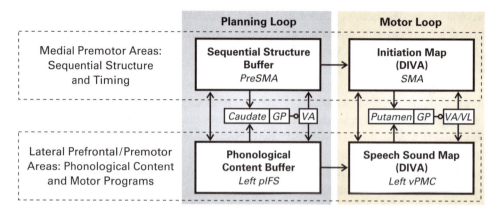

Figure 8.6
Simplified block diagram of the GODIVA model of speech sound sequencing. GP, globus pallidus; pIFS, posterior inferior frontal sulcus; PreSMA, pre-supplementary motor area; SMA, supplementary motor area; VA, ventral anterior nucleus of the thalamus; VL, ventral lateral nucleus of the thalamus; vPMC, ventral premotor cortex.

A simplified schematic of the GODIVA model is provided in figure 8.6. The model consists of two cortico–basal ganglia loops. The *planning loop* involves pIFS, preSMA, the caudate nucleus and *globus pallidus* (*GP*) of the basal ganglia, and the *ventral anterior* (*VA*) *nucleus* of the thalamus. The *motor loop*, which is an expansion of the DIVA model's speech sound map and initiation map, involves vPMC, SMA, putamen, GP, VA, and the anterior portion of the *ventral lateral* (*VL*) *nucleus* of the thalamus. Briefly, the planning loop is responsible for temporary storage, or *buffering*, of the phonological sequence to be produced whereas the motor loop is responsible for generating the motor commands for producing the current phonological unit.

The model's components can also be broken into a second dichotomy (indicated by dashed boxes in figure 8.6): the left lateral frontal areas (pIFS and vPMC) are primarily responsible for the *phonological content* of the sequence, including the associated motor programs for producing it, whereas the medial premotor areas (preSMA and SMA) are responsible for the *sequential structure* of the phonological sequence. This subdivision of phonological content and suprasegmental structure is consistent with the frame-content theories of speech planning discussed in section 8.2. *Sequential structure* is used here to refer generally to information about item order, frame structure, stress patterning, and timing in an upcoming utterance with little regard for the phonological content and associated motor programs. The term *sequential* is used here rather than *prosodic* or *suprasegmental* as the latter terms include additional aspects of speech, such as those conveying emotional state, that are controlled by different neural circuitry than the structural aspects considered here (see chapter 9 for detailed treatment of the neural bases of prosody).

According to the GODIVA model, an upcoming multisyllabic utterance is represented simultaneously in two complementary modules within the pIFS and preSMA. The sequential structure of the utterance (consisting of syllabic frame structure and syllable order information in computer implementations of the model) is represented in preSMA, whereas the phonemic content of the utterance is represented in left pIFS, organized by the location of the individual sounds within the syllable frame. Both of these representations are modeled as competitive queues that can contain multiple coactive items that represent multiple forthcoming syllables, with the order of those items represented by an activity gradient as illustrated in panels B and C of figure 8.1. This gradient representation of serial order gives rise to the model's name, gradient order DIVA (GODIVA).

The following subsections describe the neural substrates of speech sound sequencing in further detail, with reference to the GODIVA model.

Medial Premotor Areas: Sequential Structure

As discussed in chapter 2, the medial frontal cortex contains two distinct premotor cortical areas: SMA and preSMA. Tanji and colleagues have collected a wealth of data in monkeys indicating that SMA and preSMA are both crucially involved in the representation of movement sequences, with preSMA serving a higher-order role than SMA (Matsuzaka et al., 1992; Tanji & Shima, 1994; Shima et al., 1996; Shima & Tanji, 1998, 2000; Tanji, 2001). The two regions have different patterns of connectivity with cortical and subcortical areas in monkeys (Jürgens, 1984; Luppino et al., 1993), and diffusion-tensor imaging results verify these patterns in humans (Johansen-Berg et al., 2004; Lehéricy et al., 2004). Whereas preSMA is well-connected with the prefrontal cortices and the anterior striatum, SMA is more connected with the motor cortex and the posterior striatum. This suggests a role more generally associated with planning for preSMA and with motor execution for SMA.

Based in large part on these findings, the GODIVA model posits that preSMA is primarily responsible for formulating the global sequential structure of an upcoming speech utterance whereas SMA is primarily responsible for initiating the motor execution of speech articulations. At present, the neural representations utilized by speech sequencing–related neurons in SMA and preSMA are largely unknown because of the lack of single-unit recordings from these regions in humans during speech. However, insights into likely properties of these neurons can be gained from the monkey electrophysiology literature. In a seminal study of movement sequencing, Shima and Tanji (2000) trained monkeys to perform different sequences of three hand/arm movements (*push*, *pull*, or *turn* of a manipulandum). The authors identified several distinct sequence- and movement-related neuron types in SMA and preSMA, including the following:

- *Sequence-selective neurons* that fired prior to the execution of a particular movement sequence (e.g., *push-turn-pull*). These neurons were roughly equally distributed between SMA and preSMA. In the domain of speech production, one might expect analogous neurons that are *word-* or *syllable-selective neurons.*

- *Rank-order neurons* that fired prior to a particular slot or rank in the sequence (e.g., firing before the second movement regardless of the movement sequence). Rank-order neurons were more prevalent in preSMA than SMA. A plausible speech analogue would be *frame-slot* neurons, for example, neurons that represent the V slot of a CVC syllable frame or the onset slot of an ONC syllable frame.

- *Interval-selective neurons* that fired in the interval between two particular movements (e.g., between *turn* and *pull*). These neurons were more prevalent in SMA than preSMA. A possible speech analogue for these neurons would be *phoneme-transition–selective* neurons.

- *Movement-selective neurons* that fired prior to and/or during a particular movement type (e.g., *turn*). The great majority of preSMA movement-selective neurons were either sequence- or rank-order-specific neurons whereas the majority of SMA movement-selective neurons were not sequence- or rank-order-specific neurons. Possible speech analogues to movement-selective neurons would be *phoneme-*, *SSC-*, or *gesture-selective* neurons.

These considerations lead to the following conclusions: *preSMA likely represents the global structure of speech sequences*, including syllabic and/or word frame structure, whereas *SMA likely represents individual articulatory gestures and gesture transitions* within a speech sequence. Findings from the neuroimaging literature provide further support for these assertions. Using fMRI, Bohland and Guenther (2006) noted that activity in preSMA was higher for speech sequences composed of more phonologically complex syllables, as well as for complex sequences versus simple sequences (i.e., producing three unique syllables vs. producing the same syllable three times). In contrast, activity in SMA showed no such effect for syllable or sequence complexity. A word production fMRI study by Alario et al. (2006) found further evidence for preferential involvement of SMA in motor output and suggested a further functional subdivision within preSMA, specifically that anterior preSMA is more involved with lexical selection while the posterior portion is more involved in sequence encoding and execution. Tremblay and Gracco (2006) likewise observed SMA involvement across motor output tasks but found that the preSMA response increased in a word generation task (which involves lexical selection) as compared to a word reading task.

The GODIVA model formalizes these insights as follows. As in the DIVA model, SMA is hypothesized to contain an *initiation map* responsible for initiating the execution of motor programs corresponding to speech motor units, as described in chapter 7. Studies using a repetition suppression fMRI protocol[8] have found evidence that left SMA utilizes a phonemic representation (Peeva et al., 2010). In other words, there are neurons in left SMA that represent individual motor units corresponding to phonemes or articulatory gestures.[9]

As discussed in chapters 2 and 7, SMA is not likely to contain detailed motor programs for these motor units; instead, it is likely responsible for initiating or "gating on" (as represented by the GO signal in equation 7.1 of chapter 7) the readout of motor programs located in ventral premotor and motor cortical areas along with associated subcortical structures, most notably the cerebellum.

PreSMA in GODIVA represents the global sequential structure in the form of a competitive queue containing the syllable frames for the upcoming utterance. Computer simulations of GODIVA described in Bohland, Bullock, and Guenther (2010) utilized CV frames, though this was a somewhat arbitrary choice over other reasonable candidates such as ONC or OR frames. To coordinate timing, projections from preSMA to SMA activate, in the proper sequence, the appropriate neurons in the initiation map, which in turn leads to the readout of the corresponding motor programs via projections to the basal ganglia (particularly the putamen) and speech sound map in left ventral premotor cortex.

The preSMA representation proposed in Bohland, Bullock, and Guenther (2010) is completely insensitive to phonological content, in effect representing "pure" frame structure. It is likely, however, that preSMA is sensitive to both frame structure and phonological content, given the existence of both sequence-related and movement-related neurons in monkey preSMA (Shima & Tanji, 2000). Although further research is needed to decisively identify and characterize the representation of speech sequence structure in preSMA, the more general conclusion that preSMA is involved in representing the global structure of a speech utterance receives support from a wide range of experimental findings.

According to the GODIVA model, the representations of phonological units such as phonemes or SSCs in medial premotor cortical areas SMA and preSMA are highly abstracted. That is, they do not encode details regarding motor execution, acoustic structure, or linguistic properties of the encoded phonological units. Instead, they essentially act as placeholders for the phonological units within the global sequential structure of an ongoing utterance. In contrast to this abstracted representation in medial premotor areas, the model posits that details regarding the sensorimotor structure of the phonological units are encoded by lateral frontal cortical areas, as described in the next subsection.

Left Lateral Frontal Areas: Phonological Content and Motor Programs
The inferior frontal sulcus separates the inferior frontal gyrus from the middle frontal gyrus in the lateral prefrontal cortex. IFS and surrounding areas (often extending into the middle frontal gyrus and inferior frontal gyrus), particularly in the left hemisphere, have been implicated in a large number of studies of language and working memory (Gabrieli, Poldrack, & Desmond, 1998; Kerns, Cohen, Stenger, & Carter, 2004; Fiez et al., 1996; D'Esposito et al., 1998) and serial order processing (Petrides, 1991; Averbeck et al., 2002, 2003). According to the GODIVA model, the posterior portion of IFS contains a *phonological content buffer* that temporarily stores the phonological units of an upcoming utterance. This location coincides with the location of the lone activation focus from the Rottschy

et al. meta-analysis that was preferentially active in verbal versus nonverbal working memory tasks (brown sphere in figure 8.5). This region has also been shown to encode phonetic categories during speech perception (Myers et al., 2009), and Poldrack et al. (1999) implicate this region in phonological tasks without regard for semantic content. Bohland and Guenther (2006) found that left pIFS activity is sensitive both to the number of syllables in a sequence and the phonological complexity of these syllables, consistent with this area's encoding the phonological units of the sequence in a phonological content buffer.

The GODIVA model proposes that the phonological content buffer consists of nodes that each represent a different phonological content item, and that the pattern of activity across these nodes encodes the order of items to be produced, as in the competitive queuing model schematized in figure 8.1C. Based largely on the speech error data summarized above, the model further posits that distinct queues represent SSCs for each slot in a syllable's structural frame. If, for example, an ONC frame is assumed,[10] there are separate queues for onsets, nuclei, and codas. pIFS is in/near the human homologue of a region in monkey prefrontal cortex (Brodmann area 46) that has been implicated in serial working memory representations for drawing movement sequences. Averbeck et al. (2002) demonstrated with single-unit electrophysiology that, prior to initiating a planned sequence of cursor movements, there exists a parallel coactive representation for each of the component movements of the forthcoming sequence. The relative level of activity in small groups of cells that coded for the component movements corresponded to the serial order in which the movements would be produced, as in the competitive queuing model.

As in the DIVA model, left vPMC in the GODIVA model contains *speech sound map* nodes that represent the inventory of phoneme sequences with optimized motor programs for their production. Projections from left pIFS to left vPMC are responsible for choosing the next motor program to be executed based on the most highly active items in the pIFS phonological content buffer. The motor program capable of producing the most consecutive phonological items is chosen (see Bohland, Bullock, & Guenther, 2010 for details). For example, if the onset /kr/, the nucleus /i/, and the coda /k/ are the most active items in the phonological buffer, then the motor program for the syllable /krik/ will be chosen in vPMC assuming one is available. If no motor program is available for the whole syllable, then motor programs for /kr/ and /ik/ will be chosen if they are available, or for /k/, /r/, /i/, and /k/ if not. For reasons stated previously, the syllable is likely the most common motor program size in the mature system. Evidence that left vPMC utilizes a syllabic representation during speech production is provided by the repetition suppression fMRI study of Peeva et al. (2010).

Basal Ganglia Loops: Structure-Content Coordination and Chunking

In the GODIVA model, the basal ganglia is heavily involved in coordinating between medial cortical areas, which are responsible for sequence structure and timing, and lateral areas responsible for phonological content and articulatory motor programs. Following the scheme of Middleton and Strick (2000) derived from primate anatomical studies, the GODIVA model posits two largely distinct basal ganglia loops with cerebral cortex: a higher-level *planning loop*, involving projections from cortex to the caudate nucleus, and a lower-level *motor loop*, involving projections from cortex to the putamen. Such a breakdown is also apparent in humans, as determined by the study of Gil Robles et al. (2005) in which electrical stimulation of the caudate and putamen was applied during speech produced during neurosurgery. The authors differentiated the sensorimotor function of the putamen from a higher-level, more cognitive function of the caudate, in support of the Middleton and Strick (2000) view.

The GODIVA planning loop involves preSMA, left pIFS, basal ganglia (caudate nucleus and GP), and thalamus (VA). This loop is responsible for coordinating between the sequential structure and phonological content representations in preSMA and vPMC. The role of caudate nucleus and thalamus in speech sequence generation have been reviewed by Crosson (1992), who made note of the similarities between electrical stimulation effects in the caudate nucleus and anterior thalamic nuclei. Schaltenbrand (1975) reported that stimulation of the anterior nuclei of the thalamus (including VA and VL) sometimes caused compulsory speech that could not be inhibited. Stimulation of the dominant head of the caudate has also evoked word production (Van Buren, 1963), and Crosson (1992) describes the similarities in the language evoked from stimulation of the two areas as "striking." This suggests that the areas serve similar functions, and that they are involved in the generation of speech sound sequences. fMRI studies contrasting complex speech sequences with simpler sequences also support roles in speech sequencing for caudate nucleus and anterior portions of the thalamus (Sörös et al., 2006; Bohland & Guenther, 2006).

The GODIVA motor loop involves SMA, vPMC, vMC, putamen, globus pallidus, and VA/VL thalamus. The role of the cortico–basal ganglia motor loop in choosing and initiating speech motor programs was described in section 7.3 of chapter 7. We propose that learning in the cortico–basal ganglia motor loop, in concert with the cortico-cerebellar loop described in section 7.3, in effect "automates" the production of frequently produced linguistic subsequences such as syllables (cf. Alm, 2004; Redgrave et al., 2010) by encoding these subsequences as "chunks" with their own optimized motor programs, thus lowering the load on prefrontal and premotor cortical areas. Figure 8.7 schematizes this process. Panel A represents the cortical computations required to produce the word "blue" early in development. vMC contains nodes encoding articulatory gestures (labeled G) for the phonemes /b/, /l/, and /u/. Each of these phonemic gestures has a corresponding initiation map cell (labeled I) in SMA that is responsible for initiating the gesture via projections to vMC. At this early stage, vPMC does not contain a motor program for the whole syllable /blu/. Instead it has individual motor programs for each phoneme in the syllable that must be activated in sequence by projections from the phonological content buffer in pIFS.

(A) Early in Development

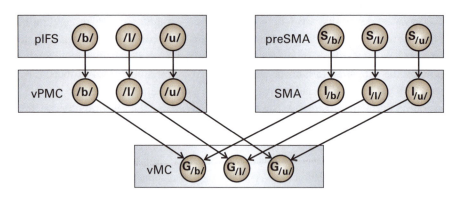

(B) Late in Development

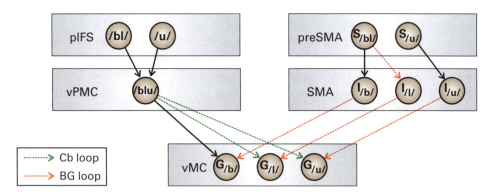

Figure 8.7
Learning of speech "chunks" according to the GODIVA model. (A) Neural processes underlying the production of the word "blue" early in development. Black arrows indicate cortico-cortical projections. (B) Neural processes underlying the production of the word "blue" late in development. Fewer cortical nodes and projections are required because of learning in cerebellar (green dashed arrows) and basal ganglia (red dashed arrows) loops. BG, basal ganglia; Cb, cerebellum; G, gestural node; I, initiation map node; pIFS, posterior inferior frontal sulcus; preSMA, pre-supplementary motor area; S, syllabic structure node; SMA, supplementary motor area; vMC, ventral primary motor cortex; vPMC, ventral premotor cortex.

Similarly, pIFS and preSMA only contain phonemic elements, not larger SSCs such as consonant clusters.

Production of /blu/ early in development proceeds as follows. The language system activates the representations for /b/, /l/, and /u/ in the pIFS phonological content buffer, as well as the structural representation for /blu/ in the preSMA sequential structure buffer. Projections from pIFS activate, in sequence, vPMC nodes representing the motor programs for /b/, /l/, and /u/, which in turn sequentially activate the corresponding gestural nodes in vMC. The timing of this sequential activation process is determined by the medial premotor areas. Projections from preSMA to SMA activate initiation map nodes for the individual phonemes in the proper order and with the proper timing. The initiation map nodes effectively "gate on" the gestural motor programs at the appropriate instant in time, as dictated by the global syllabic structure representation in preSMA. Once a motor program has been completed, the corresponding nodes in pIFS, vPMC, and pIFS are deactivated, allowing the next motor program to commence.

Panel B of figure 8.7 schematizes the production of /blu/ at a later stage of development. Now, vPMC contains a motor program for the entire syllable /blu/, with subcortical loops through the cerebellum (green dashed arrows) effectively taking over the job of coordinating and coarticulating the individual motor gestures, as described in chapter 7. Furthermore, the working memory buffers in pIFS and preSMA now contain cluster-sized SSCs, thereby reducing the number of items that have to be stored in working memory for /blu/ from three to two. The job of initiating the gesture for /l/ in /bl/ is now carried out by the basal ganglia motor loop (red dashed arrow) rather than the preSMA. The basal ganglia motor loop has also taken over the job of activating the gestural motor programs for /b/, /l/, and /u/ when the corresponding SMA initiation map cells have been activated.[11]

Comparing panels B and A, it is clear that the number of nodes that must be activated to produce /blu/ is reduced substantially in pIFS, preSMA, and vPMC, freeing up resources in these relatively high-level cortical processing areas. Similarly, the number of cortico-cortical communications (black arrows) has decreased substantially, being replaced by subcortical communications through the cerebellum (green arrows) and basal ganglia (red arrows). Among other benefits, the freed cortical resources allow larger utterance lengths to be accommodated by the speech sequencing system.

We performed two fMRI studies (Segawa et al., 2015; Beal et al., submitted) to test the account of speech sequence learning depicted in figure 8.7. In these studies, English-speaking adults were trained to produce syllables containing non-native consonant clusters. According to the GODIVA model, learning of these consonant clusters should be accompanied by decreased processing in vPMC, pIFS, and preSMA. To test these predictions, the production of novel non-native clusters was contrasted with production of non-native clusters that had been practiced prior to scanning. The results of a combined analysis of the data from these studies is shown in figure 8.8, along with spheres indicating the foci from the Rottschy et al. (2012) working memory meta-analysis as in figure 8.5.

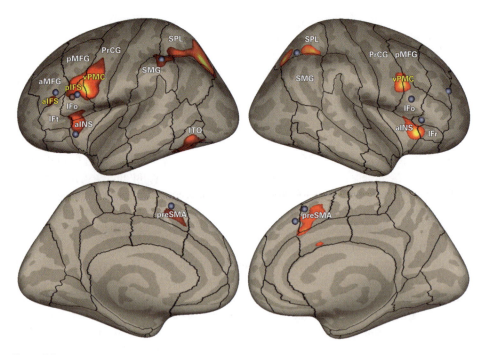

Figure 8.8
Blood-oxygen-level-dependent response in the *unlearned syllables–learned syllables* contrast from a pooled analysis (27 unique subjects) of the Segawa et al. (2015) and Beal et al. (submitted) fMRI studies. Individual vertices were thresholded at p < 0.005 uncorrected before a cluster-level analysis thresholded at a family-wise error rate of 0.05. Blue spheres indicate foci in the core working memory network identified from the Rottschy et al. (2012) meta-analysis. The brown sphere indicates the focus of increased activity for verbal versus nonverbal working memory in the Rottschy et al. (2012) meta-analysis. White labels indicate regions of interest as defined in appendix B. Yellow labels indicate additional regions discussed in the main text. aIFS, anterior inferior frontal sulcus; aINS, anterior insula; aMFG, anterior middle frontal gyrus; dPMC, dorsal premotor cortex; IFo, inferior frontal gyrus pars opercularis; IFr, inferior frontal gyrus pars orbitalis; IFt, inferior frontal gyrus pars triangularis; ITO, inferior temporal-occipital junction; pIFS, posterior inferior frontal sulcus; pMFG, posterior middle frontal gyrus; PrCG, precentral gyrus; preSMA, pre-supplementary motor area; SMG, supramarginal gyrus; SPL, superior parietal lobule; vPMC, ventral premotor cortex.

The model's predictions of higher activity for novel sequences in pIFS, preSMA, and vPMC were all upheld. Additionally, novel sequences had higher activity at almost all of the other foci corresponding to the core working memory network in the Rottschy et al. (2012) meta-analysis, including aINS and SMG bilaterally and left *anterior IFS* (*aIFS*). One possible account of the SMG activity is that reading the novel stimuli more heavily activated the phonological store in SMG (cf. Jonides et al., 1998). In support of this interpretation, Henson et al. (2000) found activity in SMG when comparing a delayed matching task involving letters to one involving nonverbal symbols (the authors suggest that this area participates in phonological recoding of visually presented verbal materials), and Crottaz-Herbette et al. (2004) found nearby areas along the left intraparietal sulcus to be more active in a verbal working memory task when stimuli were presented visually than when they were presented auditorily. Activity in aINS and left aIFS may reflect the central executive components of the Baddeley and Hitch model (as suggested by Rottschy et al., 2012). Alternatively, left aINS may constitute part of the speech sound map (see DIVA model node locations in appendix D).

Cerebellar Involvement in Speech Sequence Planning

Chapter 7 described the central role played by the cerebellum in feedforward control of speech movements, including the learning and generation of appropriate intergestural timing for the sequence of articulatory gestures within the motor program for a speech sound.[12] The cerebellar contributions discussed in chapter 7 were limited to the *execution* of speech gestural sequences. In contrast, the main focus of the current chapter, as well as the GODIVA model, is the *planning* of speech sequences. The intergestural sequencing performed during motor execution is a more local form of sequencing, limited to gestures within the same speech sound, in contrast with the global sequencing of multiple speech sounds in longer utterances addressed by the GODIVA model. Notably, disruption of the local gestural sequencing during execution of a sound, due, for example, to cerebellar damage in ataxic dysarthria, will indirectly affect the global metrical structure since the duration of sound units will be disrupted; that is, they will not properly conform to the global metrical structure intended by the speech planning mechanism. Whether the cerebellum also plays a *direct* role in global sequence planning is less clear.

Some clues regarding possible cerebellar involvement in speech sequence planning (as opposed to execution) can be gleaned from studies of covert speech (or "inner speech") and verbal working memory. Studies of covert speech have identified activation of the right cerebellar hemisphere in the absence of actual movement (e.g., Ackermann et al., 1998; Riecker et al., 2000), indicating that the cerebellum's role during speech is not exclusively related to overt execution. Ackermann et al. (2004) suggest that, through its involvement in inner speech, the cerebellum may also play a role in verbal working memory, for example, as part of a subvocal rehearsal mechanism for verbal material. This view is consistent with findings of impaired verbal working memory capacity with right-hemisphere cerebellar damage (e.g., Silveri et al., 1998). Chen and Desmond (2005) propose that verbal working memory involves two distinct cerebrocerebellar circuits: a frontal cortex/superior cerebellar network involved in articulatory processes and mental rehearsal, and a parietal cortex/inferior cerebellar network involved in phonological storage (cf. the Baddeley and Hitch model of working memory schematized in figure 8.3). Bohland and Guenther (2006) also noted functional differences between superior and inferior portions of the cerebellum in a task that required participants to temporarily store and then produce three-syllable sequences that varied in sequence complexity (three unique syllables vs. the same syllable repeated three times) as well as phonological complexity of the individual syllables. Activity in the lateral portions of the superior cerebellar cortex was sensitive to both sequence and syllable complexity, as expected if this region is part of the articulatory process component of the Baddeley and Hitch model. In contrast, the right inferior cerebellum was sensitive to sequence complexity but not the phonological complexity of the individual syllables in the sequence, suggesting that this region represents syllables as abstract chunks without regard for the internal content of these chunks. Based on these findings, Bohland and Guenther (2006) propose that the right inferior cerebellum, perhaps in concert with the left parietal lobe, was used to maintain a chunk-based working memory, whereas the lateral superior aspects of the cerebellum contribute to sequence organization in both subvocal rehearsal and overt production. The more medial (paravermal) portion of the superior cerebellar cortex was found to be more closely related to overt motor execution, as would be expected if this region is responsible for the feedforward control mechanisms treated in chapter 7.

In sum, the cerebellum appears to play roles in the generation of inner speech and verbal working memory that are above and beyond its role in motor execution. However, the degree to which these roles include direct contributions to global planning of speech sequences remains unclear.

8.6 Summary

Fluent speakers construct multiword phrases that must be temporarily stored (buffered) while the individual sound units in the phrase are produced. The processes of buffering a speech sound sequence and sequentially activating the corresponding motor programs are collectively termed *speech sound sequencing*. Current understanding of speech sound sequencing builds on prior models of serial behavior, phonology, and working memory.

Since Karl Lashley's seminal paper on serial order in 1951, two main classes of serial order models have been recognized. *Associative chain* models account for serial order by positing direct connections between item nodes that dictate item ordering. In contrast, *parallel representation* models posit that item nodes can be activated simultaneously, with serial order represented by the pattern of activity across these nodes. A refinement of the parallel representation model called *competitive queuing* further posits that the node with the highest activation wins a competition (mediated by inhibitory interconnections) with other active nodes, leading to the production of the item/action associated with that node. Upon completion, the node's activity is extinguished, the second-most-active node becomes the most active, and the process repeats until all items have been completed. A wide range of behavioral and neural evidence supports the parallel representation and competitive queuing accounts of motor sequencing, including speech sound sequencing.

Data from speech error studies imply that the speech sound sequencing system represents *phonological content* (such as the phonemes in an utterance) somewhat distinctly from *syllabic structure*. According to *frame-content* models of phonological representation, a syllable is represented by a frame with specific slots. For example, a CVC frame has slots for an initial consonant, a vowel, and a final consonant. To produce such a syllable, the brain must fill each slot with a phonological content item, for example, the onset consonant /k/, vowel /i/, and final consonant /p/ for the word "keep." The frame slots do not need to correspond to individual phonemes; for example, some researchers have proposed that initial consonant clusters act as a single item in a syllabic frame consisting of an onset, a nucleus, and a coda. Another common alternative posits an onset-rime frame structure. Although they differ regarding the size of the subsyllabic constituents that make up a syllable, these theories agree that, at some stage of the speech production process, syllabic structure is represented somewhat independently from phonological content.

In order to produce a multiword phrase generated by higher-level language centers, the brain must temporarily store the phonological items that make up the phrase in a temporary *working memory*. The most influential working memory model to date is that of Baddeley and colleagues, which posits a *central executive* that oversees performance in two slave systems: a *phonological loop* and a *visuospatial sketch pad*. Of these components, the phonological loop is the most relevant for speech sound sequencing. This loop is broken into an *articulatory process* that can covertly speak items in the sequence (inner speech) and a *phonological store* that effectively "hears" these covert productions and maintains them for a few seconds. The articulatory process is typically assumed to correspond to frontal cortex and the phonological store to parietal and/or superior temporal cortex.

The language production system is capable of generating an astronomical number of possible phoneme sequences for production. On the opposite end of the process, our brains appear to have thousands of optimized motor programs for producing syllables and other common phoneme subsequences. The working memory system must mediate between these two large-inventory systems. A key issue concerns the nature of the individual items in working memory. If the set of possible items is too large, an inordinate amount of cortical tissue would need to be devoted to this working memory representation. If the set is too small, then even short utterances will require many items to be simultaneously represented, limiting the system to very short phrases. We posit that phonological working memory uses a set of subsyllabic constituents that contains on the order of 100 distinct items (such as phonemes and consonant clusters) to interface between the near-infinite inventory of possible phoneme sequences and thousands of optimized motor programs for commonly produced subsequences, most notably common syllables from the native language.

The *GODIVA model* addresses the neural circuitry underlying speech sound sequencing, incorporating insights from the models of serial behavior, phonological representation, and working memory described above. The GODIVA model's components can be broken into two interacting cortico–basal ganglia loops: a *planning loop*, which is responsible for temporary storage of the elements of an upcoming utterance, and a *motor loop*, which is responsible for selecting and initiating the motor program for the next sound at the right instant in time. The planning loop involves a *sequence structure buffer* in pre-supplementary motor area that represents syllabic ordering and structure and a *phonological content buffer* in left posterior inferior frontal sulcus that represents phonological content in the form of subsyllabic constituents. These cortical regions interact through direct cortico-cortical projections as well as reentrant loops through the basal ganglia (including the caudate nucleus and globus pallidus) and ventral anterior thalamus. The motor loop, which overlaps with the DIVA model, involves an *initiation map* in supplementary motor area whose nodes dictate the timing of item production, and a *speech sound map* in left ventral premotor cortex that contains motor programs for common sound subsequences. These cortical regions interact through direct cortico-cortical projections as well as reentrant loops through the basal ganglia (including the putamen and globus pallidus) and ventral anterior and ventral lateral thalamus. The model's components can also be broken down into a medial premotor system that processes sequential structure and timing and a lateral prefrontal/premotor system that processes phonological content and associated motor programs.

According to the GODIVA model, production of a speech sound sequence proceeds as follows. High-level language processing regions load the phonological content of the utterance into the phonological content buffer and the global sequential structure (including syllable order and frame structure) into the sequential structure buffer. Both of these representations are modeled as competitive queues that can contain multiple coactive items representing multiple forthcoming syllables, with the order of those items represented by an activity gradient (hence the model's name, gradient order DIVA, or GODIVA). Projections from the sequential structure buffer to the initiation map indicate the next sound to be produced. The corresponding node in the initiation map monitors (via the cortico–basal ganglia motor loop) sensorimotor and cognitive context signals, becoming active when the correct context for the current sound is recognized. In parallel, the phonological content buffer signals the upcoming phonological items to left ventral premotor cortex, in the process choosing the best motor program for producing these items. For example, if the upcoming items are /pit/ and an optimized motor program exists for this syllable, then this motor program is chosen. If no motor program exists for the whole syllable, then motor programs for /p/, /i/, and /t/ will be activated in sequence. The readout of a motor program begins when the associated initiation map node is activated.

During speech development, it is postulated that learning in cortico-cerebellar and cortico–basal ganglia loops leads to the development of syllabic motor programs that require relatively little cortical involvement, thereby freeing up cortical resources. fMRI studies contrasting novel and learned syllables support the model's assertions, showing higher activity for novel syllables in brain areas associated with working memory, including the posterior inferior frontal sulcus (corresponding to the GODIVA phonological content buffer) and pre-supplementary motor area (corresponding to the GODIVA sequential structure buffer).

Notes

1. The term *subsyllabic constituent* will be used herein to refer to a phonological unit that is smaller than a syllable but no smaller than an individual phoneme. Within this definition, phonemes qualify as one type of SSC.
2. In some characterizations, the syllabic frame is represented as a hierarchy, with the nucleus and coda existing at a lower hierarchical level than the rime.
3. A later version of the model (Baddeley, 2000) added an *episodic buffer* as a third slave system.
4. The CELEX2 linguistic database identifies 22,339 distinct syllables in English.
5. For the sake of simplicity, this discussion ignores the possibility of repeated items within an utterance. See Silver et al. (2012) for a treatment of this issue.
6. The lowest number of units one might expect in this view would be equal to the number of phonemes in the language. For English, this number would be 44. The highest number would be the total number of unique SSCs in an ONC- or OR-based syllable frame, including individual phonemes as well as clusters. This number could range into the hundreds.
7. Rottschy et al. (2012) identify these locations as part of the inferior frontal gyrus rather than IFS, though both descriptions are valid since the ventral bank of the IFS is usually considered part of the inferior frontal gyrus (see the discussion in chapter 2, section 2.4).
8. Repetition suppression fMRI capitalizes on the fact that the brain's response to a repeated stimulus decreases with repetition. For example, a brain area that processes phonemes without regard to syllable structure will show suppression for the second syllable in the nonsense utterance /tid dit/ since the phonemes are the same in the two syllables, whereas a region that processes whole syllables will not since they are distinct syllables.
9. Peeva et al. (2010) could not dissociate phonemes from their component articulatory gestures, nor are there single-unit recordings from SMA during speech that could dissociate these two possibilities.
10. Computer simulations in Bohland, Bullock, and Guenther (2010) utilized a CV frame structure, but the model is compatible with other frame structures including the ONC or OR frame structures described in section 8.2. Current data do not definitively distinguish between these candidate representations in pIFS.
11. These basal ganglia loops in essence form a context-dependent version of an associative chain (see figure 8.1A) representing a highly stereotyped motor sequence, thereby "automating" production of this sequence.
12. Recall that a *speech sound* is defined herein as a speech unit with its own optimized motor program. This can be a phoneme, syllable, or word that is frequently produced and therefore has its own optimized motor program.

References

Ackermann, H., Mathiak, K., & Ivry, R. B. (2004). Temporal organization of "internal speech" as a basis for cerebellar modulation of cognitive functions. *Behavioral and Cognitive Neuroscience Reviews, 3,* 14–22.

Ackermann, H., Wildgruber, D., Daum, I., & Grodd, W. (1998). Does the cerebellum contribute to cognitive aspects of speech production? A functional magnetic resonance imaging (fMRI) study in humans. *Neuroscience Letters, 247,* 187–190.

Alario, F. X., Chainay, H., Lehericy, S., & Cohen, L. (2006). The role of the supplementary motor area (SMA) in word production. *Brain Research, 1076,* 129–143.

Alm, P. A. (2004). Stuttering and the basal ganglia circuits: a critical review of possible relations. *Journal of Communication Disorders, 37,* 325–369.

Averbeck, B. B., Chafee, M. V., Crowe, D. A., & Georgopoulos, A. P. (2002). Parallel processing of serial movements in prefrontal cortex. *Proceedings of the National Academy of Sciences of the United States of America, 99,* 13172–13177.

Averbeck, B. B., Chafee, M. V., Crowe, D. A., & Georgopoulos, A. P. (2003). Neural activity in prefrontal cortex during copying geometrical shapes: I. Single cells encode shape, sequence, and metric parameters. *Experimental Brain Research, 150,* 127–141.

Awh, E., Jonides, J., Smith, E. E., Schumacher, E. H., Koeppe, R. A., & Katz, S. (1996). Dissociation of storage and rehearsal in verbal working memory. *Psychological Science, 7,* 25–31.

Baddeley, A. D. (1986). *Working memory*. Oxford: Oxford University Press.

Baddeley, A. (1998). The central executive: a concept and some misconceptions. *Journal of the International Neuropsychological Society, 4*, 523–526.

Baddeley, A. (2000). The episodic buffer: a new component of working memory? *Trends in Cognitive Sciences, 4*, 417–423.

Baddeley, A. D., & Hitch, G. (1974). Working memory. In G. H. Bower (Ed.), *The psychology of learning and motivation: advances in research and theory* (Vol. 8, pp. 47–89). New York: Academic Press.

Beal, D. S., Segawa, J. A., Tourville, J. A., & Guenther, F. H. (submitted). The neural network for speech motor sequence learning in adults who stutter.

Bohland, J. W., Bullock, D., & Guenther, F. H. (2010). Neural representations and mechanisms for the performance of simple speech sequences. *Journal of Cognitive Neuroscience, 22*, 1504–1529.

Bohland, J. W., & Guenther, F. H. (2006). An fMRI investigation of syllable sequence production. *NeuroImage, 32*, 821–841.

Bullock, D., & Rhodes, B. (2003). Competitive queuing for serial planning and performance. In M. A. Arbib (Ed.), *Handbook of brain theory and neural networks* (pp. 241–244). Cambridge, MA: MIT Press.

Chen, S. H., & Desmond, J. E. (2005). Cerebrocerebellar networks during articulatory rehearsal and verbal working memory tasks. *NeuroImage, 24*, 332–338.

Cholin, J., Levelt, W. J. M., & Schiller, N. O. (2006). Effects of syllable frequency in speech production. *Cognition, 99*, 205–235.

Crosson, B. A. (1992). *Subcortical functions in language and memory*. New York: Guilford Press.

Crottaz-Herbette, S., Anagnoson, R. T., & Menon, V. (2004). Modality effects in verbal working memory: differential prefrontal and parietal responses to auditory and visual stimuli. *NeuroImage, 21*, 340–351.

Dell, G. S. (1988). The retrieval of phonological forms in production: tests of predictions from a connectionist model. *Journal of Memory and Language, 27*, 124–142.

Dell, G. S., Burger, L. K., & Svec, W. R. (1997). Language production and serial order: a functional analysis and a model. *Psychological Review, 104*, 123–147.

D'Esposito, M., Aguirre, G. K., Zarahn, E., Ballard, D., Shin, R. K., & Lease, J. (1998). Functional MRI studies of spatial and nonspatial working memory. *Brain Research. Cognitive Brain Research, 7*, 1–13.

Farrell, S., & Lewandowsky, S. (2004). Modelling transposition latencies: constraints for theories of serial order memory. *Journal of Memory and Language, 51*, 115–135.

Fiez, J. A., Raife, E. A., Balota, D. A., Schwarz, J. P., Raichle, M. E., & Petersen, S. E. (1996). A positron emission tomography study of the short-term maintenance of verbal information. *Journal of Neuroscience, 16*, 808–822.

Fowler, C., Treiman, R., & Gross, J. (1993). The structure of English syllables and polysyllables. *Journal of Memory and Language, 32*, 115–140.

Fromkin, V. A. (1971). The non-anomalous nature of anomalous utterances. *Language, 47*, 27–52.

Gabrieli, J. D. E., Poldrack, R. A., & Desmond, J. E. (1998). The role of left prefrontal cortex in language and memory. *Proceedings of the National Academy of Sciences of the United States of America, 95*, 906–913.

Gil Robles, S., Gatignol, P., Capelle, L., Mitchell, M.-C., & Duffau, H. (2005). The role of dominant striatum in language: a study using intraoperative electrical stimulations. *Journal of Neurology, Neurosurgery, and Psychiatry, 76*, 940–946.

Gordon, P. C., & Meyer, D. E. (1987). Hierarchical representation of spoken syllable order. In A. Allport, D. G. MacKay, & W. Prinz (Eds.), *Language perception and production* (pp. 445–462). London: Academic Press.

Grossberg, S. (1978 a). A theory of human memory: self-organization and performance of sensory-motor codes, maps, and plans. *Progress in Theoretical Biology, 5*, 233–374.

Grossberg, S. (1978 b). Behavioral contrast in short term memory: serial binary memory models or parallel continuous memory models? *Journal of Mathematical Psychology, 17*, 199–219.

Hardcastle, W. J., & Hewlett, N. (Eds.). (1999). *Coarticulation*. Cambridge, UK: Cambridge University Press.

Henson, R. N., Burgess, N., & Frith, C. D. (2000). Recoding, storage, rehearsal and grouping in verbal short-term memory: an fMRI study. *Neuropsychologia*, *38*, 426–440.

Henson, R. N., Norris, D. G., Page, M. P. A., & Baddeley, A. D. (1996). Unchained memory: error patterns rule out chaining models of immediate serial recall. *Quarterly Journal of Experimental Psychology*, *49*, 80–115.

Herman, A. B., Houde, J. F., Vinogradov, S., & Nagarajan, S. S. (2013). Parsing the phonological loop: activation timing in the dorsal speech stream determines accuracy in speech reproduction. *Journal of Neuroscience*, *33*, 5439–5453.

Houghton, G. (1990). The problem of serial order: a neural network model of sequence learning and recall. In R. Dale, C. Mellish, & M. Zock (Eds.), *Current research in natural language generation* (pp. 287–319). San Diego: Academic Press.

Johansen-Berg, H., Behrens, T. E. J., Robson, M. D., Drobnjak, I., Rushworth, M. F. S., Brady, J. M., et al. (2004). Changes in connectivity profiles define functionally distinct regions in human medial frontal cortex. *Proceedings of the National Academy of Sciences of the United States of America*, *101*, 13335–13340.

Jonides, J., Schumacher, E. H., Smith, E. E., Koeppe, R. A., Awh, E., Reuter-Lorenz, P. A., et al. (1998). The role of parietal cortex in verbal working memory. *Journal of Neuroscience*, *18*, 5026–5034.

Jürgens, U. (1984). The efferent and afferent connections of the supplementary motor area. *Brain Research*, *300*, 63–81.

Kent, R. D. (2000). Research on speech motor control and its disorders: a review and prospective. *Journal of Communication Disorders*, *33*(5), 391–427.

Kerns, J. G., Cohen, J. D., Stenger, V. A., & Carter, C. S. (2004). Prefrontal cortex guides context-appropriate responding during language production. *Neuron*, *43*, 283–291.

Klapp, S. T. (2003). Reaction time analysis of two types of motor preparation for speech articulation: action as a sequence of chunks. *Journal of Motor Behavior*, *35*, 135–150.

Lashley, K. S. (1951). The problem of serial order in behavior. In L. Jeffress (Ed.), *Cerebral mechanisms in behavior* (pp. 112–136). New York: Wiley.

Lehéricy, S., Ducros, M., Krainik, A., Francois, C., Van de Moortele, P.-F., Ugurbil, K., et al. (2004). 3-D diffusion tensor axonal tracking shows distinct SMA and pre-SMA projections to the human striatum. *Cerebral Cortex*, *14*, 1302–1309.

Levelt, W. J. M. (1989). *Speaking: from intention to articulation*. Cambridge, MA: MIT Press.

Levelt, W. J., Roelofs, A., & Meyer, A. S. (1999). A theory of lexical access in speech production. *Behavioral and Brain Sciences*, *22*, 1–38.

Levelt, W. J., & Wheeldon, L. (1994). Do speakers have access to a mental syllabary? *Cognition*, *50*, 239–269.

Loevenbruck, H., Collins, M. J., Beckman, M. E., Krishnamurthy, A. K., & Ahalt, S. C. (1999). Temporal coordination of articulatory gestures in consonant clusters and sequences of consonants. In O. Fujimura, B. D. Joseph, & B. Palek (Eds.), *Proceedings of Linguistics Phonetics 1998, Item Order in Language and Speech* (pp. 547–573). Charles University in Prague—The Karolinum Press.

Luppino, G., Matelli, M., Camarda, R., & Rizzolatti, G. (1993). Corticocortical connections of area F3 (SMA-proper) and area F6 (PreSMA) in the macaque monkey. *Journal of Comparative Neurology*, *338*, 114–140.

MacKay, D. G. (1970). Spoonerisms: the structure of errors in the serial order of speech. *Neuropsychologia*, *8*, 323–350.

MacNeilage, P. F. (1998). The frame/content theory of evolution of speech production. *Behavioral and Brain Sciences*, *21*, 499–511.

Matsuzaka, Y., Aizawa, H., & Tanji, J. (1992). A motor area rostral to the supplementary motor area (presupplementary motor area) in the monkey: neuronal activity during a learned motor task. *Journal of Neurophysiology*, *68*, 653–662.

Meijer, P. J. A. (1996). Suprasegmental structures in phonological encoding: the CV structure. *Journal of Memory and Language*, *35*, 840–853.

Middleton, F. A., & Strick, P. L. (2000). Basal ganglia and cerebellar loops: motor and cognitive circuits. *Brain Research. Brain Research Reviews*, *31*, 236–250.

Myers, E. B., Blumstein, S. E., Walsh, E., & Eliassen, J. (2009). Inferior frontal regions underlie the perception of phonetic category invariance. *Psychological Science, 20*, 895–903.

Ohman, S. (1966). Coarticulation in VCV utterances: spectrographic measurements. *Journal of the Acoustical Society of America, 39*, 151–168.

Paulesu, E., Frith, C. D., & Frackowiak, R. S. (1993). The neural correlates of the verbal component of working memory. *Nature, 362*, 342–345.

Peeva, M. G., Guenther, F. H., Tourville, J. A., Nieto-Castanon, A., Anton, J. L., Nazarian, B., et al. (2010). Distinct representations of phonemes, syllables, and supra-syllabic sequences in the speech production network. *NeuroImage, 50*, 626–638.

Petrides, M. (1991). Functional specialization within the dorsolateral frontal cortex for serial order memory. *Proceedings. Biological Sciences, 246*, 299–306.

Pierrot-Deseilligny, C., Milea, D., & Muri, R. M. (2004). Eye movement control by the cerebral cortex. *Current Opinion in Neurology, 17*, 17–25.

Poldrack, R. A., Wagner, A. D., Prull, M. W., Desmond, J. E., Glover, G. H., & Gabrieli, J. D. E. (1999). Functional specialization for semantic and phonological processing in the left inferior prefrontal cortex. *NeuroImage, 10*, 15–35.

Redgrave, P., Rodriguez, M., Smith, Y., Rodriguez-Oroz, M. C., Lehericy, S., Bergman, H., et al. (2010). Goal-directed and habitual control in the basal ganglia: implications for Parkinson's disease. *Nature Reviews. Neuroscience, 11*, 760–772.

Riecker, A., Ackermann, H., Wildgruber, D., Dogil, G., & Grodd, W. (2000). Opposite hemispheric lateralization effects during speaking and singing at motor cortex, insula, and cerebellum. *Neuroreport, 11*, 1997–2000.

Roelofs, A., & Meyer, A. S. (1998). Metrical structure in planning the production of spoken words. *Journal of Experimental Psychology: Learning, Memory, and Cognition, 24*, 922–939.

Rottschy, C., Langner, R., Dogan, I., Reetz, K., Laird, A. R., Schulz, J. B., et al. (2012). Modelling neural correlates of working memory: a coordinate-based meta-analysis. *NeuroImage, 60*, 830–846.

Schaltenbrand, G. (1975). The effects on speech and language of stereotactical stimulation in thalamus and corpus callosum. *Brain and Language, 2*, 70–77.

Segawa, J. A., Tourville, J. A., Beal, D. S., & Guenther, F. H. (2015). The neural correlates of speech motor sequence learning. *Journal of Cognitive Neuroscience, 27*, 819–831.

Sevald, C. A., Dell, G. S., & Cole, J. S. (1995). Syllable structure in speech production: are syllables chunks or schemas? *Journal of Memory and Language, 34*, 807–820.

Shattuck-Hufnagel, S. (1979). Speech errors as evidence for a serial order mechanism in sentence production. In E. Walker (Ed.), *Sentence processing: psycholinguistic studies presented to Merrill Garrett* (pp. 295–342). Hillsdale, NJ: Erlbaum.

Shattuck-Hufnagel, S. (1983). Sublexical units and suprasegmental structure in speech production planning. In P. MacNeilage (Ed.), *The production of speech* (pp. 109–136). New York: Springer-Verlag.

Shattuck-Hufnagel, S. (1992). The role of word structure in segmental serial ordering. *Cognition, 42*, 213–259.

Shattuck-Hufnagel, S. (2015). Prosodic frames in speech production. In M. A. Redford (Ed.), *The handbook of speech production* (pp. 419–444). Chichester: Wiley.

Shima, K., Mushiake, H., Saito, N., & Tanji, J. (1996). Role for cells in the presupplementary motor area in updating motor plans. *Proceedings of the National Academy of Sciences of the United States of America, 93*, 8694–8698.

Shima, K., & Tanji, J. (1998). Both supplementary and presupplementary motor areas are crucial for the temporal organization of multiple movements. *Journal of Neurophysiology, 80*, 3247–3260.

Shima, K., & Tanji, J. (2000). Neuronal activity in the supplementary and presupplementary motor areas for temporal organization of multiple movements. *Journal of Neurophysiology, 84*, 2148–2160.

Silver, M. R., Grossberg, S., Bullock, D., Histed, M. H., & Miller, E. K. (2012). A neural model of sequential movement planning and control of eye movements: Item-Order-Rank working memory and saccade selection by the supplementary eye fields. *Neural Networks, 26*, 29–58.

Silveri, M. C., Di Betta, A. M., Filippini, V., Leggio, M. G., & Molinari, M. (1998). Verbal short-term store-rehearsal system and the cerebellum: evidence from a patient with a right cerebellar lesion. *Brain*, *121*, 2175–2187.

Sörös, P., Sokoloff, L. G., Bose, A., McIntosh, A. R., Graham, S. J., & Stuss, D. T. (2006). Clustered functional MRI of overt speech production. *NeuroImage*, *32*, 376–387.

Stemberger, J. P. (1984). Length as a suprasegmental: evidence from speech errors. *Language*, *60*, 895–913.

Tanji, J. (2001). Sequential organization of multiple movements: involvement of cortical motor areas. *Annual Review of Neuroscience*, *24*, 631–651.

Tanji, J., & Shima, K. (1994). Role for supplementary motor area cells in planning several movements ahead. *Nature*, *371*, 413–416.

Treiman, R., & Danis, C. (1988). Short-term memory errors for spoken syllable are affected by the linguistic structure of the syllables. *Journal of Experimental Psychology*, *14*, 145–152.

Tremblay, P., & Gracco, V. L. (2006). Contribution of the frontal lobe to externally and internally specified verbal responses: fMRI evidence. *NeuroImage*, *33*, 947–957.

Van Buren, J. M. (1963). Confusion and disturbance of speech from stimulation in vicinity of the head of the caudate nucleus. *Journal of Neurosurgery*, *20*, 148–157.

Vousden, J. I., Brown, D. A., & Harley, T. A. (2000). Serial control of phonology in speech production: a hierarchical model. *Cognitive Psychology*, *41*, 101–175.

9
Prosody

As described in chapter 3, the speech signal consists of phonetic *segments* (e.g., phonemes) as well as suprasegmental *prosodic features* such as pitch, loudness, duration, and metrical structure that convey meaningful differences in linguistic or emotional (affective) state. Prosodic cues have two defining characteristics: (1) they co-occur with segmental units, and (2) they can extend over more than one segment (Lehiste, 1970). Prosodic control begins early in life; young children can modulate the prosody of their cries and babbles before they can produce well-formed speech sounds (de Boysson-Bardies, 1999). Prosodic control also may be more robust than segmental control as individuals with severely impaired speech can often control prosody despite little or no segmental clarity (Vance 1994; Patel, 2002, 2003, 2004).

To appreciate the influence of prosodic cues on communication, consider the following utterances involving the same short sequence of phonemes and syllables:

"*You* met him." (I didn't.)

"You *met* him." (You didn't just see him from across the room.)

"You met *him*." (Not his sister.)

Now consider that each of the above could be conveyed as a question simply by changing the pitch profile, and that all of these combinations could be produced with a sad, excited, or distressed voice. All told, these prosodic manipulations generate 18 different communicative messages from the same eight-phoneme string.

Two key acoustic features of the voice signal that convey prosodic information are the *fundamental frequency* ($F0$) of vocal fold vibration, which is perceived as *pitch*, and the *amplitude* of the acoustic signal, perceived as *loudness*. The modulation of F0 over the course of an utterance is often referred to as *intonation*, and it serves communicative purposes such as distinguishing questions from statements (the former marked by a rising intonation pattern at the end of the utterance, the latter marked by a flat or falling pattern) or conveying emotions such as excitement or sadness. F0 and amplitude increases are used together, along with increases in segment *duration*, to produce *emphatic stress* (also called *focus*) in order to highlight a particular word or words within a sentence (e.g., "That's *Tim's* car, not George's").

Beyond amplitude and F0, the voice signal contains subtle acoustic cues that collectively contribute to *voice quality*. For example, a voice may sound breathy and seductive, or it may quiver from anger or fear. Voice quality is an important cue for signaling affective state, but it does not play a significant role in signaling linguistic contrasts (Lehiste, 1970). Some aspects of voice quality can be attributed to physical properties of the vocal folds or to voice disorders such as spasmodic dysphonia or vocal hyperfunction. Here we are concerned only with those aspects that serve a communicative function such as signaling emotional state.

The speech signal also contains temporal cues that convey prosodic information, such as the relative durations of different syllables/words and the locations and durations of pauses. These cues will be referred to collectively as *metrical structure*. For example, the difference in meaning between the utterances "I went there and back; he came" and "I went there, and back he came" is conveyed by the location of the pause that corresponds to the comma or semicolon in the corresponding written sentence, coupled with lengthening the syllable just before the pause. Metrical structure can also provide cues regarding the emotional state of the speaker (Van Lancker & Sidtis, 1992).

Prosodic features can be divided into actively controlled aspects of the speech signal, such as the intonation profile, and aspects that are not willfully controlled by the speaker, such as a quivering voice. Our main concern here is with the former. As with segmental aspects of speech, these controlled aspects of prosody, which primarily involve movements of laryngeal and respiratory muscles,[1] can be viewed as movements generated by three integrated movement control subsystems (see figure 9.1): the *auditory feedback controller*, *somatosensory feedback controller*, and *feedforward controller*. For further details on these subsystems, the reader is referred to the relevant prior chapters of this book (chapters 5–7). The following sections address behavioral and neural aspects of these prosodic control processes.

9.1 Behavioral Studies of Prosodic Control

The most commonly studied prosodic cues are pitch, loudness, and duration. As described in the following paragraphs, a wide range of experimental evidence suggests that control of these cues involves both feedback and feedforward control components.

As reviewed in section 5.2 of chapter 5, experiments involving perturbations of pitch and loudness during speech have revealed evidence for auditory feedback control in the form of automatic compensations to these perturbations. For example, Lombard (1911) demonstrated that speakers increase loudness in noisy environments (the *Lombard effect*), and Elman (1981) demonstrated compensatory decreases in F0/pitch when the pitch of the talker's voice was electronically shifted upward. Hain et al. (2001) determined that

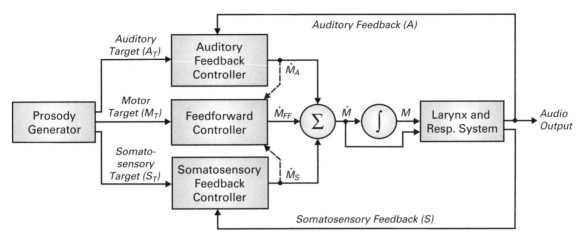

Figure 9.1
Control system for prosodic cues. The prosody generator (presumed to involve language processing regions of cortex) generates three signals: (1) a desired auditory signal (*auditory target,* denoted A_T) which includes intonation and amplitude trajectories, (2) a corresponding *somatosensory target* (S_T), and (3) a stored motor program (*motor target,* M_T) for achieving the desired intonation and amplitude trajectories. *Auditory* and *somatosensory feedback controllers* compare incoming auditory (A) and somatosensory (S) feedback to A_T and S_T, respectively. If there is a mismatch between the desired and actual sensory feedback signals, corrective movement commands ($\dot{M}_A$ and $\dot{M}_S$) are generated. The *feedforward controller* generates a feedforward movement command ($\dot{M}_{FF}$) by comparing M_T to the current motor state. The three movement commands are summed to produce the overall movement command $\dot{M}$, which is sent to the vocal tract along with a motor position command M formed by integrating $\dot{M}$ over time. The corrective motor commands $\dot{M}_A$ and $\dot{M}_S$ are also used to update the motor target for future productions (dashed arrows). Resp., respiratory.

delaying the auditory feedback of a subject's compensatory response to a pitch shift causes an increase in the duration of the initial response peak, a finding that was interpreted as strongly supporting the use of a closed-loop auditory feedback system for the control of F0, as implemented in the DIVA model (see also Larson et al., 2000, 2001). Additional studies (e.g., Hain et al., 2000; Bauer et al., 2006) indicate that compensatory responses to pitch and loudness perturbations can occur within about 100 to 150 ms of the onset of perturbed auditory feedback, providing an estimate of the delay inherent in the auditory feedback controller for pitch and loudness. This delay is approximately the same as the delay in response to auditory perturbation of segmental parameters such as formant frequencies.

In most pitch perturbation studies, subjects articulate sustained vowels in isolation, a rather unnatural speech task. Natke and Kalveram (2001) and Donath et al. (2002) extended these results by shifting pitch while subjects produced the nonsense word /tatatas/, where the first syllable could be either unstressed or stressed. If the first syllable was stressed (and therefore longer in duration), a compensatory response occurred during that syllable, but not if the first syllable was unstressed. The second syllable showed evidence of a compensatory response in both the stressed and unstressed cases. These findings indicate that talkers compensate for pitch shifts during production of syllable strings, though the compensation will occur during the first syllable only if this syllable is longer in duration than the latency of the compensatory response.

Evidence for feedforward control of pitch and loudness come from studies involving sustained perturbations of these auditory parameters during speech (see section 7.2 of chapter 7 for a review). For example, Jones and Munhall (2000) noted compensatory adaptation to sustained pitch perturbations, along with negative aftereffects when pitch feedback is returned to normal, and Patel et al. (in press) found compensatory adaptation and negative aftereffects when loudness was perturbed downward during stressed syllables. These findings implicate a feedforward control system that adapts over time by incorporating compensatory movements generated by the auditory feedback controller into the feedforward command, as detailed in chapter 7.

Although controlled by both feedback and feedforward control mechanisms much like segmental aspects of speech, prosodic cues including pitch, loudness, and duration change much more rapidly than segmental cues in response to changes in the auditory environment, for example, in response to switching off a cochlear implant (Perkell et al., 2007). Perkell et al. (2007) termed the rapidly changing cues *postural parameters* since they are quickly adjusted in response to environmental changes, contrasting them with *segmental parameters* such as formant frequencies that change much more slowly with a change in hearing status. These findings suggest that control of prosodic cues is more strongly influenced by the auditory feedback control subsystem than segmental cues, perhaps in part because of the slower timescale of the prosodic cues, which makes them less susceptible to delays in the auditory feedback control subsystem.

The control of emphatic stress involves simultaneous manipulation of pitch, loudness, and duration. It has long been known that different speakers use different combinations of these cues to signal emphatic stress, a phenomenon referred to as *cue trading* (Lieberman, 1960; Howell, 1993). For example, some speakers may rely more on duration than pitch or loudness to indicate stress, while others may use pitch or loudness more, and naive listeners appear to be able to leverage this phenomenon to perceive stress even in cases of severely dysarthric speech (Patel, 2002). These findings raise the question of whether speakers control pitch, duration, and loudness independently, as in the *independent channel model* of auditory feedback control depicted in panel A of figure 9.2, or in an integrated fashion, as in the *integrated model* in panel B. (These models are elaborations of the auditory feedback controller in figure 9.1.)

The independent channel model involves distinct auditory targets for pitch, loudness, and duration. Within the DIVA framework, these targets would be expected to project from premotor cortical areas to higher-order auditory cortical areas, where they are compared to auditory feedback of these cues. Errors are calculated for pitch, duration, and loudness separately, with each error signal leading to a corresponding corrective motor command. Using different combinations of pitch and loudness perturbations, Larson et al. (2007)

(A)

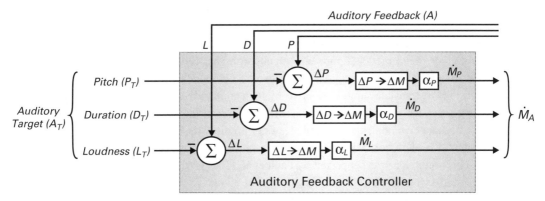

(B)

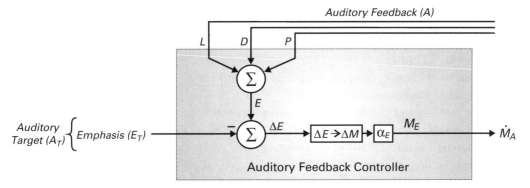

Figure 9.2
Two models for auditory feedback control of emphatic stress (cf. *auditory feedback controller* in figure 9.1). (A) In the *independent channel model*, the overall auditory target (A_T) contains distinct targets for each of the cues for emphatic stress: *pitch* (P), *duration* (D), and *loudness* (L). These targets are compared to incoming auditory feedback to detect separate pitch error (ΔP), duration error (ΔD), and loudness error (ΔL) signals. These error signals are transformed into motor coordinates and scaled by a corresponding parameter α to form corrective motor commands for each cue ($\dot{M}_P$, $\dot{M}_D$, $\dot{M}_L$). Together, these corrective motor commands make up the auditory feedback–based corrective motor command $\dot{M}_A$ in figure 9.1. (B) In the *integrated model*, the auditory target for emphatic stress consists of a single emphasis target (E_T) that characterizes the desired degree of emphatic stress for the current sound. Auditory feedback signals for pitch, duration, and loudness are combined to form an emphasis percept (E) which is compared to the emphasis target to produce an emphasis error signal (ΔE). This error is transformed into a corrective motor command for emphasis ($\dot{M}_E$), which contains commanded movements for all three emphasis cues (pitch, duration, and loudness). These movements constitute the auditory feedback–based corrective motor command $\dot{M}_A$ in figure 9.1.

demonstrated that the pitch response and loudness response are largely independent during extended vowel productions, as in the independent channel model. For example, the average response to an upward pitch shift was in the downward direction regardless of whether loudness was being perturbed upward or downward.

In the integrated model (figure 9.2B), cells in higher-order auditory cortex compare a target stress level to the perceived stress level to compute a single *emphasis* error signal which is then translated into corrective motor commands that adjust all three emphatic stress cues (pitch, duration, and loudness) to counteract the stress error. Unlike the independent channel model, the integrated model predicts that a perturbation to one stress cue should lead to compensatory responses in the other cues as well as the perturbed cue. This prediction was supported by Patel et al. (2011), who applied pitch perturbations to stressed syllables during running speech. In addition to compensatory adaptation of pitch, the authors found that talkers also adjusted the loudness and duration of stressed syllables to counteract the effect of the pitch shift on emphatic stress. Such compensations are not compatible with the independent channel model, which predicts only adjustments in the cue being perturbed. Thus, in addition to individual targets for pitch, loudness, and duration, the auditory feedback control subsystem appears to use an emphatic stress target that represents the combined effects of these cues, as in the integrated model of figure 9.2B.

Studies of somatosensory feedback control of prosodic cues have primarily investigated control of F0/pitch. Leonard and Ringel (1979) demonstrated that application of a topical anesthetic to the laryngeal mucosa led to impairments in performance of a pitch matching task, including slower responses and more oscillations around the target pitch. Sorenson, Horii, and Leonard (1980) noted increased jitter of F0 during sustained vowels when topical anesthesia was applied to the larynx. Larson et al. (2008) demonstrate that response to a pitch perturbation is larger if the vocal folds are anesthetized, presumably because this eliminates somatosensory error signals that would otherwise arise for large compensatory pitch adjustments (see related discussion in section 7.2 of chapter 7). Together, these studies indicate that somatosensory feedback control is involved in maintaining stability of F0 during speech.

Regarding the control of metrical timing in speech, a largely feedforward control process is suggested by the results of Gammon et al. (1971), who found no effects on stress pattern or juncture when topical anesthesia of the oral cavity was combined with masking noise during speech. However, metrical timing is not entirely insensitive to the influence of sensory feedback. For example, delaying auditory feedback by approximately 200 ms can result in severe disruption to the metrical timing of an utterance (e.g., Lee, 1950), and manipulations of auditory feedback that make an ongoing formant trajectory seem slower have the effect of delaying the onset of subsequent phonemes in the utterance (Cai et al., 2011).

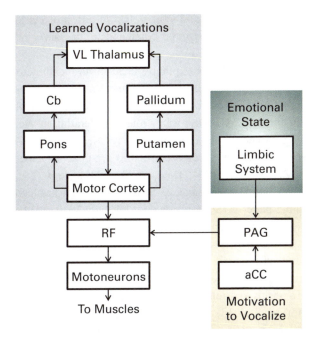

Figure 9.3
Schematic of the Jürgens (2009) primate vocalization model modified to account for emotional influences on voice quality. aCC, anterior cingulate cortex; Cb, cerebellum; PAG, periaqueductal gray matter; RF, reticular formation; VL, ventral lateral nucleus.

9.2 Neural Bases of Prosodic Control

In this section we will review the neuroimaging and lesion literatures related to prosody in order to identify the brain regions most likely to be responsible for the generation of prosody in speech production. For the purposes of this discussion, we will distinguish between areas involved in motor execution of prosodic cues (in particular the voice signal) from those involved in the planning or generation of prosody.

Motor Execution of Prosodic Cues

The primary articulator for generating prosodic information is the larynx, acting in concert with the respiratory system. In chapter 2, the Jürgens (2009) model of the neural circuit underlying primate vocalization was introduced. A modified version of this model is shown in figure 9.3. Starting at the output stage, motoneurons for the laryngeal muscles (located in the nucleus ambiguus of the medulla) project to the laryngeal musculature via cranial nerves X and XI (see section 2.2 of chapter 2 for further detail). The *anterior cingulate cortex* (*aCC*) is thought to signal the *motivation to vocalize* via projections to the *periaqueductal gray matter* (*PAG*) located in the midbrain, which in turn projects to the reticular formation where it modulates descending motor commands to the laryngeal muscles.

The *emotional state* component of figure 9.3 is not explicitly included in the Jürgens (2009) model but is added here to address a key component of affective prosody. As mentioned above, *voice quality* is a highly informative source regarding the affective state of the speaker. According to the Jürgens model, PAG plays a key role in generating innate, emotive vocalizations and modulating learned vocalizations in monkeys (see also Larson, 1985). Similar roles for PAG in vocalization have been identified in a number of other vertebrate species, including cats (Zhang et al., 1994; Davis et al., 1996), dogs (Liu, Solomon, & Luschei, 1998), and fish (Kittelberger, Land, & Bass, 2006). PAG has also been implicated in control of breathing patterns (Subramanian & Holstege, 2010, 2014), which can impart emotive qualities to the voice signal. Finally, PAG is a major processing center of signals related to pain (e.g., Basbaum & Fields, 1978), another factor that can affect voice quality. At present the human literature on the role of PAG in vocalization is sparse, but findings to date appear in line with the primate findings. The neuroimaging study of Dietrich et al. (2012) found a positive correlation between stress reaction and PAG activity during speech, and Wattendorf et al. (2013) noted PAG activity during the production of laughter. Collectively, these observations point to an important role for PAG in affective modulation of the human voice during speech. Although it is not currently clear what upstream emotional brain regions may be involved, PAG has bilateral connections with several limbic and paralimbic structures implicated in processing emotions (lumped together as *limbic system* in figure 9.3), including the hypothalamus, amygdala, insula, and cingulate cortex[2] (Vianna & Brandão, 2003).

The blue shaded portion of figure 9.3 indicates the core neural circuitry for generating learned vocalizations in primates. Motor commands for learned vocalizations project from motor cortex to brain stem motoneurons via the reticular formation.[3] In essence, chapters 5 through 7 of this book provide elaborations of this primate learned vocalization circuit to include sensory feedback control mechanisms in addition to feedforward control mechanisms and to account for learning and execution of segmental aspects of speech in humans. Motor execution of the laryngeal and respiratory movements underlying prosody appears to be carried out by largely the same brain networks that control articulations for segmental aspects of speech, including the auditory feedback control network (chapter 5), somatosensory feedback control network (chapter 6), and feedforward control network (chapter 7). This conclusion is supported by behavioral findings described in the previous section that indicate strong similarities in responses to unpredictable as well as sustained perturbations of prosodic and segmental cues.

Further support for this view comes from neuroimaging studies of prosody-related motor systems. The motor cortical and cerebellar foci for laryngeal and respiratory movements from the meta-analyses of simple articulator movements in appendix A are shown in figure

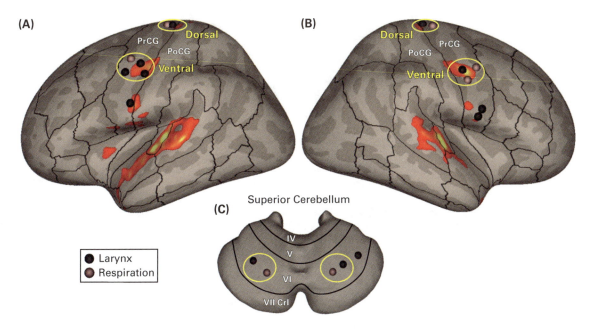

Figure 9.4
Motor cortical and cerebellar activation foci for simple respiratory movements (brown spheres) and laryngeal movements (black spheres) from the articulatory meta-analyses of appendix A plotted on the (A) inflated left lateral cortical surface, (B) inflated right lateral cortical surface, and (C) superior cerebellar surface. Hot colormap indicates cortical activity from 13 subjects performing a humming task with no supralaryngeal articulatory movement (contrasted with controlled breathing). Yellow circles indicate regions involved in both respiratory and laryngeal control, the two primary articulations required for execution of prosody. White labels in (A) and (B) indicate regions of interest as defined in appendix B. Roman numerals in (C) indicate cerebellar lobules. CrI, crus 1; PoCG, postcentral gyrus; PrCG, precentral gyrus.

9.4, along with cortical fMRI activity associated with pure phonation (humming) without movement of the supralaryngeal articulators. Two motor cortical locations (yellow circles, labeled *dorsal* and *ventral* in figure 9.4) contain foci for both laryngeal and respiratory movements. To date, only the ventral region, which includes the larynx motor area identified by Brown, Ngan, and Liotti (2008), has been identified in neuroimaging studies of prosodic control, primarily for linguistic prosodic manipulations (Aziz-Zadeh et al., 2010; Golfinopoulos et al., 2015). Subcortically, the paravermal portion of cerebellar lobule VI contains foci for both laryngeal and respiratory movements and has been identified in at least one study involving the generation of affective prosody (Mayer et al., 2002). It is noteworthy that both the ventral motor cortical region and the cerebellar lobule VI region contain representations of multiple speech articulators in addition to the larynx, including the jaw, lips, and tongue (see appendix A for details), suggesting that the same motor cortical and cerebellar regions responsible for executing segmental aspects of the speech signal

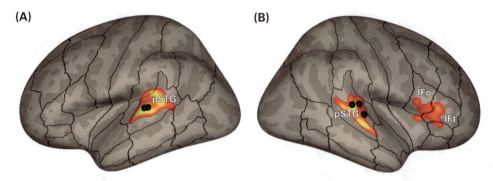

Figure 9.5
Comparison of auditory cortical activation peaks for pitch-perturbed speech (black spheres; peak coordinates from Parkinson et al., 2012) and formant-perturbed speech (hot colormap; see caption of figure 5.4 in chapter 5 for details) compared to unperturbed speech plotted on inflated (A) left and (B) right lateral cortical surfaces. IFo, inferior frontal gyrus pars opercularis; IFt, inferior frontal gyrus pars triangularis; pSTG, posterior superior temporal gyrus.

are also responsible for execution of prosodic aspects, at least when considering these regions at a macroscopic level.

The available neuroimaging evidence also suggests that auditory feedback control of F0 involves bilateral auditory error maps in the same portion of the *posterior superior temporal gyrus* (*pSTG*) that contains auditory error maps for formant frequencies. This is illustrated in figure 9.5, which shows auditory cortical activation peaks in response to a perturbation of pitch during speech (black spheres; Parkinson et al., 2012) overlaid on brain activity associated with perturbations of F1 and F2 during speech (Tourville et al., 2008; Niziolek & Guenther, 2013).

Although neuroimaging evidence suggests bilateral auditory and motor representations for the execution of prosody, there is evidence from the lesion literature that certain aspects of prosody may rely more heavily on right-hemisphere mechanisms compared to segmental aspects of speech, a topic we will return to below.

Perceiving and Generating Prosodic Messages
The neuroimaging literature indicates that perception and generation of prosodic messages involves regions beyond the core areas for speech motor control described in the previous subsection. To date, neuroimaging studies that aim to separate prosodic mechanisms from other speech and language mechanisms have largely focused on the perception of prosodic cues rather than prosody generation (e.g., George et al., 1996; Buchanan et al., 2000; Meyer et al., 2002; Kotz et al., 2003; Mitchell et al., 2003; Doherty et al., 2004). Figure 9.6 illustrates the cortical activity foci derived from two ALE meta-analyses of prosody perception (Witteman et al., 2012; Belyk & Brown, 2013). Red spheres indicate foci from studies of affective prosody, and green spheres indicate foci from studies of linguistic prosody. A

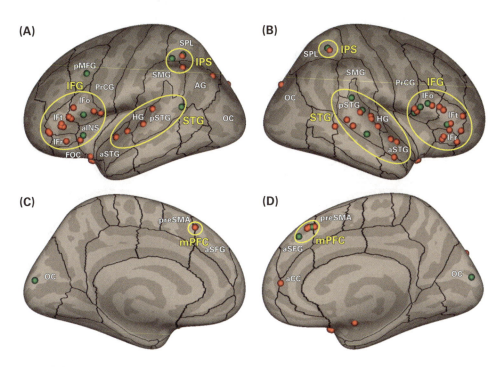

Figure 9.6
Activation foci from ALE meta-analyses of neuroimaging studies of linguistic prosody perception (green spheres; Belyk & Brown, 2013) and affective prosody perception (red spheres; Witteman et al., 2012; Belyk & Brown, 2013) plotted on inflated (A) left lateral, (B) right lateral, (C) left medial, and (D) right medial cortical surfaces. White labels indicate regions of interest as defined in appendix B. Yellow circles and labels indicate additional regions discussed in the main text. aCC, anterior cingulate cortex; AG, angular gyrus; aINS, anterior insula; aSFG, anterior superior frontal gyrus; aSTG, anterior superior temporal gyrus; HG, Heschl's gyrus; FOC, frontal orbital cortex; IFG, inferior frontal gyrus; IFo, inferior frontal gyrus pars opercularis; IFr, inferior frontal gyrus pars orbitalis; IFt, inferior frontal gyrus pars triangularis; IPS, intraparietal sulcus; mPFC, medial prefrontal cortex; OC, occipital cortex; pMFG, posterior middle frontal gyrus; PrCG, precentral gyrus; preSMA, pre-supplementary motor area; pSTG, posterior superior temporal gyrus; SMG, supramarginal gyrus; SPL, superior parietal lobule; STG, superior temporal gyrus; vPMC, ventral premotor cortex.

heavy concentration of foci is found in the *inferior frontal gyrus* (*IFG*) bilaterally. This is true for both affective and linguistic prosody, though the foci for affective prosody extend further anteriorly than those for linguistic prosody. Another heavy concentration of foci is found in auditory cortical regions in the *superior temporal gyrus* (*STG*), with a rightward bias for both affective and linguistic prosody. This may reflect a right-hemisphere bias for representing relatively slowly varying aspects of the acoustic signal such as the global pitch profile (e.g., Poeppel, 2003; Sidtis & Van Lancker Sidtis, 2003). Affective prosody also involves left-hemisphere auditory cortical areas, perhaps because of timing cues that signal emotional state (Van Lancker & Sidtis, 1992). Foci also appear in the right medial prefrontal cortex, including the *pre-supplementary motor area* (*preSMA*) and *anterior*

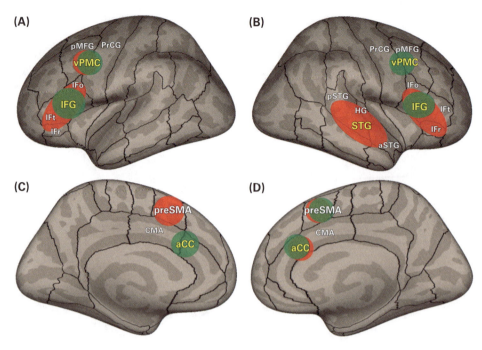

Figure 9.7
Schematic of cortical regions implicated in the generation of affective (red) and linguistic (green) prosody in fMRI studies by Mayer et al. (2002), Aziz-Zadeh et al. (2010), Pichon and Kell (2013), and Golfinopoulos et al. (2015) plotted on inflated (A) left lateral, (B) right lateral, (C) left medial, and (D) right medial cortical surfaces. White labels indicate regions of interest as defined in appendix B. Yellow labels indicate additional regions discussed in the main text. aCC, anterior cingulate cortex; aSTG: anterior superior temporal gyrus; CMA, cingulate motor area; IFG, inferior frontal gyrus; IFo, inferior frontal gyrus pars opercularis; IFr, inferior frontal gyrus pars orbitalis; IFt, inferior frontal gyrus pars triangularis; pMFG, posterior middle frontal gyrus; PrCG, precentral gyrus; preSMA, pre-supplementary motor area; pSTG, posterior superior temporal gyrus; STG, superior temporal gyrus; vPMC, ventral premotor cortex.

superior frontal gyrus (*aSFG*), and bilateral *intraparietal sulcus* (*IPS*) for both types of prosody perception.

Very few neuroimaging studies have directly investigated prosody production to date. A schematic compilation of the results of four such fMRI studies of prosody production (Mayer et al., 2002; Aziz-Zadeh et al., 2010; Pichon & Kell, 2013; Golfinopoulos et al., 2015) is provided in figure 9.7. Overall, the regions identified in these studies of prosody production overlap substantially with those involved in prosody perception. The following paragraphs discuss these studies in further detail.

Three neuroimaging studies have involved comparison of speech produced with an emotional affect (e.g., *happy*, *sad*, *angry*) to neutral speech. Mayer et al. (2002) noted right-lateralized activity in IFG for the emotive conditions, particularly in the *pars orbitalis* (*IFo*; Brodmann area 47). Aziz-Zadeh et al. (2010) found that *happy* speech yielded activity peaks in preSMA (with a left-hemisphere bias), aCC, left posterior IFG, and left *ventral premotor cortex* (*vPMC*) near the laryngeal motor area. Notably, there is no overlap in the regions identified by these two studies of emotional prosody generation. Pichon and Kell (2013) contrasted the production of sentences under *fearful, sad, angry*, and *happy* conditions to producing the same sentences with neutral prosody. The emotionally charged speech was accompanied by increased bilateral activity in IFG, consistent with the studies of prosody perception summarized in figure 9.6, as well as the Mayer et al. (2002) and Aziz-Zadeh et al. (2010) studies of prosody production. Additional activity for emotionally charged speech was found in auditory cortical areas in STG with a rightward bias. Subcortically, emotional prosody production was associated with increased activity bilaterally in the cerebellum, basal ganglia (including the globus pallidus and substantia nigra), and ventral lateral thalamus. One important caveat about these studies of emotional prosody generation is that *acting like* you are happy or sad is different than truly *being* happy or sad, suggesting that the brain regions identified in figure 9.6 may have as much or more to do with acting than with generation of true affective prosody. This concern is somewhat alleviated by the observation that the same regions appear to be involved in the perception of affective prosody (figure 9.6), which does not require the subject to artificially generate an affective state.

Linguistic prosody has also been studied in three studies. Mayer et al. (2002) had subjects produce nonsense syllable strings in two linguistic prosody conditions, a *focus* condition involving emphatic stress on one or more words, and a *modus* condition involving the use of boundary tones such as a rising pitch at the end of an utterance to indicate that it is a question. The prosodic conditions were contrasted with a *mono* condition involving the production of nonsense syllables in a monotone voice. Both *focus–mono* and *modus–mono* contrasts yielded activity in left IFG. The *focus–mono* contrast also identified activity in left vPMC. When contrasting production of sentences involving emphatic stress to those with no emphatic stress, Golfinopoulos et al. (2015) found activity in right ventral motor/premotor cortex (approximately in the right-hemisphere homologue of the region identified by Mayer et al., 2002) as well as right SMA. Aziz-Zadeh et al. (2010) contrasted the speaking of nonsense syllable strings with a *question* intonation pattern to producing the same strings with *neutral* prosody. This contrast yielded activity peaks in bilateral vPMC, left posterior IFG, right preSMA, and right aCC. Subcortically, the *question–neutral* contrast yielded activity peaks in right cerebellum and right putamen.

In sum, neuroimaging studies of prosody indicate possible roles in the generation of prosody for IFG, vPMC, SMA/preSMA, and aCC as schematized in figure 9.7, with most of these regions showing evidence of bilateral involvement. One limitation of these studies is that they conflate different aspects of prosodic control. For example, emphatic stress involves manipulations of both intonation and metrical timing, as do affect-related production tasks. The following paragraphs attempt to tease apart the different elements of prosodic control, particularly with regard to cortical laterality, through review of the neurological literature on prosody.

Hemispheric Lateralization of Prosodic Processing

Much of the lesion literature on prosody has focused on left-hemisphere versus right-hemisphere specialization, particularly in the auditory, premotor, and prefrontal regions of the cerebral cortex. Accordingly, a number of so-called *hemispheric models* of prosodic function have been proposed. For example, in their *dynamic dual pathway* model of auditory language comprehension, Friederici and Alter (2004) posit that the left hemisphere primarily processes syntactic and semantic information whereas the right hemisphere primarily processes sentence-level prosody. A different type of hemispheric model suggests that the difference between hemispheres has to do with acoustic processing; the left hemisphere is specialized for the processing of timing (rate) whereas the right hemisphere is specialized for pitch processing (e.g., Sidtis & Van Lancker Sidtis, 2003). Variants of this view suggest that the left hemisphere is specialized for analyzing sounds using a short temporal window (as needed to distinguish rapidly changing segmental cues) while the right hemisphere preferentially processes sound using a longer temporal window, making it more suitable for spectral rather than temporal processing (Poeppel, 2003; Zatorre, Belin, & Penhune, 2002).

One common theme of these models concerns a *right-hemisphere preference for processing of pitch and intonation*. Evidence from the lesion literature indicates right-hemisphere dominance for perceptual processing of pitch as well as for the production of intonation patterns during speech. On the perceptual side, Zatorre (1988) compared left temporal lobectomy patients to right temporal lobectomy patients in a task involving pitch perception from complex tones. Only right-hemisphere patients with lobectomies involving Heschl's gyrus and surrounding areas were impaired on this task. Similarly, Sidtis and Volpe (1988) found that only right-hemisphere stroke patients were impaired on a complex-pitch perception task whereas left-hemisphere patients were impaired on a speech perception task. Right-hemisphere damage also appears to impair the perception of emotional prosody (see Baum & Pell, 1999, and Ackermann, Hertrich, & Ziegler, 1993, for reviews), and Van Lancker and Sidtis (1992) provide evidence that this impairment is greatest for pitch-based cues as opposed to temporal cues (which are more affected by left-hemisphere damage).

On the production side, speakers with right-hemisphere damage use less variation of F0 than neurologically normal speakers, and they are less able to convey emphasis or emotional contrasts (Pell, 1999). Ross and Mesulam (1979) and Ghacibeh and Heilman (2003) report patients with right-hemisphere damage who lost the ability to impart affective qualities to their speech. Van Lancker Sidtis et al. (2006) report a patient with right-hemisphere basal ganglia damage who lost the ability to mimic other dialects and sing, and whose speech was monotone. The fMRI study of Riecker et al. (2000) found that reproducing a nonlyrical tune through voice intonation elicited activity in right-hemisphere motor cortex and anterior insula and left-hemisphere cerebellum[4] whereas speaking produced an opposite pattern of hemispheric dominance (see also Wildgruber et al., 1996). In their review of cases of dysprosodic speech during epileptic seizures, Peters et al. (2011) note that dysprosody, primarily in the form of disrupted intonation, was associated with epileptic seizures in the right (nondominant) hemisphere but never during left-hemisphere seizures.

The neurology and neuroimaging literature on singing, which shares with prosodic intonation the requirement of generating a well-learned pitch contour while articulating, provides additional insight into the neural mechanisms of intonation. It has been noted that aphasics with left-hemisphere damage can often sing lyrics of a song better than they speak the same words (e.g., Albert, Sparks, & Helm; 1973; Schlaug et al., 2010; Yamadori et al., 1977). The fMRI study of Ozdemir, Norton, and Schlaug (2006) found a large degree of overlap in brain mechanisms underlying singing and speaking. Overall singing was accompanied by more activity than speaking the same words in several regions of the speech network bilaterally, including ventral Rolandic cortical areas and auditory cortical areas (Heschl's gyrus as well as anterior and middle portions of STG), with a stronger response in the right hemisphere. Singing also evoked more activity than speaking in the most inferior portion of the IFG bilaterally.

A second theme common to many hemispheric models is a *left-hemisphere preference for processing segmental and/or lexical information*, including prosodic cues that signal differences in word meaning (e.g., raising pitch to stress the first syllable in *off*ense vs. the second syllable in *off*ence). A review of the lesion literature related to lexical stress by Häuser and Domahs (2014) found that impaired generation of lexical stress was nearly always associated with damage to the left (language-dominant) hemisphere, and the fMRI study by Gandour et al. (2003) found that lexical pitch in a tonal language such as Chinese appears to be predominantly processed by the left hemisphere. These lexical uses of stress and intonation are likely more appropriately considered part of a word's motor program than as suprasegmental or prosodic parameters.

A third common theme in hemispheric models is a *left-hemisphere preference for processing timing information*. When reviewing the lesion literature on metrical structure and timing-related aspects of prosody, it is important, but not always possible, to distinguish between brain areas whose damage can *impact* metrical structure from the smaller set of regions that *generate* metrical structure. Damage to almost any part of the feedforward control system for speech will impact the metrical structure of a spoken utterance because of poor execution of speech motor programs, including poor intergestural coordination and timing. Similarly, an impaired ability of the language network to fluently generate longer utterances, for example, because of Broca's aphasia, will necessarily result in abnormal metrical structure characterized by long pauses between words (Ackermann, Hertrich, & Ziegler, 1993). In these cases, impaired metrical structure does not imply that the damaged brain regions are responsible for *generating* metrical structure.

Nonetheless, insights into the generation of metrical structure in speech can be gleaned from the neurology literature. One frequent finding is that patients with left-hemisphere damage appear to have larger deficits in perceiving and generating timing cues than right-hemisphere patients. For example, Van Lancker and Sidtis (1992) report that left-hemisphere patients relied mostly on F0 rather than temporal cues when perceiving affective prosody, consistent with the view that the intact right-hemisphere auditory areas are specialized for detecting global intonation profiles (which change relatively slowly) while the damaged left-hemisphere auditory areas are specialized for more precise temporal processing.

With regard to generation of metrical structure, Ouellette and Baum (1994) reported that left-hemisphere patients were impaired in their of use temporal cues to mark emphatic stress correctly compared to patients with right-hemisphere damage whereas measures of F0 were normal for the left-hemisphere patients. Gandour and Baum (2001) identified a deficit in patients with left-hemisphere damage in generating global stress patterns that conform to norms for the language. In English, speakers prefer to alternate stress and will modify a word's stress depending on the word immediately following a word that carries stress, a process called *stress retraction*. For example, the word "thirteen," which is usually stressed on the second syllable, becomes "*thir*teen" if followed by a word like "women," which is stressed on the first syllable. Compared to neurotypical controls and right-hemisphere patients, the left-hemisphere patients were impaired particularly in their generation of the temporal cues that signal stressed syllables.

To date, the lesion-based findings of left-hemisphere cortical dominance for production of timing and metrical structure have received little support from the neuroimaging literature, perhaps in part because of a lack of studies that distinguish metrical structure from other aspects of speech prosody. In one of the very few neuroimaging studies to explicitly target metrical structure in speech, Riecker et al. (2002) contrasted the production of short syllable sequences with a rhythmic pattern (e.g., short-short-long) to producing the same syllables isochronously. The rhythmic production condition yielded increased activity in left putamen and thalamus and in right STG, IFo, and premotor cortex. This right-hemisphere cortical bias appears to contradict the lesion findings of left-hemisphere bias for generation and perception of timing cues described above. Note, however, that isochronous syllable production (which, like the rhythmic condition, is also an "unnatural" timing pattern) might also involve regions responsible for generation of metrical structure, possibly explaining the lack of left-hemisphere cortical activity when contrasting the rhythmic and isochronous production conditions in Riecker et al. (2002).

Disturbances to global metrical structure resulting from subcortical damage are typically, though not always, attributed to impaired motor execution rather than impaired planning of metrical structure. For example, reduced prosodic modulation often occurs in both Parkinson's disease and Huntington's disease (Ackermann et al., 1993). In contrast, Parkinson's patients are largely unimpaired in terms of speech tempo (Ackermann &

Ziegler, 1991), prompting Ackermann et al. (1993) to conclude that reduced prosodic modulation in basal ganglia diseases likely results from impaired motor execution. However, Van Lancker Sidtis et al. (2006) report a patient with a lesion confined to globus pallidus and putamen that had preserved ability to sing but had monotonous speech, suggestive of a deficit at the level of prosodic planning rather than a motor execution deficit. The lesion literature concerning the role of the cerebellum in controlling metrical structure is confounded by the fact that the cerebellum is a major contributor to gestural timing at the motor program level, which would impact metrical structure independent of any potential role in planning or controlling it. Nonetheless, at least one finding is suggestive of a possible role in generating metrical timing in speech above any role in motor execution. Ataxic dysarthria due to cerebellar damage impairs the temporal organization of speech articulations and can result in so-called *scanning speech* in which syllables tend to all have similar durations of about 300 to 350 ms (Ackermann, 2008). This finding of constant duration across syllables hints that the cerebellum may be at least partly responsible for generating durational cues that distinguish stressed from unstressed syllables.

9.3 Summary

Prosodic cues such as pitch, loudness, duration, and metrical structure shape the meaning of speech utterances. Intonation, which involves variations of pitch and loudness over the duration of an utterance, serves linguistic purposes such as distinguishing questions and statements. Emphatic stress puts focus on particular words within a sentence through increases in pitch, duration, and loudness. Voice quality conveys information about the speaker's emotional state.

Behavioral studies suggest that pitch, loudness, and duration are controlled via feedforward and feedback control mechanisms similar to those described in chapters 5 through 7. Compensatory adjustments in response to pitch, loudness, and duration perturbations provide evidence of an auditory feedback controller for these prosodic parameters with a delay and gain similar to those found in the auditory feedback controller for segmental parameters such as formant frequencies. In addition to individual targets for pitch, loudness, and duration, the auditory feedback control subsystem appears to use an integrated stress target that represents the combined effects of these cues. This results in compensatory cue trading—for example, a downward perturbation of pitch during a stressed syllable results in compensatory adjustments to loudness and duration in addition to pitch. Evidence for somatosensory feedback control of pitch comes from studies involving the application of a topical anesthetic to the larynx, which results in increased pitch variation due to the elimination of somatosensory feedback control. Feedforward control mechanisms for pitch and loudness are revealed by studies of sustained pitch and loudness perturbations, which result in sensorimotor adaptations that only gradually return to baseline when feedback is returned to normal. Feedforward control of duration and metrical structure is revealed by studies involving simultaneous removal of auditory and somatosensory feedback; despite the near-complete lack of relevant sensory feedback, speakers show no impairment of metrical structure during speech.

With the exception of affective aspects of voice quality, which likely involve limbic and paralimbic structures acting on the periaqueductal gray matter, motor execution of prosodic movements appears to involve the same cortical and subcortical regions that control articulations for segmental aspects of speech. In Rolandic cortex, distinct dorsal and ventral locations may be involved in coordinating respiratory and laryngeal movements for generating prosodic cues. At the motor planning and linguistic levels, prosody generation involves cortical regions outside the speech motor network, including bilateral inferior frontal gyrus, pre-supplementary motor area, and anterior cingulate cortex. A number of hemispheric models of prosodic function have been proposed to account for neurological findings regarding hemispheric specialization of prosodic processing. Common themes in these models include a right-hemisphere bias for the processing of pitch and intonation, a left-hemisphere bias for processing lexical information, including lexical stress and lexical tones in tonal languages, and a left-hemisphere bias for processing timing-related prosodic cues.

Notes

1. Though the larynx and respiratory system are the primary articulators for generating prosodic cues, other articulators such as the jaw, tongue, and lips also contribute to some aspects of prosody. For example, emphatic stress of a syllable containing a low vowel such as /a/ involves increased lowering of the tongue and opening of the jaw and lips, manipulations that increase the sound intensity and provide visual cues to the listener regarding emphasis.

2. It is possible that aCC is involved in signaling both the motivation to vocalize and the emotional state to PAG, and further that these two aspects of speech are not really dissociable, in which case the *emotional state* and *motivation to vocalize* components of figure 9.3 could be collapsed into a single component. They are treated separately in figure 9.3 to highlight aspects of affective prosody related to voice quality and to convey the possibility that limbic areas beyond aCC may be responsible for emotional aspects of voice quality.

3. Humans also have direct corticobulbar projections to motoneurons not shown in figure 9.3; see chapter 2 for details.

4. Recall from chapter 2 that the cerebellar hemispheres interact primarily with the contralateral hemispheres of the cerebral cortex.

References

Ackermann, H. (2008). Cerebellar contributions to speech production and speech perception: psycholinguistic and neurobiological perspectives. *Trends in Neurosciences, 31*, 265–272.

Ackermann, H., Hertrich, I., & Ziegler, W. (1993). Prosodic disorders in neurologic diseases—a review of the literature. *Fortschritte der Neurologie-Psychiatrie, 61*, 241–253.

Ackermann, H., & Ziegler, W. (1991). Articulatory deficits in parkinsonian dysarthria: an acoustic analysis. *Journal of Neurology, Neurosurgery, and Psychiatry, 54*, 1093–1098.

Albert, M. L., Sparks, R. W., & Helm, N. A. (1973). Melodic intonation therapy for aphasia. *Archives of Neurology, 29*, 130–131.

Aziz-Zadeh, L., Sheng, T., & Gheytanchi, A. (2010). Common premotor regions for the perception and production of prosody and correlations with empathy and prosodic ability. *PLoS One, 20*, e8759.

Basbaum, A. I., & Fields, H. L. (1978). Endogenous pain control mechanisms: review and hypothesis. *Annals of Neurology*, *4*, 451–462.

Bauer, J. J., Mittal, J., Larson, C. R., & Hain, T. C. (2006). Vocal responses to unanticipated perturbations in voice loudness feedback: an automatic mechanism for stabilizing voice amplitude. *Journal of the Acoustical Society of America*, *119*, 2363–2371.

Baum, S. R., & Pell, M. D. (1999). The neural bases of prosody: insights from lesion studies and neuroimaging. *Aphasiology*, *13*, 581–608.

Belyk, M., & Brown, S. (2013). Perception of affective and linguistic prosody: an ALE meta-analysis of neuroimaging studies. *Social Cognitive and Affective Neuroscience*, *9*, 1395–1403.

Brown, S., Ngan, E., & Liotti, M. (2008). A larynx area in the human motor cortex. *Cerebral Cortex*, *18*, 837–845.

Buchanan, T. W., Lutz, K., Mirzazade, S., Specht, K., Shah, N. J., Zilles, K., et al. (2000). Recognition of emotional prosody and verbal components of spoken language: an fMRI study. *Brain Research. Cognitive Brain Research*, *9*, 227–238.

Cai, S., Ghosh, S. S., Guenther, F. H., & Perkell, J. S. (2011). Focal manipulations of formant trajectories reveal a role of auditory feedback in the online control of both within-syllable and between-syllable speech timing. *Journal of Neuroscience*, *31*, 16483–16490.

Davis, P. J., Zhang, S. P., Winkworth, A., & Bandler, R. (1996). Neural control of vocalization: respiratory and emotional influences. *Journal of Voice*, *10*, 23–38.

de Boysson-Bardies, B. (1999). *How language comes to children* (M. B. DeBevoise, Trans.). Cambridge, MA: MIT Press.

Dietrich, M., Andreatta, R. D., Jiang, Y., Joshi, A., & Stemple, J. C. (2012). Preliminary findings on the relation between the personality trait of stress reaction and the central neural control of human vocalization. *International Journal of Speech-Language Pathology*, *14*, 377–389.

Doherty, C. P., West, W. C., Dilley, L. C., Shattuck-Hufnagel, S., & Caplan, D. (2004). Question/statement judgments: an fMRI study of intonation processing. *Human Brain Mapping*, *23*, 85–98.

Donath, T. M., Natke, U., & Kalveram, K. T. (2002). Effects of frequency-shifted auditory feedback on voice F0 contours in syllables. *Journal of the Acoustical Society of America*, *111*, 357–366.

Elman, J. L. (1981). Effects of frequency-shifted feedback on the pitch of vocal productions. *Journal of the Acoustical Society of America*, *70*, 45–50.

Friederici, A. D., & Alter, K. (2004). Lateralization of auditory language functions: a dynamic dual pathway model. *Brain and Language*, *89*, 267–276.

Gammon, S. A., Smith, P. J., Daniloff, R. G., & Kim, C. W. (1971). Articulation and stress-juncture production under oral anesthetization and masking. *Journal of Speech and Hearing Research*, *14*, 271–282.

Gandour, J., & Baum, S. R. (2001). Production of stress retraction by left- and right-hemisphere-damaged patients. *Brain and Language*, *79*, 482–494.

Gandour, J., Dzemidzic, M., Wong, D., Lowe, M., Tong, Y., Hsieh, L., et al. (2003). Temporal integration of speech prosody is shaped by language experience: an fMRI study. *Brain and Language*, *84*, 318–336.

George, M. S., Parekh, P. I., Rosinsky, N., Ketter, T. A., Kimbrell, T. A., Heilman, K. M., et al. (1996). Understanding emotional prosody activates right hemisphere regions. *Archives of Neurology*, *53*, 665–670.

Ghacibeh, G. A., & Heilman, K. M. (2003). Progressive affective aprosodia and prosoplegia. *Neurology*, *60*, 1192–1194.

Golfinopoulos, E., Cai, S., Blood, A., Burns, J., Noordzij, J. P., & Guenther, F. H. (2015). Resting and task-based functional neuroimaging of adductor spasmodic dysphonia. *Proceedings of the 2015 Human Brain Mapping Meeting, Honolulu, Hawaii*.

Hain, T. C., Burnett, T. A., Kiran, S., Larson, C. R., Singh, S., & Kenney, M. K. (2000). Instructing subjects to make a voluntary response reveals the presence of two components to the audio-vocal reflex. *Experimental Brain Research*, *130*, 133–134.

Hain, T. C., Burnett, T. A., Larson, C. R., & Kiran, S. (2001). Effects of delayed auditory feedback (DAF) on the pitch-shift reflex. *Journal of the Acoustical Society of America*, *109*, 2146–2152.

Häuser, K., & Domahs, F. (2014). Functional lateralization of lexical stress representation: a systematic review of patient data. *Frontiers in Psychology*, *5*, 317.

Howell, P. (1993). Cue trading in the production and perception of vowel stress. *Journal of the Acoustical Society of America*, *94*, 2063–2073.

Jones, J. A., & Munhall, K. G. (2000). Perceptual calibration of F0 production: evidence from feedback perturbation. *Journal of the Acoustical Society of America*, *108*, 1246–1251.

Jürgens, U. (2009). The neural control of vocalization in mammals: a review. *Journal of Voice*, *23*, 1–10.

Kittelberger, J. M., Land, B. R., & Bass, A. H. (2006). Midbrain periaqueductal gray and vocal patterning in a teleost fish. *Journal of Neurophysiology*, *96*, 71–85.

Kotz, S. A., Meyer, M., Alter, K., Besson, M., von Cramon, D. Y., & Friederici, A. D. (2003). On the lateralization of emotional prosody: an event-related functional MR investigation. *Brain and Language*, *86*, 366–376.

Larson, C. R. (1985). The midbrain periaqueductal gray: a brainstem structure involved in vocalization. *Journal of Speech and Hearing Research*, *28*, 241–249.

Larson, C. R., Altman, K. W., Liu, H., & Hain, T. C. (2008). Interactions between auditory and somatosensory feedback for voice F0 control. *Experimental Brain Research*, *187*, 613–621.

Larson, C. R., Burnett, T. A., Bauer, J. J., Kiran, S., & Hain, T. C. (2001). Comparison of voice F0 responses to pitch-shift onset and offset conditions. *Journal of the Acoustical Society of America*, *110*, 2845–2848.

Larson, C. R., Burnett, T. A., Kiran, S., & Hain, T. C. (2000). Effects of pitch-shift velocity on voice F0 responses. *Journal of the Acoustical Society of America*, *107*, 559–564.

Larson, C. R., Sun, J., & Hain, T. C. (2007). Effects of simultaneous perturbations of voice pitch and loudness feedback on voice F0 and amplitude control. *Journal of the Acoustical Society of America*, *12*, 2862–2872.

Lee, B. (1950). Some effects of side-tone delay. *Journal of the Acoustical Society of America*, *22*, 639–640.

Lehiste, I. (1970). *Suprasegmentals*. Cambridge, MA: MIT Press.

Leonard, R. J., & Ringel, R. L. (1979). Vocal shadowing under conditions of normal and altered laryngeal sensation. *Journal of Speech and Hearing Research*, *22*, 794–817.

Lieberman, P. (1960). Some acoustic correlates of word stress in American English. *Journal of the Acoustical Society of America*, *32*, 451–454.

Liu, K., Solomon, N. P., & Luschei, E. S. (1998). Midbrain regions for eliciting vocalization by electrical stimulation in anesthetized dogs. *Annals of Otology, Rhinology, and Laryngology*, *107*, 977–986.

Lombard, E. (1911). Le signe de l'elevation de la voix. *Annales des Maladies de l'Oreille du Larynx*, *37*, 101–119.

Mayer, J., Wildgruber, D., Reicker, A., Dogil, G., Ackermann, H., & Grodd, W. (2002). Prosody production and perception: converging evidence from fMRI studies. *International Symposium on Computer Architecture*, Aix-en-Provence, France.

Meyer, M., Alter, K., Friederici, A. D., Lohmann, G., & von Cramon, D. Y. (2002). FMRI reveals brain regions mediating slow prosodic modulations in spoken sentences. *Human Brain Mapping*, *17*, 73–88.

Mitchell, R. L. C., Elliott, R., Barry, M., Cruttenden, A., & Woodruff, P. W. R. (2003). The neural response to emotional prosody, as revealed by functional magnetic resonance imaging. *Neuropsychologia*, *41*, 1410–1421.

Natke, U., & Kalveram, K. T. (2001). Effects of frequency-shifted auditory feedback on fundamental frequency of long stressed and unstressed syllables. *Journal of Speech, Language, and Hearing Research*, *44*, 577–584.

Niziolek, C., & Guenther, F. H. (2013). Vowel category boundaries enhance cortical and behavioral responses to speech feedback alterations. *Journal of Neuroscience*, *33*, 12090–12098.

Ouellette, G., & Baum, S. (1994). Acoustic analysis of prosodic cues in left- and right-hemisphere damaged patients. *Aphasiology*, *8*, 257–283.

Ozdemir, E., Norton, A., & Schlaug, G. (2006). Shared and distinct neural correlates of singing and speaking. *NeuroImage*, *33*, 628–635.

Parkinson, A. L., Flagmeier, S. G., Manes, J. L., Larson, C. R., Rogers, B., & Robin, D. A. (2012). Understanding the neural mechanisms involved in sensory control of voice production. *NeuroImage, 61*, 314–322.

Patel, R. (2002). Prosodic control in severe dysarthria: preserved ability to mark the question-statement contrast. *Journal of Speech, Language, and Hearing Research, 45*, 858–870.

Patel, R. (2003). Acoustic characteristics of the question-statement contrast in severe dysarthria due to cerebral palsy. *Journal of Speech, Language, and Hearing Research, 46*, 1401–1415.

Patel, R. (2004). The acoustics of contrastive prosody in adults with cerebral palsy. *Journal of Medical Speech-Language Pathology, 12*, 189–193.

Patel, R., Niziolek, C., Reilly, K., & Guenther, F. H. (2011). Prosodic adaptations to pitch perturbation in running speech. *Journal of Speech, Language, and Hearing Research, 54*, 1051–1059.

Patel, R., Reilly, K. J., Archibald, E., Cai, S., & Guenther, F. H. (in press). Responses to intensity-shifted auditory feedback during running speech. *Journal of Speech, Language, and Hearing Research*.

Pell, M. D. (1999). Fundamental frequency encoding of linguistic and emotional prosody by right hemisphere-damaged speakers. *Brain and Language, 69*, 161–192.

Perkell, J. S., Lane, H., Denny, M., Matthies, M. L., Tiede, M., Zandipour, M., et al. (2007). Time course of speech changes in response to unanticipated short-term changes in hearing state. *Journal of the Acoustical Society of America, 121*, 2296–2311.

Peters, A. S., Rémi, J., Vollmar, C., Gonzalez-Victores, J. A., Cunha, J. P., & Noachtar, S. (2011). Dysprosody during epileptic seizures lateralizes to the nondominant hemisphere. *Neurology, 77*, 1482–1486.

Pichon, S., & Kell, C. A. (2013). Affective and sensorimotor components of emotional prosody generation. *Journal of Neuroscience, 23*, 1640–1650.

Poeppel, D. (2003). The analysis of speech in different temporal integration windows: cerebral lateralization as "asymmetric sampling in time." *Speech Communication, 41*, 245–255.

Riecker, A., Achermann, H., Widgruber, D., Dogil, G., & Grodd, W. (2000). Opposite hemispheric lateralization effects during speaking and singing at motor cortex, insula and cerebellum. *Neuroreport, 26*, 1997–2000.

Riecker, A., Wildgruber, D., Dogil, G., Grodd, W., & Ackermann, H. (2002). Hemispheric lateralization effects of rhythm implementation during syllable repetitions: an fMRI study. *NeuroImage, 16*, 169–176.

Ross, E. D., & Mesulam, M. M. (1979). Dominant language functions of the right hemisphere? Prosody and emotional gesturing. *Archives of Neurology, 36*, 144–148.

Schlaug, G., Norton, A., Marchina, S., Zipse, L., & Wan, C. Y. (2010). From singing to speaking: facilitating recovery from nonfluent aphasia. *Future Neurology, 5*, 657–665.

Sidtis, J. J., & Van Lancker Sidtis, D. (2003). A neurobehavioral approach to dysprosody. *Seminars in Speech and Language, 24*, 93–105.

Sidtis, J. J., & Volpe, B. T. (1988). Selective loss of complex-pitch or speech discrimination after unilateral lesion. *Brain and Language, 34*, 235–245.

Sorenson, D., Horii, Y., & Leonard, R. (1980). Effects of laryngeal topical anesthesia on voice fundamental frequency perturbation. *Journal of Speech and Hearing Research, 23*, 274–283.

Subramanian, H. H., & Holstege, G. (2010). Periaqueductal gray control of breathing. *Advances in Experimental Medicine and Biology, 669*, 353–358.

Subramanian, H. H., & Holstege, G. (2014). The midbrain periaqueductal gray changes the eupneic respiratory rhythm into a breathing pattern necessary for survival of the individual and of the species. *Progress in Brain Research, 212*, 351–384.

Tourville, J. A., Reilly, K. J., & Guenther, F. H. (2008). Neural mechanisms underlying auditory feedback control of speech. *NeuroImage, 39*, 1429–1443.

Vance, J. E. (1994). Prosodic deviation in dysarthria: a case study. *European Journal of Disorders of Communication, 29*, 61–76.

Van Lancker, D., & Sidtis, J. J. (1992). The identification of affective-prosodic stimuli by left- and right-hemisphere-damaged subjects: all errors are not created equal. *Journal of Speech and Hearing Research, 35*, 963–970.

Van Lancker Sidtis, D., Pachana, N., Cummings, J. L., & Sidtis, J. J. (2006). Dysprosodic speech following basal ganglia insult: toward a conceptual framework for the study of the cerebral representation of prosody. *Brain and Language, 97*, 135–153.

Vianna, D. M., & Brandão, M. L. (2003). Anatomical connections of the periaqueductal gray: specific neural substrates for different kinds of fear. *Brazilian Journal of Medical and Biological Research, 36*, 557–566.

Wattendorf, E., Westermann, B., Fiedler, K., Kaza, E., Lotze, M., & Celio, M. R. (2013). Exploration of the neural correlates of ticklish laughter by functional magnetic resonance imaging. *Cerebral Cortex, 23*, 1280–1289.

Wildgruber, D., Ackermann, H., Klose, U., Kardatzki, B., & Grodd, W. (1996). Functional lateralization of speech production at primary motor cortex: a fMRI study. *Neuroreport, 7*, 2791–2795.

Witteman, J., Can Heuven, V. J., & Schiller, N. O. (2012). Hearing feelings: a quantitative meta-analysis on the neuroimaging literature of emotional prosody perception. *Neuropsychologia, 50*, 2752–2763.

Yamadori, A., Osumi, Y., Masuhara, S., & Okubo, M. (1977). Preservation of singing in Broca's aphasia. *Journal of Neurology, Neurosurgery, and Psychiatry, 40*, 221–224.

Zatorre, R. J. (1988). Pitch perception of complex tones and human temporal-lobe function. *Journal of the Acoustical Society of America, 84*, 556–572.

Zatorre, R. J., Belin, P., & Penhune, V. B. (2002). Structure and function of auditory cortex: music and speech. *Trends in Cognitive Sciences, 6*, 37–46.

Zhang, S. P., Davis, P. J., Bandler, R., & Carrive, P. (1994). Brain stem integration of vocalization: role of the midbrain periaqueductal gray. *Journal of Neurophysiology, 72*, 1337–1356.

10

Neurological Disorders of Speech Production

Prior chapters have touched on a number of different neurological disorders that provide clues to the neural control of speech. In this chapter we will investigate such disorders in more depth. Our goal is not to provide a comprehensive coverage of motor speech disorders; several excellent sources for this are available (see, e.g., Darley, Aronson, & Brown, 1975; Duffy, 1995; Kent, 2004). Instead, our aim is to use the neurocomputational modeling framework for speech production developed in prior chapters to gain mechanistic insights into a number of disorders that affect the neural circuits described by the DIVA and GODIVA models.

Figure 10.1 illustrates the components of the DIVA model that are associated with particular speech disorders. These disorders will be treated in detail in the following subsections. Before proceeding, it is useful to consider that the severity of the effects of neural damage on speech output depends heavily on whether the damage affects feedforward control mechanisms, feedback control mechanisms, or both. In earlier chapters, we described how the primary role of the feedback control system was for tuning feedforward motor commands (or motor programs) and keeping them tuned in the face of growth and other changes that can occur in the vocal tract. Once tuned, the feedforward commands are capable of generating speech with little or no input from the feedback control system. This account is in keeping with findings of only minor impairments to speech output when auditory and/or somatosensory feedback is removed in mature speakers (see chapters 5 and 6). Thus, *damage to the feedback control system in mature speakers will typically have only minor effects*[1] *on speech output. Damage to feedback control mechanisms in the early years of life is expected to impair development of fluent speech* since the tuning of feedforward commands required for fluent speech is dependent on intact sensory feedback control mechanisms. The impaired ability to develop fluent spoken language with profound prelingual deafness provides strong evidence for this view. *Regardless of developmental stage, damage to the feedforward control system is expected to cause significant motor impairment.* This is apparent from figure 10.1 since damage to any portion of the feedforward control network is associated with a speech motor disorder whereas damage to the feedback control system is only associated with speech motor disorders if the

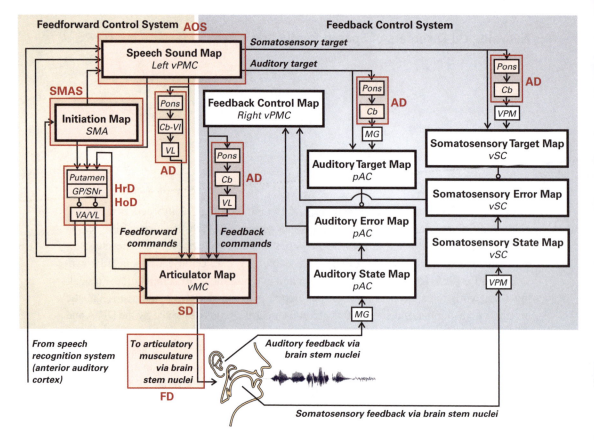

Figure 10.1
Sites of neural impairment (red boxes) associated with various speech disorders (red text labels) within the context of the DIVA model. These disorders include flaccid dysarthria (FD), spastic dysarthria (SD), ataxic dysarthria (AD), hyperkinetic dysarthria (HrD), hypokinetic dysarthria (HoD), apraxia of speech (AOS), and supplementary motor area syndrome (SMAS). Cb, cerebellum (specific lobule unknown); Cb-VI, cerebellum lobule VI; GP, globus pallidus; MG, medial geniculate nucleus of the thalamus; pAC, posterior auditory cortex; SMA, supplementary motor area; SNr, substantia nigra pars reticula; VA, ventral anterior nucleus of the thalamus; VL, ventral lateral nucleus of the thalamus; vMC, ventral motor cortex; VPM, ventral posterior medial nucleus of the thalamus; vPMC, ventral premotor cortex; vSC, ventral somatosensory cortex.

damage also extends to regions involved in feedforward control. This is because the feedforward system is specialized for generating properly timed and coarticulated movements, capabilities that cannot be subsumed by the feedback control system (see chapters 5 through 7 for details).

10.1 Dysarthria

The term *dysarthria* refers to a disorder of speech of neurological origin (as distinct from a language disorder, or *aphasia*) that is characterized by poor articulation, respiration, and/or phonation, including slurred, slow, effortful, and prosodically abnormal speech (for detailed treatments, see Darley, Aronson, & Brown, 1969; Netsell, 1991; Duffy, 2005). Generally speaking, dysarthrias are characterized by weakness and/or abnormal muscle tone of the speech musculature and are often accompanied by diminished amplitude and velocity of speech movement (Turner & Weismer, 1993).

The main dysarthria types, illustrated by location of neural impairment within the DIVA model in figure 10.1, are (1) *flaccid dysarthria* due to lesions of cranial nerves and/or associated nuclei in the brain stem and midbrain (also called *lower motor neurons*), (2) *spastic dysarthria* due to bilateral lesions of motor cortical neurons (or *upper motor neurons*) and/or their projections to the motor periphery, (3) *ataxic dysarthria* due to cerebellum or cortico-cerebellar pathway lesions, (4) *hypokinetic dysarthria* due to Parkinson's disease, and (5) *hyperkinetic dysarthria* due to lesions or diseases of the basal ganglia (cf. Darley, Aronson, & Brown, 1969).

Flaccid Dysarthria

Damage to the cranial motor nuclei, nerves, or neuromuscular junctions involved in speech results in flaccid dysarthria, characterized by weakness in the afflicted muscles. This damage can occur as the result of stroke or neurological diseases such as neuromuscular junction disease (including *myasthenia gravis*), demyelinating disease (including *Guillain-Barré syndrome*), muscular disease (including *muscular dystrophy*), and motor neuron diseases (including *progressive bulbar palsy* and *amyotrophic lateral sclerosis; ALS*). Impairments from unilateral damage tend to be mild whereas bilateral damage involving multiple nerves can be devastating (Duffy, 2004). The articulators affected depend on the nuclei/nerves that are impaired, with trigeminal nerve damage impacting jaw movement, facial nerve damage impacting lip and facial movement, and hypoglossal nerve damage impacting tongue movement. Impairments in speech articulation associated with flaccid dysarthria result directly from weakness in speech-related muscles innervated by the damaged nuclei or nerves. This disorder thus sheds relatively little light on the neural control mechanisms of speech. However, it is the most illuminating dysarthria regarding the muscular innervations of the cranial nerves and the roles particular muscles play in speech articulation (Duffy, 1995).

Spastic Dysarthria
Bilateral damage[2] to motor cortex and/or its descending projections leads to spastic dysarthria, whose characteristic features include strained voice quality, slowed speaking rate, and restricted pitch and loudness variability. Spastic dysarthria can occur as the result of stroke, neurological diseases such as *amyotrophic lateral sclerosis*, or inflammatory diseases such as *leukoencephalitis* (Duffy, 2004). Unlike flaccid dysarthria, spastic dysarthria involves a combination of muscle hypoactivity resulting from damage to upper motor neurons in motor cortex and hyperactivity resulting from release of lower motor neurons in the brain stem from cortical modulation (Darley, Aronson, & Brown, 1975). The latter phenomenon highlights the fact that motor cortex provides both inhibitory and excitatory input to the lower motor neurons. Because motor cortex is involved in generating both feedforward- and feedback-based motor commands, spastic dysarthria at any age would be expected to impair both motor execution (because of impaired feedforward motor commands normally represented in motor cortex) and motor learning (because of impaired ability to generate corrective motor commands when sensory errors are detected).

Ataxic Dysarthria
Damage to the cerebellum often results in *ataxia* (from the Greek *a taxis*, "without order"), characterized by uncoordinated movements with inappropriate range (*dysmetria*), poor balance, and low muscle tone (*hypotonia*). Ataxia can also result from damage to the *ventral lateral (VL) nucleus* of the thalamus (Solomon et al., 1994) and to the *pons* (Fisher, 1989; Schmahmann, Ko, & MacMore, 2004; Varsou et al., 2014), likely through impairment of cortico-pontine-cerebellar-thalamo-cortical motor loops (see box labeled HoD in figure 10.1). Ataxic dysarthria involves uncoordinated, poorly timed articulations, including inappropriate variation in prosodic cues such as pitch, loudness, and duration (Ackermann et al., 1992; Duffy, 2004; Spencer and Slocomb, 2007). Many cases involve equal stress and duration across syllables, a property referred to as *scanning speech* (Charcot, 1877; Spencer & Slocomb, 2007), and slowed speech rate is also common. Resting muscle tone is low in ataxia, and muscular activity during voluntary movements is poorly controlled and coordinated (Darley, Aronson, & Brown, 1975; Duffy, 1995).

Ataxic dysarthria often results from strokes involving the superior cerebellar artery (Ackermann et al., 1992). It can also arise from degenerative diseases such as *Friedreich's ataxia* (a common cause of ataxia in children, given the disease's typical time of onset before adolescence) and *multiple sclerosis*, traumatic brain injury, or toxicity/nutrition deficiency, for example, due to severe alcoholism. Figure 10.2 provides a schematic summary of lesion sites that have been frequently associated with ataxic dysarthria in the research literature. A large amount of evidence points to the superior cerebellar cortex, particularly the paravermal and lateral portions of lobules V and VI, as key cerebellar cortical site for lesions that induce ataxic dysarthria. Dysarthria is most commonly associated

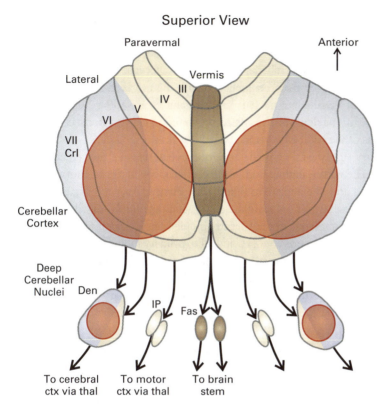

Figure 10.2
Schematic of locations where cerebellar damage has been strongly associated with ataxic dysarthria (red). Roman numerals refer to cerebellar lobules (Schmahmann et al., 2000). CrI, crus I; ctx, cortex; Den, dentate nucleus; Fas, fastigial nucleus; IP, interpositus nucleus; thal, thalamus.

with cerebellar strokes of the superior cerebellar artery, which supplies the superior cerebellar cortex, and bilateral lesions are associated with the most severe articulatory deficits (Ackermann et al., 1992), though it can occur with unilateral damage to either hemisphere (Lechtenberg & Gilman, 1978; Ackermann et al., 1992; Urban et al., 2003). This is in keeping with neuroimaging studies showing bilateral cerebellar activation during speech articulation, most prominently in cerebellar lobule VI (see section 2.3 in chapter 2). Although there are occasional reports of dysarthria with strokes involving the posterior inferior cerebellar artery or the anterior cerebellar artery, these cases generally also include brain stem damage that may be responsible for dysarthric symptoms (Ackermann et al., 1992; Urban et al., 2003). Schoch et al. (2006) used voxel-based lesion symptom mapping in an attempt to more accurately localize motor functions within the cerebellum. The authors noted a somatotopic representation of the superior cerebellar cortex, with vermal lobules II–III associated with gait and posture, vermal and paravermal lobules III–IV with lower limb ataxia, lobules IV–VI with upper limb ataxia, and paravermal and lateral lobules V–VI with ataxic dysarthria. Richter et al. (2005) noted that speech impairments are rare in children after surgery to remove cerebellar tumors, likely because lobules VI and VII Crus I are spared.[3] Regarding the deep cerebellar nuclei, dysarthria is most commonly associated with lesions of the dentate nucleus (Ackermann et al., 1992; Schoch et al., 2006), as expected since this nucleus receives its cerebellar cortical input from the paravermal and lateral portions of the cerebellar hemispheres.

As illustrated in figure 10.1, the DIVA model ascribes several roles for the cerebellum in speech motor control. The corresponding cortico-cerebellar loops in the feedforward and feedback control systems are schematized in figure 10.3. First, as shown in figure 10.3A, the cerebellum plays a crucial role in learning and generating finely timed, smoothly coarticulated feedforward commands to the speech articulators via a loop from premotor cortical areas located in *Brodmann areas* (*BA*) 6 and 44 to primary motor cortex (BA 4) via the pons (not shown), cerebellum (especially paravermal lobule VI), and ventral lateral thalamic nucleus (not shown). Since this cerebellar loop is a core component of the feedforward control system, damage to this loop is most likely the primary cause of ataxic dysarthria (see also Spencer & Slocomb, 2007). In addition to the symptoms described

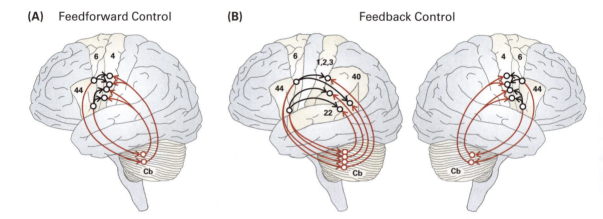

Figure 10.3
Different roles assigned to the cerebellum (red pathways) in the DIVA model. Relay nuclei in the pons and thalamus omitted for clarity. Although projections to contralateral cortical areas are not shown, all projections to sensory and primary motor cortical areas are bilateral. (A) Feedforward motor commands from left-hemisphere premotor cortical regions (Brodmann areas 6, 44) to bilateral primary motor cortex (Brodmann area 4) include a transcerebellar component responsible for learning and generating finely timed motor commands. (B) The feedback control system involves two transcerebellar loops: one from left-hemisphere premotor cortical areas to bilateral auditory (Brodmann area 22) and somatosensory (Brodmann areas 1, 2, 3, 40) cortical areas that is responsible for generating the sensory targets (or expectations) associated with the current sound (left panel), and one from right-hemisphere premotor areas to bilateral primary motor cortex that is responsible for generating corrective motor commands (right panel). Cb, cerebellum.

previously, ataxia is characterized by an inability to perform components of a movement sequence at the proper time (*dyssynergia*), as expected given the key role of the cortico-cerebellar loop in coarticulation described in section 7.3 of chapter 7.

Figure 10.3B illustrates two roles for the cerebellum in feedback control. The left panel shows pathways involved in generating the auditory and somatosensory targets for the current sound. The right panel shows pathways involved in generating corrective motor commands in response to sensory errors. Damage to either of these pathways will impede the correction of auditory and somatosensory errors during speech. If this damage occurs in the mature system, relatively little impairment of speech would be expected.[4] If the damage occurs early in development, it would result in learning of poorly timed and coarticulated feedforward commands, that is, ataxic symptoms. The application of unexpected auditory or somatosensory perturbation during speech would provide a means to test for impairment in the feedback control system in ataxic dysarthria, which would be evident if patients do not show compensatory motor responses to perturbed feedback conditions, but no such studies have been published to date.

Although ataxic dysarthria is typically considered to be an impairment in motor execution, some cases may also involve impairments to higher-level aspects of motor programming, working memory, and language capacities. Spencer and Rogers (2005) found that participants with ataxic dysarthria showed diminished effects of sequence length on reaction time when reading multisyllabic utterances, evidence of a reduced ability to preprogram future syllables in an upcoming utterance. Marien et al. (2001a) suggest a "lateralized linguistic cerebellum" in which the right cerebellar hemisphere is crucially involved in working memory and language processes through its indirect interconnections with language areas in left cerebral cortex. In this view, damage to the right cerebellar hemisphere would, through *diaschisis* (in which damage to one brain structure impairs performance of a distant structure that is not itself damaged), result in impaired functioning in left cerebral cortical language areas, much like damage to lobule VI impairs performance of motor cortex in the DIVA model. In keeping with this view, the voxel-based lesion-symptom mapping study of cerebellar patients by Richter et al. (2007) found that only right-hemispheric lesions to lobule VIIA Crus II were associated with impaired performance on a verb generation task that required participants to generate verbs related to an object presented visually (e.g., "drive" for a picture of a car). However, the notion of a lateralized linguistic cerebellum has been challenged by Cook et al. (2004), who found language deficits in patients with left-lateralized cerebellar lesions.

Hypokinetic Dysarthria

Damage to the cortico-basal ganglia-thalamo-cortical motor loop (*basal ganglia motor loop*, sometimes called the *extrapyramidal system*)[5] can result in either reduced movement (*hypokinesia*) or excessive movement (*hyperkinesia*), in keeping with its key roles in selecting between competing motor programs and initiating execution of the current motor program discussed in chapters 7 and 8. Behavioral characteristics of hypokinetic dysarthria include reduced vocal loudness, monotone voice, and a breathy, hoarse voice quality (collectively termed *hypophonia*; Ramig et al., 2004); reduced articulatory movement speeds (*bradykinesia*) and extents; delays in initiating movement (*akinesia*) or ending an ongoing movement; imprecise and/or "blurred" articulations; and bursts of rapid, accelerating speech (Darley, Aronson, & Brown, 1975; Duffy, 2004). Hypokinetic dysarthria is primarily associated with *Parkinson's disease*,[6] though it can also be caused by other basal ganglia–related conditions (Duffy, 2004).

Figure 10.4 illustrates the effects of Parkinson's disease on function in the basal ganglia motor loop. In the neurologically normal brain (panel A), there is a balance between two pathways through the basal ganglia: the *direct pathway* (green), which has a net excitatory effect on motor/premotor cortex, and the *indirect pathway* (red), which has a net inhibitory effect on motor/premotor cortex. In Parkinson's disease (panel B), there is a loss of dopaminergic input to the basal ganglia due to neuronal death within the *substantia nigra pars compacta* (*SNc*), which normally provides dopamine to the striatum of the basal ganglia via the *nigrostriatal pathway*. This reduction of striatal dopamine, along with concomitant reductions in metabolic measures, has been verified in a number of neuroimaging studies (see Niethammer, Feigin, & Eidelber, 2012, for a review) and is most prominent in the *putamen*, which is the portion of the striatum most heavily involved in motor control. Recall from chapter 7 that dopaminergic inputs to putamen have opposite effects on the direct and indirect pathways in the basal ganglia. In the direct pathway, dopamine binding to D1 receptors has an excitatory effect on striatal neurons. A reduction of dopaminergic input would thus be expected to weaken the influence of the direct pathway, resulting in a reduction of excitatory input to cortex from the basal ganglia loop. In the indirect pathway, dopamine binding to D2 receptors inhibits striatal neurons; a decrease of dopamine would thus lead to excess activity in the indirect pathway, which also decreases excitation of cortex via the basal ganglia motor loop. Current treatments for Parkinson's disease are aimed at restoring the balance between the underactivated direct pathway and the overactivated indirect pathway. The most commonly used drug treatment is *L-dopa* (also known as *levodopa*), a dopamine precursor that, among other actions, increases dopamine levels in the striatum (Duffy, 1995), thereby strengthening the direct pathway and weakening the indirect pathway. The most common surgical intervention is *deep brain stimulation*, in which an electric current is applied to *subthalamic nucleus* (*ST*) or the *internal segment of the globus pallidus* (*GPi*; see Da Cunha et al., 2015, for a review). Deep brain stimulation is generally thought to act as a "virtual lesion" to these areas, effectively weakening the indirect pathway to restore balance with the weakened direct pathway. However, as discussed in the next subsection, the effects of deep brain stimulation are complex and not entirely consistent with a simple virtual lesion interpretation. Although there are many reports of successful treatments of speech difficulties with L-dopa or deep brain stimulation, there

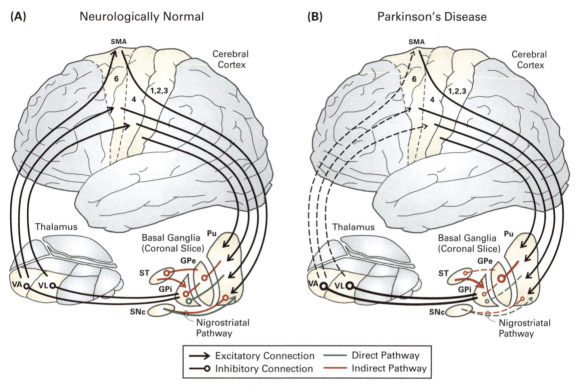

Figure 10.4
The basal ganglia motor loop in (A) the neurologically normal brain and (B) Parkinson's disease. Dashed lines in (B) indicate pathways weakened by Parkinson's disease; thick lines indicate pathways that are strengthened (disinhibited) in Parkinson's disease. Numbers indicate Brodmann areas. See the text for details. GPe, external segment of the globus pallidus; GPi, internal segment of the globus pallidus; Pu, putamen; SMA, supplementary motor area; SNc, substantia nigra pars compacta; ST, subthalamic nucleus; VA, ventral anterior thalamic nucleus; VL, ventral lateral thalamic nucleus.

are also a number of cases in which speech has been resistant to these treatments (see Skodda, 2012, and Sapir, 2014, for reviews), raising the possibility that speech difficulties in Parkinson's may also involve mechanisms beyond the basal ganglia dopaminergic system.

The most common behavioral therapy for hypophonia is *Lee Silverman voice treatment* (*LSVT*; Ramig, Pawlas, & Countryman, 1995), which essentially consists of training patients to speak louder and with maximum effort, with frequent reminders to monitor their loudness and the effort it takes to produce it. Although the neural mechanisms underlying successful LSVT treatment are not well understood at present, it appears to involve increased involvement of right-hemisphere cortical regions after therapy (Narayana et al., 2010), perhaps reflecting increased use of feedback monitoring mechanisms located in the right hemisphere (cf. the right-hemisphere bias for sensory feedback control mechanisms discussed in chapters 5 and 6).

According to the DIVA model, the cortico–basal ganglia motor loop forms a critical component of the feedforward control system for speech, the *initiation circuit* (see chapter 7 for details). The production of a sound begins with activation of a cell in an *initiation map* in the *supplementary motor area* (*SMA*; the dorsomedial portion of BA 6), which in turn sends a *GO signal* (see equation 7.1 in chapter 7) to *ventral premotor cortex* (*vPMC*) and *ventral motor cortex* (*vMC*) nodes that, along with associated subcortical structures, contain motor programs for the current sound. The loss of cortical excitation through the basal ganglia motor loop in Parkinson's disease is expected to impair the ability to activate cells in the SMA initiation map, thereby accounting for the commonly noted difficulties in initiating motor programs (resulting from an impaired ability to activate initiation map nodes), slow movements of reduced extent (resulting from weaker than normal GO signals), and increased rate of speech (resulting from an inability to keep motor programs active for their full duration). In support of this view, one of the more commonly reported findings from functional neuroimaging studies of brain activity during motor tasks in Parkinson's patients is decreased activity in the supplementary motor areas compared to neurotypical controls (see Ceballos-Baumann, 2003, for a review), though some studies indicate this hypoactivity may be primarily in the *preSMA* rather than the SMA proper (e.g., Sabatini et al., 2000).

Another consistent finding from the Parkinson's neuroimaging literature is hyperactivity of the lateral premotor cortical areas (Ceballos-Baumann, 2003). This finding is also consistent with the DIVA/GODIVA modeling framework as illustrated in figure 10.5, which schematizes the premotor and motor cortical processes involved in production of the word "blue" in three different scenarios: (A) very early in development, before a motor program for the whole syllable /blu/ has developed; (B) later in development, after a motor program for /blu/ has been learned; and (C) in Parkinson's disease. For simplicity, we assume that each phoneme is associated with a single gesture in ventral motor cortex. Early in development (panel A), vPMC is responsible for activating motor programs for each of the three phonemic gestures in sequence. After repeated productions of the sequence (panel B), learning in the basal ganglia motor loop produces a single motor program for /blu/, represented by its own node in vPMC; this motor learning process is described in chapter 8. The vPMC node for /blu/ is responsible for activating the first phonemic gesture whereas the SMA/basal ganglia initiation circuit is now responsible for initiating the remaining gestures in the sequence, thus freeing up resources in vPMC by effectively *automating* the readout of these gestures (cf. Alm, 2004; Redgrave et al., 2010). In a speaker with Parkinson's disease (panel C), the SMA/basal ganglia initiation circuit is no longer capable of activating these subsequent gestures, forcing vPMC to activate all of the gestures in the syllable individually, thereby increasing the computational load in vPMC relative to a neurotypical speaker.

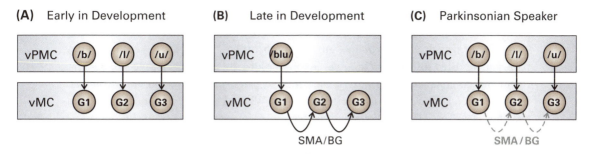

Figure 10.5
Schematized DIVA/GODIVA account of production of the word "blue" in three scenarios: (A) early in development, (B) later in development, after a motor program for the word "blue" has been learned, and (C) in Parkinson's disease. BG, basal ganglia; G, gesture; SMA, supplementary motor area; vMC, ventral motor cortex; vPMC, ventral premotor cortex.

The DIVA/GODIVA modeling framework attributes a major role in learning new speech sequences to the basal ganglia motor loop (see chapter 8). Because Parkinson's disease generally occurs late in life (with the vast majority of cases occurring after age 50), individuals with the disease generally have well-developed motor programs for commonly occurring syllables in the language(s) they speak. However, one would expect an impaired ability in Parkinson's patients for learning new speech sequences, such as in a foreign language. In support of this, Smits-Bandstra and Gracco (2013) found impaired implicit learning of nonsense syllable sequences in Parkinson's patients, extending prior findings of impaired nonspeech motor sequence learning in Parkinson's disease (e.g., Ferraro, Balota, & Connor, 1993; Jackson et al., 1995; Helmuth, Mayr, & Daum, 2000; Vakil et al., 2000) to the speech domain. Furthermore, Mure et al. (2012) report improved motor sequence learning with deep brain stimulation to ST in Parkinson's patients.

Hyperkinetic Dysarthria

The terms *hyperkinesia* and *dyskinesia*[7] refer to abnormal involuntary movements (in contrast to normal involuntary movements, such as rapid withdrawal of the hand upon touching a hot object). *Hyperkinesia* is usually used to describe cases presumed to involve basal ganglia impairment (as opposed to, say, tremor in cerebellar ataxia), thereby acting as a complement to *hypo*kinesia resulting from Parkinson's disease.

A number of different hyperkinesias can affect speech. These are usually characterized by the type of movement elicited, and they may co-occur. The most common hyperkinesia is *tremor*, which involves rhythmic, small amplitude movements of a body part. *Spasm* is a general term for abnormal muscle contractions that may result in movement or impede normal movement (as in back spasms that prevent normal back movements). *Myoclonus* (from the Greek *myo-*, "muscle," and *klonos*, "tumult") consists of sudden, short muscle contractions in groups of muscles that cause one or more body parts to jerk, possibly repetitively. *Tics* are rapid, transient, nonrhythmic contractions that form patterned movements, such as exaggerated eye blinks. *Chorea* (from the Greek *khoreia*, "dance") consists of slower contractions, lasting between 100 ms and 1 second (Darley, Aronson, & Brown, 1975). *Dystonia* involves slow, involuntary muscle contractions (including co-contractions of antagonistic muscle pairs) that delay and slow voluntary movements and often result in distorted postures. When acting on the orofacial or laryngeal muscles, any of these hyperkinesias can impact speech production by interfering with voluntary speech movements.

Hyperfunction of the striatal dopaminergic system is a likely cause of many hyperkinesias. The previous subsection detailed how an imbalance in the direct and indirect pathways of the basal ganglia, specifically a weakening of the direct pathway relative to the indirect pathway due to a reduction of dopamine in the striatum, has an inhibitory effect on movement. Conversely, strengthening of the direct pathway and/or weakening of the indirect pathway, due, for example, to excessive striatal dopamine or hypersensitivity of dopamine receptors in striatal neurons, would be expected to result in excessive movement, including generation of unwanted movements that would normally be inhibited by the indirect pathway. Figure 10.6 schematizes this situation. Dopaminergic hyperfunction in the putamen leads to increased excitation of putamen neurons in the direct pathway, in turn leading to increased inhibition of GPi, indicated by the thick green path between putamen and GPi in figure 10.6B. Dopaminergic hyperfunction has an inhibitory effect on striatal neurons in the indirect pathway, thereby reducing inhibition of the *external segment of the globus pallidus* (*GPe*), as indicated by the dashed red path between putamen and GPe in figure 10.6B. This in turn increases the inhibitory influence of GPe on GPi, both directly and via ST. The net effect on GPi from both the direct and indirect pathways is increased inhibition, which decreases GPi's inhibitory influence on the VL and *ventral anterior* (*VA*) nuclei of the thalamus, leading to increased excitation of motor and premotor cortical regions via thalamocortical projections.

Several hyperkinesias appear to arise, at least in part, in this fashion. Prolonged use of neuroleptic drugs (often used to treat schizophrenia) can result in *tardive dyskinesia*, characterized by unwanted stereotyped movements such as lip smacking, tongue protrusion, or jaw opening/closing. Tardive dyskinesia is often attributed to hypersensitivity of dopamine receptors in the striatum (e.g., Calabrese et al., 1992). Hyperfunction of the striatal dopaminergic system is also believed to play a key role in *Tourette's syndrome*, a hereditary disease characterized primarily by tics that can include involuntary production of inappropriate words or sounds (Buse et al., 2013). Excessive dosage of the drug L-dopa, a dopamine precursor used to treat Parkinson's disease, can result in chorea (Darley, Aronson, & Brown, 1975), and overactivity in the basal ganglia dopamine system (as well as the acetylcholine system) has been suggested as a primary cause of dystonia (Duffy, 1995). Although the exact mechanisms may vary (e.g., excessive dopamine vs. hypersensitive dopaminergic receptors), these disorders all appear to involve an imbalance in which the

Neurological Disorders of Speech Production

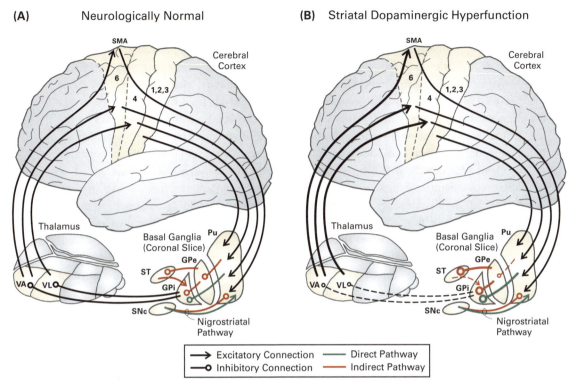

Figure 10.6
The basal ganglia motor loop in (A) the neurologically normal brain and (B) hyperkinesias thought to involve striatal dopaminergic hyperfunction (e.g., tardive dyskinesia, Tourette's syndrome, and some dystonias). Dashed lines in (B) indicate pathways weakened by dopaminergic hyperfunction; thick lines indicate pathways strengthened (disinhibited) by dopaminergic hyperfunction. See the text for details. GPe, external segment of the globus pallidus; GPi, internal segment of the globus pallidus; Pu, putamen; SMA, supplementary motor area; SNc, substantia nigra pars compacta; ST, subthalamic nucleus; VA, ventral anterior thalamic nucleus; VL, ventral lateral thalamic nucleus.

indirect pathway is weak relative to the direct pathway in part because of hyperfunction in the striatal dopaminergic system, thus disinhibiting inappropriate movements.

Other forms of basal ganglia impairment can also cause dyskinesia. *Huntington's disease*, a hereditary disease that constitutes the most common cause of chorea, involves severe loss of neurons in the caudate and putamen (Darley, Aronson, & Brown, 1975), likely impacting both the direct and indirect pathways. Exactly how these pathways are impacted is not entirely clear at present. However, the presence of chorea and other dyskinesias indicates that the net effect of basal ganglia damage in Huntington's disease is to decrease GPi inhibition of thalamus (and thus motor cortex), thereby disinhibiting unwanted movements. Increased GPe activity and decreased GPi activity in Huntington's disease compared to Parkinson's disease has been demonstrated using electrode recordings performed during deep brain stimulator implantation surgeries (e.g., Starr et al., 2008).

The neural mechanisms underlying focal dystonias[8] of speech, in particular *oromandibular dystonia* and *spasmodic dysphonia* (also called *laryngeal dystonia*) remain an active topic of discussion in the speech motor control literature. Many researchers consider these focal dystonias to be a separate category of disorder from hyperkinetic dysarthria rather than a subtype, and it is not clear whether the basal ganglia are the locus of the primary deficit in these disorders. Oromandibular dystonia is characterized by involuntary contractions in muscles of the face, jaw, and tongue, often resulting in difficulty opening or closing the mouth, swallowing, or speaking. Spasmodic dysphonia is characterized by spasms of the intrinsic muscles of the larynx during speech,[9] often accompanied by tremor. Cases that involve muscles that close the glottis are referred to as *adductor* spasmodic dysphonia while those involving muscles that open the glottis are called *abductor* spasmodic dysphonia. In *mixed* spasmodic dysphonia, both sets of muscles are involved. The most common treatment for spasmodic dysphonia and oromandibular dystonia is *botulinum toxin* (often referred to by the brand name *Botox*), injected into the affected muscles. This treatment is purely palliative, essentially reducing muscle spasms by partially denervating (and thereby weakening) the afflicted muscles (Ludlow, 2004). The palliative effects of botulinum toxin extend beyond partial muscle paralysis of injected muscles, as Bielamowicz and Ludlow (2000) identified reduced spasmodic bursting in noninjected muscles as well as injected muscles. The authors hypothesized that such changes are indicative of modification to the somatosensory feedback control system after botulinum toxin injection. Over a period of months, however, the effects of botulinum toxin wear off, necessitating continued treatments.

To date, far more research has been done on spasmodic dysphonia than oromandibular dystonia. Nonetheless, the neuromotor mechanisms underlying spasmodic dysphonia remain a topic of debate. spasmodic dysphonia is more common in women than men and usually occurs later in life, at an average age of approximately 45 to 50 years (Epidemiological Study of Dystonia in Europe Collaborative Group, 1999). Based in part by association to other dystonias, one leading theory regarding the neural impairment underlying spasmodic dysphonia is malfunction of the cortico–basal ganglia motor loop. Such a malfunction might cause overactivation in the feedforward control system for laryngeal movements. A substantial body of evidence of basal ganglia involvement in spasmodic dysphonia has accumulated (e.g., Liberman & Reif, 1989; Schaefer et al., 1985; Lee, Lee, & Kim, 1996; Simonyan et al., 2013; Walter et al., 2014) though the exact nature of this impairment remains unclear and likely varies substantially from case to case.

A second theoretical view suggests that an impaired somatosensory system may be responsible for spasmodic dysphonia. Perhaps the most striking neurological aspect of focal dystonias induced in monkeys is the presence of highly aberrant somatosensory cortical representations, wherein neural receptive fields (i.e., the portion of the skin surface that affects neural firing) are an order of magnitude larger than normal (e.g., Blake et al., 2002). Abnormal somatosensory activity has also been noted during speech in spasmodic dysphonia participants (Simonyan & Ludlow, 2010). Simulations of the DIVA model indicate that hyperfunction of the somatosensory feedback control subsystem for the laryngeal muscles can produce spasmodic dysphonia-like symptoms such as voice breaks (Golfinopoulos, Bullock, & Guenther, 2015).

The hyperactive feedforward control and hyperactive somatosensory feedback control hypotheses are not mutually exclusive—for example, somatosensory hyperfunction could impair function in the basal ganglia motor loop via projections from somatosensory cortex to the striatum, thereby impairing feedforward control—and current data are not yet sufficient to favor one over the other. As illustrated in figure 10.7, individuals with spasmodic dysphonia have excessive brain activity during speech in core regions of the speech motor network with a left-hemisphere bias. Hyperactivity in the ventral portion of the precentral gyrus is consistent with the idea of excessive feedforward commands in spasmodic dysphonia since the feedforward control system is thought to be left lateralized. Hyperactivity in the ventral portion of the postcentral gyrus is consistent with a hyperactive somatosensory feedback control system, but these might arise as a secondary consequence of excessive feedforward commands, which would result in somatosensory error signals, rather than being the cause of excessive motor commands. Hyperactivity in in the posterior superior temporal gyrus may reflects auditory error signals generated when auditory feedback of voice quality does not match the auditory target. The left-hemisphere bias may reflect the fact that voice quality cues occur on a fast time-scale, and left-hemisphere auditory cortex is thought to be specialized for processing rapidly changing stimuli.

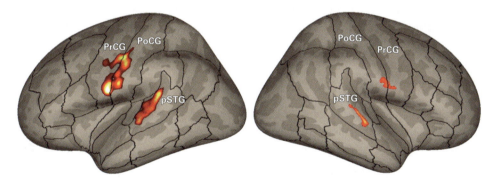

Figure 10.7
Regions of higher activity for individuals with spasmodic dysphonia compared to neurotypical controls when speaking short sentences (Golfinopoulos et al., 2015; fixed effects analysis with a voxel-wise threshold of $p < 0.001$ and a cluster-level family-wise error threshold of $p < 0.05$). PoCG, postcentral gyrus; PrCG, precentral gyrus; pSTG, posterior superior temporal gyrus.

As with hypokinetic dysarthria due to Parkinson's disease, deep brain stimulation applied to the basal ganglia can improve symptoms of a number of hyperkinetic disorders, including Huntington's chorea, dystonia, and Tourette's syndrome (see Da Cunha, 2015, and Smith & Spindler, 2015, for reviews). The most effective stimulation site for hyperkinetic disorders appears to be GPi,[10] which (seemingly paradoxically) is also one of the most effective stimulation sites for *hypo*kinetic dysarthria due to Parkinson's disease. This situation highlights the inadequacy of the "virtual lesion" account of the effects of deep brain stimulation. In fact, the effects (both positive and negative) of deep brain stimulation in a particular patient depend heavily on the precise location of the stimulating electrode as well as a number of stimulation parameters, including frequency, pulse width, and amplitude of electrical stimulation (Volkmann, Moro, & Pahwa, 2006). The mechanism by which deep brain stimulation applied to GPi alleviates hyperkinetic disorders remains unclear at present.

The dysarthrias involve weakness, spasticity, and/or abnormal tone in the articulatory musculature. In the following sections, we will address disorders of speech production that occur in the absence of weakness or abnormal muscle tone, that is, disorders due to impairment of the premotor and prefrontal cortical regions and associated subcortical circuitry responsible for speech initiation, motor programming, and sequencing.

10.2 Apraxia of Speech

The term *apraxia* is used to describe the inability to perform particular purposeful actions despite normal muscle strength and tone. *Apraxia of speech* (*AOS*) involves an impaired ability to plan or program speech movements. It is differentiated from dysarthrias in that it is not due to problems in strength, speed, and coordination of the articulatory musculature (Darley, 1968; Darley, Aronson, & Brown, 1975; Duffy, 2005) and from aphasias in that it is not an impairment of linguistic capabilities but instead lies at the interface between phonological and motor systems (Ziegler, 2002). In the words of McNeil, Robin, and Schmidt (1997), AOS is a "phonetic-motoric disorder of speech production caused by inefficiencies in the translation of a well-formed and filled phonological frame to previously learned kinematic parameters assembled for carrying out the intended movement" (p. 329). AOS can be divided into two types: *childhood apraxia of speech* and *acquired apraxia of speech*. The main difference between the two lies in the underlying cause: acquired AOS typically results from an acute brain insult in mature speakers whereas developmental AOS begins very early in life and in many cases may be genetic (Kent, 2000; Maassen, 2002). As a result, developmental AOS is characterized not only by symptoms seen in acquired AOS, but also by a multitude of other developmental issues (Maassen, 2002). As these developmental issues are beyond the scope of the current book, we focus here on adult-onset acquired AOS.

The primary behavioral characteristics of AOS are overall slowed speech, abnormal prosody, sound distortions and substitutions, and consistent error types from trial to trial, although the presence of error is inconsistent (Wambaugh et al., 2006). Individuals with AOS often appear to be groping for the right articulation patterns. Distortions of speech sounds are commonly regarded as the predominant segmental error type in AOS (McNeil, Robin, & Schmidt, 1997; Odell et al., 1990). Phoneme substitutions may also occur, but these also tend to be distorted, differing from the target in only one or two articulatory features (e.g., Odell et al., 1990; Sugishita et al., 1987; Trost & Canter, 1974). Reduced speech rate in AOS does not appear to be caused by reduction in peak velocities of movements (McNeil & Adams, 1991; McNeil, Caligiuri, & Rosenbek, 1989; Robin, Bean, & Folkins, 1989; Van Lieshout, Bose, Square, & Steele, 2007), instead arising primarily from sound prolongations and/or increased pauses between sounds. Attempts to increase speech rate lead to increased error rates (McNeil, Doyle, & Wambaugh, 2000; Robin, Bean, & Folkins, 1989). Transitioning from one sound/syllable to the next appears to be particularly difficult for individuals with AOS (e.g., Kent & Rosenbek, 1983). Abnormal prosody is often present in the form of equalized stress on successive syllables, for example, by stressing syllables that should be unstressed (Kent & Rosenbek, 1983; Masaki, Tatsumi, & Sasanuma, 1991).

At present, the diagnosis of AOS remains purely behavioral, based primarily on perceptual evaluation of speech characteristics without regard to neural lesions (McNeil, Robin, & Schmidt, 1997; Strand et al., 2014). This situation may confound disorders that have different neural substrates but share some surface behavioral features. Since the spared brain mechanisms differ by lesion site, the residual capabilities of an individual with AOS that might be leveraged to improve speech will depend on lesion site. It is thus imperative that, as the field moves forward, the location of neural damage is taken into account when treating individual cases of AOS.

Although damage to a wide range of brain structures has been associated with apraxia of speech at one time or another,[11] in recent years most of the focus has been on a contiguous set of left-hemisphere inferior frontal cortical areas, illustrated in figure 10.8, including the posterior inferior frontal gyrus, rostral precentral gyrus, and anterior insula (e.g., Dronkers, 1996; Kuriki, Mori, & Hirata, 1999; Fox et al., 2001; Hillis et al., 2004; Marien et al., 2001b; Ziegler, 2002; New et al., 2015). This region largely coincides with the hypothesized location of the DIVA model's *speech sound map* (red box labeled AOS in figure 10.1; see appendix D for DIVA model stereotactic coordinates), which we have referred to as vPMC herein.[12] This map, along with associated subcortical structures, is the primary source of the feedforward motor commands that constitute the motor programs for commonly produced speech strings, as detailed in chapters 3 and 7. Generally speaking, damage to premotor cortex can lead to a dissolution of learned, skilled, purposeful movements (Rothi & Heilman, 1997). Regarding speech production, a number of authors have hypothesized that left vPMC is involved in speech motor programming (e.g., Kuriki, Mori, &

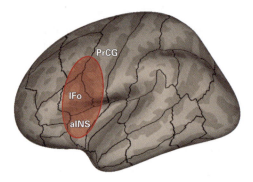

Figure 10.8
Approximate portion of left-hemisphere cerebral cortex most commonly associated with apraxia of speech (red). This region includes left inferior frontal gyrus pars opercularis (IFo), anterior insula (aINS), and the ventral rostral portion of the precentral gyrus (PrCG).

Hirata, 1999; Ziegler, 2002; Guenther, Ghosh, & Tourville, 2006), consistent with the DIVA model and the apraxia findings referenced above.

The effects of damage to the speech sound map on feedforward and feedback control mechanisms in the DIVA model are schematized in figure 10.9. The speech impairments characteristic of AOS are due primarily to impaired feedforward control, as in panel A (e.g., Jacks, 2008; Maas, Mailend, & Guenther, 2015). The speech sound map is the source of projections (red arrows) to articulator maps in primary motor cortex, both directly and via the cortico-cerebellar loop. These projections encode the feedforward commands, or motor programs, for speech sound chunks (such as syllables) represented in the speech sound map. Damage to these projections impairs the generation of feedforward commands, resulting in impaired articulation, as well as *groping*, in which the speaker appears unable to articulate an intended utterance because the appropriate motor programs cannot be accessed. This account can be thought of as a neural instantiation of earlier proposals that associate AOS with impaired syllabic motor programs (e.g., Aichert & Ziegler, 2004). Damage to the speech sound map will also impair feedback control because of damage of projections from vPMC to the higher-order auditory and somatosensory cortical areas that carry sensory targets for speech sounds (panel B of figure 10.9). Without these targets, errors in production cannot be detected via auditory or somatosensory feedback, thus eliminating sensory feedback-based corrective motor commands. Relatively little disruption of speech would be expected in the mature system from impaired feedback control alone since the impact of the feedback control system is small when properly tuned feedforward commands are available. If it occurs in childhood, it would impair tuning of feedforward commands for speech since this tuning process depends on the generation of corrective motor commands for speech errors, as discussed in chapters 3 and 7.

(A) Feedforward Control   **(B)** Feedback Control

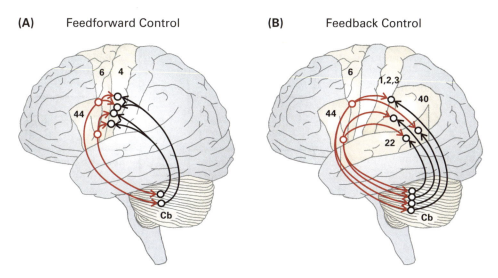

Figure 10.9
Effects of damage to the DIVA model speech sound map. Numbers indicate Brodmann areas. (A) Projections from speech sound map nodes in left premotor cortical regions (Brodmann areas 6 and 44) to primary motor cortex (Brodmann area 4) encode feedforward commands for speech sounds (syllables and phonemes). Damage to these projections (indicated in red) results in an inability to access the feedforward commands (or motor programs) for proper articulation of speech. (B) Projections from speech sound map cells in left premotor cortex to higher-order auditory and somatosensory cortices (Brodmann areas 40 and 22) encode the sensory expectations for these sounds. Damage to these projections impairs the auditory and somatosensory feedback control subsystems for speech by eliminating the ability to detect production errors via sensory feedback. Cb, cerebellum.

It is noteworthy that damage to right vPMC in the DIVA model will not significantly impact feedforward control mechanisms since this region is not part of the feedforward control system, instead constituting a *feedback control map* involved in transforming sensory errors into corrective motor commands (see chapters 5 and 6 for details). This view is consistent with the fact that AOS is far more likely to occur with damage to the left hemisphere. The model predicts that damage to right vPMC would impair feedback control mechanisms such as the ability to compensate for auditory or somatosensory perturbations. This prediction has not been tested to date.

The neural impairments described thus far can account for many aspects of AOS, most notably the sound distortions that are central to the disorder. However, the diagnostic criteria for AOS include impairments at a level higher than the syllabic motor program, including increased duration of pauses between syllables and abnormal prosody at the suprasyllabic level. Insight into these impairments, which likely involve regions of the brain beyond the core speech motor control network embodied by the DIVA model, can be gained by considering the effects of damage to different components of the GODIVA model, schematized in figure 10.10.

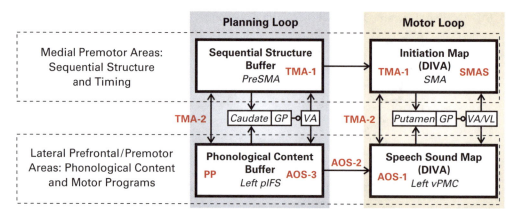

Figure 10.10
Locations of damage to the GODIVA model associated with speech disorders (red text). AOS, apraxia of speech; GP, globus pallidus; pIFS, posterior inferior frontal sulcus; PP, phonemic paraphasia; PreSMA, pre-supplementary motor area; SMA, supplementary motor area; SMAS, supplementary motor area syndrome; TMA, transcortical motor aphasia; VA, ventral anterior nucleus of the thalamus; VL, ventral lateral nucleus of the thalamus; vPMC, ventral premotor cortex.

Recall from chapter 8 that, in the GODIVA model, a *phonological content buffer* in left *posterior inferior frontal sulcus* (*pIFS*) temporarily stores the phonological content items (e.g., phonemes and/or consonant clusters) for an upcoming utterance, with earlier items in the utterance having higher activity than later items. Projections from left pIFS to left vPMC are responsible for choosing the next motor program to be executed based on the most highly active items in the pIFS phonological content buffer. If a syllable- or word-sized motor program exists in vPMC for the most active items in the buffer, it is chosen over smaller, phoneme-sized motor programs (see Bohland, Bullock, & Guenther, 2010 for details). If the upcoming syllable/word is novel to the speaker, then the motor programs for the individual phonemes are activated in sequence. This aspect of the model is in essence a neurocomputational implementation of the *dual-route theory* of word production (e.g., Levelt, 1989; Coltheart et al., 2001). According to dual-route models, production of a high-frequency word involves a *direct route* for accessing efficient and stereotyped movement sequences (i.e., motor programs) for producing the whole word. Articulatory plans for low-frequency words, in contrast, must be constructed by accessing motor programs for each of the phonemes in the word individually (the *indirect route*).

Whiteside and Varley (1998; see also Varley & Whiteside, 2001) proposed an account of AOS based on dual-route theory. In their view, the direct route is damaged in AOS, forcing the speaker to utilize the indirect route to produce all words, not just low-frequency words. The indirect route is more resource intensive and possibly more error prone, resulting in the greater number of speech errors, loss of automaticity, prolongations of segments and pauses, reduced coarticulation, and reductions in speech rate that characterize AOS. The GODIVA model provides a means for interpreting the Whiteside and Varley account of AOS in neurocomputational terms. Earlier in this subsection we described how damage to the speech sound map in left vPMC (labeled AOS-1 in figure 10.10) would impair syllable-level motor programs and produce many apraxic symptoms. In addition, damage to the projections between left pIFS and left vPMC (AOS-2 in figure 10.10) would eliminate access to these motor programs even if the motor programs themselves were spared. In both cases, access to word- and syllable-level motor programs would be impaired or eliminated, as in the Whiteside and Varley (1998) account of AOS (see also Mailend & Maas, 2013).

An alternative proposal to the dual-route account of AOS is the *reduced buffer capacity hypothesis* (Rogers & Storkel, 1999), in which AOS results from a reduced capacity to hold multiple syllables in a processing buffer that underlies motor planning. The reduced buffer size would account for a number of AOS symptoms, including pauses between syllables and a lack of intersyllable dependency of the F0 contour. Interpreted within the GODIVA model, the reduced buffer capacity hypothesis would correspond to damage to the phonological content buffer (AOS-3 in figure 10.10), thereby limiting the buffer to containing only a few phonological units at a time. The GODIVA model further predicts that, if the phonological content buffer is damaged but the speech sound map is intact, speech errors may include nondistorted productions of incorrect syllables/words that differ from the intended syllables/words by one or more phonological units (e.g., "paker" for "paper"). Such errors have been noted in Broca's aphasics and are termed *phonemic paraphasias* (Canter, Trost, & Burns, 1985; Goodglass, 1993; labeled PP in figure 10.10).

10.3 Medial Premotor Syndromes

As described in the previous section, AOS typically results from damage to the lateral premotor cortical areas involved in speech, in particular vPMC. Damage to the medial premotor cortical areas, specifically the SMA and preSMA, leads to a very different type of disorder, characterized by impairments in the ability to initiate speech rather than an inability to articulate it.

Unilateral resection of SMA typically results in *SMA syndrome* (labeled SMAS in figures 10.1 and 10.10), which is characterized by reduction in spontaneous movements of body parts contralateral to the resection (Bannur & Rajshekhar, 2000; Abel et al., 2015). This is consistent with the DIVA/GODIVA characterization of an *initiation map* in this region. If the resection is in the language-dominant (typically left) hemisphere, transient mutism or loss of spontaneous speech can occur (Laplane et al., 1977; Jonas, 1981; Pai, 1999; Abel et al., 2015). Interestingly, SMA syndrome typically resolves almost completely within a few weeks of resection (Zentner et al., 1996; Endo et al., 2014), indicating that the original function of the removed cortex is taken over by other parts of the brain, most likely including surrounding ipsilateral cortical areas such as preSMA, cingulate cortex, and superior frontal cortex, as well as the contralateral SMA.

Transcortical motor aphasia (*TMA*; also referred to as *dynamic aphasia*) is a disorder characterized by greatly reduced spontaneous speech output with a spared ability to repeat speech, often including long sentences. Transcortical motor aphasia is commonly associated with strokes that damage either the supplementary motor areas[13] (labeled TMA-1 in figure 10.10; see Bogousslavsky, Assal, & Regli, 1987; Mochizuki & Saito, 1990; Pai, 1999; Krainik et al., 2003; Satoer et al., 2014) and/or projections from these areas to the lateral premotor and prefrontal cortical areas involved in speech and language (labeled TMA-2 in figure 10.10; see Enokido et al., 1984; Friedmann, Alexander, & Naeser, 1984). Some authors differentiate subtypes of TMA at the two lesion sites. For example, Pai (1999) refers to a case of lost spontaneous speech with spared repetition after damage to the supplementary motor areas (TMA-1 in figure 10.10) as *supplementary motor area aphasia*, and Enokido et al. (1984) suggest that dynamic aphasia is a subtype of transcortical motor aphasia related to lesions in the posterior left middle frontal gyrus, which may cut off communication between the medial and lateral premotor/prefrontal cortical areas involved in speech and language (TMA-2 in figure 10.10). The spared ability to repeat speech in transcortical motor aphasia suggests that the SMA *initiation map* is either bilateral or is only involved in initiating self-generated speech, not in externally triggered speech. Large bilateral lesions that involve the supplementary motor areas and cingulate cortex have also been associated with *akinetic mutism* (e.g., Németh, Hegedüs, & Molnár, 1988; Mochizuki & Saito, 1990; Nagaratnam et al., 2004), characterized by a lack of voluntary movement (sometimes described as lack of the will to move), including speech, although individuals with this condition nonetheless appear alert to their surroundings.

10.4 Stuttering

Stuttering (also referred to as *stammering*) is a speech disorder characterized by intermittent disruptions of the fluent flow of speech that take one of three forms: (1) involuntary *repetitions* of one or more speech sounds (e.g., "ta-ta-ta-take"); (2) *blocks*, in which an involuntary hesitation occurs prior to producing a sound (usually the first sound in a phrase or sentence); and (3) *prolongations*, in which production of a sound is inappropriately extended (e.g., "rrrrrake"). The individuals know exactly what they want to say, but their ability to say it is disrupted by one or more of the behaviors listed above. Repetitions, blocks, and prolongations are referred to as *primary* or *core behaviors* and are differentiated from a wide variety of *secondary behaviors* (e.g., eye blinking, extraneous body movements) that can develop over time, presumably as learned reactions to the core behaviors. We restrict our attention here to core deficits in the neural circuitry responsible for speech production, while acknowledging that other factors may contribute to stuttering dysfluencies, for example, emotional effects on the respiratory pattern (Denny & Smith,

1997), monitoring loops in the higher-level linguistic system (Postma & Kolk, 1993), and dyssynchrony between the linguistic and paralinguistic systems (Perkins, Kent, & Curlee, 1991).

Stuttering is often referred to as a *fluency disorder* and is generally considered to be a distinct disorder from dysarthrias or AOS. Stuttering differs from dysarthrias in that there is no underlying weakness or paralysis of the articulatory musculature and no inappropriate muscle tone at rest. Although stuttering can co-occur with dysarthric symptoms, a distinction between the two can be justified by noting that most cases of stuttering do not involve such symptoms. AOS and stuttering can be differentiated according to the pattern of speech errors that occur: stuttering involves the interruptions of flow described above with otherwise intact productions of intended sounds whereas AOS involves uncoordinated movements, distorted productions, omissions, and substitutions of sounds.

Neurogenic stuttering can arise through stroke, neurological disease, or as a result of treatments for neurological diseases (see Lundgren, Helm-Estabrooks, & Klein, 2010, and Theys, 2012, for reviews). When neural damage is unilateral, the large majority of cases involve left-hemisphere cortical or subcortical structures. Using voxel-based lesion symptom mapping, Theys et al. (2013) identified nine left-hemisphere regions associated with neurogenic stuttering, including the inferior frontal cortex, superior temporal cortex, inferior parietal cortex, caudate nucleus, and putamen. This wide range of possible damage sites indicates that stuttering is a *system-level disorder* that can arise from different sites of damage within a multiregional brain network. This view is also supported by the fact that drug-induced stuttering can be caused by medications affecting a number of neurotransmitter systems, including cholinergic, dopaminergic, noradrenergic, or serotonergic systems (Lundgren et al., 2010). The neural system most commonly damaged in neurogenic stuttering appears to be the *left-hemisphere basal ganglia motor loop* (Theys, 2012; Theys et al., 2013), which is associated with the initiation of speech motor programs in the DIVA model. We will return to this topic below.

Developmental stuttering is by far the most prevalent form of the disorder, typically emerging at 2 to 5 years of age in approximately 3–5% of the general population, with approximately 75% of cases spontaneously resolving within about 2 years (Yaruss, 2004; Curlee, 2004). Those cases that do not resolve are referred to as *persistent developmental stuttering* (*PDS*), which affects approximately 1% of the population and occurs in nearly all cultures (Van Riper, 1982). Although the male/female ratio among *people who stutter* (*PWS*) is approximately 1:1 at onset, females recover at a higher rate than males; as a result, PDS is more prevalent in males than females, with roughly a 4:1 male-female ratio among adults who stutter (Yaruss, 2004). PDS is much more common in those with a familial history of stuttering, and if one of a pair of identical twins stutters, the other is very likely to stutter, though they may stutter differently (Van Riper, 1982).

Van Riper (1982) provides a detailed and eloquent description of the developmental trajectory of PDS, briefly summarized here. Typically the first sign is an excessive amount of repetition in speech, primarily of syllables. Prolongations and blocks can also occur at the age of onset, though their frequency increases over time. Children who show complete blockages at a young age are less likely to recover than those that produce primarily repetitions. Tension (muscular co-contraction, often occurring with the articulators set to their position for the first sound in the word to be spoken) is usually not present early on but increases over time, possibly because the individual gradually becomes aware of his or her stutter. Speech tempo is normal at first, may become faster than normal for a period of time, and then becomes slower than normal as tension and prolongation/blocks increase. Despite these common tendencies, considerable variation is seen across individuals in the development of PDS, particularly as they begin to develop compensatory strategies. For example, many PWS develop *interrupter reactions,* such as a sudden head or body jerk, that occur during blocks. If such a movement accompanies successful release of the blocked sound (which is a rewarding situation), motor learning can occur, making the reaction more likely to arise during future blocks. The order in which such behaviors are learned over time is unique to an individual, thereby contributing to the wide variety of behaviors found in adult PWS.

In this section we will take the view that the *core deficit in PDS is an impaired ability to initiate, sustain, and/or terminate motor programs for phonemic/gestural units within a speech sequence due to impairment of the left-hemisphere basal ganglia motor loop*[14] (cf. Alm, 2004). Recall from chapters 5 and 8 that in the DIVA and GODIVA models this *initiation circuit* is responsible for sequentially initiating phonemic gestures within a (typically syllabic) motor program by activating nodes for each phoneme in an *initiation map* in SMA. Figure 10.11 schematizes this situation. Very early in development (prior to the age of 2–3 years), initiation of the phonemes in a CVC syllable requires preSMA input to sequence through the SMA initiation map nodes for the three phonemes in the proper order

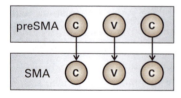 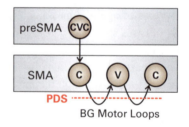

Figure 10.11
Schematized view of the process of sequencing through three phonemes in a CVC utterance (A) early in development, when preSMA involvement is required to sequentially activate nodes in SMA for initiating each phoneme, and (B) later in development, when the basal ganglia motor loop has taken over sequential activation of the SMA nodes. BG, basal ganglia; C, consonant; PDS, persistent developmental stuttering; preSMA, pre-supplementary motor area; SMA, supplementary motor area; V, vowel.

(panel A). Later in development (panel B), the basal ganglia motor loop has taken over the job of activating the three phoneme nodes in SMA when producing the syllable. The dashed red line in panel B of figure 10.11 indicates the proposed locus of impairment in PDS, namely, the basal ganglia motor loops normally responsible for initiating motor programs for the second and subsequent phonemes are impaired such that they cannot reliably activate *initiation map* nodes in SMA for these phonemes.

Figure 10.12 provides an expanded view of the basal ganglia motor loop. The basal ganglia in essence performs a pattern matching operation in which it monitors the current cognitive context (as represented by activity in prefrontal cortical areas including preSMA and pIFS), motor context (represented in vPMC, SMA, and vMC), and sensory context (represented in *posterior auditory cortex*, or *pAC*, and *ventral somatosensory cortex, vSC*). When the proper context is detected, the basal ganglia signal to SMA that it is time to terminate the ongoing phoneme (*termination signal*) and/or initiate the next phoneme of the speech sequence (*initiation signal*). Failure to recognize the sensory, motor, and cognitive context for terminating the current phoneme would result in a prolongation stutter since activity of the SMA *initiation map* node for the current sound will not be terminated at the right time. Failure to recognize the context for initiating the next phoneme would result in a block stutter since the *initiation map* node for the next phoneme in SMA will not

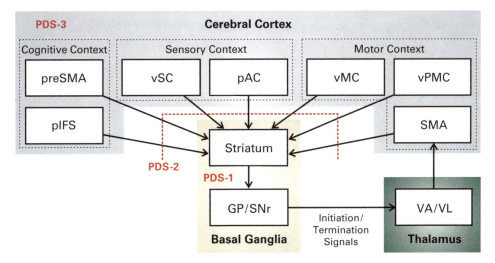

Figure 10.12
Potential impairments of the basal ganglia motor loop that may contribute to persistent developmental stuttering (PDS), specifically the basal ganglia (PDS-1), corticostriatal projections from cerebral cortex to the basal ganglia (PDS-2), and the network of cortical regions involved in speech (PDS-3). GP, globus pallidus; pAC, posterior auditory cortex; pIFS, posterior inferior frontal sulcus; preSMA, pre-supplementary motor area; SMA, supplementary motor area; SNr, substantia nigra pars reticulata; VA, ventral anterior thalamic nucleus; VL, ventral lateral thalamic nucleus; vMC, ventral motor cortex; vPMC, ventral premotor cortex; vSC, ventral somatosensory cortex.

be activated at the right time. If the initiation signal "drops out" momentarily, production of the next phoneme might begin but prematurely terminate and then restart, as in a repetition stutter.

Alm (2004) refers to signals such as the initiation and termination signals illustrated in figure 10.12 as *timing signals* since they indicate the right time to terminate/initiate movements. This characterization provides insight into the frequent observation that stuttering is often greatly reduced or eliminated in situations where external timing cues are available, such as choral reading and metronome-timed speech (Bloodstein, 1995). Interpreted within the DIVA/GODIVA framework, these tasks involve timing signals that are perceived by sensory cortical areas, which then relay the signals to SMA, thereby reducing dependence on the basal ganglia motor loop for generating initiation/termination signals. Similarly, singing, which also increases fluency in PDS (Starkweather, 1987), likely involves different mechanisms for generating phonemic timing than the basal ganglia motor loop used for initiating propositional speech.

Impairment to several different parts of the basal ganglia motor loop could lead to the disruptions just described, including the basal ganglia proper (labeled PDS-1 in figure 10.12), projections from cerebral cortex to the basal ganglia that provide cognitive and sensorimotor contextual information (PDS-2), and the network of cortical regions that process cognitive and sensorimotor aspects of speech (PDS-3).

The basal ganglia are frequently associated with stuttering in the speech production literature. For example, stuttering often develops or reemerges in Parkinson's disease (Koller, 1983; Leder, 1996; Benke et al., 2000; Shahed & Jankovic, 2001; Lim et al., 2005), the motor components of which are thought to arise from impairment of function within basal ganglia structures as described earlier in this chapter. Deep brain stimulation applied to the ST of the basal ganglia can relieve acquired stuttering in some Parkinson's disease patients (Walker et al., 2009; Thiriez et al., 2013) while in others it seems to exacerbate stuttering (Burghaus et al., 2006; Toft & Dietrichs, 2011). Levodopa treatment, aimed at increasing dopamine levels in the striatum of the basal ganglia, can also exacerbate stuttering (Anderson et al., 1999; Louis et al., 2001; Tykalová et al., 2015).

This last finding fits well with one popular hypothesis regarding the neurogenesis of PDS, namely *dopamine excess theory*, which is based on the Wu et al. (1997) finding of excessive dopamine in the striatum of 3 PWS compared to 6 nonstuttering control participants.[15] Computer simulations performed by Civier et al. (2013) verified that an increased level of dopamine in the striatum can lead to stuttering behaviors in the GODIVA model. To understand how this can occur, it is useful to consider the role of the basal ganglia motor loop in performing a *winner-take-all competition* between competing motor programs, such as the set of phonemic motor programs that make up a larger syllabic motor program. When the winning motor program has been selected, an initiation signal is transmitted to SMA (see figure 10.12) and the corresponding *initiation map* node is activated, causing the readout of the chosen motor program (see chapter 7 for details).

Within this process, the striatum can be viewed as a *pattern detector* that monitors cortical inputs, such that striatal neurons associated with a particular motor program become active when the cognitive and sensorimotor context (represented in cerebral cortex) matches the context appropriate for that motor program. The motor program that best matches the current context will have the highest striatal activity. This situation is illustrated in panel A of figure 10.13; in this case, striatal neurons corresponding to MP1 are the most active, followed by MP2 and MP3. The larger the difference in striatal activity between MP1 and competing motor programs, the faster and more reliable the winner-take-all choice will be.

Increasing striatal dopamine has the effect of exciting neurons with D1 receptors, most notably those in the direct pathway, while inhibiting those with D2 receptors, which are predominant in the indirect pathway. This latter action has the effect of reducing the inhibition of competing motor programs via the indirect pathway, as illustrated in panel B of figure 10.13, where the desired motor program (MP1) has a greatly decreased activity advantage over competing motor programs compared to panel A. Such a situation could delay the choice of the desired motor program, leading to a block or prolongation stutter, or it may lead to an unstable initiation signal that starts to increase but suffers drop-outs that result in repetition stutters.

In support of the dopamine excess theory of stuttering, it has been noted that antipsychotic drugs such as haloperidol and risperidone that block dopamine D2 striatal receptors are effective in treating symptoms of stuttering[16] (see Bothe et al., 2006, for a review). These D2 antagonists increase the efficacy of the indirect pathway by removing it from dopaminergic inhibition, thus correcting the hypothesized direct/indirect imbalance and increasing inhibition of competing actions. A weakened indirect pathway and concomitant inability to maintain the chosen action over competing actions is also supported by the study of Webster (1989) demonstrating that PWS are particularly impaired in initiating and progressing through sequences in the presence of competing tasks. Alm (2004) suggests that developmental changes in dopamine receptor density in the putamen

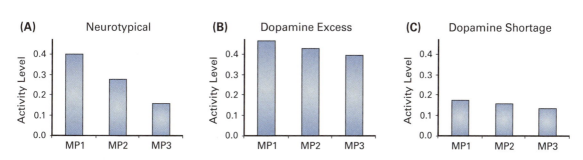

Figure 10.13
Schematic of striatal activity for three motor programs (MP1, MP2, and MP3) under (A) normal conditions, (B) excess striatal dopamine, and (C) insufficient striatal dopamine.

could also explain the pattern of early childhood onset and recovery, including gender differences.

Notably, the computer simulations of Civier et al. (2013) also indicate that *decreased* dopamine levels in the striatum could lead to stuttering dysfluencies, which would account for the onset of stuttering in individuals with Parkinson's disease noted above. In this scenario, illustrated in panel C of figure 10.13, reduced excitation of the desired motor program through the direct pathway leads to a decrease in the competitive advantage of this motor program, which in turn leads to a delayed, weakened, and/or unstable initiation signal.

These considerations suggest that there may be (at least) two subtypes of PDS: one characterized by an underactive indirect pathway and another characterized by an underactive direct pathway. Behaviorally, the former might be characterized by a tendency toward excessive motor activity due to reduced inhibition of movement from the indirect pathway whereas the latter might be characterized by a reduced level of motor activity due to reduced excitation of movement from the direct pathway. This is similar to the proposal put forth by Alm (2004), who proposed a breakdown into *D2-responsive* and *stimulant-responsive* subgroups of PWS. It should be noted, however, that our treatment of basal ganglia anatomy and physiology has been highly schematic, and that the actual situation is very complex, involving many neurotransmitter types and axonal pathways in addition to those discussed herein. Nonetheless, there is sufficient evidence for the differentiation of stuttering subtypes involving different malfunctions of the basal ganglia to merit increased research on this topic, including large-sample studies investigating striatal dopamine levels in PDS.

The view of stuttering as an inability to quickly and reliably resolve the competition between motor programs for upcoming speech sounds can shed light on additional aspects of PDS. In the GODIVA model's *competitive queue* (*CQ*) representation of a speech sequence, the total activity of nodes in the phonological buffer in pIFS and the sequential structure buffer in preSMA is normalized. Panel A of figure 10.14 illustrates normalized activities for a four-item sequence. After the motor program for the first item of the sequence has been produced, the activity of the corresponding node in the CQ buffer is extinguished, as illustrated in panel B. Because activity in the buffer is then renormalized, the difference in activity for the first and second items (labeled $\Delta_{1,2}$ in panel A) is smaller than the difference between the second and third items (labeled $\Delta_{2,3}$ in panel B). This makes the winner-take-all competition process more difficult for the first item in a sequence compared to subsequent items, providing a potential explanation for the observation that stuttering occurs far more frequently at the beginning of an utterance (see Van Riper, 1982 for a review). Figure 10.14 also illustrates the fact that differences in activity between items are larger for shorter sequences (panel B) than longer sequences (panel A), thus accounting for the frequently reported finding that stuttering is more prevalent for

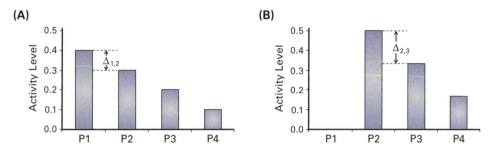

Figure 10.14
(A) Activity levels in a competitive queue (CQ) buffer for a four-phoneme sequence. (B) Activity levels in the CQ buffer after the first phoneme has been completed and its corresponding activity in the buffer extinguished. P, phoneme.

longer speech sequences (Yaruss, 1999; Zackheim & Conture, 2003; see Bloodstein, 1995, for review).

The second potential source of impairment in the basal ganglia motor loop of PDS identified in figure 10.12, labeled PDS-2, is the set of projections from cerebral cortex to striatum that convey the current sensorimotor and cognitive context to the basal ganglia. Using *diffusion-tensor imaging (DTI)* performed on children who stutter and age-matched controls, Chang and Zhu (2013) found that children who stutter have weaker white matter tracts between left putamen and several left-hemisphere cortical regions involved in speech, including IFo and SMA. Computer simulations of the GODIVA model by Civier et al. (2013) indicate that impaired corticostriatal connectivity can result in poor detection of the cognitive and sensorimotor context for initiating the next sound by the basal ganglia motor loop, thereby impairing the generation of initiation/termination signals to SMA. It is thus tempting to conclude that impaired left-hemisphere corticostriatal connectivity may be a root cause of stuttering. To date, however, clear evidence of impaired corticostriatal connectivity in adults who stutter remains scarce, raising the question of whether the weakened corticostriatal connectivity found in children who stutter may be a secondary consequence of impairment elsewhere rather than a root cause of stuttering.

The third possible source of impairment in the basal ganglia motor loop of PDS is the network of cerebral cortical regions involved in speech production (PDS-3 in figure 10.12). As noted above, neurogenic stuttering is generally associated with damage to the left (language-dominant) hemisphere. This suggests that *stuttering involves prefrontal and/or premotor cortical mechanisms for speech*, which, unlike primary sensory and motor cortical areas that show relatively little hemispheric differentiation, are predominantly located in the left hemisphere, as detailed in chapters 7 and 8.

Structural and functional neuroimaging studies of PDS support this assertion. For example, Cai et al. (2014b) used DTI and probabilistic tractography to identify correlations between stuttering severity and white matter tract strengths in PDS. It is commonly believed that, all else being equal, stronger white matter tracts are associated with better performance. According to this view, if a particular tract is part of the underlying cause of stuttering, we would expect that the weaker the tract, the more severe the stuttering, that is, tract strength should be negatively correlated with stuttering severity. Conversely, the strength of a tract that is forced into action to (incompletely) compensate for the core neural impairment should be positively correlated with severity. Figure 10.15 indicates all intrahemispheric tracts between inferior frontal cortical ROIs and sensorimotor (Rolandic) cortical ROIs that were significantly correlated with severity in Cai et al. (2014b). Strikingly, all such tracts in the left hemisphere were negatively correlated with stuttering, while all right-hemisphere tracts were positively correlated.[17] This finding suggests that *impaired performance in the left-hemisphere cortical network for speech in PDS forces reliance on right-hemisphere homologues*, leading to increased right-hemisphere white matter tract strengths due to additional use. This interpretation is a variant of the *atypical cerebral laterality* view of stuttering that dates as far back as Orton (1927). Interpreted within the DIVA/GODIVA framework, the left-hemisphere white matter impairments are indicative of impaired function of the left-lateralized feedforward control system, resulting in sensory errors that must be corrected by the right-lateralized auditory and somatosensory feedback control subsystems.

Further support for the view that left-hemisphere impairments in PDS result in increased right-hemisphere involvement during speech comes from functional neuroimaging studies.

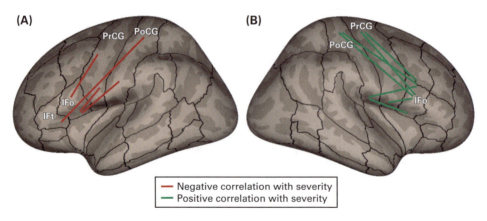

Figure 10.15
Schematic of intrahemispheric white matter tracts between inferior frontal cortical regions and Rolandic cortical regions whose strengths are significantly correlated with stuttering severity (Cai et al., 2014b) plotted on (A) left and (B) right lateral inflated cortical surfaces. Red tracts indicate a negative correlation with severity (i.e., weaker tracts are associated with higher severity); green tracts indicate a positive correlation. IFo, inferior frontal gyrus pars opercularis; IFt, inferior frontal gyrus pars triangularis; PoCG, postcentral gyrus; PrCG, precentral gyrus.

Anomalous functioning in left-hemisphere inferior frontal cortex of PDS during single-word production was identified using magnetoencephalography by Salmelin et al. (2000), who also noted that suppression of motor rhythms (which reflect task-related processing) was right-dominant in PDS but left-dominant in fluent speakers. Hyperactivity in right-hemisphere cerebral cortex of PWS has been noted in a number of prior PET and fMRI studies (e.g., Fox et al., 1996; Braun et al., 1997; De Nil et al., 2000). Figure 10.16 illustrates the results of an activation likelihood estimate (ALE) meta-analysis of neuroimaging studies contrasting brain activity during fluent speech in PWS to activity in nonstuttering control participants speaking the same utterances (Belyk, Kraft, & Brown, 2015). PWS have higher activity in a number of right-hemisphere regions involved in speech, including prefrontal, sensorimotor, and parietal regions of cortex. The view that right-hemisphere cortical hyperactivity results from impaired left-hemisphere function is also

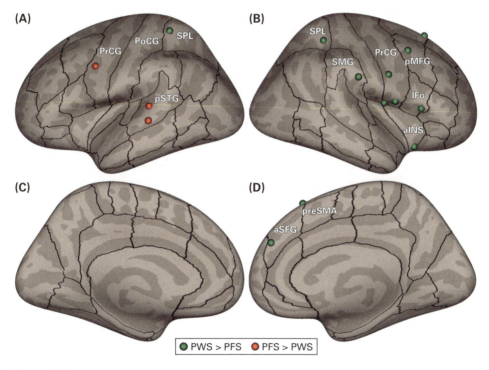

Figure 10.16
Cortical activation foci from the Belyk, Kraft, and Brown (2015) meta-analysis of speech functional neuroimaging studies comparing brain activity during fluently produced speech of people who stutter (PWS) to brain activity in fluent speakers (PFS) producing the same speech stimuli, displayed on the (A) left lateral, (B) right lateral, (C) left medial, and (D) right medial inflated cortical surfaces. aINS, anterior insula; aSFG, anterior superior frontal gyrus; IFo, inferior frontal gyrus pars opercularis; pMFG, posterior middle frontal gyrus; PoCG, postcentral gyrus; PrCG, precentral gyrus; preSMA, pre-supplementary motor area; pSTG, posterior superior temporal gyrus; SMG, supramarginal gyrus; SPL, superior parietal lobule.

consistent with the effects of fluency-inducing therapy on BOLD responses; successful treatment has been associated with a shift toward more normal, left-lateralized frontal activation (De Nil et al., 2003; Neumann et al., 2005).

Another interesting aspect of figure 10.16 is reduced activity in left-hemisphere auditory cortex of the posterior superior temporal gyrus in PDS compared to fluent controls, a finding reported in numerous prior neuroimaging studies (e.g., Braun et al., 1997; Fox et al., 1996, 2000; De Nil et al., 2000). Auditory cortical activity impacts motor actions via corticostriatal projections (Znamenskiy & Zador, 2013). Thus, if auditory feedback of one's own speech does not match the expected pattern for the current sound (e.g., because of subtle errors in articulation), the striatum may detect a mismatch between the current sensorimotor context and the context needed for initiating the next motor program, thus reducing its competitive advantage over competing motor programs. As described above in relation to figure 10.13, this reduced competitive advantage could result in the impaired generation of initiation signals by the basal ganglia and a concomitant stutter.

This view receives support from a number of findings. First, it has long been known that there is a very low rate of stuttering in congenitally deaf individuals (e.g., Backus, 1938; Harms & Malone, 1939; Van Riper, 1982). Furthermore, a number of manipulations that interfere with normal auditory feedback processing of one's own speech can alleviate stuttering, including noise masking (Maraist & Hutton, 1957; Adams & Hutchison, 1974), chorus reading (Barber, 1939; Kalinowski & Saltuklaroglu, 2003), pitch-shifted auditory feedback (Macleod et al., 1995), and delayed auditory feedback (Stephen & Haggard, 1980). These conditions may have the effect of eliminating the detection of small errors in articulation that would otherwise reduce the match between expected and actual sensorimotor context for the next motor program in striatum. In light of these considerations, the reduced activity in auditory cortex of PWS in figure 10.16 may reflect a compensatory mechanism involving inhibition of auditory feedback of one's own speech to avoid detection of minor errors in production. This conjecture receives some support from findings of reduced responses to auditory perturbations during speech in adult PWS compared to age-matched controls (e.g., Cai et al., 2012, 2014a; Daliri et al., submitted). Interestingly, children who stutter do not show a reduction in adaptation to auditory perturbations compared to nonstuttering children (Daliri et al., submitted), suggesting that increased inhibition of auditory feedback during speech may develop gradually in PWS as a means to reduce dysfluency.

Further insight into cortical processing in stuttering can be gained from comparing brain activity during moments of stuttering to activity during fluent speech. Figure 10.17 illustrates activation foci from the Belyk, Kraft, and Brown (2015) meta-analysis of neuroimaging studies that contrasted brain activity during stuttered speech with activity during fluent speech in PWS. Several factors are expected to contribute to higher brain activity during stuttered speech, including (1) conscious detection of major speech errors (as opposed to the subconscious detection of relatively small sensory errors during fluent speech), (2)

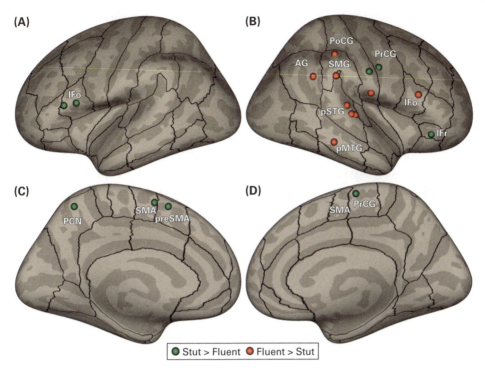

Figure 10.17
Cortical activation foci of brain activity differences between stuttered and fluent speech of people who stutter in the Belyk, Kraft, and Brown (2015) meta-analysis displayed on the (A) left lateral, (B) right lateral, (C) left medial, and (D) right medial inflated cortical surfaces. AG, angular gyrus; IFo, inferior frontal gyrus pars opercularis; IFr, inferior frontal gyrus pars orbitalis; PCN, precuneus; pMTG, posterior middle temporal gyrus; PoCG, postcentral gyrus; PrCG, precentral gyrus; preSMA, pre-supplementary motor area; pSTG, posterior superior temporal gyrus; SMA, supplementary motor area; SMG, supramarginal gyrus; Stut, stuttered speech.

increased effort to generate speech output, and (3) longer overall motor execution duration due to repetitions and prolongations. Nonetheless, stuttering is accompanied by an activity *decrease* in auditory cortical areas in posterior superior temporal gyrus and higher-order somatosensory cortical areas in supramarginal gyrus compared to fluent speech. This may reflect increased inhibition of sensory feedback during a stutter in an attempt to eliminate auditory and somatosensory error signals that are impeding the recognition of the sensorimotor context for initiating the next motor program in striatum. Alternatively, it may reflect the "shutting down" of feedback control mechanisms upon detection of sensory errors so large that feedback control is no longer appropriate for error correction. Increased activity in left inferior frontal gyrus and preSMA may reflect increased reliance on cognitive mechanisms to overcome the moment of stuttering by activating the *initiation map* node for the next sound via cortico-cortical means rather than through the basal ganglia motor loop (see related discussion in chapter 7, section 7.3). At present, however, these interpretations remain speculative.

10.5 Future Directions

Prior sections have illustrated how the DIVA and GODIVA models act as a framework for better understanding the primary disorders of speech motor control and motor programming—namely, the dysarthrias, apraxia of speech, and stuttering. The same neurocomputational framework can also provide valuable insights into a much wider range of cognitive and motor disorders that involve a speech component. For example, approximately 25% of individuals with *autism spectrum disorder* (*ASD*) do not develop functional language skills (Mody et al., 2013). Using DTI, Peeva et al. (2013) investigated the speech network in individuals with ASD and age-matched neurotypical controls and found abnormally weak white matter tracts between left SMA and left vPMC in the ASD participants (see figure 10.18). Interpreted within the DIVA/GODIVA framework, the expected behavioral correlate would be reduced speech output since this pathway lies between the initiation map in SMA and the speech sound map in left vPMC. This hypothesis is being further tested in ongoing studies and may at least partially account for reduced propositional speech in many individuals with ASD.

The DIVA/GODIVA modeling framework can also serve as a guide for the development of speech restoration technologies. For example, Guenther et al. (2009) developed a *brain-computer interface* (*BCI*) involving a neurotrophic electrode (Kennedy, 1989;

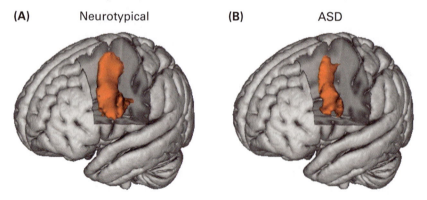

Figure 10.18
Magnetic resonance image with cutout to reveal canonical white matter tracts between left supplementary motor area (SMA) and left ventral premotor cortex (vPMC) in (A) neurotypical individuals and (B) individuals with autism spectrum disorder (ASD), generated from diffusion-weighted magnetic resonance images (see Peeva et al., 2013, for details). The volume of the left SMA-vPMC tract was found to be 32% smaller in ASD participants than in neurotypical participants. According to the DIVA model, this tract is involved in the initiation of speech output.

Bartels et al., 2008) implanted in the left ventral precentral gyrus of an individual suffering from *locked-in syndrome*, characterized by near-complete loss of voluntary movement with intact cognition and sensation. According to the DIVA model, neurons in the vicinity of the electrode encode speech motor programs, in the form of intended formant frequency trajectories (see chapters 3 and 4 for details). This led to the insight that it should be possible to decode the formant frequencies of intended speech movements from the electrode signals, a prediction that was verified in offline analyses of electrode signals collected during attempted speech (see figure 2.18 in chapter 2). This same insight was central to the design of a brain-computer interface that utilized real-time decoding of intended formant frequency trajectories to drive a formant synthesizer, thereby generating an audible speech signal from the decoded intentions of the locked-in BCI user (see figure 10.19). The total delay from neural firing to acoustic output was 50 ms, which approximates the delay from motor cortex neural activity to corresponding acoustic output in the intact vocal tract. The user was able to use the system to successfully produce vowels, and his performance improved substantially with practice. This latter finding indicates that the user's auditory feedback control subsystem was intact (as illustrated in figure 10.19), and commands generated by the auditory feedback controller were used to update the intact feedforward control system (see chapter 7 for details). The Guenther et al. (2009) BCI involved only two electrode recording channels, thereby limiting recording to a small number of neurons, in turn limiting performance to the production of vowel sounds. It is anticipated that future systems involving 100 or more electrode channels (such as the Utah array; Maynard, Nordhausen, & Normann, 1997) will allow real-time decoding and synthesis of a full set of speech sounds, thereby restoring conversational speech capabilities to those with locked-in syndrome.

The examples in this section illustrate the utility of the DIVA/GODIVA modeling framework for guiding research and treatment of speech disorders (for additional examples, see Perkell et al., 2000, 2007; Max et al., 2004; Lane et al., 2007; Loui et al., 2008; Terband et al., 2009, 2014; Beal et al., 2010; Civier, Tasko, & Guenther, 2010; Terband & Maassen, 2010; Cai et al., 2012, 2014; Civier et al., 2013; Croot, 2014; Maas, Mailend, & Guenther, 2015). As new data accumulate, it is important that the framework continues to evolve in order to account for these data, thereby forming an increasingly accurate and detailed account of the neural computations underlying speech production.

10.6 Summary

The DIVA and GODIVA models provide a theoretical framework for understanding the core neural deficits responsible for disordered speech production, including impairments of motor execution, articulatory motor programming, and initiation of speech output. Generally speaking, speech motor disorders involve impairment to the feedforward control system, which is essential for the production of fluent speech. Damage to feedback

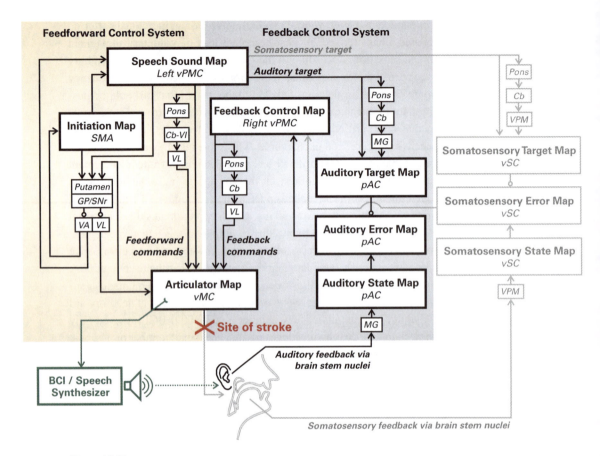

Figure 10.19
Schematic illustrating the replacement of normal vocal tract function with a brain-computer interface (BCI) that generates synthetic speech output in a locked-in patient whose motor output pathways have been abolished because of a brain stem stroke (indicated in red; see Guenther et al., 2009, for details). Portions of the speech motor control system that are unusable because of the stroke and resulting paralysis are indicated in gray. The BCI (shown in green) involved electrodes implanted in ventral motor cortex that transmit signals to a decoder which translated these signals into intended formant frequencies. The decoded frequencies were then fed to a formant-based speech synthesizer to generate an acoustic signal. The total delay from neural firing to acoustic output was 50 ms, which is the approximate delay from motor cortex neural activity to corresponding acoustic output in the intact speech motor system. The generated acoustic signal is used by the auditory feedback control subsystem in place of feedback from the paralyzed vocal tract. Corrective motor commands generated by the auditory feedback control subsystem become incorporated into the feedforward commands for subsequent productions, allowing the BCI user to improve performance with practice. Cb, cerebellum (specific lobule unknown); Cb-VI, cerebellum lobule VI; GP, globus pallidus; MG, medial geniculate nucleus of the thalamus; pAC, posterior auditory cortex; SMA, supplementary motor area; SNr, substantia nigra pars reticula; VA, ventral anterior nucleus of the thalamus; VL, ventral lateral nucleus of the thalamus; vMC, ventral motor cortex; vPMC, ventral premotor cortex.

control mechanisms can impair development of fluent speech if they occur early in life, but in the mature system the effects of impaired feedback control on intelligibility are likely to be minor.

Dysarthrias are disorders of motor execution characterized by weakness, abnormal muscle tone, and impaired articulation. Damage to the cranial nerves or associated motor nuclei that provide input to the articulatory musculature results in flaccid dysarthria, characterized by weakness in the musculature. Impairments from unilateral damage are typically relatively mild, whereas bilateral damage can be devastating. Spastic dysarthria, characterized by a strained voice, slow speech, and restricted pitch and loudness variation, typically occurs with damage to the motor cortex or its descending projections to the motor periphery. Because each hemisphere of motor cortex provides output to articulatory muscles on both sides of the body, spastic dysarthria typically involves bilateral damage, which can be caused by diseases such as amyotrophic lateral sclerosis. Weakness due to decreased motor cortical input co-occurs with muscle hyperactivity due to the release of brain stem motor neurons from cortical inhibition. Because they involve damage to the motor output end of the speech network, flaccid and spastic dysarthria affect both feedforward and feedback control mechanisms for speech.

Ataxic dysarthria, characterized by uncoordinated movements with inappropriate extent and low muscle tone, typically results from damage to the cerebellum, though it can also occur with damage to the pons (which relays signals from cortex to the cerebellum) or the ventral lateral thalamic nucleus (which relays cerebellar outputs to motor cortex). The most commonly reported damage sites are the superior paravermal cerebellar cortex (in or near lobule VI) and the dentate nucleus. Although the cerebellum plays roles in both feedforward and feedback control, the articulation deficits in ataxic dysarthria are primarily due to its role in generating highly coordinated and finely timed feedforward motor commands required for fluent speech.

Hypokinetic and hyperkinetic dysarthria generally involve impaired function of the basal ganglia. Hypokinetic dysarthria is most often associated with Parkinson's disease, which involves a depletion of dopamine in the striatum due to damage to the substantia nigra pars compacta. Insufficient striatal dopamine effectively weakens the direct pathway through the basal ganglia (which facilitates motor output) and strengthens the indirect pathway (which inhibits motor output), resulting in reductions in articulatory movement speeds and extents, decreased loudness and pitch range, and delays in initiating or ending movements. According to the DIVA model, these symptoms arise because of underactivation in the initiation circuit, resulting in initiation difficulties and a decreased GO signal that controls movement speed. The model further posits that the initiation circuit impairment forces reliance on premotor cortical regions for initiating the individual motor programs in a speech sequence. Hyperkinetic dysarthria involves abnormal involuntary movements of the speech articulators that can result from hyperfunction of the striatal dopaminergic system or other conditions that strengthen the direct pathway and/or weaken the indirect pathway, corresponding to an overactive initiation circuit in the DIVA model. Although dystonias are often categorized as hyperkinetic disorders, focal dystonias such as oromandibular dystonia and spasmodic dysphonia may involve mechanisms outside the basal ganglia, such as a hyperactive somatosensory control system.

Apraxia of speech is a disorder of speech motor planning or programming without comprehension impairment and without weakness in the speech musculature. It can occur developmentally (typically in combination with a number of other developmental issues) or as the result of stroke or other neurological insult to the left (language-dominant) inferior frontal gyrus, anterior insula, and/or precentral gyrus. According to the DIVA model, such damage impairs the speech sound map, which contains representations of frequently produced sound sequences such as syllables. This map is a central component of the motor programs for these sequences, so damage to it will impair the feedforward commands for articulating them. Such damage might also impair the readout of sensory expectations, or targets, for these sounds to higher-order auditory and somatosensory cortical areas, thereby impairing auditory and somatosensory feedback control mechanisms that rely on comparing these sensory targets to incoming sensory information. Apraxia of speech also involves impairments at a suprasyllabic level, including increased pausing between syllables and abnormal prosody. According to the GODIVA model, these impairments may occur because of damage to the phonological content buffer that temporarily stores upcoming phonological items for production.

The DIVA model distinguishes mechanisms for initiating a speech motor program from the mechanisms that generate the pattern of articulatory movements that make up the motor program. Whereas apraxia of speech primarily involves the latter, damage to the medial premotor cortex, particularly the supplementary motor area, affects the initiation circuit. Unilateral damage to this region is associated with supplementary motor area syndrome, characterized by a reduction in spontaneous movements of the body. If the damage is in the language-dominant hemisphere, transient mutism and loss of spontaneous speech can occur. According to the DIVA model, these symptoms arise from damage to the initiation map responsible for releasing speech motor programs for upcoming sounds. If the damage is unilateral, mutism resolves within a few weeks, indicating that functions performed in language-dominant supplementary motor area can be taken over by surrounding cortical regions and/or the supplementary motor area of the opposite hemisphere. Damage to language-dominant medial premotor and prefrontal cortex or projections from these regions to lateral speech and language areas can also result in transcortical motor aphasia, characterized by greatly reduced spontaneous speech output with a spared ability to repeat speech. The spared ability to repeat speech in transcortical motor aphasia suggests that the initiation map is either bilateral or is only involved in initiating self-generated speech and is not needed for externally triggered speech. If damage to medial premotor cortex is bilateral and spreads into the cingulate cortex, akinetic mutism, characterized by a near-complete lack of voluntary movement or speech, can occur, indicating that these areas are involved in processing the "motivation to speak."

Stuttering, which is often categorized as a fluency disorder, can also be considered a disorder of speech initiation. Although there is no current consensus regarding the root neural cause(s) of stuttering, the most likely locus of neural impairment is the left-hemisphere basal ganglia motor loop that constitutes the initiation circuit in the DIVA model. Impairment to several different parts of the basal ganglia motor loop have been associated with stuttering. Within the basal ganglia, excess dopamine in striatum has been identified in people who stutter; this may have the effect of decreasing inhibition of competing motor programs mediated by the indirect pathway, making it more difficult for the proper motor program to be chosen. Insufficient dopamine, as in Parkinson's disease, may lead to a different subtype of stuttering in which insufficient activation of the proper motor program occurs through the direct pathway. In order to initiate an upcoming motor program, the basal ganglia in effect monitors the cognitive, sensory, and motor context as represented in cerebral cortex via corticostriatal projections; abnormalities in these projections have been identified in children who stutter. Neural abnormalities in motor planning areas of the left inferior frontal cortex have also been noted in people who stutter; impaired functioning in this left-hemisphere network appears to force reliance on right-hemisphere homologues in stuttering. Reduced auditory cortical activity in people who stutter may reflect a compensatory strategy of inhibiting incoming auditory information, which might otherwise signal auditory production errors that could exacerbate stuttering.

The DIVA and GODIVA models can also provide insights into medical conditions beyond the classic motor speech disorders. For example, the DIVA model has been used to interpret findings regarding speech network abnormalities in autism and to guide the design of a brain-computer interface for restoring rudimentary speech capabilities to an individual suffering from locked-in syndrome. With time it is expected that details of this neurocomputational framework will change as new data accumulate, thereby providing an ever-refined account of the neural computations underlying speech production.

Notes

1. One exception would be a case where a hyperactive sensory feedback control system generated aberrant "corrective" motor commands that are strong enough to effectively override feedforward commands, resulting in aberrant articulations. This will be discussed further with respect to spasmodic dysphonia below.

2. Recall from chapter 2 that descending projections to the speech articulators from motor cortex are largely bilateral, and as a result unilateral damage of motor cortex has only mild effects on speech capabilities.

3. Childhood cerebellar tumors tend to be located in and around the posterior-inferior vermis and not in the superior cerebellum (Richter et al., 2005).

4. Based on studies of auditory and somatosensory deprivation described in chapters 5 and 6, the most likely sounds to be impaired would be fricative consonants, which require a precise constriction size to produce frication.

5. In addition to the basal ganglia motor loop, the extrapyramidal system also includes projections from the pallidum to the reticular formation and on to peripheral motor neurons. See footnote 5 in chapter 2 for a

discussion of different uses of the term *extrapyramidal system* by different authors, in particular inclusion of the cerebellum and related structures by some authors but not others (particularly in older publications).

6. Although the motor symptoms of Parkinson's disease are attributed to basal ganglia malfunction, the disease also involves damage to other brain regions; see Braak et al. (2004) for a review.

7. These terms are not used in the same way by all authors. Here we use *dyskinesia* to refer to any abnormal involuntary movement whereas *hyperkinesia* is limited to those abnormal involuntary movements originating from basal ganglia impairment (cf. Duffy, 1995).

8. Dystonia can be *generalized*, affecting many/most parts of the body, or *focal*, affecting only a particular neuromuscular system, with *segmental* or *multifocal dystonia* describing intermediate cases involving two or more neuromuscular systems (often involve neighboring body parts). Here we focus on focal dystonias affecting speech.

9. Interestingly, the intrinsic laryngeal musculature of individuals with spasmodic dysphonia often do not spasm during nonspeech vocalizations, such as laughing.

10. An exception is Tourette's syndrome, where thalamic stimulation is most commonly reported (Smith & Spindler, 2015).

11. These disparate findings likely arise in large part because of a lack of consensus on the precise definition of AOS (Miller, 2002), compounded by the fact that brain imaging was rarely used in diagnosis.

12. The ventral rostral precentral gyrus (BA 6) is often referred to as ventral premotor cortex. However, in this book the term *ventral premotor cortex* is used to refer to a larger functional region that also includes portions of BA 44 and aINS, which fit the definition of premotor cortex by virtue of the fact that lesions here can result in deficits in motor programming, typically considered a function performed by premotor cortex, that are central to AOS.

13. Most studies of transcortical motor aphasia do not differentiate between the SMA and preSMA; thus the umbrella term *supplementary motor areas* is used here.

14. Impaired initiation and/or termination of movements in PDS has also been reported in a number of nonspeech tasks such as auditory tracking (Nudelman et al., 1992) or producing a minimal displacement of the fingers or speech articulators (De Nil & Abbs, 1991; Howell, Sackin, & Rustin, 1995; Loucks & De Nil, 2001).

15. The small sample of PWS in this study, coupled with the lack of published follow-up studies, suggests caution in interpreting these experimental findings. In particular, excess striatal dopamine might be representative of only one subtype of stuttering, as discussed later in this section.

16. Unfortunately, these drugs typically have serious side effects that often outweigh improvements in fluency in most study participants, as detailed in Bothe et al. (2006).

17. Positive correlations were also found for tracts from right-hemisphere inferior frontal ROIs to left-hemisphere Rolandic ROIs (not shown in figure).

References

Abel, T. J., Buckley, R. T., Morton, R. P., Gabikian, P., & Silbergeld, D. L. (2015). Recurrent supplementary motor area syndrome following repeat brain tumor resection involving supplementary motor cortex. *Neurosurgery, 11*, 447–456.

Ackermann, H., Vogel, M., Petersen, D., & Poremba, M. (1992). Speech deficits in ischaemic cerebellar lesions. *Journal of Neurology, 239*, 223–227.

Adams, M. R., & Hutchison, J. (1974). The effects of three levels of auditory masking on selected vocal characteristics and the frequency of disfluency of adult stutterers. *Journal of Speech and Hearing Research, 17*, 682–688.

Aichert, I., & Ziegler, W. (2004). Syllable frequency and syllable structure in apraxia of speech. *Brain and Language, 88*, 148–159.

Alm, P. A. (2004). Stuttering and the basal ganglia circuits: a critical review of possible relations. *Journal of Communication Disorders, 37*, 325–369.

Anderson, J. M., Hughes, J. D., Rothi, L. J., Crucian, G. P., & Heilman, K. M. (1999). Developmental stuttering and Parkinson's disease: the effects of levodopa treatment. *Journal of Neurology, Neurosurgery, and Psychiatry, 66*, 776–778.

Backus, O. (1938). Incidence of stuttering among the deaf. *Annals of Otology, Rhinology, and Laryngology, 47*, 632–635.

Bannur, U., & Rajshekhar, V. (2000). Post operative supplementary motor area syndrome: clinical features and outcome. *British Journal of Neurosurgery, 14*, 204–210.

Barber, V. (1939). Studies in the psychology of stuttering: XV. Chorus reading as a distraction in stuttering. *Journal of Speech Disorders, 4*, 371–383.

Bartels, J., Andreasen, D., Ehirim, P., Mao, H., Siebert, S., Wright, E. J., et al. (2008). Neurotrophic electrode: method of assembly and implantation into human motor speech cortex. *Journal of Neuroscience Methods, 174*, 168–176.

Beal, D. S., Cheyne, D. O., Gracco, V. L., Quraan, M. A., Taylor, M. J., & De Nil, L. F. (2010). Auditory evoked fields to vocalization during passive listening and active generation in adults who stutter. *NeuroImage, 52*, 1645–1653.

Belyk, M., Kraft, S. J., & Brown, S. (2015). Stuttering as a trait or state—an ALE meta-analysis of neuroimaging studies. *European Journal of Neuroscience, 41*, 275–284.

Benke, T., Hohenstein, C., Poewe, W., & Butterworth, B. (2000). Repetitive speech phenomena in Parkinson's disease. *Journal of Neurology, Neurosurgery, and Psychiatry, 69*, 319–324.

Bielamowicz, S., & Ludlow, C. L. (2000). Effects of botulinum toxin on pathophysiology in spasmodic dysphonia. *Annals of Otology, Rhinology, and Laryngology, 109*, 194–203.

Blake, D. T., Byl, N. N., Cheung, S., Bedenbaugh, P., Nagarajan, S., Lamb, M., et al. (2002). Sensory representation abnormalities that parallel focal hand dystonia in a primate model. *Somatosensory & Motor Research, 19*, 347–357.

Bloodstein, O. (1995). *A handbook of stuttering* (5th ed.). London: Singular.

Bogousslavsky, J., Assal, F., & Regli, F. (1987). Infarct in the area of the left anterior cerebral artery: II. Language disorders. *Revue Neurologique, 143*, 121–127.

Bohland, J. W., Bullock, D., & Guenther, F. H. (2010). Neural representations and mechanisms for the performance of simple speech sequences. *Journal of Cognitive Neuroscience, 22*, 1504–1529.

Bothe, A. K., Davidow, J. H., Bramlett, R. E., Franic, D. M., & Ingham, R. J. (2006). Stuttering treatment research 1970–2005: II. Systematic review incorporating trial quality assessment of pharmacological approaches. *American Journal of Speech-Language Pathology, 15*, 342–352.

Braak, H., Ghebremedhim, E., Rub, U., Bratzke, H., & Del Tredici, K. (2004). Stages in the development of Parkinson's disease-related pathology. *Cell and Tissue Research, 318*, 121–134.

Braun, A. R., Varga, M., Stager, S., Schulz, G., Selbie, S., Maisog, J. M., et al. (1997). Altered patterns of cerebral activity during speech and language production in developmental stuttering: an H2(15)O positron emission tomography study. *Brain, 120*, 761–784.

Burghaus, L., Hilker, R., Thiel, A., Galldiks, N., Lehnhardt, F. G., Zaro-Weber, O., et al. (2006). Deep brain stimulation of the subthalamic nucleus reversibly deteriorates stuttering in advanced Parkinson's disease. *Journal of Neural Transmission (Vienna), 113*, 625–631.

Buse, J., Schoenefeld, K., Munchau, A., & Roessner, V. (2013). Neuromodulation in Tourette syndrome: dopamine and beyond. *Neuroscience and Biobehavioral Reviews, 37*, 1069–1084.

Cai, S., Beal, D. S., Ghosh, S. S., Guenther, F. H., & Perkell, J. S. (2014 a). Impaired timing adjustments in response to time-varying auditory perturbation during connected speech production in persons who stutter. *Brain and Language, 129*, 24–29.

Cai, S., Beal, D. S., Ghosh, S. S., Tiede, M. K., Guenther, F. H., & Perkell, J. S. (2012). Weak responses to auditory feedback perturbation during articulation in persons who stutter: evidence for abnormal auditory-motor transformation. *PLoS One, 7*, e41830.

Cai, S., Tourville, J. A., Beal, D. S., Perkell, J. S., Guenther, F. H., & Ghosh, S. S. (2014 b). Diffusion imaging of cerebral white matter in persons who stutter: evidence for network-level anomalies. *Frontiers in Human Neuroscience, 8*, 54.

Calabrese, P., De Murtas, M., Mercuri, N. B., & Bernardi, G. (1992). Chronic neuroleptic treatment: D2 dopamine receptor supersensitivity and striatal glutamatergic transmission. *Annals of Neurology, 31*, 366–373.

Canter, G. J., Trost, J. E., & Burns, M. S. (1985). Contrasting speech patterns in apraxia of speech and phonemic paraphasia. *Brian and Language*, *24*, 204–222.

Ceballos-Baumann, A. O. (2003). Functional imaging in Parkinson's disease: activation studies with PET, fMRI and SPECT. *Journal of Neurology*, *250*, 115–123.

Chang, S.-E., & Zhu, D. C. (2013). Neural network connectivity differences in children who stutter. *Brain*, *136*, 3709–3726.

Charcot, J. M. (1877). *Lectures on the diseases of the nervous system*. London: New Sydenham Society.

Civier, O., Bullock, D., Max, L., & Guenther, F. H. (2013). Computational modeling of stuttering caused by impairments in a basal ganglia thalamo-cortical circuit involved in syllable selection and initiation. *Brain and Language*, *126*, 263–278.

Civier, O., Tasko, S. M., & Guenther, F. H. (2010). Overreliance on auditory feedback may lead to sound/syllable repetitions: simulations of stuttering and fluency-inducing conditions with a neural model of speech production. *Journal of Fluency Disorders*, *35*, 246–279.

Coltheart, M., Rastle, K., Perry, C., Langdon, R., & Ziegler, J. (2001). DRC: a dual route cascaded model of visual word recognition and reading aloud. *Psychological Review*, *108*, 204–256.

Cook, M., Murdoch, B., Cahill, L., & Whelan, B. M. (2004). Higher-level language deficits resulting from left primary cerebellar lesions. *Aphasiology*, *18*, 771–784.

Croot, K. (2014). Motor speech disorders and models of speech production. In L. Cummings (Ed.), *The Cambridge handbook of communication disorders* (pp. 501–523). Cambridge, UK: Cambridge University Press.

Curlee, R. F. (2004). Stuttering. In R. D. Kent (Ed.), *The MIT encyclopedia of communication disorders* (pp. 220–222). Cambridge, MA: MIT Press.

Da Cunha, C., Boschen, S. L., Gomez-A, A., Ross, E. K., Gibson, W. S. J., Min, H.-K., et al. (2015). Toward sophisticated basal ganglia neuromodulation: review on basal ganglia deep brain stimulation. *Neuroscience & Biobehavioral Reviews*, *15*. doi: 10.1016.

Daliri, A., Wieland, E. A., Cai, S., Guenther, F. H., & Chang, S. (submitted). Sensorimotor adaptation is reduced in adults who stutter but not in children who stutter.

Darley, F. L. (1968). Apraxia of speech: 107 years of terminological confusion. Paper presented to the American Speech and Hearing Association, Denver.

Darley, F. L., Aronson, A. E., & Brown, J. R. (1969). Clusters of deviant speech dimensions in the dysarthrias. *Journal of Speech and Hearing Research*, *12*, 462–497.

Darley, F. L., Aronson, A. E., & Brown, J. R. (1975). *Motor speech disorders*. Philadelphia: Saunders.

De Nil, L. F., & Abbs, J. H. (1991). Kinaesthetic acuity of stutterers and non-stutterers for oral and non-oral movements. *Brain*, *114*, 2145–2158.

De Nil, L. F., Kroll, R. M., Kapur, S., & Houle, S. (2000). A positron emission tomography study of silent and oral single word reading in stuttering and nonstuttering adults. *Journal of Speech, Language, and Hearing Research*, *43*, 1038–1053.

De Nil, L. F., Kroll, R. M., Lafaille, S. J., & Houle, S. (2003). A positron emission tomography study of short- and long-term treatment effects on functional brain activation in adults who stutter. *Journal of Fluency Disorders*, *28*, 357–379.

Denny, M., & Smith, A. (1997). Respiratory and laryngeal control in stuttering. In R. F. Curlee & G. M. Siegel (Eds.), *Nature and treatment of stuttering: new directions* (2nd ed., pp. 128–142). Boston, MA: Allyn & Bacon.

Dronkers, N. F. (1996). A new brain region for coordinating speech articulation. *Nature*, *384*, 159–161.

Duffy, J. R. (1995). *Motor speech disorders: substrates, differential diagnosis, and management*. St. Louis: Mosby.

Duffy, J. R. (2004). Dysarthrias: characteristics and classification. In R. D. Kent (Ed.), *The MIT encyclopedia of communication disorders* (pp. 126–128). Cambridge, MA: MIT Press.

Duffy, J. R. (2005). *Motor speech disorders: substrates, differential diagnosis, and management* (2nd ed.). St. Louis: Mosby-Year Book.

Endo, Y., Saito, Y., Otsuki, T., Takahashi, A., Nakata, Y., Okada, K., et al. (2014). Persistent verbal and behavioral deficits after resection of the left supplementary motor area in epilepsy surgery. *Brain & Development, 36,* 74–79. doi: 10.1016.

Enokido, H., Torri, H., Ainoda, N., Hanyu, T., & Omori, S. (1984). Spontaneous speech disturbances in so called transcortical motor aphasia—comparison of 3 cases with the different lesion sites. *No To Shinkei, 36,* 895–902.

Epidemiological Study of Dystonia in Europe Collaborative Group. (1999). Sex-related influences on the frequency and age of onset of primary dystonia. *Neurology, 53,* 1871–1873.

Ferraro, F. R., Balota, D. A., & Connor, L. T. (1993). Implicit memory and the formation of new associations in nondemented Parkinson's disease individuals and individuals with senile dementia of the Alzheimer type: a serial reaction time (SRT) investigation. *Brain and Cognition, 21,* 163–180.

Fisher, C. M. (1989). Lacunar infarct of the tegmentum of the lower lateral pons. *Archives of Neurology, 46,* 566–567.

Fox, P. T., Ingham, R. J., Ingham, J. C., Hirsch, T. B., Downs, J. H., Martin, C., et al. (1996). A PET study of the neural systems of stuttering. *Nature, 382,* 158–161.

Fox, P. T., Ingham, R. J., Ingham, J. C., Zamarripa, F., Xiong, J.-H., & Lancaster, J. (2000). Brain correlates of stuttering and syllable production: A PET performance-correlation analysis. *Brain, 123,* 1985–2004.

Fox, R. J., Kasner, S. E., Chatterjee, A., & Chalela, J. A. (2001). Aphemia: an isolated disorder of articulation. *Clinical Neurology and Neurosurgery, 103,* 123–126.

Friedmann, M., Alexander, M. P., & Naeser, M. A. (1984). Anatomic basis of transcortical motor aphasia. *Neurology, 34,* 409–417.

Golfinopoulos, E., Bullock, D., & Guenther, F. H. (2015). Neurocomputational modeling of impaired sensory feedback control in spasmodic dysphonia. *Proceedings of the 2015 Human Brain Mapping Meeting, Honolulu, Hawaii.*

Goodglass, H. (1993). *Understanding aphasia.* New York: Academic Press.

Guenther, F. H., Brumberg, J. S., Wright, E. J., Nieto-Castanon, A., Tourville, J. A., Panko, M., et al. (2009). A wireless brain-machine interface for real-time speech synthesis. *PLoS One, 4,* e8218.

Guenther, F. H., Ghosh, S. S., & Tourville, J. A. (2006). Neural modeling and imaging of the cortical interactions underlying syllable production. *Brain and Language, 96,* 280–301.

Harms, H. A., & Malone, J. Y. (1939). The relationship of hearing acuity to stammering. *Journal of Speech Disorders, 4,* 363–370.

Helmuth, L., Mayr, U., & Daum, I. (2000). Sequence learning in patients with Parkinson's disease: a comparison between spatial-attention and number-response sequences. *Neuropsychologia, 38,* 1443–1451.

Hillis, A. E., Work, M., Barker, P. B., Jacobs, M. A., Breese, E. L., & Maurer, K. (2004). Re-examining the brain regions crucial for orchestrating speech articulation. *Brain, 127,* 1479–1487.

Howell, P., Sackin, S., & Rustin, L. (1995). Comparison of speech motor development in stutterers and fluent speakers between 7 and 12 years old. *Journal of Fluency Disorders, 20,* 243–255.

Jacks, A. (2008). Bite block vowel production in apraxia of speech. *Journal of Speech, Language, and Hearing Research, 51,* 898–913.

Jackson, G. M., Jackson, S. R., Harrison, J., Henderson, L., & Kennard, C. (1995). Serial reaction time learning in Parkinson's disease: evidence for a procedural learning deficit. *Neuropsychologia, 33,* 577–593.

Jonas, S. (1981). The supplementary motor region and speech emission. *Journal of Communication Disorders, 14,* 349–373.

Kalinowski, J., & Saltuklaroglu, T. (2003). Choral speech: the amelioration of stuttering via imitation and the mirror neuronal system. *Neuroscience and Biobehavioral Reviews, 27,* 339–347.

Kennedy, P. R. (1989). The cone electrode: a long-term electrode that records from neurites grown onto its recording surface. *Journal of Neuroscience Methods, 29,* 181–193.

Kent, R. D. (2000). Research on speech motor control and its disorders: a review and prospective. *Journal of Communicable Diseases, 33,* 391–428.

Kent, R. D. (2004). *The MIT encyclopedia of communication disorders.* Cambridge, MA: MIT Press.

Kent, R. D., & Rosenbek, J. C. (1983). Acoustic patterns of apraxia of speech. *Journal of Speech and Hearing Research, 26,* 231–249.

Koller, W. C. (1983). Dysfluency (stuttering) in extrapyramidal disease. *Archives of Neurology, 40,* 175–177.

Krainik, A., Lehericy, S., Duffau, H., Capelle, L., Chainay, H., Cornu, P., et al. (2003). Postoperative speech disorder after medial frontal surgery: role of the supplementary motor area. *Neurology, 60,* 587–594.

Kuriki, S., Mori, T., & Hirata, Y. (1999). Motor planning center for speech articulation in the normal human brain. *Neuroreport, 17,* 765–769.

Lane, H., Denny, M., Guenther, F. H., Hanson, H., Marrone, N., Matthies, M. L., et al. (2007). On the structure of phoneme categories in listeners with cochlear implants. *Journal of Speech, Language, and Hearing Research, 50,* 2–14.

Laplane, D., Talairach, J., Meininger, V., Bancaud, J., & Orgogozo, J. M. (1977). Clinical consequences of corticectomies involving the supplementary motor area in man. *Journal of the Neurological Sciences, 34,* 301–314.

Lechtenberg, R., & Gilman, S. (1978). Speech disorders in cerebellar disease. *Annals of Neurology, 3,* 285–290.

Leder, S. B. (1996). Adult onset of stuttering as a presenting sign in a parkinsonian-like syndrome: a case report. *Journal of Communication Disorders, 29,* 471–477.

Lee, M. S., Lee, S. B., & Kim, W. C. (1996). Spasmodic dysphonia associated with a left ventrolateral putaminal lesion. *Neurology, 47,* 827–828.

Levelt, W. J. M. (1989). *Speaking: from intention to articulation.* Cambridge, MA: MIT Press.

Liberman, J. A., & Reif, R. (1989). Spastic dysphonia and denervation signs in a young man with tardive dyskinesia. *British Journal of Psychiatry, 154,* 105–109.

Lim, E. C., Wilder-Smith, E., Ong, B. K., & Seet, R. C. (2005). Adult-onset re-emergent stuttering as a presentation of Parkinson's disease. *Annals of the Academy of Medicine, Singapore, 34,* 579–581.

Loucks, T. M., & De Nil, L. F. (2001). Oral kinesthetic deficit in adults who stutter: a target-accuracy study. *Journal of Motor Behavior, 38,* 238–246.

Loui, P., Guenther, F. H., Mathys, C., & Schlaug, G. (2008). Action-perception mismatch in tone-deafness. *Current Biology, 18,* R331–R332.

Louis, E. D., Winfield, L., Fahn, S., & Ford, B. (2001). Speech dysfluency exacerbated by levodopa in Parkinson's disease. *Movement Disorders, 16,* 562–565.

Ludlow, C. L. (2004). Laryngeal movement disorders: treatment with botulinum toxin. In R. D. Kent (Ed.), *The MIT encyclopedia of communication disorders* (pp. 38–40). Cambridge, MA: MIT Press.

Lundgren, K., Helm-Estabrooks, N., & Klein, R. (2010). Stuttering following acquired brain damage: a review of the literature. *Journal of Neurolinguistics, 23,* 447–454.

Maas, E., Mailend, M., & Guenther, F. H. (2015). Feedforward and feedback control in apraxia of speech (AOS): effects of noise masking on vowel production. *Journal of Speech, Language, and Hearing Research, 58,* 185–200.

Maassen, B. (2002). Issues contrasting adult acquired versus developmental apraxia of speech. *Seminars in Speech and Language, 23,* 257–266.

Macleod, J., Kalinowski, J., Stuart, A., & Armson, J. (1995). Effect of single and combined altered auditory feedback on stuttering frequency at two speech rates. *Journal of Communication Disorders, 28,* 217–228.

Mailend, M. L., & Maas, E. (2013). Speech motor programming in apraxia of speech: evidence from a delayed picture-word interference task. *American Journal of Speech-Language Pathology, 22,* S380–S396.

Maraist, J. A., & Hutton, C. (1957). Effects of auditory masking upon the speech of stutterers. *Journal of Speech and Hearing Disorders, 22,* 385–389.

Marien, P., Engelborghs, S., Fabbro, F., & De Deyn, P. P. (2001 a). The lateralized linguistic cerebellum: a review and a new hypothesis. *Brain and Language, 79,* 580–600.

Marien, P., Pickut, B. A., Engelborghs, S., Martin, J. J., & De Dyn, P. P. (2001 b). Phonological agraphia following a focal anterior insula-opercular infarction. *Neuropsychologia, 39,* 845–855.

Masaki, S., Tatsumi, I. F., & Sasanuma, S. (1991). Analysis of the temporal relationship between pitch control and articulatory movements in the realization of Japanese word accent by a patient with apraxia of speech. *Clinical Aphasiology*, *19*, 307–316.

Max, L., Guenther, F. H., Gracco, V. L., Ghosh, S. S., & Wallace, M. E. (2004). Unstable or insufficiently activated internal models and feedback-biased motor control as sources of dysfluency: a theoretical model of stuttering. *Contemporary Issues in Communication Science and Disorders*, *31*, 105–122.

Maynard, E. M., Nordhausen, C. T., & Normann, R. A. (1997). The Utah Intracortical Electrode Array: a recording structure for potential brain-computer interfaces. *Electroencephalography and Clinical Neurophysiology*, *102*, 228–239.

McNeil, M. R., & Adams, S. (1991). A comparison of speech kinematics among apraxic, conduction aphasic, ataxic dysarthric, and normal geriatric speakers. *Clinical Aphasiology*, *19*, 279–294.

McNeil, M. R., Caligiuri, M., & Rosenbek, J. C. (1989). A comparison of labiomandibular kinematic durations, displacements, velocities, and dysmetrias in apraxic and normal adults. *Clinical Aphasiology*, *18*, 173–193.

McNeil, M. R., Doyle, P. J., & Wambaugh, J. (2000). Apraxia of speech: a treatable disorder of motor planning and programming. In S. E. Nadeau, L. J. Gonzalez-Rothi, & B. Crosson (Eds.), *Aphasia and language: theory to practice* (pp. 221–266). New York: Guilford Press.

McNeil, M. R., Robin, D. A., & Schmidt, R. A. (1997). Apraxia of speech: definition, differentiation, and treatment. In M. R. McNeil (Ed.), *Clinical management of sensorimotor speech disorders* (pp. 311–344). Stuttgart: Thieme.

Mochizuki, H., & Saito, H. (1990). Mesial frontal lobe syndromes: correlations between neurological deficits and radiological localizations. *Tohoku Journal of Experimental Medicine*, *161*, 231–239.

Mody, M., Manoach, D. S., Guenther, F. H., Tenet, K., Bruno, K. A., McDougle, C. J., et al. (2013). Speech and language in autism spectrum disorder: a view through the lens of behavior and brain imaging. *Neuropsychiatry*, *3*, 223–232.

Mure, H., Tang, C. C., Argyelan, M., Ghilardi, M. F., Kaplitt, M. G., Dhawan, V., et al. (2012). Improved sequence learning with subthalamic nucleus deep brain stimulation: evidence for treatment-specific network modulation. *Journal of Neuroscience*, *32*, 2804–2813.

Nagaratnam, N., Nagaratnam, K., Ng, K., & Diu, P. (2004). Akinetic mutism following stroke. *Journal of Clinical Neuroscience*, *11*, 25–30.

Narayana, S., Fox, P. T., Zhang, W., Franklin, C., Robin, D. A., Vogel, D., et al. (2010). Neural correlates of efficacy of voice therapy in Parkinson's disease identified by performance-correlation analysis. *Human Brain Mapping*, *31*, 222–236.

Németh, G., Hegedüs, K., & Molnár, L. (1988). Akinetic mutism associated with bicingular lesions: clinicopathological and functional anatomical correlates. *European Archives of Psychiatry and Neurological Sciences*, *237*, 218–222.

Netsell, R. (1991). *A neurobiologic view of speech production and the dysarthrias*. San Diego: Singular.

Neumann, K., Preibisch, C., Euler, H. A., von Guderberg, A. W., Lanfermann, H., Gall, V., et al. (2005). Cortical plasticity associated with stuttering therapy. *Journal of Fluency Disorders*, *30*, 23–39.

New, A. B., Robin, D. A., Parkinson, A. L., Duffy, J. R., McNeil, M. R., Piguet, O., et al. (2015). Altered resting-state network connectivity in stroke patients with and without apraxia of speech. *NeuroImage. Clinical*, *25*, 429–439.

Niethammer, M., Feigin, A., & Eidelber, D. (2012). Functional neuroimaging in Parkinson's disease. *Cold Spring Harbor Perspectives in Medicine*, *2*, a009274.

Nudelman, H. B., Herbrich, K. E., Hess, K. R., Hoyt, B. D., & Rosenfield, D. B. (1992). A model of the phonatory response time of stutterers and fluent speakers to frequency-modulated tones. *Journal of the Acoustical Society of America*, *92*, 1882–1888.

Odell, K., McNeil, M. R., Rosenbek, J. C., & Hunter, L. (1990). Perceptual characteristics of consonant production by apraxic speakers. *Journal of Speech and Hearing Disorders*, *55*, 345–359.

Orton, S. T. (1927). Studies in stuttering. *Archives of Neurology and Psychiatry*, *18*, 671–672.

Pai, M. C. (1999). Supplementary motor area aphasia: a case report. *Clinical Neurology and Neurosurgery*, *101*, 29–32.

Peeva, M. G., Tourville, J. A., Agam, Y., Holland, B., Manoach, D. S., & Guenther, F. H. (2013). White matter impairment in the speech network of individuals with autism spectrum disorder. *NeuroImage. Clinical, 3*, 234–241.

Perkell, J. S., Denny, M., Lane, H., Guenther, F. H., Matthies, M. L., Tiede, M., et al. (2007). Effects of masking noise on vowel and sibilant contrasts in normal-hearing speakers and postlingually deafened cochlear implant users. *Journal of the Acoustical Society of America, 121*, 505–518.

Perkell, J. S., Guenther, F. H., Lane, H., Matthies, M. L., Perrier, P., Vick, J., et al. (2000). A theory of speech motor control and supporting data from speakers with normal hearing and profound hearing loss. *Journal of Phonetics, 28*, 233–272.

Perkins, W. H., Kent, R. D., & Curlee, R. F. (1991). A theory of neuropsycholinguistic function in stuttering. *Journal of Speech and Hearing Research, 34*, 734–752.

Postma, A., & Kolk, H. (1993). The covert repair hypothesis: prearticulatory repair processes in normal and stuttered disfluencies. *Journal of Speech and Hearing Research, 36*, 472–487.

Ramig, L. O., Brin, M. F., Velickovic, M., & Fox, C. (2004). Hypokinetic laryngeal movement disorders. In R. D. Kent (Ed.), *The MIT encyclopedia of communication disorders* (pp. 30–31). Cambridge, MA: MIT Press.

Ramig, L., Pawlas, A., & Countryman, S. (1995). *The Lee Silverman voice treatment (LSVT®): a practical guide to treating the voice and speech disorders in Parkinson disease.* Iowa City, IA: National Center for Voice and Speech.

Redgrave, P., Rodriguez, M., Smith, Y., Rodrigues-Oroz, M. C., Lehericy, S., Bergman, H., et al. (2010). Goal-directed and habitual control in the basal ganglia: implications for Parkinson's disease. *Nature, 11*, 760–772.

Richter, S., Gerwig, M., Aslan, B., Wilhelm, H., Schoch, B., Dimitrova, A., et al. (2007). Cognitive functions in patients with MR-defined chronic focal cerebellar lesions. *Journal of Neurology, 254*, 1193–1203.

Richter, S., Schoch, B., Ozimek, A., Gorissen, B., Hein-Kropp, C., Kaiser, O., et al. (2005). Incidence of dysarthria in children with cerebellar tumors: a prospective study. *Brain and Language, 92*, 153–167.

Robin, D. A., Bean, C., & Folkins, J. W. (1989). Lip movement in apraxia of speech. *Journal of Speech and Hearing Research, 32*, 512–523.

Rogers, M. A., & Storkel, H. L. (1999). Planning speech one syllable at a time: the reduced buffer capacity hypothesis in apraxia of speech. *Aphasiology, 13*, 793–805.

Rothi, L. J., & Heilman, K. M. (1997). *Apraxia: the neuropsychology of action.* Hove, UK: Psychology Press.

Sabatini, U., Boulanouar, K., Fabre, N., Martin, F., Carel, C., Colonnese, C., et al. (2000). Cortical motor reorganization in akinetic patients with Parkinson's disease: a functional MRI study. *Brain, 123*, 394–403.

Salmelin, R., Schnitzler, A., Schmitz, F., & Fruend, H. J. (2000). Single word reading in developmental stutterers and fluent speakers. *Brain, 123*, 1184–1202.

Sapir, S. (2014). Multiple factors are involved in the dysarthria associated with Parkinson's disease: a review with implications for clinical practice and research. *Journal of Speech and Hearing Research, 57*, 1330–1343.

Satoer, D., Kloet, A., Vincent, A., Dirven, C., & Visch-Brink, E. (2014). Dynamic aphasia following low-grade glioma surgery near the supplementary motor area: a selective spontaneous speech deficit. *Neurocase, 20*, 704–716. doi: 10.1080.

Schaefer, S., Freeman, F., Finitzo, T., Close, L., Cannito, M., Ross, E., et al. (1985). Magnetic resonance imaging findings and correlations in spasmodic dysphonia patients. *Annals of Otology, Rhinology, and Laryngology, 94*, 595–601.

Schmahmann, J. D., Doyon, J., Toga, A. W., Petrides, M., & Evans, A. C. (2000). *MRI atlas of the human cerebellum.* San Diego: Academic Press.

Schmahmann, J. D., Ko, R., & MacMore, J. (2004). The human basis pontis: motor syndromes and topographic organization. *Brain, 127*, 1269–1291.

Schoch, B., Dimitrova, A., Gizewski, E. R., & Timmann, D. (2006). Functional localization in the human cerebellum based on voxelwise statistical analysis: a study of 90 patients. *NeuroImage, 30*, 36–51.

Shahed, J., & Jankovic, J. (2001). Re-emergence of childhood stuttering in Parkinson's disease: a hypothesis. *Movement Disorders, 16*, 114–118.

Simonyan, K., Berman, B. D., Herscovitch, P., & Hallett, M. (2013). Abnormal striatal dopaminergic neurotransmission during rest and task production in spasmodic dysphonia. *Journal of Neuroscience, 33*, 14705–14714.

Simonyan, K., & Ludlow, C. L. (2010). Abnormal activation of the primary somatosensory cortex in spasmodic dysphonia: an fMRI study. *Cerebral Cortex, 20*, 2749–2759.

Skodda, S. (2012). Effect of deep brain stimulation on speech performance in Parkinson's disease. *Parkinson's Disease, 2012*, 850596.

Smith, K. M., & Spindler, M. A. (2015). Uncommon applications of deep brain stimulation in hyperkinetic movement disorders. *Tremor and Other Hyperkinetic Movements (New York, N.Y.), 5*, 278.

Smits-Bandstra, S., & Gracco, V. (2013). Verbal implicit sequence learning in persons who stutter and persons with Parkinson's disease. *Journal of Motor Behavior, 45*, 381–393.

Solomon, D. H., Barohn, R. J., Bazan, C., & Grisson, J. (1994). The thalamic ataxia syndrome. *Neurology, 44*, 810–814.

Spencer, K. A., & Rogers, M. A. (2005). Speech motor programming in hypokinetic and ataxic dysarthria. *Brain and Language, 94*, 347–366.

Spencer, K. A., & Slocomb, D. L. (2007). The neural basis of ataxic dysarthria. *Cerebellum (London, England), 6*, 58–65.

Starkweather, C. W. (1987). *Fluency and stuttering*. Englewood Cliffs, NJ: Prentice-Hall.

Starr, P. A., Kang, G. A., Heath, S., Shimamoto, S., & Turner, R. S. (2008). Pallidal neuronal discharge in Huntington's disease: support for selective loss of striatal cells originating the indirect pathway. *Experimental Neurology, 211*, 227–233.

Stephen, S. C., & Haggard, M. P. (1980). Acoustic properties of masking/delayed feedback in the fluency of stutterers and controls. *Journal of Speech and Hearing Research, 23*, 527–538.

Strand, E. A., Duffy, J. R., Clark, H. M., & Josephs, K. (2014). The apraxia of speech rating scale: a tool for diagnosis and description of apraxia of speech. *Journal of Communication Disorders, 51*, 43–50.

Sugishita, M., Konno, K., Kabe, S., Yunoki, K., Togashi, O., & Kawamura, M. (1987). Electropalatographic analysis of apraxia of speech in a left hander and in a right hander. *Brain, 110*, 1393–1417.

Terband, H., & Maassen, B. (2010). Speech motor development in childhood apraxia of speech: generating testable hypotheses by neurocomputational modeling. *Folia Phoniatrica et Logopaedica, 62*, 134–142.

Terband, H., Maassen, B., Guenther, F. H., & Brumberg, J. (2009). Computational neural modeling of speech motor control in childhood apraxia of speech. *Journal of Speech, Language, and Hearing Research, 52*, 1595–1609.

Terband, H., Maassen, B., Guenther, F. H., & Brumberg, J. (2014). Auditory-motor interactions in pediatric motor speech disorders: neurocomputational modeling of disordered development. *Journal of Communicable Diseases, 47*, 17–33.

Theys, C. (2012). Unraveling the enigma of neurogenic stuttering: prevalence, behavioral characteristics and neural correlates. PhD dissertation, Katholieke Universiteit Leuven.

Theys, C., De Nil, L., Thijs, V., van Wieringen, A., & Sunaert, S. (2013). A crucial role for the cortico-striato-cortical loop in the pathogenesis of stroke-related neurogenic stuttering. *Human Brain Mapping, 34*, 2103–2112.

Thiriez, C., Roubeau, B., Ouerchefani, N., Gurruchaga, J. M., Palfi, S., & Fenelon, G. (2013). Improvement in developmental stuttering following deep brain stimulation for Parkinson's disease. *Parkinsonism & Related Disorders, 19*, 383–384.

Toft, M., & Dietrichs, E. (2011). Aggravated stuttering following subthalamic deep brain stimulation in Parkinson's disease—two cases. *BMC Neurology, 11*, 44. doi: 10.1186.

Trost, J. E., & Canter, G. J. (1974). Apraxia of speech in patients with Broca's aphasia: a study of phoneme production accuracy and error patterns. *Brain and Language, 1*, 63–79.

Turner, G. S., & Weismer, G. (1993). Characteristics of speaking rate in the dysarthria associated with amyotrophic lateral sclerosis. *Journal of Speech and Hearing Research, 36*, 1134–1144.

Tykalová, T., Rusz, J., Čmejla, R., Klempíř, J., Růžičková, H., Roth, J., et al. (2015). Effect of dopaminergic medication on speech dysfluency in Parkinson's disease: a longitudinal study. *Journal of Neural Transmission, 122*, 1135–1142.

Urban, P. P., Marx, J., Hunsche, S., Gawehn, J., Vucurevic, G., Wicht, S., et al. (2003). Cerebellar speech representation: lesion topography in dysarthria as derived from cerebellar ischemia and functional magnetic resonance imaging. *Archives of Neurology, 60*, 965–972.

Vakil, E., Kahan, S., Huberman, M., & Osimani, A. (2000). Motor and non-motor sequence learning in patients with basal ganglia lesions: the case of serial reaction time (SRT). *Neuropsychologia, 38*, 1–10.

Van Lieshout, P. H. H. M., Bose, A., Square, P. A., & Steele, C. M. (2007). Speech motor control in fluent and dysfluent speech production of an individual with apraxia of speech and Broca's aphasia. *Clinical Linguistics & Phonetics, 21*, 159–188.

Van Riper, C. (1982). *The nature of stuttering* (2nd ed.). Prospect Heights, IL: Waveland Press.

Varley, R., & Whiteside, S. P. (2001). What is the underlying impairment in acquired apraxia of speech? *Aphasiology, 15*, 39–84.

Varsou, O., Stringer, M. S., Fernandes, C. D., Schwarzbauer, C., & MacLeod, M. J. (2014). Stroke recovery and lesion reduction following acute isolated bilateral ischaemic pontine infarction: a case report. *BMC Research Notes, 7*, 728.

Volkmann, J., Moro, E., & Pahwa, R. (2006). Basic algorithms for the programming of deep brain stimulation in Parkinson's disease. *Movement Disorders, 21*, S284–S289.

Walker, H. C., Phillips, D. E., Boswell, D. B., Guthrie, B. L., Guthrie, S. L., Nicholas, A. P., et al. (2009). Relief of acquired stuttering associated with Parkinson's disease by unilateral left subthalamic brain stimulation. *Journal of Speech, Language, and Hearing Research, 52*, 1652–1657.

Walter, U., Blitzer, A., Benecke, R., Grossman, A., & Dressler, D. (2014). Sonographic detection of basal ganglia abnormalities in spasmodic dysphonia. *European Journal of Neurology, 21*, 349–352.

Wambaugh, J., Duffy, J. R., McNeil, M. R., Robin, D. A., & Rogers, M. A. (2006). Treatment guidelines for acquired apraxia of speech: a synthesis and evaluation of the evidence. *Journal of Medical Speech-Language Pathology, 14*, xv–xxxiii.

Webster, W. G. (1989). Sequence initiation performance by stutterers under conditions of response competition. *Brain and Language, 36*, 286–300.

Whiteside, S. P., & Varley, R. A. (1998). A reconceptualisation of apraxia of speech: a synthesis of evidence. *Cortex, 34*, 221–231.

Wu, J. C., Maguire, G. A., Riley, G. D., Lee, A., Keator, D., Tang, C., et al. (1997). Increased dopamine activity associated with stuttering. *Neuroreport, 8*, 767–770.

Yaruss, J. S. (1999). Utterance length, syntactic complexity, and childhood stuttering. *Journal of Speech, Language, and Hearing Research, 42*, 329–344.

Yaruss, J. S. (2004). Speech disfluency and stuttering in children. In R. D. Kent (Ed.), *The MIT encyclopedia of communication disorders* (pp. 180–183). Cambridge, MA: MIT Press.

Zackheim, C. T., & Conture, E. G. (2003). Childhood stuttering and speech disfluencies in relation to children's mean length of utterance: a preliminary study. *Journal of Fluency Disorders, 28*, 115–141.

Zentner, J., Hufnagel, A., Pechstein, U., Wolf, H. K., & Schramm, J. (1996). Functional results after respective procedures involving the supplementary motor area. *Journal of Neurosurgery, 85*, 542–549.

Ziegler, W. (2002). Psycholinguistic and motor theories of apraxia of speech. *Seminars in Speech and Language, 23*, 231–243.

Znamenskiy, P., & Zador, A. M. (2013). Corticostriatal neurons in auditory cortex drive decisions during auditory discrimination. *Nature, 497*, 482–485.

Appendix A: Articulator Meta-analyses

This appendix describes a set of *activation likelihood estimate* (*ALE*) meta-analyses of studies that involve isolated nonspeech movements of speech articulators. Five separate meta-analyses were performed, one for each of the following articulators: jaw, lips, larynx, tongue, and the respiratory mechanism. Typical tasks included raising/lowering of the jaw (including gum chewing), side-to-side movement or protrusion/retraction of the tongue, protrusion/spreading of the lips, voicing with no articulator movements (e.g., humming), and consciously controlled inspiration/expiration. The ALE meta-analysis for each articulator identifies the locations of peak activation foci that are reliably found across studies (see Eickhoff et al., 2009, for details). These foci are specified as three-dimensional coordinates within the *Montreal Neurological Institute* (*MNI*) *stereotactic coordinate frame*. The origin of this coordinate frame is located at the anterior commissure (a relatively small fiber bundle connecting the left and right hemispheres) at the midline between the hemispheres, and each coordinate specifies a distance in millimeters[1] from the origin: the x-coordinate indicates left/right location relative to the midline, with negative values for left and positive values for right; the y-coordinate indicates anterior/posterior location relative to the anterior commissure, with positive values for anterior and negative values for posterior; and the z-coordinate represents superior/anterior position, with positive values for superior and negative values for inferior.

A.1 Methods

Study Selection

Table A.1 lists the studies included for each meta-analysis. For all meta-analyses, only studies that met the following criteria were considered:

1. The study utilized *functional magnetic resonance imaging* (*fMRI*) or *positron emission tomography* (*PET*).
2. The study reported activation peaks in stereotactic coordinates.

Table A.1
Studies included in the articulator meta-analyses (listed by first author and year), along with movement tasks studied

Articulator	Study	Task	No. of Contrasts
Jaw			
	Grabski (2012)	Jaw lowering	1
	Onozuka (2002)	Gum chewing	2
	Onozuka (2003)	Gum chewing	3
	Takahashi (2007)	Gum chewing	2
Larynx			
	Brown (2008)	Glottal stop; intonation	2
	Olthoff (2008)	Phonation; intonation	2
	Terumitsu (2006)	Intonation	1
Lip			
	Brown (2008)	Lip protrusion	1
	Grabski (2012)	Lip protrusion	1
	Gerardin (2003)	Lip contraction	1
	Hanakawa (2005)	Lips right/left	2
	Hesselmann (2004)	Lip pursing	1
	Lotze (2000a)	Lip pursing	1
	Lotze (2000b)	Lip pursing	1
	Pulvermuller (2006)	Lip up/down	1
	Rotte (2002)	Lip pursing	1
Respiratory system			
	Evans (1999)	Inspiration	1
	Isaev (2002)	Inspiration	1
	McKay (2003)	Hypercapnia breathing	1
	Colebatch (1991)	Inspiration	1
	Fink (1996)	Inspiration w/resistance	1
	Nakayama (2004)	Diaphragmatic motion	1
	Ramsay (1993)	Inspiration; expiration	2
Tongue			
	Alkadhi (2002a)	Tongue right/left	1
	Alkadhi (2002b)	Tongue right/left	1
	Brown (2008)	Tongue tip up/down	1
	Corfield (1999)	Tongue forward	1
	Curt (2002)	Tongue right/left	1
	Fesl (2003)	Tongue right/left	1
	Grabski (2012)	Tongue retraction	1
	Hauk (2004)	Tongue movement	1
	Hesselmann (2004)	Tongue right/left	1
	Lotze (2000b)	Vertical tongue movement	1
	Martin (2004)	Tongue elevation	1
	Pardo (1997)	Tongue protrusion	1
	Pulvermuller (2006)	Tongue up/down	1
	Riecker (2000)	Tongue right/left	1
	Rotte (2002)	Tongue right/left	1
	Shinagawa (2003)	Tongue protrusion; right; left	3
	Shinagawa (2004)	Tongue protrusion; right; left	9
	Stippich (2002)	Tongue up/down	1
	Stippich (2007)	Tongue up/down	1
	Terumitsu (2006)	Silent articulation of syllables	1
	Vincent (2006)	Tongue up/down/left/right	1
	Watanabe (2004)	Tongue right; left; retraction	3
	Zald (1999)	Tongue left/right	1

3. The study involved neurotypical adult participants. For studies that involved both neurotypical and disordered participants, only results from the neurotypical participants were included.

Additional criteria were used for each meta-analysis as follows. For the jaw meta-analysis, contrasts that involved isolated movements of the jaw or gum chewing were included (8 contrasts from 4 studies).[2] For the larynx meta-analysis, contrasts that involved phonation without any other articulator movements were included (5 contrasts from 3 studies). For the lip meta-analysis, contrasts involving lip movements (e.g., pursing) without other articulator movements were included (10 contrasts from 9 studies). Studies involving whistling were not included as this task requires substantial articulatory coordination (cf. the lip movement meta-analysis of Takai et al., 2010, which included whistling). For the respiration meta-analysis, contrasts involving consciously controlled inspiration, expiration, hypercapnia breathing, and breathing against a resistive load were included (8 contrasts from 7 studies). For the tongue meta-analysis, contrasts involving simple tongue movements (leftward, rightward, upward, downward, or forward movements within the mouth, or protrusion) with no other articulator movements were included (35 contrasts from 23 studies).

ALE Meta-analyses
Five ALE meta-analyses were performed using the quantitative method of Eickhoff et al. (2009) as implemented in the GingerALE analysis software (http://brainmap.org; Research Imaging Center of the University of Texas Health Science Center, San Antonio, Texas). Briefly, an activation likelihood estimate is calculated at every location (voxel) in the brain from a pooled set of activation foci from neuroimaging studies involving a similar task. For each focus in the pooled data set, an ALE score is calculated at every voxel as a function of the voxel's distance from the focus using a 3-D Gaussian probability density function centered on the focus. This is repeated for every focus, and the final estimate for each voxel is the sum of the ALE scores across all foci. To test for statistical significance, the computed ALEs are compared to a null distribution created by repeatedly generating ALE statistics from randomly placed activation foci, with the same number of foci as the pooled studies. The null distribution is used to estimate an ALE statistic threshold resulting in a given false discovery rate (0.05 in the current study), and a minimum spatial extent is applied to eliminate very small clusters of contiguous significant voxels (100 mm^2 in the current study). The peaks of surviving clusters are reported in MNI coordinates, and their volumes are reported in mm^3. Studies utilizing Talairach coordinates were transformed into MNI space by the tal2icbm transformation implemented in the GingerALE toolbox (Lancaster et al., 2007). Anatomical labels for subcortical structures were determined using the Automated Anatomical Labeling atlas (Tzourio-Mazoyer et al., 2002) from the MRIcron software package (author Chris Rorden). Cortical anatomical labels were derived from the cortical

parcellation system described in Appendix B after ALE foci were projected to the cortical surface.

Identifying Regions of Articulatory Convergence

After completion of the five meta-analyses, visual inspection was used to identify candidate areas that contained foci of three or more articulators within close proximity to each other. The identified groupings were then quantitatively analyzed by (1) calculating the MNI location of the center of the cluster (i.e., the averaged x-, y-, and z-coordinates for the foci in the cluster) and (2) calculating the distance from the cluster center to each of the individual foci within the cluster. Only clusters that met the criterion of having foci for three or more articulators within 1 cm of the cluster center were considered to demonstrate articulatory convergence. The 1-cm threshold corresponds approximately to the average full width half maximum of a Gaussian distribution describing the uncertainty in spatial locations of activity foci determined empirically by Eickhoff et al. (2009), which was 10.2 mm.

A.2 Results

Table A.2 through table A.6 provide the results of ALE analyses for jaw, larynx, lip, respiratory, and tongue movements, respectively. The MNI coordinates, activation likelihood estimates, cluster volumes, and brain regions are listed for all foci identified by the analyses.

Table A.2
Results of ALE analysis for jaw movements

Region	Hemisphere	X	Y	Z	ALE	Cluster No.	Cluster Volume (mm^3)
PrCG	Left	−50	−10	32	0.026	1	2,496
PrCG	Right	54	−8	36	0.024	2	2,376
SMA	Left	−2	−6	58	0.017	3	1,976
Thalamus	Right	14	−18	2	0.023	4	1,336
PoCG	Left	−40	−12	16	0.024	5	1,328
Cerebellum	Right	20	−66	−26	0.017	6	1,112
Cerebellum	Left	−10	−66	−20	0.016	7	1,056
Thalamus	Left	−10	−20	2	0.017	8	808
PrCG	Right	42	−6	12	0.013	9	448
FP	Right	36	56	16	0.009	10	200

Note. For each significant cluster, the Montreal Neurological Institute coordinate of the peak voxel is given along with the associated brain region label. ALE, activation likelihood estimate; FP, frontal pole; PoCG, postcentral gyrus; PrCG, precentral gyrus; SMA, supplementary motor area.

Table A.3
Results of ALE analysis for larynx movements

Region	Hemisphere	X	Y	Z	ALE	Cluster No.	Cluster Volume (mm³)
PrCG	Left	−54	0	22	0.020	1	2,440
PoCG	Left	−52	−12	12	0.018	1	—
PrCG	Right	48	−10	40	0.020	2	2,088
SMG	Left	−52	−32	16	0.017	3	1,456
HG	Right	54	−10	4	0.017	4	1,304
Cerebellum	Right	26	−62	−24	0.013	5	912
Cerebellum	Right	34	−58	−30	0.011	5	—
PrCG	Left	−46	−14	44	0.014	6	728
PrCG	Right	54	2	20	0.015	7	560
PrCG	Left	−18	−30	58	0.015	8	520
SMA	Right	4	−8	62	0.015	9	496
preSMA	Left	−4	6	56	0.014	10	488
Cerebellum	Left	−22	−72	−50	0.014	11	480
PrCG	Right	60	8	10	0.014	12	480
Cerebellum	Left	−26	−60	−28	0.013	13	456
PrCG	Left	−52	0	44	0.013	14	424
PrCG	Right	16	−32	56	0.015	15	424
PrCG	Left	−38	−12	32	0.012	16	400
CMA	Right	6	4	38	0.012	17	344
pSTG	Right	52	−30	16	0.012	18	304

Note. For each significant cluster, the Montreal Neurological Institute coordinate of the peak voxel is given along with the associated brain region label. ALE, activation likelihood estimate; CMA, cingulate motor area; HG, Heschl's gyrus; PoCG, postcentral gyrus; PrCG, precentral gyrus; preSMA, pre-supplementary motor area; pSTG, posterior superior temporal gyrus; SMA, supplementary motor area; SMG, supramarginal gyrus.

Table A.4
Results of ALE analysis for lip movements

Region	Hemisphere	X	Y	Z	ALE	Cluster No.	Cluster Volume (mm³)
PoCG	Left	−54	−14	40	0.032	1	5,376
PrCG	Left	−58	6	32	0.016	1	—
PrCG	Right	54	−6	38	0.028	2	3,512
PrCG	Right	60	6	32	0.015	2	—
Globus pallidus	Left	−24	−2	−4	0.025	3	2,224
Putamen	Right	26	0	4	0.011	4	912
Globus pallidus	Right	22	−4	−4	0.010	4	—
Cerebellum	Right	20	−60	−20	0.014	5	880
SMA	Right	2	−4	62	0.014	6	768
PoCG	Right	58	−16	20	0.013	7	456
Cerebellum	Left	−22	−60	−22	0.010	8	304
PrCG	Right	62	10	14	0.010	9	168

Note. For each significant cluster, the Montreal Neurological Institute coordinate of the peak voxel is given along with the associated brain region label. ALE, activation likelihood estimate; PoCG, postcentral gyrus; PrCG, precentral gyrus; SMA, supplementary motor area.

Table A.5
Results of ALE analysis for respiratory movements

Region	Hemisphere	X	Y	Z	ALE	Cluster No.	Cluster Volume (mm³)
PrCG	Right	18	−20	60	0.014	1	4,384
SMA	Right	2	−8	60	0.009	1	—
PrCG	Left	−18	−24	64	0.013	2	3,072
Thalamus	Left	−12	−10	0	0.009	3	1,256
Globus pallidus	Left	−24	−8	−2	0.008	3	—
pCC	Right	2	−26	46	0.009	4	1,136
PrCG	Left	−48	−4	46	0.010	5	856
PrCG	Right	48	4	46	0.007	6	704
PrCG	Right	54	−4	36	0.006	6	—
SMG	Right	66	−32	28	0.009	7	616
SMG	Left	−44	−46	40	0.009	8	560
Thalamus	Right	2	−16	−2	0.006	9	504
Thalamus	Right	16	−10	0	0.006	9	—
Globus pallidus	Right	22	−6	−2	0.006	9	—
SMG	Left	−46	−22	18	0.009	10	496
CMA	Right	14	−4	42	0.007	11	288
Putamen	Left	−26	10	−6	0.006	12	200
Putamen/caud.	Right	22	18	8	0.006	13	168

Note. For each significant cluster, the Montreal Neurological Institute coordinate of the peak voxel is given along with the associated brain region label. ALE, activation likelihood estimate; Caud., caudate nucleus; CMA, cingulate motor area; pCC, posterior cingulate cortex; PoCG, postcentral gyrus; PrCG, precentral gyrus; SMA, supplementary motor area; SMG, supramarginal gyrus.

Table A.6
Results of ALE analysis for tongue movements

Region	Hemisphere	X	Y	Z	ALE	Cluster No.	Cluster Volume (mm³)
PrCG	Left	−56	−2	24	0.105	1	11,488
PrCG	Right	60	−2	22	0.096	2	10,288
PrCG	Right	56	−6	28	0.086	2	—
Cerebellum	Right	18	−64	−24	0.038	3	2,408
SMA	Left	−4	−2	56	0.039	4	2,408
Cerebellum	Left	−16	−62	−22	0.038	5	1,984
Putamen	Left	−26	−2	−2	0.024	6	1,696
Thalamus	Right	14	−18	2	0.026	7	1,312
Globus pallidus	Right	24	2	2	0.022	8	1,240
Thalamus	Left	−14	−18	2	0.019	9	720
Cerebellum	Right	14	−70	−50	0.019	10	528
SMA	Right	6	−2	72	0.015	11	464
SPL	Left	−36	−42	46	0.018	12	456
PrCG	Right	40	−8	58	0.012	13	160

Note. For each significant cluster, the Montreal Neurological Institute coordinate of the peak voxel is given along with the associated brain region label. ALE, activation likelihood estimate; PrCG, precentral gyrus; SMA, supplementary motor area; SPL, superior parietal lobule.

A.3 Discussion

The following subsections discuss the findings organized by brain region. This discussion will also treat the activity foci identified in two prior meta-analyses of speech production: the Turkeltaub et al. (2002) meta-analysis of single-word production and the Brown et al. (2005) meta-analysis of word and sentence production.

Subcortical Structures

Basal Ganglia and Thalamus The ALE analysis identified bilateral foci of thalamic activity for jaw, tongue, and respiratory movements (panel A of figure A.1) and bilateral foci of basal ganglia activity for lip, tongue, and respiratory movements (panel B of figure A.1). With the exception of one medial focus for respiration, the thalamic foci were in lateral portions, in or near the ventral lateral nucleus. These foci were located slightly medially relative to the speech-related focus found by Turkeltaub et al. (2002) in the left thalamus; no other speech-related foci were identified for the basal ganglia or thalamus in the two speech meta-analyses. Basal ganglia foci were clustered in the putamen and globus pallidus, with the exception of one respiratory focus in each hemisphere that was located anterior to these clusters, near the border between the caudate and putamen.

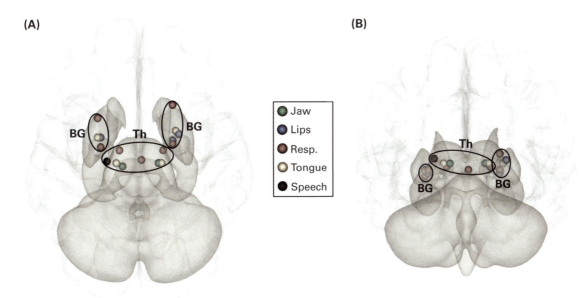

Figure A.1
(A) Superior and (B) posterior views of activity foci (color-coded spheres) from the articulatory and speech meta-analyses located in the basal ganglia and thalamus viewed through a transparent brain (subcortical structures shown in darker shade). BG, basal ganglia; Resp., respiratory system; Th, thalamus.

Cerebellum As illustrated in figure A.2, the ALE analyses identified bilateral foci of superior cerebellar activity for jaw, larynx, lip, and tongue movements, as well as one right inferior cerebellar focus for tongue movements and one left inferior cerebellar focus for larynx movements. No cerebellar foci were found for respiratory movements. However, it is noteworthy that many of the respiration studies are older and used PET, and some older PET studies did not capture the cerebellum because of limitations in the field of view of the scanner. Inspection of the individual respiration studies yielded 5 respiration contrasts showing left-hemisphere cerebellar activity peaks (Colebatch et al., 1991, 1 peak; Isaev et al., 2002, 3 peaks; McKay et al., 2003, 1 peak) and 5 showing right-hemisphere cerebellar peaks (Isaev et al., 2002, 3 peaks; McKay et al., 2003, 1 peak; Ramsay et al., 1993, 1 peak). The average locations of these peaks (left-hemisphere coordinate [–17,–66,–27]; right-hemisphere coordinate [18,–64,–24]) are included in figure A.2 with the caveat that these foci were not identified by the GingerALE analysis.

All but two of the cerebellar foci are in lobule VI of the superior cerebellar cortex, which also contains all of the cerebellar activity foci from the Turkeltaub et al. (2002) and Brown et al. (2005) speech meta-analyses. The paravermal region of lobule VI includes foci from all of the articulators as well as the speech meta-analyses bilaterally. Additional speech-related foci are found near the vermis in lobule VI.

Lateral Cortical Areas

Figure A.3 illustrates the ALE results for the lateral aspect of the cerebral cortex plotted on inflated cortical surfaces. The ALE analyses yielded bilateral activity foci in the lateral

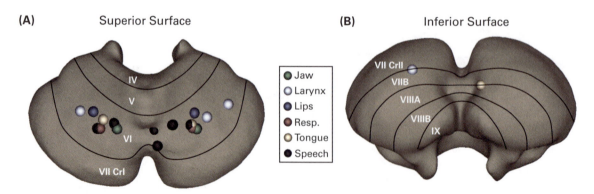

Figure A.2
Activity foci (color-coded spheres) from the articulatory and speech meta-analyses projected onto the (A) superior and (B) inferior cerebellar cortical surfaces. The foci for respiratory movement were formed by averaging activity peaks found in individual studies rather than from the ALE meta-analysis (see the text for details). Roman numerals indicate cerebellar lobules (Schmahmann et al., 2000), and black lines indicate approximate lobule boundaries. Cr, crus; Resp., respiratory system.

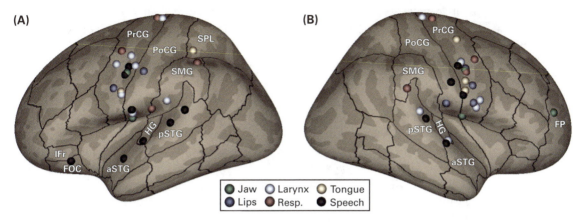

Figure A.3
(A) Left- and (B) right-hemisphere lateral cortical activity foci (color-coded spheres) from the articulatory and speech meta-analyses projected onto inflated cortical surfaces. aSTG, anterior superior temporal gyrus; FOC, frontal orbital cortex; FP, frontal pole; HG, Heschl's gyrus; IFr, inferior frontal gyrus pars orbitalis; PoCG, postcentral gyrus; PrCG, precentral gyrus; pSTG, posterior superior temporal gyrus; Resp., respiratory system; SMG, supramarginal gyrus; SPL, superior parietal lobule.

cerebral cortex for all articulator movements; these foci are described in the following paragraphs.

Frontal Lobe Articulator-related activity foci in the left-hemisphere frontal lobe were limited to the precentral gyrus, which contains the primary motor cortex and portions of premotor cortex. Right frontal lobe foci were also primarily located in *precentral gyrus* (*PrCG*), with the exception of one focus for jaw movements that was located in the frontal pole, far anterior to the other foci.

In both hemispheres, speech-related foci in PoCG and PrCG from the Turkeltaub et al. (2002) and Brown et al. (2005) studies are found in the vicinity of clusters of articulatory foci in PrCG or PoCG. One speech-related focus is located in the left *frontal orbital cortex* (*FOC*), very near the border with the *inferior frontal gyrus pars orbitalis* (*IFr*) and far away from any articulator foci. This location is likely involved in speech- or language-specific processing.

Parietal Lobe The ALE analyses identified activity foci for jaw, larynx, and lip movements in left-hemisphere *postcentral gyrus* (*PoCG*), which contains the primary somatosensory cortex. The lip ALE analysis identified an activity focus in right-hemisphere PoCG. Activity foci were found in the ventral *supramarginal gyrus* (*SMG*), in or near the parietal operculum, bilaterally for respiratory movements and in the left hemisphere for laryngeal movements. One focus each for tongue and respiratory movements was found in/near the intraparietal sulcus, which separates SMG from the *superior parietal lobule* (*SPL*). As discussed in chapter 2, this region may be involved in reading of stimuli during the neuroimaging experiment.

Speech-related foci in the parietal lobe were limited to one focus in the PoCG of each hemisphere, both occurring in the vicinity of foci from several speech articulators.

Temporal Lobe No articulatory foci were found in the left temporal lobe, and only two were reported in the right temporal lobe, both for laryngeal movements. One of these is located in *Heschl's gyrus* (*HG*), which contains primary auditory cortex, and the other is located in the *posterior superior temporal gyrus* (*pSTG*), which is a higher-order auditory cortical area that is heavily involved in speech processing. It is not surprising that movements of nonlaryngeal articulators do not activate the auditory cortical areas in the temporal lobes since these movements are essentially silent.

In contrast to the articulator meta-analyses, several speech-related foci are found in the temporal lobes, including foci in higher-order auditory areas of the *anterior superior temporal gyrus* (*aSTG*) and pSTG bilaterally as well as a focus in left HG. Left-hemisphere superior temporal cortex, which contains four speech-related foci, has long been known to play an important role in speech and language processing. The right-hemisphere speech foci are located very near the foci for laryngeal movements.

Medial Cortical Areas
Figure A.4 illustrates the foci identified by the ALE analyses in the medial cortical areas.

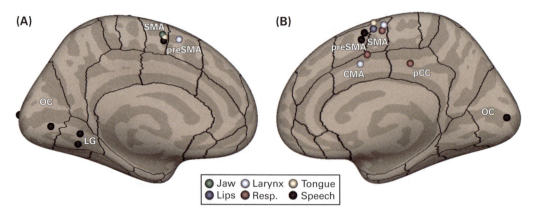

Figure A.4
(A) Left- and (B) right-hemisphere medial cortical activity foci (color-coded spheres) from the articulator and speech meta-analyses projected onto inflated cortical surfaces. CMA, cingulate motor area; LG, lingual gyrus; OC, occipital cortex; pCC, posterior cingulate cortex; preSMA, pre-supplementary motor area; Resp., respiratory system; SMA, supplementary motor area.

Supplementary Motor Areas The ALE meta-analyses identified activation foci in medial frontal cortex for all articulators. Seven of these foci localized to the *supplementary motor area* (*SMA*), and one focus for laryngeal movements localized to the left *pre-supplementary motor area* (*preSMA*). Bilateral activation of the SMA was found for larynx, tongue, and respiratory movements. Left-lateralized SMA activity was found for jaw movements and right-lateralized SMA activity for lip movements. It should be noted that identifying hemispheric laterality of neuroimaging activity in the supplementary motor areas can be difficult. Because it lies very near the midline of the brain, bilateral SMA activity can appear as a single activity focus rather than separate foci for left and right hemispheres. Furthermore, SMA from one hemisphere may cross over the $x = 0$ plane of the MNI coordinate frame; for example, left-hemisphere SMA may have a positive x value in some subjects and studies. It is thus possible that SMA activity is bilateral for all articulators.

The speech meta-analyses yielded bilateral foci near the border between SMA and preSMA, all located in the vicinity of foci from multiple articulators.

Cingulate Cortex The ALE meta-analyses yielded three significant foci of activity in right-hemisphere cingulate cortex. Two of these foci are in the *cingulate motor area* (*CMA*), one for laryngeal movements and one for respiratory movements. The third focus, for respiratory movements, is located in the *posterior cingulate cortex* (*pCC*). No speech-related foci were found in cingulate cortex by Turkeltaub et al. (2002) or Brown et al. (2005).

Occipital Cortex and Lingual Gyrus The speech meta-analyses identified a total of five foci in *occipital cortex* (*OC*) and *lingual gyrus* (*LG*). No foci of high articulatory convergence are located in these regions. These foci are likely the result of reading of stimuli in the speech conditions of the studies from the meta-analysis.

Regions of High Articulatory Convergence
Table A.7 and figures A.5 and A.6 provide the mean locations of the clusters of high articulatory convergence as defined in section A.1, along with activity foci from the Turkeltaub et al. (2002) and Brown et al. (2005) speech meta-analyses.

Several noteworthy observations can be made regarding the clusters of articulatory convergence. First, they are largely bilaterally symmetrical; for every left-hemisphere cluster there is a right-hemisphere cluster in the same region, with the exception of the cluster in left ventral PoCG, which is somewhat anterior and superior to its closest right-hemisphere counterpart in the ventral PrCG. Second, they are confined to core sensorimotor control areas: PoCG, PrCG, SMA, cerebellum, basal ganglia, and thalamus. Third, there are no articulator foci in the temporal or occipital lobes, prefrontal cortex, or posterior parietal cortex.

Table A.7
Comparison of locations of high articulatory convergence to activity foci from meta-analyses of simple speech tasks

Region	Hemisphere	Articulatory Convergence			Turkeltaub et al.				Brown et al.			
		X	Y	Z	X	Y	Z	Distance	X	Y	Z	Distance
Cerebellum	Left	-18	-63	-24	-14	-65	-15	10	-22	-67	-16	10
Cerebellum	Right	21	-63	-25	12	-65	-15	14	20	-66	-15	11
GP/Putamen	Left	-25	-4	-3								
GP/Putamen	Right	24	-2	0								
Thalamus	Left	-12	-16	1	-20	-17	5	9				
Thalamus	Right	15	-15	1								
PrCG	Left	-50	-4	35	-48	-12	36	8	-51	-6	33	3
PrCG	Right	53	-7	36	44	-10	34	10	59	-6	34	6
PrCG	Right	59	3	19					61	-5	18	8
PoCG	Left	-46	-15	15					-61	-12	10	16
SMA	Left	-3	-1	57	0	1	52	6				
SMA	Right	4	-6	64	0	1	52	15	7	4	59	12

Note. For each articulatory convergence location, the closest focus from the Turkeltaub et al. (2002) and Brown et al. (2005) meta-analyses of speech production is shown if there is such a focus within 2 cm. Coordinates are in Montreal Neurological Institute space. Distance, distance between the articulatory convergence location and the meta-analysis focus in millimeters; GP, globus pallidus; PoCG, postcentral gyrus; PrCG, precentral gyrus; SMA, supplementary motor area.

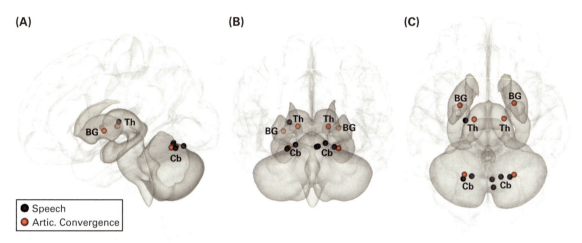

Figure A.5
Comparison of subcortical locations of high articulatory convergence (red spheres) to activity foci from the Brown et al. (2005) and Turkeltaub et al. (2002) meta-analyses of speech production (black spheres), shown in (A) left, (B) posterior, and (C) superior views of a transparent brain. Artic., articulatory; BG, basal ganglia; Cb, cerebellum; Th, thalamus.

All foci from the speech meta-analyses that are located in core sensorimotor areas (PoCG, PrCG, thalamus, and cerebellum) occur near foci of high articulatory convergence (see table A.7), with the exception of speech-related foci near the vermis of lobule VI in the cerebellar cortex. The speech meta-analyses revealed 3 left-hemisphere and 2 right-hemisphere foci in the superior temporal cortex not found in the articulatory convergence analysis; these foci are likely due to processing of auditory signals during speech. Two speech-related foci fall in left prefrontal cortical areas (FOC and FP) that are located far anterior to any of the foci of high articulatory convergence, and five speech-related foci are found in OC and LG (likely related to reading rather than articulation of speech), with no nearby foci of articulatory convergence.

Also notable is that the speech meta-analyses did not find any foci in the basal ganglia, unlike the articulatory convergence analysis. This appears to be a limitation of ALE meta-analysis since activity in basal ganglia and thalamus is frequently reported in speech neuroimaging experiments and is clearly visible in the fMRI activity from 116 subjects producing short speech utterances shown in figure 2.8 of chapter 2.

This comparative analysis indicates that speech-related regions of the core sensorimotor network overlap highly with those used for nonspeech movements of the speech articulators. Regions specialized for speech production appear to be limited to the temporal lobe and left prefrontal cortex, at least when viewed from the perspective provided by ALE meta-analyses.

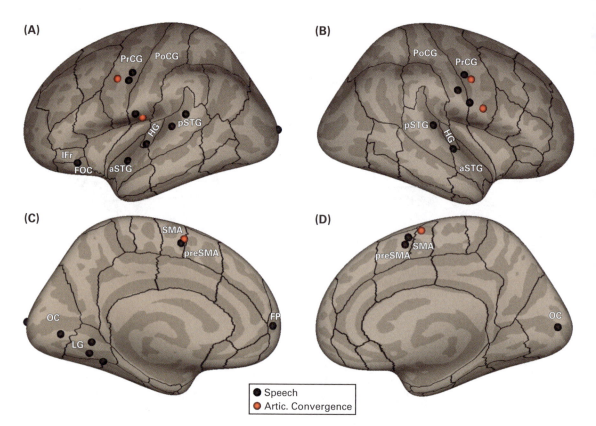

Figure A.6
Comparison of cortical foci of high articulatory convergence (red spheres) with activity foci from the Brown et al. (2005) and Turkeltaub et al. (2002) meta-analyses of speech production (black spheres) plotted on (A) left lateral, (B) right lateral, (C) left medial, and (D) right medial inflated cortical surfaces. Artic., articulatory; aSTG, anterior superior temporal gyrus; FOC, frontal orbital cortex; FP, frontal pole; HG, Heschl's gyrus; IFr, inferior frontal gyrus pars orbitalis; LG, lingual gyrus; OC, occipital cortex; PoCG, postcentral gyrus; PrCG, precentral gyrus; preSMA, pre-supplementary motor area; pSTG, posterior superior temporal gyrus; SMA, supplementary motor area.

Notes

1. The interpretation of the coordinate values as millimeters is only approximate because of the registration process that deforms each brain to match a standard template.

2. The reliability of ALE meta-analyses decreases for smaller numbers of studies. In particular, the likelihood of obtaining false positives based on the results of a single study increases when only a few studies are considered. This factor should be kept in mind when interpreting the results of the jaw and larynx meta-analyses, which involve only 4 and 3 studies, respectively.

References

Alkadhi, H., Crelier, G. R., Boendermaker, S. H., Golay, X., Hepp-Reymond, M. C., & Kollias, S. S. (2002 a). Reproducibility of primary motor cortex somatotopy under controlled conditions. *AJNR. American Journal of Neuroradiology, 23,* 1524–1532.

Alkadhi, H., Crelier, G. R., Boendermaker, S. H., Hepp-Reymond, M. C., & Kollias, S. S. (2002 b). Somatotopy in the ipsilateral primary motor cortex. *Neuroreport, 13*, 2065–2070.

Brown, S., Ingham, R. J., Ingham, J. C., Laird, A., & Fox, P. T. (2005). Stuttered and fluent speech production: an ALE meta-analysis of functional neuroimaging studies. *Human Brain Mapping, 25*, 105–117.

Brown, S., Ngan, E., & Liotti, M. (2008). A larynx area in the human motor cortex. *Cerebral Cortex, 18*, 837–845.

Colebatch, J. G., Adams, L., Murphy, K., Martin, A. J., Lammertsma, A. A., Tochon-Danguy, H. J., et al. (1991). Regional cerebral blood flow during volitional breathing in man. *Journal of Physiology, 443*, 91–103.

Corfield, D. R., Murphy, K., Josephs, O., Fink, G. R., Frackowiak, R. S., Guz, A., et al. (1999). Cortical and subcortical control of tongue movement in humans: a functional neuroimaging study using fMRI. *Journal of Applied Physiology (Bethesda, Md.), 86*, 1468–1477.

Curt, A., Alkadhi, H., Crelier, G. R., Boendermaker, S. H., Hepp-Reymond, M. C., & Kollias, S. S. (2002). Changes of non-affected upper limb cortical representation in paraplegic patients as assessed by fMRI. *Brain, 125*, 2567–2578.

Eickhoff, S. B., Laird, A. R., Grefkes, C., Wang, L. E., Zilles, K., & Fox, P. T. (2009). Coordinate-based activation likelihood estimation meta-analysis of neuroimaging data: a random-effects approach based on empirical estimates of spatial uncertainty. *Human Brain Mapping, 30*, 2907–2926.

Evans, K. C., Shea, S. A., & Saykin, A. J. (1999). Functional MRI localisation of central nervous system regions associated with volitional inspiration in humans. *Journal of Physiology, 520*, 383–392.

Fesl, G., Moriggl, B., Schmid, U. D., Naidich, T. P., Herholz, K., & Yousry, T. A. (2003). Inferior central sulcus: variations of anatomy and function on the example of the motor tongue area. *NeuroImage, 20*, 601–610.

Fink, G. R., Corfield, D. R., Murphy, K., Kobayashi, I., Dettmers, C., Adams, L., et al. (1996). Human cerebral activity with increasing inspiratory force: a study using positron emission tomography. *Journal of Applied Physiology (Bethesda, Md.), 81*, 1295–1305.

Gerardin, E., Lehericy, S., Pochon, J. B., Tezenas du Montcel, S., Mangin, J. F., Poupon, F., et al. (2003). Foot, hand, face and eye representation in the human striatum. *Cerebral Cortex, 13*, 162–169.

Grabski, K., Lamalle, L., Vilian, C., Schwartz, J. L., Vallee, N., Tropres, I., et al. (2012). Functional MRI assessment of orofacial articulators: neural correlates of lip, jaw, larynx, and tongue movements. *Human Brain Mapping, 33*, 2306–2321.

Hanakawa, T., Parikh, S., Bruno, M. K., & Hallett, M. (2005). Finger and face representations in the ipsilateral precentral motor areas in humans. *Journal of Neurophysiology, 93*, 2950–2958.

Hauk, O., Johnsrude, I., & Pulvermuller, F. (2004). Somatotopic representation of action words in human motor and premotor cortex. *Neuron, 41*, 301–307.

Hesselmann, V., Sorger, B., Lasek, K., Guntinas-Lichius, O., Krug, B., Sturm, V., et al. (2004). Discriminating the cortical representation sites of tongue and up movement by functional MRI. *Brain Topography, 16*, 159–167.

Isaev, G., Murphy, K., Guz, A., & Adams, L. (2002). Areas of the brain concerned with ventilatory load compensation in awake man. *Journal of Physiology, 539*, 935–945.

Lancaster, J. L., Tordesillas-Gutierrez, D., Martinez, M., Salinas, F., Evans, A., Zilles, K., et al. (2007). Bias between MNI and Talairach coordinates analyzed using the ICBM-152 brain template. *Human Brain Mapping, 28*, 1194–1205.

Lotze, M., Erb, M., Flor, H., Huelsmann, E., Godde, B., & Grodd, W. (2000 a). FMRI evaluation of somatotopic representation in human primary motor cortex. *NeuroImage, 11*, 473–481.

Lotze, M., Seggewies, G., Erb, M., Grodd, W., & Birbaumer, N. (2000 b). The representation of articulation in the primary sensorimotor cortex. *Neuroreport, 11*, 2985–2989.

Martin, R. E., MacIntosh, B. J., Smith, R. C., Barr, A. M., Stevens, T. K., Gati, J. S., et al. (2004). Cerebral areas processing swallowing and tongue movement are overlapping but distinct: a functional magnetic resonance imaging study. *Journal of Neurophysiology, 92*, 2428–2443.

McKay, L. C., Evans, K. C., Frackowiak, R. S., & Corfield, D. R. (2003). Neural correlates of voluntary breathing in humans. *Journal of Applied Physiology (Bethesda, Md.), 95*, 1170–1178.

Nakayama, T., Fujii, Y., Suzuki, K., Kanazawa, I., & Nakada, T. (2004). The primary motor area for voluntary diaphragmatic motion identified by high field fMRI. *Journal of Neurology, 251*, 730–735.

Onozuka, M., Fujita, M., Watanabe, K., Hirano, Y., Niwa, M., Nishiyama, K., et al. (2002). Mapping brain region activity during chewing: a functional magnetic resonance imaging study. *Journal of Dental Research, 81*, 743–746.

Onozuka, M., Fujita, M., Watanabe, K., Hirano, Y., Niwa, M., Nishiyama, K., et al. (2003). Age-related changes in brain regional activity during chewing: a functional magnetic resonance imaging study. *Journal of Dental Research, 82*, 657–660.

Olthoff, A., Baudewig, J., Kruse, E., & Dechent, P. (2008). Cortical sensorimotor control in vocalization: a functional magnetic resonance imaging study. *Laryngoscope, 118*, 2091–2096.

Pardo, J. V., Wood, T. D., Costello, P. A., Pardo, P. J., & Lee, J. T. (1997). PET study of the localization and laterality of lingual somatosensory processing in humans. *Neuroscience Letters, 234*, 23–26.

Pulvermuller, F., Huss, M., Kherif, F., Moscoso del Prado Martin, F., Hauk, O., & Shtyrov, Y. (2006). Motor cortex maps articulatory features of speech sounds. *Proceedings of the National Academy of Sciences of the United States of America, 103*, 7865–7870.

Ramsay, S. C., Adams, L., Murphy, K., Corfield, D. R., Grootoonk, S., Bailey, D. L., et al. (1993). Regional cerebral blood flow during volitional expiration in man: a comparison with volitional inspiration. *Journal of Physiology, 461*, 85–101.

Riecker, A., Ackermann, H., Wildgruber, D., Meyer, J., Dogil, G., Haider, H., et al. (2000). Articulatory/phonetic sequencing at the level of the anterior perisylvian cortex: a functional magnetic resonance imaging (fMRI) study. *Brain and Language, 75*, 259–276.

Rotte, M., Kanowski, M., & Heinze, H. J. (2002). Functional magnetic resonance imaging for the evaluation of the motor system: primary and secondary brain areas in different motor tasks. *Stereotactic and Functional Neurosurgery, 78*, 3–16.

Schmahmann, J. D., Doyon, J., Toga, A. W., Petrides, M., & Evans, A. C. (2000). *MRI atlas of the human cerebellum*. San Diego: Academic Press.

Shinagawa, H., Ono, T., Honda, E., Sasaki, T., Taira, M., Iriki, A., et al. (2004). Chewing-side preference is involved in differential cortical activation patterns during tongue movements after bilateral gum-chewing: a functional magnetic resonance imaging study. *Journal of Dental Research, 83*, 762–766.

Shinagawa, H., Ono, T., Ishiwata, Y., Honda, E., Sasaki, T., Taira, M., et al. (2003). Hemispheric dominance of tongue control depends on the chewing-side preference. *Journal of Dental Research, 82*, 278–283.

Stippich, C., Blatow, M., Durst, A., Dreyhaupt, J., & Sartor, K. (2007). Global activation of primary motor cortex during voluntary movements in man. *NeuroImage, 34*, 1227–1237.

Stippich, C., Ochmann, H., & Sartor, K. (2002). Somatotopic mapping of the human primary sensorimotor cortex during motor imagery and motor execution by functional magnetic resonance imaging. *Neuroscience Letters, 331*, 50–54.

Takahashi, T., Miyamoto, T., Terao, A., & Yokoyama, A. (2007). Cerebral activation related to the control of mastication during changes in food hardness. *Neuroscience, 145*, 791–794.

Takai, O., Brown, S., & Liotti, M. (2010). Representation of the speech effectors in the human motor cortex: somatotopy or overlap? *Brain and Language, 113*, 39–44.

Terumitsu, M., Fujii, Y., Suzuki, K., Kwee, I. L., & Nakada, T. (2006). Human primary motor cortex shows hemispheric specialization for speech. *Neuroreport, 17*, 1091–1095.

Turkeltaub, P. E., Eden, G. F., Jones, K. M., & Zeffiro, T. A. (2002). Meta-analysis of the functional neuroanatomy of single-word reading: method and validation. *NeuroImage, 16*, 765–780.

Tzourio-Mazoyer, N., Landeau, B., Papathanassiou, D., Crivello, F., Etard, O., Delcroix, N., et al. (2002). Automated anatomical labeling of activations in SPM using a macroscopic anatomical parcellation of the MNI MRI single-subject brain. *NeuroImage, 15*, 273–289.

Vincent, D. J., Bloomer, C. J., Hinson, V. K., & Bergmann, K. J. (2006). The range of motor activation in the normal human cortex using bold FMRI. *Brain Topography, 18*, 273–280.

Watanabe, J., Sugiura, M., Miura, N., Watanabe, Y., Maeda, Y., Matsue, Y., et al. (2004). The human parietal cortex is involved in spatial processing of tongue movement—an fMRI study. *NeuroImage, 21*, 1289–1299.

Zald, D. H., & Pardo, J. V. (1999). The functional neuroanatomy of voluntary swallowing. *Annals of Neurology, 46*, 281–286.

Appendix B: Cortical Parcellation Scheme

This appendix describes the parcellation scheme used to define cerebral cortical *regions of interest* (*ROIs*) in this book. The boundaries of these ROIs are indicated by black lines in figure B.1 and in figures showing cortical activity throughout the book. The region boundaries are a modification of the protocol originally described by Caviness et al. (1996) to focus on anatomical distinctions relevant for studies of speech (Tourville & Guenther, 2003). Region definitions were also informed by the Petrides (2012) atlas of human gyri and sulci. The following paragraphs describe the anatomical landmarks used to define each ROI. For sulci that act as boundaries, it is the fundus of the sulcus that demarcates the boundary. Boundaries defined by points (such as the anterior or posterior limit of a sulcus or other structure) are formed by the coronal plane passing through the point. Note that the cortical inflation process transforms coronal planes into irregular boundaries in figure B.1.

Angular gyrus (AG). Bordered anteriorly by the first segment of the caudal superior temporal sulcus, dorsally by the intraparietal sulcus, ventrally by the superior temporal sulcus, and posterolaterally by the third segment of the caudal superior temporal sulcus.

Anterior cingulate cortex (aCC). The portion of the cingulate gyrus (cortex between the cingulate sulcus and the callosal sulcus) that lies anterior to the genu of the corpus callosum.

Anterior inferior temporal gyrus (aITG). Bordered dorsally by the inferior temporal sulcus and ventrally by the occipitotemporal sulcus; bounded anteriorly by the anterior limit of the superior temporal sulcus and posteriorly by the anterior limit of Heschl's sulcus.

Anterior insula (aINS). The portion of the insular cortex (which is encircled by the circular sulcus of the insula) that lies dorsorostral to the central sulcus of the insula.

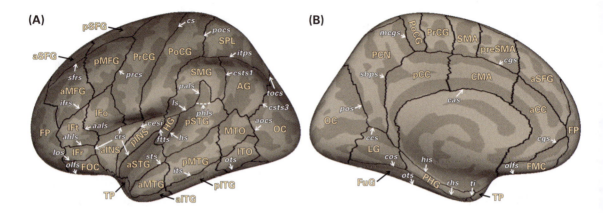

Figure B.1
Inflated view of the (A) left lateral and (B) left medial cerebral cortex with region boundaries and labels indicated by black lines and beige labels. Major sulci are indicated by white arrows and labels. Darker gray shades of the cortical surface indicate sulci; lighter shades indicate gyri. See main text for region abbreviations. Sulcus abbreviations: aals, anterior ascending ramus of the lateral sulcus; ahls, anterior horizontal ramus of the lateral sulcus; aocs, anterior occipital sulcus; cas, callosal sulcus; ccs, calcarine sulcus; cesi, central sulcus of the insula; cgs, cingulate sulcus; cos, collateral sulcus; crs, circular sulcus of the insula; cs, central sulcus; csts1, first segment of the caudal superior temporal sulcus; csts3, third segment of the caudal superior temporal sulcus; ftts, first transverse temporal sulcus; his, hippocampal sulcus; hs, Heschl's sulcus; ifrs, inferior frontal sulcus; itps, intraparietal sulcus; its, inferior temporal sulcus; los, lateral orbital sulcus; ls, lateral sulcus (Sylvian fissure); mcgs, marginal ramus of the cingulate sulcus; olfs, olfactory sulcus; ots, occipitotemporal sulcus; pals, posterior ascending ramus of the lateral sulcus; phls, posterior horizontal ramus of the lateral sulcus; pocs, postcentral sulcus; pos, parieto-occipital sulcus; prcs, precentral sulcus; rhs, rhinal sulcus; sbps, subparietal sulcus; sfrs, superior frontal sulcus; sts, superior temporal sulcus; ti, temporal incisor; tocs, transverse occipital sulcus.

Anterior middle frontal gyrus (aMFG). Bordered dorsally by the superior frontal sulcus and ventrally by the inferior frontal sulcus; bounded anteriorly by the anterior limit of the inferior frontal sulcus and posteriorly by the genu of the corpus callosum.

Anterior middle temporal gyrus (aMTG). Bordered dorsally by the superior temporal sulcus and ventrally by the inferior temporal sulcus; bounded anteriorly by the anterior limit of the superior temporal sulcus and posteriorly by the anterior limit of Heschl's sulcus.

Anterior superior frontal gyrus (aSFG). Extends from the cingulate sulcus on the medial surface laterally to the superior frontal sulcus; bounded anteriorly by the anterior limit of the inferior frontal sulcus and posteriorly by the genu of the corpus callosum.

Anterior superior temporal gyrus (aSTG). Bordered medially by the circular sulcus of the insula and laterally by the superior temporal sulcus; bounded anteriorly by the anterior limit of the superior temporal sulcus and posteriorly by first transverse temporal sulcus (on the supratemporal plane) and the anterior limit of Heschl's sulcus (on the lateral surface). Note that this definition includes the planum polare (the portion of the supratemporal plane anterior to the first transverse temporal sulcus) as part of aSTG.

Cingulate motor area (*CMA*). The portion of the cingulate gyrus (cortex between the cingulate sulcus and the callosal sulcus) that is bounded anteriorly by the genu of the corpus callosum and posteriorly by the caudomedial limit of the precentral sulcus.

Frontal medial cortex (*FMC*). Bordered medially by the cingulate sulcus and laterally by the olfactory sulcus; bounded anteriorly by the anterior limit of the inferior frontal sulcus and posteriorly by the posterior limit of the orbitofrontal cortex (emergence of the septal nuclei).

Frontal orbital cortex (*FOC*). Bordered laterally by the olfactory sulcus and medially by the lateral orbital sulcus; bounded anteriorly by the anterior limit of the inferior frontal sulcus and posteriorly by the circular sulcus of the insula.

Frontal pole (*FP*). The portion of the frontal lobe located anterior to the anterior limit of inferior frontal sulcus.

Fusiform gyrus (*FuG*). Bordered medially by the rhinal sulcus anteriorly and by the collateral sulcus posteriorly and bordered laterally by the occipitotemporal sulcus; bounded anteriorly by the anterior limit of the occipitotemporal sulcus and posteriorly by the intersection of the parieto-occipital sulcus and the calcarine sulcus.

Heschl's gyrus (*HG*). The transverse temporal gyrus on the ventral bank of the lateral sulcus (supratemporal plane) that is bordered anteriorly by the first transverse temporal sulcus, posteriorly by Heschl's sulcus, medially by the circular sulcus of the insula and, more posteriorly, the lateral sulcus, and bordered laterally by the hemispheric margin of the ventral bank of the lateral sulcus.

Inferior frontal gyrus pars opercularis (*IFo*). Bordered anteriorly by the anterior ascending ramus of the lateral sulcus, posteriorly by the precentral sulcus, dorsally by the inferior frontal sulcus, and ventrally by the circular sulcus of the insula.

Inferior frontal gyrus pars orbitalis (*IFr*). Bordered dorsally by the anterior horizontal ramus of the lateral sulcus, ventrally by the lateral orbital sulcus, and posteriorly by the lateral sulcus; bounded anteriorly by the anterior limit of the inferior frontal sulcus.

Inferior frontal gyrus pars triangularis (*IFt*). Bordered anterodorsally by the inferior frontal sulcus, ventrolaterally by the anterior horizontal ramus of the lateral sulcus, and posteriorly by the anterior ascending ramus of the lateral sulcus.

Inferior temporal-occipital junction (*ITO*). Bordered dorsally by the inferior temporal sulcus, ventrally by the occipitotemporal sulcus, and posteriorly by anterior occipital sulcus; bounded anteriorly by the anterior limit of the posterior ascending ramus of the lateral sulcus.

Lingual gyrus (*LG*). Bordered dorsally by the calcarine sulcus, ventrally by the collateral sulcus, and posteriorly by the intersection of the parieto-occipital sulcus; bounded anteriorly by the anterior limit of the calcarine sulcus.

Middle temporal-occipital junction (*MTO*). Bordered dorsally by the superior temporal sulcus, ventrally by the inferior temporal sulcus, and posteriorly by the anterior occipital sulcus; bounded anteriorly by the posterior extent of the lateral sulcus.

Occipital cortex (*OC*). Cortex posterior to the parieto-occipital sulcus on the medial surface and ventral surfaces, posterior to the anterior occipital sulcus on the inferior lateral surface, and posterior to the transverse occipital sulcus and the third segment of the caudal superior temporal sulcus on the superior lateral surface.

Parahippocampal gyrus (*PHG*). Bordered laterally by the rhinal sulcus anteriorly and the collateral sulcus posteriorly; bordered medially by dorsomedial margin of the temporal lobe anteriorly, by the amygdala centrally, and by the hippocampus more posteriorly; bounded anteriorly by the temporal incisure and posteriorly by the anterior limit of the calcarine sulcus.

Postcentral gyrus (*PoCG*). Extends from the cingulate sulcus on the medial surface laterally to the circular sulcus of the insula; bordered anteriorly by the central sulcus and posteriorly by the postcentral sulcus.

Posterior cingulate cortex (*pCC*). The portion of the cingulate gyrus (cortex between the cingulate sulcus and the callosal sulcus) that lies posterior to the caudomedial limit of the precentral sulcus.

Posterior inferior temporal gyrus (*pITG*). Bordered dorsally by the inferior temporal sulcus, ventrally by the occipitotemporal sulcus; bounded anteriorly by the anterior limit of Heschl's sulcus and posteriorly by the posterior extent of the lateral sulcus.

Posterior insula (pINS). The portion of the insular cortex (which is encircled by the circular sulcus of the insula) that lies anterior to the central sulcus of the insula.

Posterior middle frontal gyrus (pMFG). Bordered dorsally by the superior frontal sulcus and ventrally by the inferior frontal sulcus; bounded anteriorly by the genu of the corpus callosum and posteriorly by the precentral sulcus.

Posterior middle temporal gyrus (pMTG). Bordered dorsally by the superior temporal sulcus and ventrally by the inferior temporal sulcus; bounded anteriorly by the anterior limit of Heschl's sulcus and posteriorly by the posterior extent of the lateral sulcus.

Posterior superior frontal gyrus (pSFG). Bordered dorsally by the dorsal hemispheric margin, ventrally by the superior frontal sulcus, and posteriorly by the precentral sulcus; bounded anteriorly by the genu of the corpus callosum.

Posterior superior temporal gyrus (pSTG). Bordered dorsally by the lateral sulcus anteriorly and by the posterior horizontal ramus of the lateral sulcus posteriorly, ventrally by the superior temporal sulcus, anteriorly by Heschl's sulcus (and its projection onto the lateral surface), and posteriorly by the first segment of the caudal superior temporal sulcus. Note that this definition includes the planum temporale (the portion of the supratemporal plane posterior to Heschl's sulcus) as part of pSTG.

Precentral gyrus (PrCG). Extends from the cingulate sulcus on the medial surface laterally to the circular sulcus of the insula; bordered anteriorly by the precentral sulcus and posteriorly by the central sulcus.

Precuneus (PCN). Bordered dorsally by the dorsal hemispheric margin, ventrally by the subparietal sulcus anteriorly and the calcarine sulcus posteriorly, anteriorly by the marginal ramus of the cingulate sulcus, and posteriorly by the parieto-occipital sulcus.

Pre-supplementary motor area (preSMA). The portion of medial prefrontal cortex extending from the dorsal hemispheric margin ventrally to the cingulate sulcus; bounded anteriorly by the genu of the corpus callosum and posteriorly by the decussation of the anterior commissure.

Superior parietal lobule (SPL). Bordered anteriorly by the precentral sulcus, ventrally by the intraparietal sulcus, dorsally by the dorsal hemispheric margin, and posteriorly by the transverse occipital sulcus.

Supplementary motor area (*SMA*). The portion of medial prefrontal cortex extending from the dorsal hemispheric margin ventrally to the cingulate sulcus; bounded anteriorly by the decussation of the anterior commissure and posteriorly by the caudomedial limit of the precentral sulcus.

Supramarginal gyrus (*SMG*). Bordered anteriorly by the postcentral sulcus, dorsally by the intraparietal sulcus, posteriorly by the first segment the caudal superior temporal sulcus, and ventromedially by the lateral sulcus anteriorly and the posterior horizontal ramus of the lateral sulcus posteriorly.

Temporal pole (*TP*). The portion of the temporal lobe located anterior to the anterior limit of the superior temporal sulcus laterally and the temporal incisure medially.

References

Caviness, V. S., Meyer, J., Makris, N., & Kennedy, D. N. (1996). MRI-based topographic parcellation of human neocortex: an anatomically specified method with estimation of reliability. *Journal of Cognitive Neuroscience, 8*, 566–587.

Petrides, M. (2012). *The human cerebral cortex: an MRI atlas of the sulci and gyri in MNI stereotaxic space.* London: Academic Press.

Tourville, J. A., & Guenther, F. H. (2003). *A cortical and cerebellar parcellation system for speech studies.* Boston University Technical Report CAS/CNS-2003-022.

Appendix C: Speech Network Cortical Connectivity Maps

This appendix provides maps of cortico-cortical functional and structural connectivity within the brain regions that make up the cortical speech network.

Functional connectivity maps were derived from statistical analysis of resting state *functional magnetic resonance imaging* (*fMRI*) data collected from 497 healthy adult participants (293 female) as part of the *Human Connectome Project* (*HCP*; http://www.humanconnectome.org). Maps were derived from the HCP minimally processed data included in the WU-Minn Q6 500 + MEG2 release. The HCP fMRI image preprocessing steps included image distortion correction, motion correction, within-subject registration to a structural volume, and nonlinear registration to the MNI152 standard brain template space. The CONN toolbox (http://www.nitrc.org/projects/conn/) was used to further preprocess the data, including artifact rejection/scrubbing, detrending, aCompCor correction for removal of physiological noise and residual movement artifacts, and band-pass filtering (0.01Hz to 0.10Hz), as well as for functional connectivity analyses (Whitfield-Gabrieli & Nieto-Castanon, 2012). FreeSurfer (http://freesurfer.net/) was used to segment a T1 structural image of each subject's brain and to generate a representation of the cortical surface. The preprocessed *blood-oxygen-level-dependent* (*BOLD*) responses from functional volume cortical voxels were then mapped to the corresponding vertex of the cortical surface. The cortical speech network was broken into a set of anatomically defined seed regions, indicated by white patches in the figures that follow. Individual functional connectivity maps for each seed region were determined by calculating the Pearson's correlation between the mean BOLD time series in the seed region to that of all cortical vertices. Group-level statistical tests of connectivity strength were done by transforming the resulting surface correlation maps to an approximate normal distribution using the Fisher Z transform, then performing vertex-wise two-tailed t tests to determine whether the correlation at each voxel differed significantly from 0. The t statistical map for each seed was first thresholded at the vertex level, then an additional correction was done to ensure that the family-wise probability of any cluster of suprathreshold vertices being a false positive was less than 5%. In the figures below, areas shown in red survived the t = 5 voxel-level threshold and areas in yellow survived the more stringent t = 15 voxel-level threshold.

Structural connectivity maps were derived from the diffusion-weighted imaging data of 83 healthy adult participants. Motion artifacts were removed and eddy-current distortions were corrected with DTIPrep (Oguz et al., 2014). Using the FMRIB Diffusion Toolbox (http://www.fmrib.ox.ac.uk/fsl/fslwiki), diffusion tensors were then fitted at each voxel within a cortical mask. DTI volumes were coregistered with subjects' anatomical T1-weighted volumes using the FreeSurfer image analysis suite (Dale, Fischl, & Sereno, 1999; Fischl, Sereno, & Dale, 1999a; Fischl et al., 1999b). FreeSurfer was also used to identify white matter voxels that lie 2 mm below the cortical gray-white boundary of the anatomically defined seed regions (Kang et al., 2012). For each subject, probablistic tractography was performed from each seed region to the rest of the cortical surface using the FMRIB diffusion toolbox. Five thousand streamlines traced probabilistic paths from each seed region voxel to the brain surface. Streamlines were terminated if they exceeded 2,000 steps (corresponding to a 1-m-long tract). To compare connectivity across individuals, a surface voxel was considered "connected" to a seed region if at least 10 streamlines from the seed region reached that voxel in a given subject. These data were projected to an average brain surface using FreeSurfer and averaged across subjects. At each vertex, figures show the percentage of subjects for whom that vertex is connected to the seed region; blue represents 50% of subjects connected to the seed region at that vertex, and green represents 80% of subjects.

C.1 Rolandic Cortex

Rostral Precentral Gyrus
Figures C.1 through C.3 provide functional and structural connectivity maps for three subregions of the rostral half of the precentral gyrus, commonly considered to be premotor cortex (Brodmann area 6): *rostral ventral precentral gyrus*, *rostral middle precentral gyrus*, and *rostral dorsal precentral gyrus*.

Caudal Precentral Gyrus
Figures C.4 through C.6 provide functional and structural connectivity maps for three subregions of the caudal half of the precentral gyrus, commonly considered to be primary motor cortex (Brodmann area 4): *caudal ventral precentral gyrus*, *caudal middle precentral gyrus*, and *caudal dorsal precentral gyrus*.

Postcentral Gyrus
Figures C.7 through C.9 provide functional and structural connectivity maps for three subregions of the postcentral gyrus, commonly considered to be primary somatosensory cortex (Brodmann areas 1–3): *ventral postcentral gyrus*, *middle postcentral gyrus*, and *dorsal postcentral gyrus*.

Speech Network Cortical Connectivity Maps 347

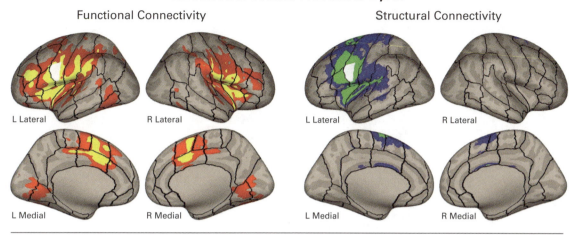

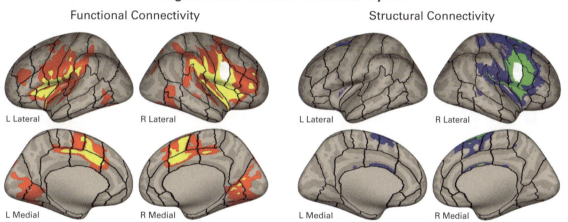

Figure C.1
Functional and structural connectivity maps for rostral ventral precentral gyrus. Seed region for connectivity analyses indicated in white.

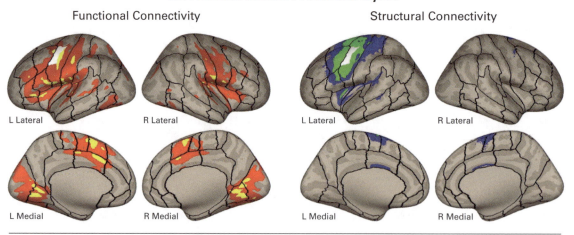

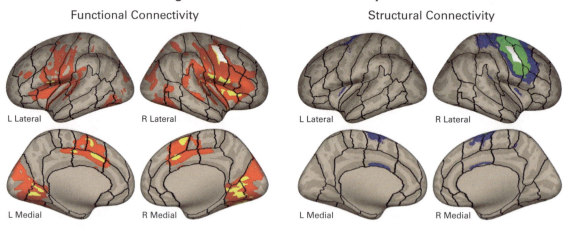

Figure C.2
Functional and structural connectivity maps for rostral middle precentral gyrus. Seed region for connectivity analyses indicated in white.

Left Rostral Dorsal Precentral Gyrus

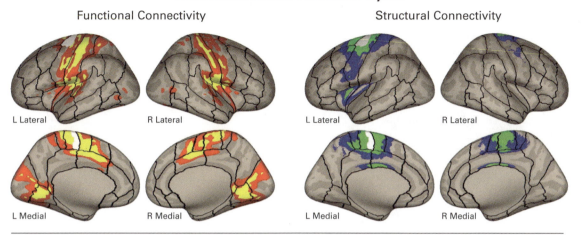

Right Rostral Dorsal Precentral Gyrus

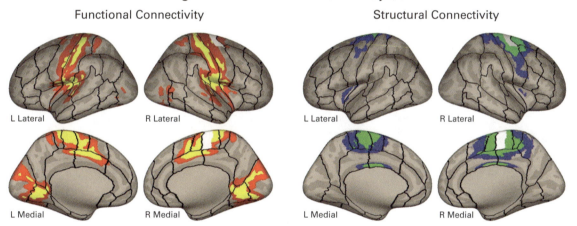

Figure C.3
Functional and structural connectivity maps for rostral dorsal precentral gyrus. Seed region for connectivity analyses indicated in white.

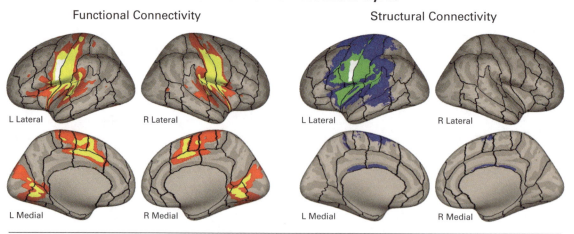

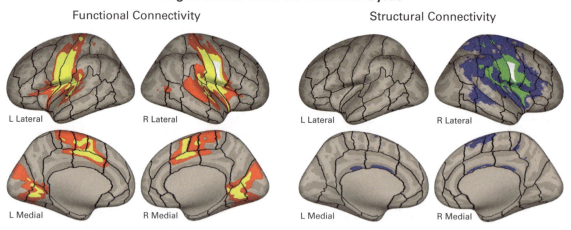

Figure C.4
Functional and structural connectivity maps for caudal ventral precentral gyrus. Seed region for connectivity analyses indicated in white.

Left Caudal Middle Precentral Gyrus

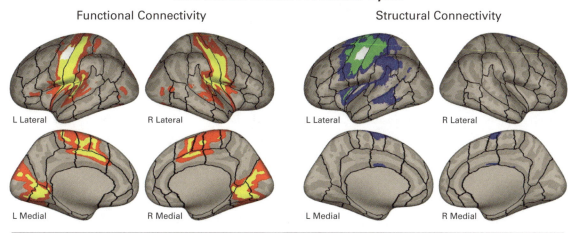

Right Caudal Middle Precentral Gyrus

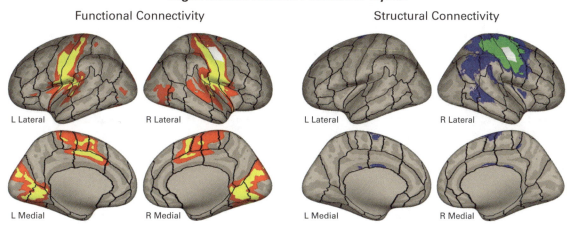

Figure C.5
Functional and structural connectivity maps for caudal middle precentral gyrus. Seed region for connectivity analyses indicated in white.

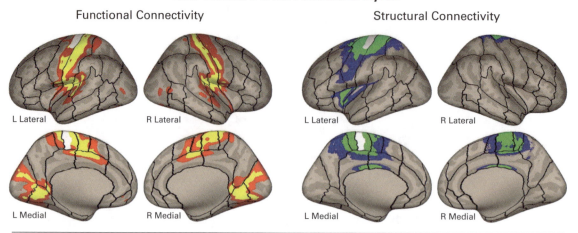

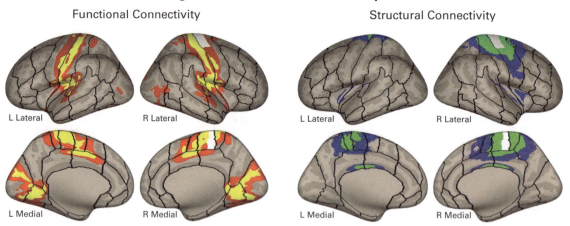

Figure C.6
Functional and structural connectivity maps for caudal dorsal precentral gyrus. Seed region for connectivity analyses indicated in white.

Left Ventral Postcentral Gyrus

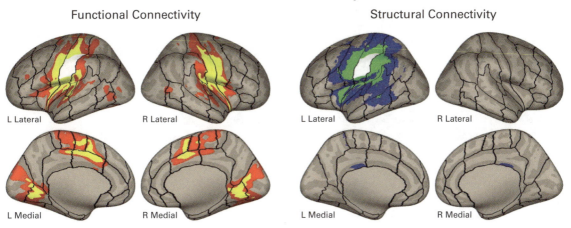

Right Ventral Postcentral Gyrus

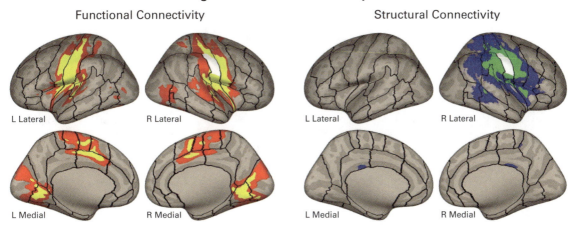

Figure C.7
Functional and structural connectivity maps for ventral postcentral gyrus. Seed region for connectivity analyses indicated in white.

Left Middle Postcentral Gyrus

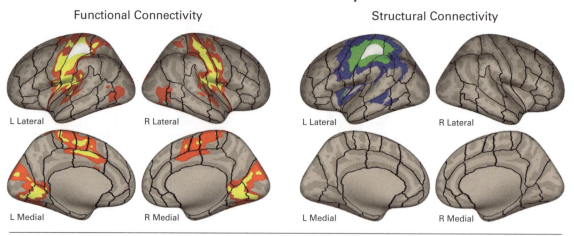

Right Middle Postcentral Gyrus

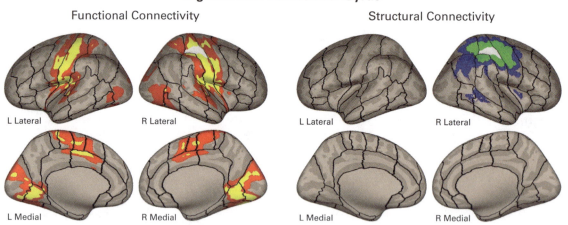

Figure C.8
Functional and structural connectivity maps for middle postcentral gyrus. Seed region for connectivity analyses indicated in white.

Left Dorsal Postcentral Gyrus

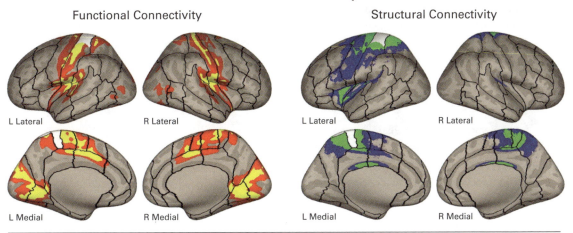

Right Dorsal Postcentral Gyrus

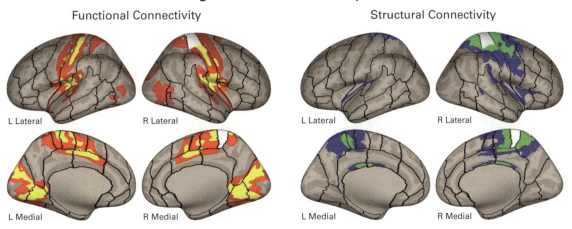

Figure C.9
Functional and structural connectivity maps for dorsal postcentral gyrus. Seed region for connectivity analyses indicated in white.

Central Operculum

Figures C.10 and C.11 provide functional and structural connectivity maps for two subregions of the central (Rolandic) operculum, commonly considered to include Brodmann area 43, a relatively poorly understood region that may be involved in gustatory and somatic sensation: *anterior central operculum* and *posterior central operculum*.

C.2 Medial Frontal Cortex

Supplementary Motor Areas

Figures C.12 and C.13 illustrate functional and structural connectivity maps for the supplementary motor areas in the medial frontal cortex, commonly considered to be premotor cortex (Brodmann area 6): *supplementary motor area* and *pre-supplementary motor area*.

Cingulate Motor Area

Figures C.14 and C.15 provide functional and structural connectivity maps for two subregions of the cingulate motor area, which is commonly considered to be paralimbic cortex: *dorsal cingulate motor area* (corresponding approximately to Brodmann area 32) and *ventral cingulate motor area* (Brodmann area 24).

C.3 Inferior Parietal Cortex

Supramarginal Gyrus

Figures C.16 through C.18 provide functional and structural connectivity maps for three subregions of the supramarginal gyrus (Brodmann area 40) in the inferior parietal lobule, which is commonly considered to include higher-level somatosensory and association cortex: *anterior supramarginal gyrus*, *posterior supramarginal gyrus*, and *parietal operculum*.

C.4 Inferior Frontal Cortex

Inferior Frontal Gyrus Pars Opercularis

Figures C.19 through C.21 illustrate functional and structural connectivity maps for three subregions of the *inferior frontal gyrus pars opercularis* (*IFo*, Brodmann area 44), commonly considered to be association cortex and a core part of Broca's language production area in the left hemisphere: *dorsal IFO* (corresponding approximately to the cytoarchitectonic subregion denoted *44d* by Amunts & Zilles, 2012), *ventral IFO* (subregion *44v* of Amunts and Zilles, 2012), and *posterior frontal operculum* (subregion *op8* of Amunts & Zilles, 2012).

Left Anterior Central Operculum

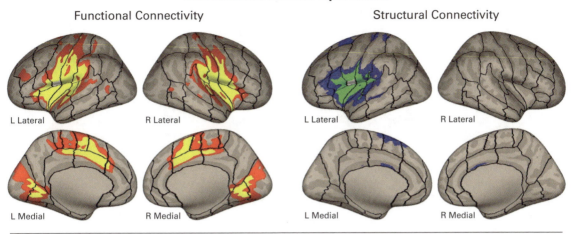

Right Anterior Central Operculum

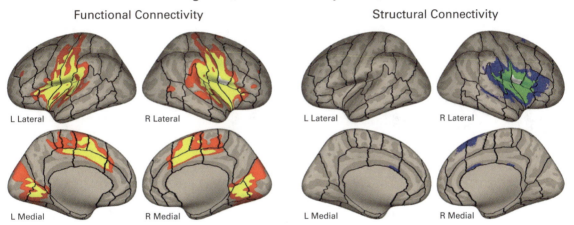

Figure C.10
Functional and structural connectivity maps for anterior central operculum. Seed region for connectivity analyses indicated in white.

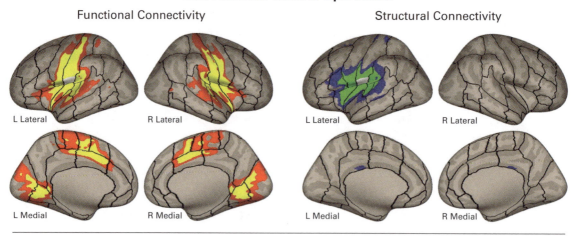
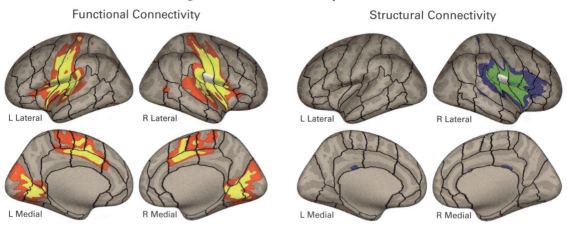

Figure C.11
Functional and structural connectivity maps for posterior central operculum. Seed region for connectivity analyses indicated in white.

Left Supplementary Motor Area

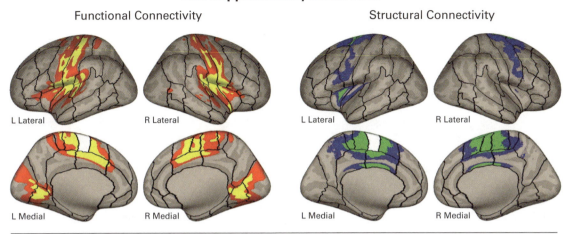

Right Supplementary Motor Area

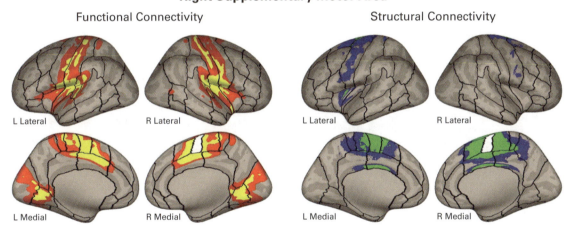

Figure C.12
Functional and structural connectivity maps for supplementary motor area. Seed region for connectivity analyses indicated in white.

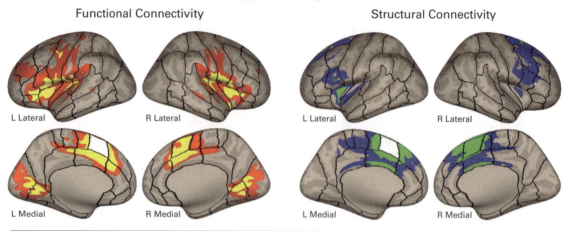
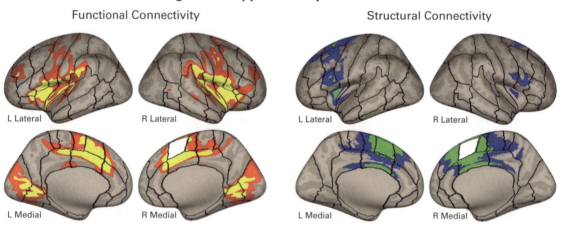

Figure C.13
Functional and structural connectivity maps for pre-supplementary motor area. Seed region for connectivity analyses indicated in white.

Left Dorsal Cingulate Motor Area

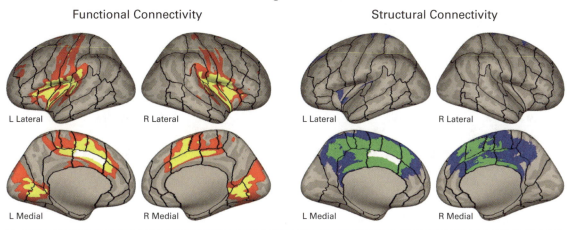

Right Dorsal Cingulate Motor Area

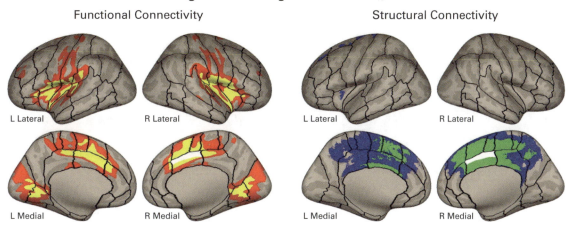

Figure C.14
Functional and structural connectivity maps for dorsal cingulate motor area. Seed region for connectivity analyses indicated in white.

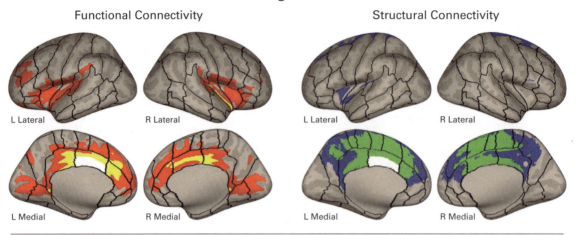

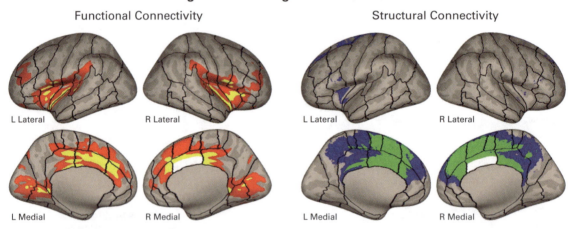

Figure C.15
Functional and structural connectivity maps for ventral cingulate motor area. Seed region for connectivity analyses indicated in white.

Left Anterior Supramarginal Gyrus

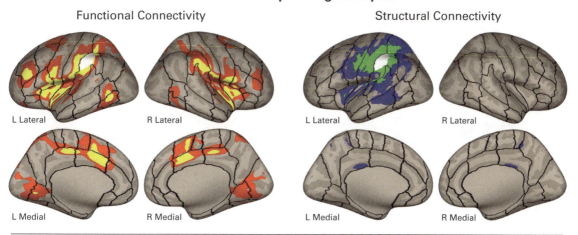

Right Anterior Supramarginal Gyrus

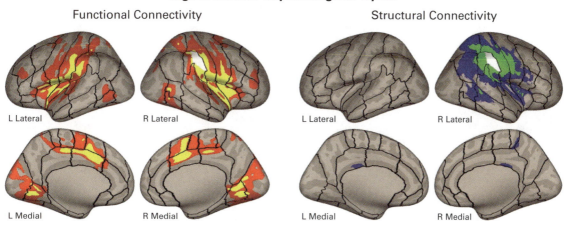

Figure C.16
Functional and structural connectivity maps for anterior supramarginal gyrus. Seed region for connectivity analyses indicated in white.

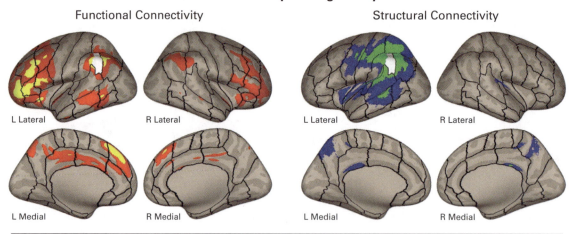
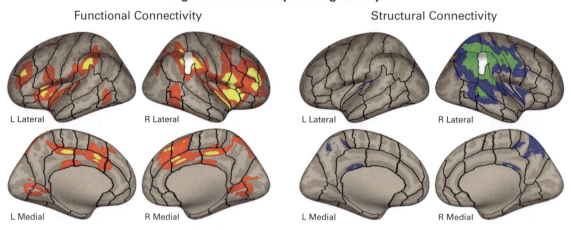

Figure C.17
Functional and structural connectivity maps for posterior supramarginal gyrus. Seed region for connectivity analyses indicated in white.

Left Parietal Operculum

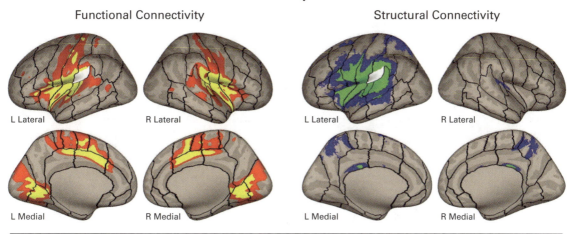

Right Parietal Operculum

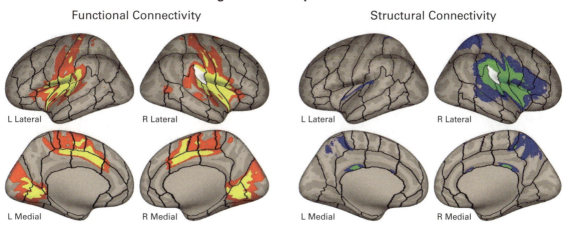

Figure C.18
Functional and structural connectivity maps for parietal operculum. Seed region for connectivity analyses indicated in white.

Left Dorsal Inferior Frontal Gyrus Pars Opercularis

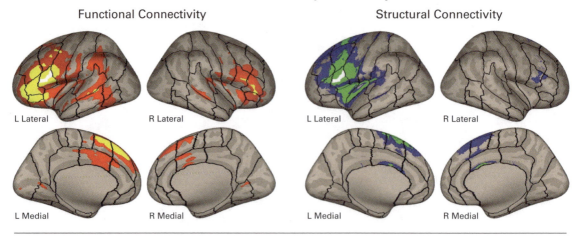

Right Dorsal Inferior Frontal Gyrus Pars Opercularis

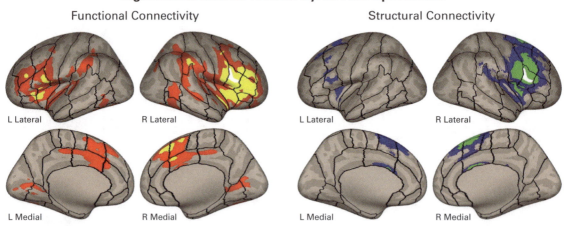

Figure C.19
Functional and structural connectivity maps for dorsal inferior frontal gyrus pars opercularis. Seed region for connectivity analyses indicated in white.

Speech Network Cortical Connectivity Maps

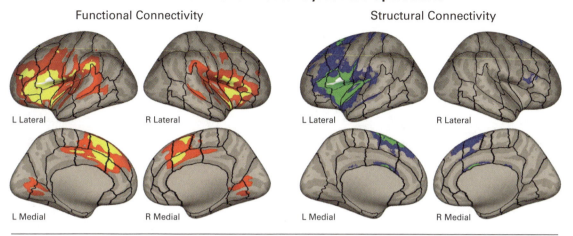

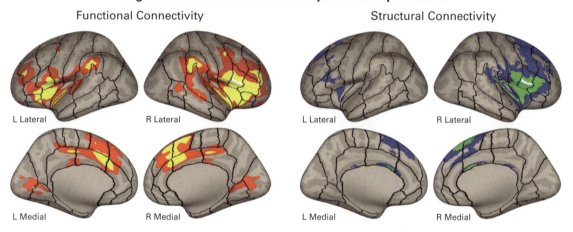

Figure C.20
Functional and structural connectivity maps for ventral inferior frontal gyrus pars opercularis. Seed region for connectivity analyses indicated in white.

Left Posterior Frontal Operculum

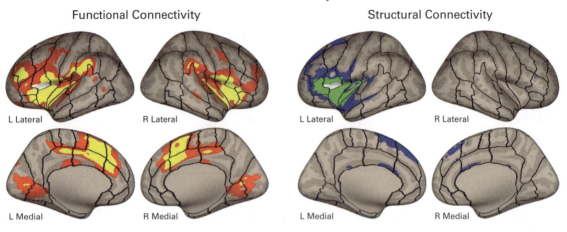

Right Posterior Frontal Operculum

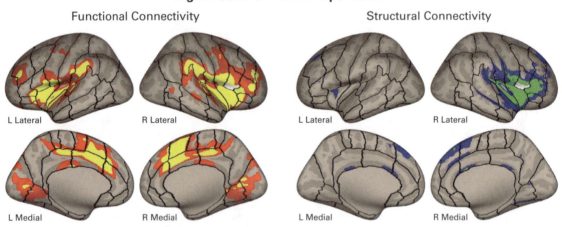

Figure C.21
Functional and structural connectivity maps for posterior frontal operculum. Seed region for connectivity analyses indicated in white.

Inferior Frontal Gyrus Pars Triangularis

Figures C.22 through C.24 illustrate functional and structural connectivity maps for three subregions of the *inferior frontal gyrus pars triangularis* (*IFt*, Brodmann area 45) commonly considered to be association cortex and a core part of Broca's language production area in the left hemisphere: *anterior IFt* (corresponding to the cytoarchitectonic subregion denoted *45a* by Amunts and Zilles, 2012), *posterior IFt* (subregion *45p* of Amunts & Zilles, 2012), and *anterior frontal operculum* (subregion *op9* of Amunts & Zilles, 2012).

Inferior Frontal Gyrus Pars Orbitalis

Figure C.25 provides functional and structural connectivity maps for *inferior frontal gyrus pars orbitalis* (Brodmann area 47, also called area 47/12 since it subsumes monkey area 12), which is association cortex. In the left hemisphere it is commonly associated with language processing, including syntax.

Inferior Frontal Sulcus

Figures C.26 and C.27 illustrate functional and structural connectivity maps for the inferior frontal sulcus: *anterior inferior frontal sulcus* (which includes portions of Brodmann areas 9, 45, and 46) and *posterior inferior frontal sulcus* (which includes portions of Brodmann areas 9 and 44). These regions consist of association cortex; in the left hemisphere they coincide approximately with the dorsal border of Broca's language production area.

C.5 Insular Cortex

Insula

Figures C.28 and C.29 illustrate functional and structural connectivity maps for *anterior insula* and *posterior insula*. These regions are commonly considered to be paralimbic structures, though they contain subregions with very diverse cytoarchitecture (see the discussion in chapter 2, section 2.4).

C.6 Superior Temporal Cortex

Heschl's Gyrus

Figure C.30 provides functional and structural connectivity maps for *Heschl's gyrus* (Brodmann area 41), commonly considered to be primary auditory cortex.

Posterior Superior Temporal Cortex

Figures C.31 through C.34 provide functional and structural connectivity maps for four subregions of posterior superior temporal cortex (Brodmann areas 22 and 42, along with

Left Anterior Inferior Frontal Gyrus Pars Triangularis

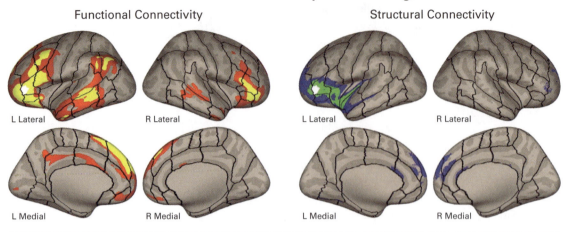

Right Anterior Inferior Frontal Gyrus Pars Triangularis

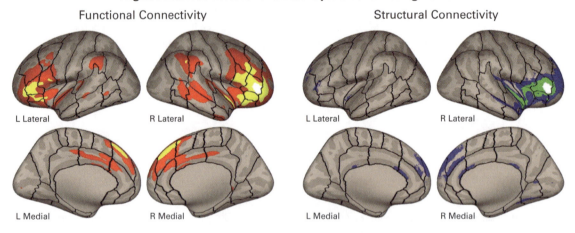

Figure C.22
Functional and structural connectivity maps for anterior inferior frontal gyrus pars triangularis. Seed region for connectivity analyses indicated in white.

Speech Network Cortical Connectivity Maps

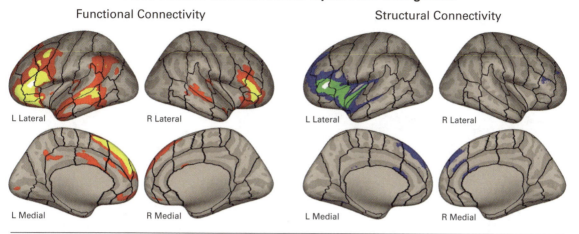

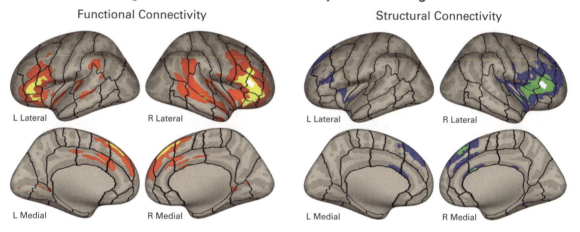

Figure C.23
Functional and structural connectivity maps for posterior inferior frontal gyrus pars triangularis. Seed region for connectivity analyses indicated in white.

Left Anterior Frontal Operculum

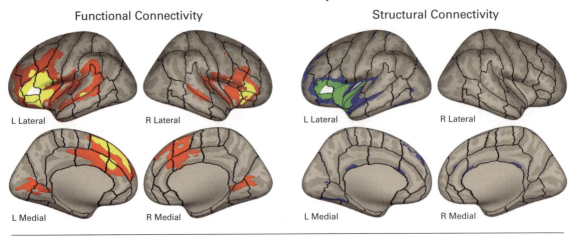

Right Anterior Frontal Operculum

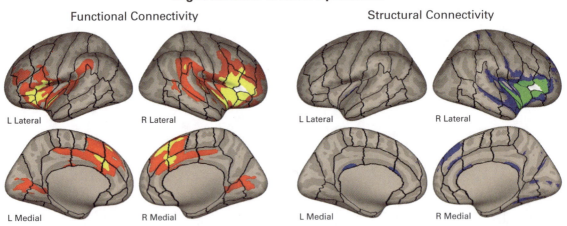

Figure C.24
Functional and structural connectivity maps for anterior frontal operculum. Seed region for connectivity analyses indicated in white.

Speech Network Cortical Connectivity Maps

Left Inferior Frontal Gyrus Pars Orbitalis

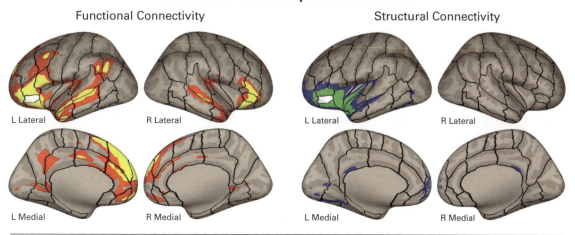

Right Inferior Frontal Gyrus Pars Orbitalis

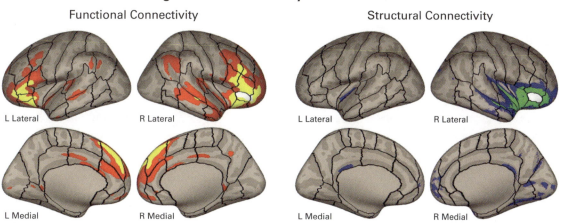

Figure C.25
Functional and structural connectivity maps for inferior frontal gyrus pars orbitalis. Seed region for connectivity analyses indicated in white.

Left Anterior Inferior Frontal Sulcus

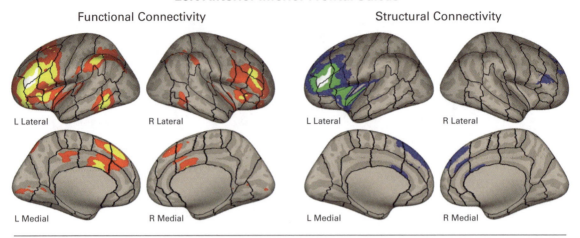

Right Anterior Inferior Frontal Sulcus

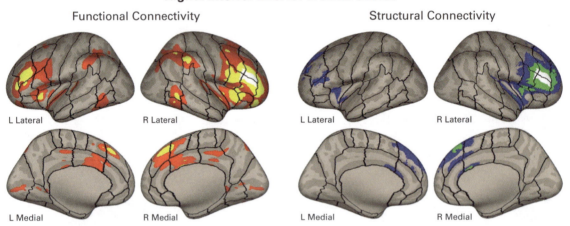

Figure C.26
Functional and structural connectivity maps for anterior inferior frontal sulcus. Seed region for connectivity analyses indicated in white.

Left Posterior Inferior Frontal Sulcus

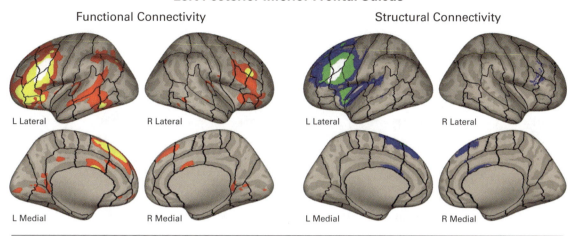

Right Posterior Inferior Frontal Sulcus

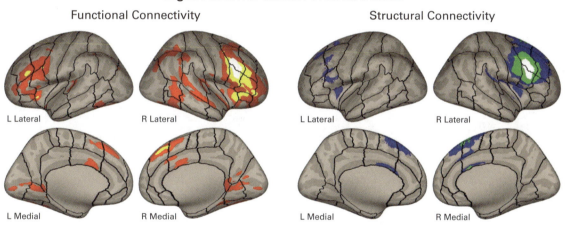

Figure C.27
Functional and structural connectivity maps for posterior inferior frontal sulcus. Seed region for connectivity analyses indicated in white.

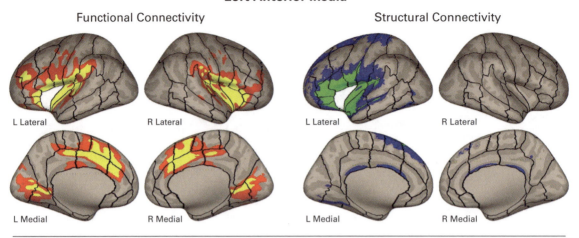
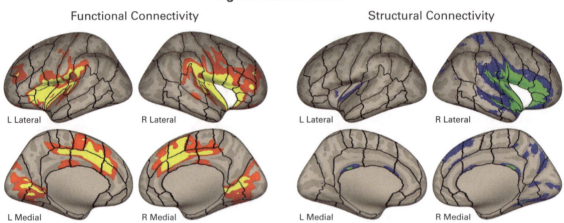

Figure C.28
Functional and structural connectivity maps for anterior insula. Seed region for connectivity analyses indicated in white.

Left Posterior Insula

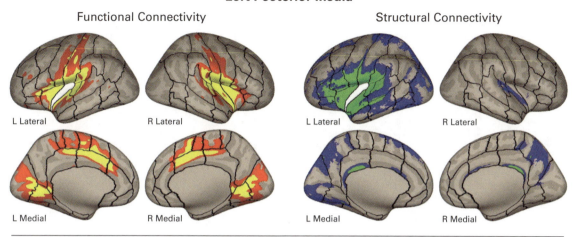

Right Posterior Insula

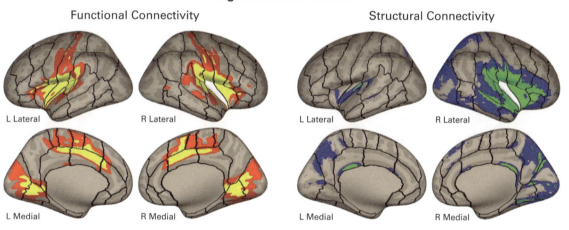

Figure C.29
Functional and structural connectivity maps for posterior insula. Seed region for connectivity analyses indicated in white.

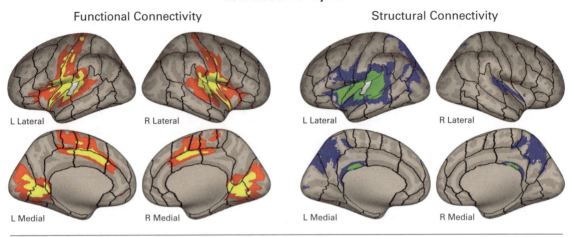

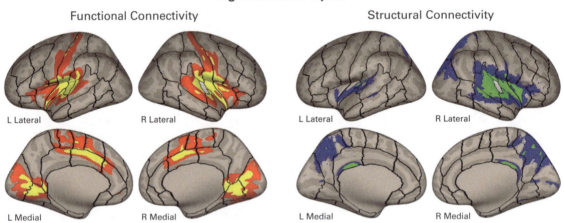

Figure C.30
Functional and structural connectivity maps for Heschl's gyrus. Seed region for connectivity analyses indicated in white.

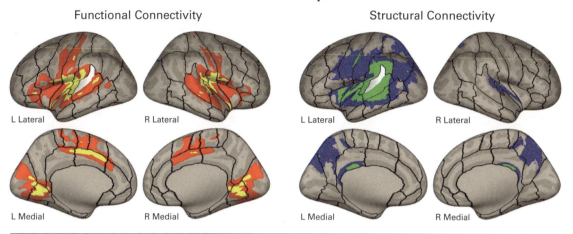
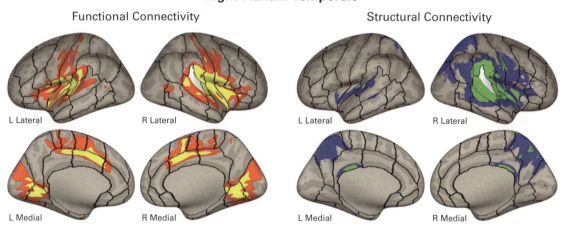

Figure C.31
Functional and structural connectivity maps for planum temporale. Seed region for connectivity analyses indicated in white.

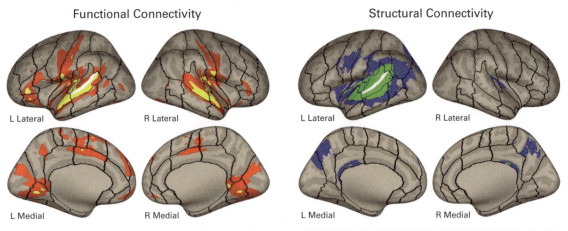

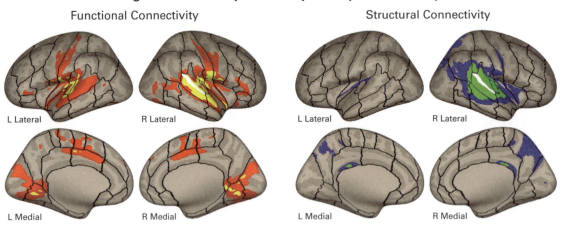

Figure C.32
Functional and structural connectivity maps for posterior superior temporal gyrus convexity. Seed region for connectivity analyses indicated in white.

Left Posterior Dorsal Superior Temporal Sulcus

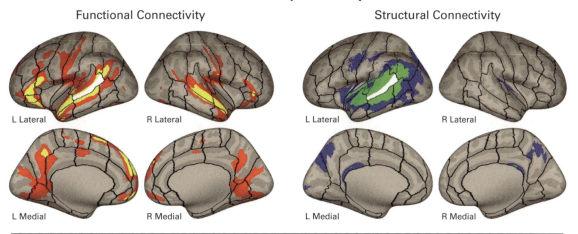

Right Posterior Dorsal Superior Temporal Sulcus

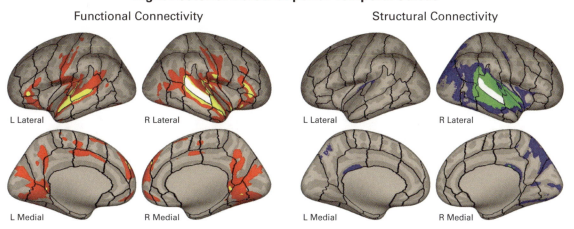

Figure C.33
Functional and structural connectivity maps for posterior dorsal superior temporal sulcus. Seed region for connectivity analyses indicated in white.

Left Posterior Ventral Superior Temporal Sulcus

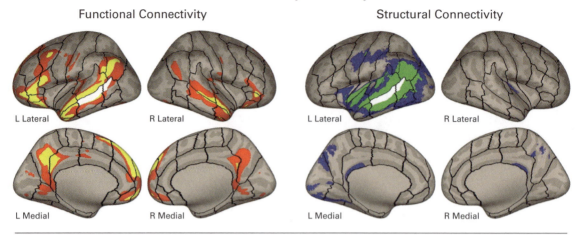

Right Posterior Ventral Superior Temporal Sulcus

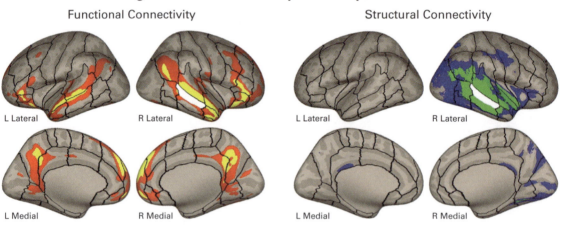

Figure C.34
Functional and structural connectivity maps for posterior ventral superior temporal sulcus. Seed region for connectivity analyses indicated in white.

dorsal portions of Brodmann area 21 in the ventral bank of the inferior frontal sulcus), which is commonly considered to include higher-order auditory cortex and constitutes part of Wernicke's language reception area: *planum temporale*, *posterior superior temporal gyrus convexity* (i.e., the portion of the posterior superior temporal gyrus that is not in the Sylvian fissure or superior temporal sulcus), *posterior dorsal superior temporal sulcus*, and *posterior ventral superior temporal sulcus*.

Anterior Superior Temporal Cortex

Figures C.35 through C.38 provide functional and structural connectivity maps for four subregions of anterior superior temporal cortex (Brodmann area 22, along with dorsal portions of Brodmann area 21 in the ventral bank of the inferior temporal sulcus), which is commonly considered to include higher-order auditory cortex and association cortex: *planum polare, anterior superior temporal gyrus convexity* (i.e., the portion of the anterior superior temporal gyrus that is not in the Sylvian fissure or superior temporal sulcus), *anterior dorsal superior temporal sulcus*, and *anterior ventral superior temporal sulcus*.

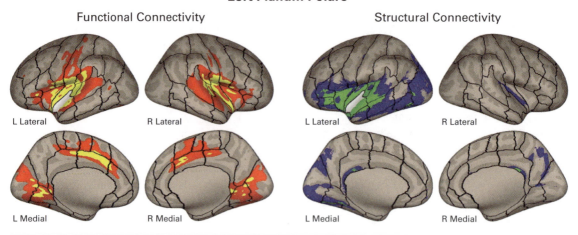

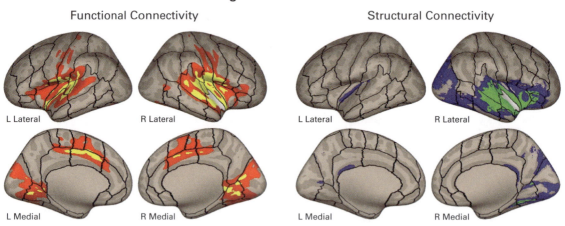

Figure C.35
Functional and structural connectivity maps for planum polare. Seed region for connectivity analyses indicated in white.

Speech Network Cortical Connectivity Maps

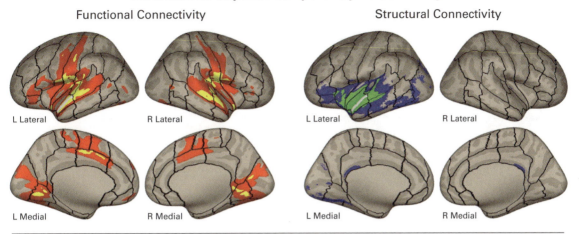

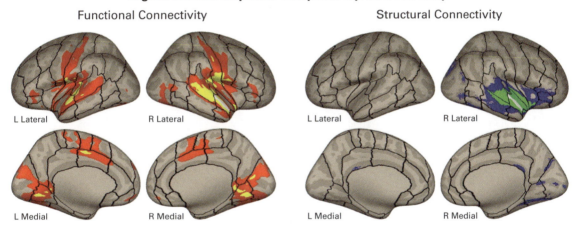

Figure C.36
Functional and structural connectivity maps for anterior superior temporal gyrus convexity. Seed region for connectivity analyses indicated in white.

Left Anterior Dorsal Superior Temporal Sulcus

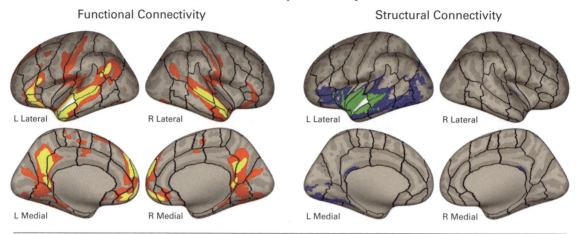

Right Anterior Dorsal Superior Temporal Sulcus

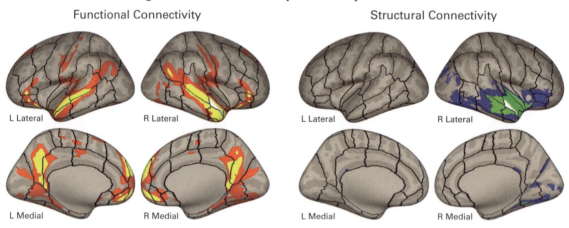

Figure C.37
Functional and structural connectivity maps for anterior dorsal superior temporal sulcus. Seed region for connectivity analyses indicated in white.

Left Anterior Ventral Superior Temporal Sulcus

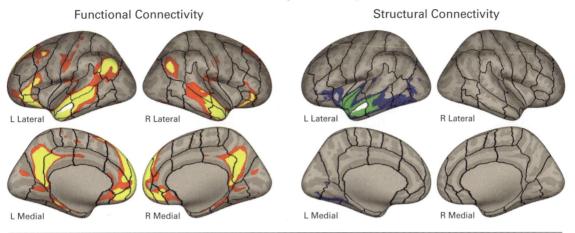

Right Anterior Ventral Superior Temporal Sulcus

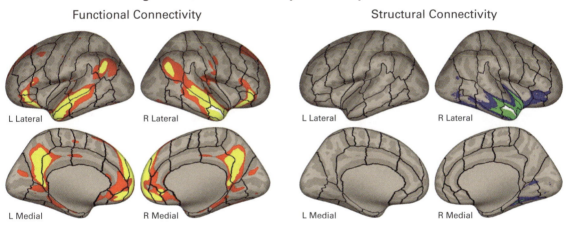

Figure C.38
Functional and structural connectivity maps for anterior ventral superior temporal sulcus. Seed region for connectivity analyses indicated in white.

References

Amunts, K., & Zilles, K. (2012). Architecture and organizationl principles of Broca's region. *Trends in Cognitive Sciences*, *16*, 418–426.

Dale, A. M., Fischl, B., & Sereno, M. I. (1999). Cortical surface-based analysis: I. Segmentation and surface reconstruction. *NeuroImage*, *9*, 179–194.

Fischl, B., Sereno, M. I., & Dale, A. M. (1999 a). Cortical surface-based analysis: II. Inflation, flattening, and a surface-based coordinate system. *NeuroImage*, *9*, 195–207.

Fischl, B., Sereno, M. I., Tootell, R. B. H., & Dale, A. M. (1999 b). High-resolution intersubject averaging and a coordinate system for the cortical surface. *Human Brain Mapping*, *8*, 272–284.

Kang, X., Herron, T. J., Turken, U., & Woods, D. L. (2012). Diffusion properties of cortical and pericortical tissue: regional variations, reliability and methodological issues. *Magnetic Resonance Imaging*, *30*, 1111–1122.

Oguz, I., Farzinfar, M., Matsui, J., Budin, F., Liu, Z., Gerig, G., et al. (2014). DTIPrep: quality control of diffusion-weighted images. *Frontiers in Neuroinformatics*, *8*(4).

Whitfield-Gabrieli, S., & Nieto-Castanon, A. (2012). Conn: a functional connectivity toolbox for correlated and anticorrelated brain networks. *Brain Connectivity*, *2*, 125–141.

Appendix D: DIVA Brain Activity Simulations

This appendix describes the generation of simulated brain activity from computer simulations of the DIVA model performing speaking tasks. Each neural component of the model (e.g., the *articulator map*) is associated with one or more nodes, each centered at a particular location in the *Montreal Neurological Institute* (*MNI*) coordinate frame. Node locations for specific articulators in the *articulator map* and *somatosensory state map*, as well as the cerebellum, were chosen to be consistent with the distribution of Rolandic cortex and cerebellar activation foci for the corresponding articulators in the simple articulator movement studies summarized in Appendix A. Locations for the remaining foci were guided by the results of three contrasts from our database of speech fMRI studies: (1) the *speech–baseline* contrast from 116 subjects producing mono- and bisyllabic utterances illustrated in figure 1.8 in chapter 1, (2) the *auditory-perturbed speech–unperturbed speech* contrast from 25 subjects illustrated in figure 5.4 in chapter 5, and (3) the *jaw-perturbed speech–unperturbed speech* contrast from 13 subjects[1] illustrated in figure 6.5 in chapter 6 (see figure captions for fMRI data analysis details).

Computer simulations of the model were performed for five monosyllabic utterances ("bed," "but," "peek," "cat," and "mat") under three speech conditions: no perturbation, auditory perturbation, and jaw perturbation. During a simulation, each node is associated with a *computational load function*, denoted $L_n(t)$, where the index n identifies the particular node and t denotes the particular time step within the simulation. This load function relates activity of the node to variables from the computer simulation, that is, the variables in equations 3.1, 5.1, 6.1, and 7.1, which can be combined into the following equation characterizing the model's core dynamics:

$$\dot{M} = \alpha_{FF}G[M_T - M] + \alpha_A Z_{AM}[A - A_T] + \alpha_S Z_{SM}[S - S_T]. \tag{D.1}$$

where $\dot{M}$ is the overall motor command to the speech articulators in vector form; α_{FF} is the feedforward control gain factor; G is a GO signal that controls speech rate; M_T is the target motor state (or motor program) for the current sound in vector form; M is the current motor state in vector form; α_A is the auditory feedback control gain factor; Z_{AM} is a matrix of synaptic weights representing the pseudoinverse of the Jacobian matrix relating the auditory and motor spaces; A is the current auditory state in vector form; A_T is the auditory target for the current sound in vector form; α_S is the somatosensory feedback control gain factor; Z_{SM} is a matrix of synaptic weights representing the pseudoinverse of the Jacobian matrix relating the somatosensory and motor spaces; S is the somatosensory state in vector form; and S_T is the somatosensory target for the current sound in vector form.

Once the instantaneous neural activity of all nodes has been calculated for all time steps in a simulation, the following equation is used to generate a simulated BOLD signal at each cortical location:

$$\hat{B}_i = \sum_t \sum_n \alpha_n L_n e^{-(D_{i,n}/\beta_n)^2}, \tag{D.2}$$

where $\hat{B}_i$ is the simulated BOLD signal (corresponding to a t statistic in the fMRI analyses) at vertex i, α_n is a scaling parameter for model node n, $D_{i,n}$ is the distance along the cortical surface between vertex i and model node n, and β_n is an activation spread parameter for model node n. In words, each model node produces a Gaussian of activity on the cortical surface, with the height of the Gaussian proportional to the instantaneous activity of the node scaled by the weighting parameter α_n and the spread of the Gaussian determined by the parameter β_n. The BOLD signal is then estimated at each vertex by the sum of all of the superimposed activity Gaussians over all time steps in the simulation. Activity at subcortical locations is determined in the same way except that $D_{i,n}$ represents distance in 3-D stereotactic space rather than distance along the cortical surface. The parameters α_n and β_n were optimized to produce the best fit to the measured brain activity across all three contrasts, using the same parameter values for all contrasts. A trust-region-reflective algorithm for minimization of nonlinear least squares (Coleman & Li, 1994, 1996) was employed to optimize the parameters α_n and β_n.

The following sections provide the node locations and computational load functions for the cortical and subcortical model components in the simulations reported herein.

D.1 Cortical Components

Table D.1 provides MNI coordinates and computational load functions for the cortical nodes of the DIVA model. Note that the *motivation map* in *cingulate motor area* (*CMA*) is not included in the model descriptions in earlier chapters as it is simply set to 1 during all simulations, meant to roughly correspond to having the "will to speak" (see the discussion of CMA in chapter 2).

Figure D.1 illustrates the DIVA cortical node locations on inflated lateral and medial surfaces, color coded by model component. Figure D.2 illustrates nodes in the *articulator map* and *somatosensory state map* color coded by articulator (cf. figure 2.16 in chapter 2).

Table D.1

Stereotactic coordinates and computational load functions for cortical nodes in the DIVA model brain activity simulations reported herein

Model Component	MNI Left			MNI Right			Computational Load
	X	Y	Z	X	Y	Z	
Motivation map	−8	13	36	6	15	37	$L(t) = 1$
Initiation map	−3	−1	59	4	2	60	$L(t) = G(t)$
Speech sound map	−57	−1	40				$L(t) = G(t)$
Speech sound map	−55	9	0				$L(t) = G(t)$
Speech sound map	−37	19	8				$L(t) = G(t)$
Speech sound map	−34	13	4	39	12	0	$L(t) = G(t)$
Feedback control map—auditory				46	23	6	$L(t) = |\dot{M}_A(t)|$
Feedback control map—somatosensory				41	18	−2	$L(t) = |\dot{M}_S(t)|$
Feedback control map—somatosensory				53	23	6	$L(t) = |\dot{M}_S(t)|$
Feedback control map—somatosensory				50	8	13	$L(t) = |\dot{M}_S(t)|$
Articulator map—jaw	−51	−9	33	49	−10	33	$L(t) = |\dot{M}_{Jaw}(t)|$
Articulator map—jaw	−59	−4	17	59	−3	9	$L(t) = |\dot{M}_{Jaw}(t)|$
Articulator map—larynx	−48	−12	38	53	−3	50	$L(t) = |\dot{M}_{Larynx}(t)|$
Articulator map—larynx	−58	−4	23	63	0	21	$L(t) = |\dot{M}_{Larynx}(t)|$
Articulator map—lip	−40	−17	38	58	−2	41	$L(t) = |\dot{M}_{Lip}(t)|$
Articulator map—lip	−62	4	23	63	2	14	$L(t) = |\dot{M}_{Lip}(t)|$
Articulator map—respiratory	−21	−28	55	21	−28	55	$L(t) = |\dot{M}_{Resp}(t)|$
Articulator map—respiratory	−50	−1	50	43	−13	38	$L(t) = |\dot{M}_{Resp}(t)|$
Articulator map—tongue	−56	−5	31	58	−3	29	$L(t) = |\dot{M}_{Tongue}(t)|$
Articulator map—tongue	−61	2	13	60	2	6	$L(t) = |\dot{M}_{Tongue}(t)|$
Somatosensory state map—jaw	−51	−11	29	50	−15	43	$L(t) = |\dot{S}_{Jaw}(t)|$
Somatosensory state map—jaw	−53	−12	11	53	−13	12	$L(t) = |\dot{S}_{Jaw}(t)|$
Somatosensory state map—larynx	−47	−17	35	49	−15	39	$L(t) = |\dot{S}_{Larynx}(t)|$
Somatosensory state map—larynx	−66	−11	35	67	−11	12	$L(t) = |\dot{S}_{Lip}(t)|$
Somatosensory state map—lip	−62	−16	34	52	−10	28	$L(t) = |\dot{S}_{Lip}(t)|$
Somatosensory state map—lip	−65	−12	28	66	−16	18	$L(t) = |\dot{S}_{Lip}(t)|$
Somatosensory state map—respiratory	−22	−30	56	19	−30	56	$L(t) = |\dot{S}_{Resp}(t)|$
Somatosensory state map—tongue	−59	−9	21	61	−6	25	$L(t) = |\dot{S}_{Tongue}(t)|$
Somatosensory target map	−65	−16	21	66	−16	24	$L(t) = |\dot{S}_T(t)|$
Somatosensory target map	−59	−19	30	60	−35	20	$L(t) = |\dot{S}_T(t)|$
Somatosensory error map	−65	−16	21	66	−16	24	$L(t) = |S(t) - S_T(t)|$
Somatosensory error map	−54	−25	33	57	−22	34	$L(t) = |S(t) - S_T(t)|$
Somatosensory error map				58	−31	34	$L(t) = |S(t) - S_T(t)|$
Somatosensory error map				64	−41	28	$L(t) = |S(t) - S_T(t)|$
Somatosensory error map	−60	−36	31	60	−35	20	$L(t) = |S(t) - S_T(t)|$
Somatosensory error map	−61	−24	23	60	−25	25	$L(t) = |S(t) - S_T(t)|$
Somatosensory error map	−39	−9	−8	41	−10	−7	$L(t) = |S(t) - S_T(t)|$
Somatosensory error map	−43	−11	2	40	−7	9	$L(t) = |S(t) - S_T(t)|$
Auditory state map	−65	−22	−2	64	−22	−4	$L(t) = |\dot{A}(t)|$
Auditory state map	−68	−30	7	58	−34	1	$L(t) = |\dot{A}(t)|$
Auditory state map	−51	−37	15	52	−28	10	$L(t) = |\dot{A}(t)|$
Auditory state map	−61	−12	5	65	−15	6	$L(t) = |\dot{A}(t)|$
Auditory state map	−55	1	−3	62	2	−3	$L(t) = |\dot{A}(t)|$
Auditory state map	−57	−42	4	49	−33	−5	$L(t) = |\dot{A}(t)|$
Auditory state map	−45	−30	6	40	−23	3	$L(t) = |\dot{A}(t)|$
Auditory state map	−37	−25	3	61	−28	10	$L(t) = |\dot{A}(t)|$
Auditory target map	−68	−31	7	69	−30	2	$L(t) = |\dot{A}_T(t)|$
Auditory target map	−66	−38	15	60	−39	6	$L(t) = |\dot{A}_T(t)|$

Table D.1 (continued)

Model Component	MNI Left			MNI Right			Computational Load
	X	Y	Z	X	Y	Z	
Auditory target map	−56	−26	6	56	−21	−5	$L(t) = |\dot{A}_T(t)|$
Auditory target map	−46	−39	19	54	−30	12	$L(t) = |\dot{A}_T(t)|$
Auditory error map	−66	−38	15	69	−30	2	$L(t) = |A(t) - A_T(t)|$
Auditory error map	−56	−26	6	60	−39	6	$L(t) = |A(t) - A_T(t)|$
Auditory error map	−46	−39	19	56	−21	−5	$L(t) = |A(t) - A_T(t)|$
Auditory error map	−68	−31	7	54	−30	12	$L(t) = |A(t) - A_T(t)|$

Note. MNI, Montreal Neurological Institute stereotactic coordinate.

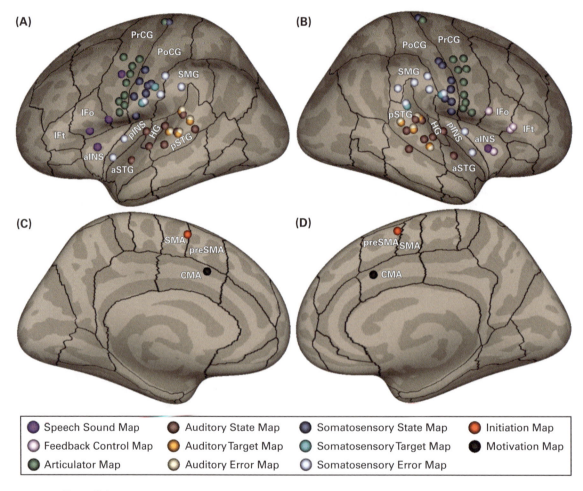

Figure D.1
Anatomical locations of the cortical components of the DIVA model on the (A) left lateral, (B) right lateral, (C) left medial, and (D) right medial inflated cortical surfaces. aINS, anterior insula; aSTG, anterior superior temporal gyrus; CMA, cingulate motor area; HG, Heschl's gyrus; IFo, inferior frontal gyrus pars opercularis; IFt, inferior frontal gyrus pars triangularis; pINS, posterior insula; PoCG, postcentral gyrus; PrCG, precentral gyrus; preSMA, pre-supplementary motor area; pSTG, posterior superior temporal gyrus; SMA, supplementary motor area; SMG, supramarginal gyrus.

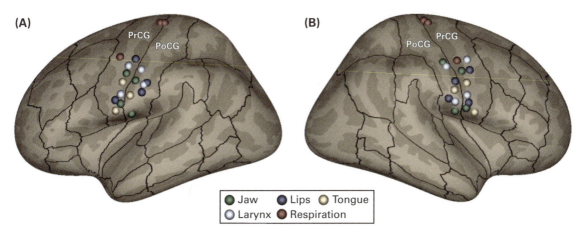

Figure D.2
Speech articulator representations in the DIVA model *articulator map* in *precentral gyrus* (PrCG) and *somatosensory state map* in *postcentral gyrus* (PoCG) plotted on (A) left lateral and (B) right lateral inflated cortical surfaces. For comparison, see articulator foci from fMRI studies of single articulator movements summarized in figure 2.16 of chapter 2.

D.2 Subcortical Components

Table D.2 provides MNI coordinates for the subcortical components of the DIVA model. Figures D.3, D.4, and D.5 illustrate the thalamic, basal ganglia, and cerebellar components, respectively, within the MNI152 template brain.

Table D.2
MNI locations and computational load functions for subcortical nodes in the DIVA model brain activity simulations reported herein

Model Component	MNI Left			MNI Right			Computational Load
	X	Y	Z	X	Y	Z	
Thalamus—VL	−12	−15	5	13	−15	3	$L(t) = |\dot{M}(t)|$
Thalamus—VA	−8	−3	6	10	−2	8	$L(t) = G(t)$
Thalamus—VPM	−13	−20	1	12	−20	−3	$L(t) = |\dot{S}(t)|$
Thalamus—MG	−12	−28	−8	13	−28	−7	$L(t) = |\dot{A}(t)|$
Basal ganglia—anterior putamen	−22	7	2				$L(t) = G(t)$
Basal ganglia—putamen	−23	−2	8	22	3	5	$L(t) = G(t)$
Basal ganglia—GPe	−25	−14	−3	26	−12	−3	$L(t) = G(t)$
Basal ganglia—Gpi	−19	−10	−4	17	−5	−4	$L(t) = G(t)$
Basal ganglia—SNr	−8	−20	−14	10	−20	−14	$L(t) = G(t)$
Cerebellum—lob. VI jaw	−10	−66	−20	20	−66	−26	$L(t) = |\dot{M}_{Jaw}(t)|$
Cerebellum—lob. VI larynx	−26	−60	−28	26	−62	−24	$L(t) = |\dot{M}_{Larynx}(t)|$
Cerebellum—lob. VI larynx				34	−58	−30	$L(t) = |\dot{M}_{Larynx}(t)|$
Cerebellum—lob. VI lips	−22	−60	−22	20	−60	−20	$L(t) = |\dot{M}_{Lip}(t)|$
Cerebellum—lob. VI respiratory	−17	−66	−27	20	−66	−21	$L(t) = |\dot{M}_{Resp}(t)|$
Cerebellum—lob. VI tongue	−16	−62	−22	18	−64	−24	$L(t) = |\dot{M}_{Tongue}(t)|$
Cerebellum—lob. VI speech	−14	−65	−15	12	−65	−15	$L(t) = |\dot{M}(t)|$
Cerebellum—lob. VI speech				2	−65	−17	$L(t) = |\dot{M}(t)|$

Note. GPe, external segment of the globus pallidus; GPi, internal segment of the globus pallidus; Lob., cerebellar cortex lobule; MG, medial geniculate nucleus; MNI, Montreal Neurological Institute stereotactic coordinate; VA, ventral anterior nucleus; VL, ventral lateral nucleus; VPM, ventral posterior medial nucleus; SNr, substantia nigra pars reticulata.

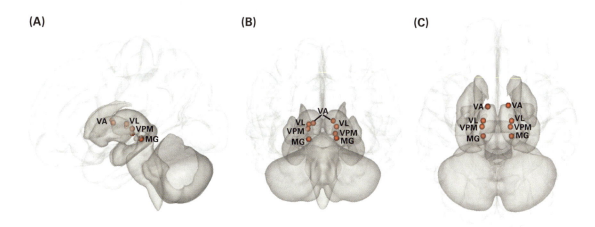

Figure D.3
Locations of the thalamic components of the DIVA model within the MNI152 template brain. (A) Left view. (B) Posterior view. (C) Superior view. MG, medial geniculate nucleus; VA, ventral anterior nucleus; VL, ventral lateral nucleus; VPM, ventral posterior medial nucleus.

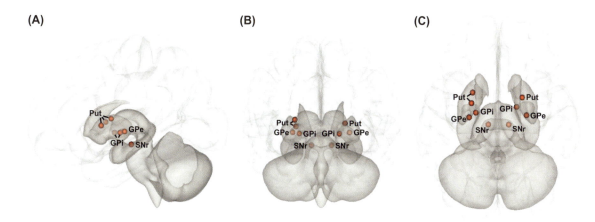

Figure D.4
Locations of the basal ganglia components of the DIVA model within the MNI152 template brain. (A) Left view. (B) Posterior view. (C) Superior view. GPe, external segment of the globus pallidus; GPi, internal segment of the globus pallidus; Put, putamen; SNr, substantia nigra pars reticulata.

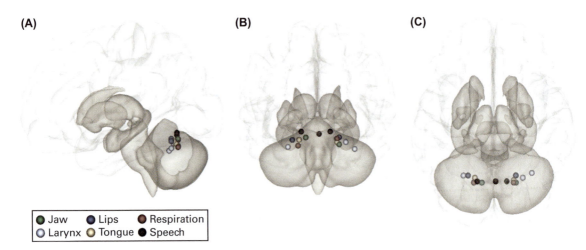

Figure D.5
Locations of the cerebellar components of the DIVA model within the MNI152 template brain, color coded by speech articulator. (A) Left view. (B) Posterior view. (C) Superior view. For comparison, see articulator foci from fMRI studies of single articulator movements summarized in figure 2.12 of chapter 2.

Note

1. The relatively low number of subjects in this study suggests caution when interpreting node locations that were based primarily on its results, namely, those in the *somatosensory error map* and the somatosensory components of the *feedback control map*.

References

Coleman, T. F., & Li, Y. (1994). On the convergence of reflective Newton methods for large-scale nonlinear minimization subject to bounds. *Mathematical Programming, 67*, 189–224.

Coleman, T. F., & Li, Y. (1996). An interior, trust region approach for nonlinear minimization subject to bounds. *SIAM Journal on Optimization, 6*, 418–445.

Index

Accessory nerve. *See* Cranial nerves, accessory (XIth)
Acoustic phonetics, 3
Acoustic reference frame, 8
Adaptive control, 98
Affective prosody. *See* Prosody, affective (emotional)
Agranular cortex, 77
Akinesia, 280
Akinetic mutism. *See* Mutism
Allocortex, 57, 77
Alpha motor neuron, 6, 42
Amyotrophic lateral sclerosis, 275–276, 309
Angular gyrus, 14, 16, 73, 261, 305, 339–340
Anterior cingulate cortex, 71–73. *See also* Cingulate motor area
  and affect, 49, 73, 258
  anatomical definition of, 339–340
  cortical connectivity of, 72
  electrical stimulation of, 72–73
  and motivation (will) to vocalize, 38–39, 49, 72, 257, 294, 311
  and mutism, 72–73, 82, 294, 310–311
  in the primate vocalization circuit, 38–39, 257
  and prosodic control, 257–258, 261–263, 268
  subcortical connectivity of, 38–39, 257–258
Anterior insula. *See* Insula
Anticipatory coarticulation. *See* Coarticulation, anticipatory
Aphasia, 9, 14–16, 78, 275, 288
  Broca's (motor), 9, 14, 16, 79, 265
  conduction, 16
  supplementary motor area, 294
  transcortical motor, 16, 71, 80, 82, 206, 292, 294, 310
  transcortical sensory, 16
  Wernicke's (sensory), 14, 16, 79
Apraxia of speech, 76, 78, 80, 82, 169, 210, 274, 288–293, 295, 306, 310
Arcuate fasciculus, 164
Articulation circuit. *See* Feedforward control, articulation circuit
Articulatory convergence, 49–50, 331–334

Articulatory gestures. *See also* Gestural score; Gestural target model
  and buccofacial apraxia, 76
  feedforward control of, 193–194, 199, 205, 209, 212–213, 215–216
  language specificity of, 109, 111
  and Parkinson's disease, 282–283
  as phonological units, 27
  sequencing of, 228, 234, 237–239, 228, 234, 237–239, 241, 245n9
  somatosensory feedback control of, 181–182
  in speech motor programs, 107
  as speech targets, 123
  and stuttering, 296
  timing of, 188
  visual perception of, 124–125
Articulatory phonetics, 3
Articulatory process. *See* Baddeley working memory model
Articulatory redundancy. *See* Motor redundancy
Articulatory reference frame, 7, 13, 29
Articulatory synthesizer, 26–27, 99–100, 127–128
Articulatory trade—offs/trading relations, 126, 129, 145
Articulatory training, 134
Articulatory variability, 126, 129, 131, 134–135, 145
Articulometry, 13
Artificial neural networks, 27–29, 30, 99
Artificial palate. *See* Somatosensory perturbation, palate
Associative chain model of serial order, 221–223, 242, 245n11
Ataxic dysarthria. *See* Dysarthria, ataxic
Auditory cortex, 78–83. *See also* Heschl's gyrus; Planum polare; Planum temporale; Superior temporal gyrus; Superior temporal sulcus
  activity when viewing speaking faces, 143
  and auditory feedback control, 79–80, 82, 162–168, 171
  and covert speech, 79
  cytoarchitecture of, 84n17

Auditory cortex (cont.)
　in the DIVA model, 101, 103–106, 108, 114, 163–167, 171, 392 (*see also* DIVA model, auditory error map; DIVA model, auditory state map; DIVA model, auditory target map)
　excision of, 78
　lateralization of function in, 80, 83, 287
　M100 in, 20
　and metrical structure, 266–267
　and non-speech articulator movements, 74, 330
　and prosodic control, 254, 256, 260–263, 265–266
　and singing, 265
　sound representation in, 8–9, 11, 29
　and spasmodic dysphonia, 287
　speech-induced suppression of, 79–80
　and speech sound targets, 143–145, 163–166, 171
　and stuttering, 297, 3305, 311
　and Wernicke's aphasia, 14, 79
Auditory error, 108, 114, 153–154, 158, 162, 166–168, 170–171, 196–197, 287, 305. *See also* DIVA model, auditory error map
Auditory feedback control, 153–172. *See also* DIVA model, auditory feedback control subsystem
　behavioral studies of, 155–163
　in a brain-computer interface, 307–308
　neural mechanisms of, 79, 82, 105–106, 114, 163–171
　in the overall control system for speech, 27, 100–102, 108, 114, 153–155
　and prosodic control, 252–256, 258, 260, 267
　relation to somatosensory feedback control, 162, 181–182, 188
　in the servosystem model of speech, 27, 95–97
　and speech development, 108, 114, 116, 131, 193–195, 214
　and tuning of feedforward commands, 193–195, 197, 202–204, 214–215, 307
Auditory nerve. *See* Cranial nerves, cochleovestibular (VIIIth)
Auditory perturbation. *See also* Sensorimotor adaptation
　and apraxia of speech, 291
　and auditory acuity, 200–202
　and the cerebellum, 56, 279
　combined with jaw perturbation, 202–203
　compensation generalization, 203
　conscious versus subconscious compensation, 200
　delayed auditory feedback, 79, 157, 166, 253, 256, 304
　duration, 256, 267
　emphatic stress, 254–256
　formant, 79, 158–162, 167–169, 170, 200–203, 217n8, 253, 256, 260
　fricative, 200
　incomplete compensation to, 162–163, 181, 202–203, 215
　loudness (intensity), 157, 159, 162, 170–171, 252–256, 267

masking noise, 157–158, 166, 171, 180, 197, 214, 256, 304
negative aftereffects, 198–202, 215, 217n7, 254
pitch, 158–159, 161–162, 166, 169, 170, 200, 203, 252–256, 260, 267
and speech targets, 145
and stuttering, 304
subphonemic, 160–162
unexpected versus sustained, 158
Auditory reference frame, 8, 11, 29, 100, 113
Auditory state, 101, 153. *See also* DIVA model, auditory state map
Auditory target. *See also* DIVA model, auditory target map
　and ataxic dysarthria, 279
　and auditory feedback control, 153–155, 156–157, 164–166, 170–171
　in the DIVA model, 100–103, 106, 108–109
　and feedforward control, 196–197, 202, 204, 214–215
　learning of, 108–109, 113–116, 140, 143–145
　as the primary goal of speech, 121–123, 125–129, 134
　and prosodic control, 253–255
　relation to mirror neuron system, 143–144
　and spasmodic dysphonia, 287
　in speech models, 27, 95, 123, 126 (*see also* Auditory target, in the DIVA model)
　time-varying, 139–140
Autism spectrum disorder, 306, 311

Babbling, 109–112, 124, 154–155, 179. *See also* DIVA model, babbling phase
Baddeley working memory model, 73, 226–227, 230–231, 240–243, 245n3
Basal ganglia, 46–50. *See also* Cortico-basal ganglia motor loop; Cortico-basal ganglia planning loop
　caudate nucleus, 17–18, 45, 47, 49–50, 71, 105, 232, 237, 244, 285, 292, 295, 326–327
　cortical connectivity of, 46, 48, 62
　direct and indirect pathways, 207–209, 216, 280–281, 284–285, 299–300, 309–311
　electrical stimulation of, 17–18, 47, 237, 280, 283, 286, 288, 298
　globus pallidus, 39, 47–50, 83n5, 103, 105, 205–208, 232, 237, 244, 257, 263, 267, 274, 280–281, 284, 285, 292, 297, 308, 311n5, 325–327, 332
　hyperdirect pathway, 209
　lateralization of function in, 80, 83
　and non-speech articulator movements, 49–50, 326–327, 331, 333
　nucleus accumbens, 47
　and prosody, 263–264, 266–267
　putamen, 39, 45, 47–48, 50, 62, 71, 80, 105, 206–207, 232, 235, 237, 244, 263, 266–267, 280–281, 284–285, 295, 299, 301, 325–327, 332

striatum, 47–48, 62, 83n5, 207–208; 233; 280, 284, 287, 298–301, 304–305, 309, 311 (*see also* Basal ganglia, caudate nucleus; Basal ganglia, putamen)
 subcortical connectivity of, 45–48
 substantia nigra, 47–50, 103, 105, 205, 207–208, 216, 263, 274, 280–281, 285, 297, 308–309
 subthalamic nucleus, 47, 50, 207–209, 280–281, 283–285, 298
 timing signals in, 298
 ventral striatum, 47
Basal ganglia motor loop. *See* Cortico-basal ganglia motor loop
Basal ganglia planning loop. *See* Cortico-basal ganglia planning loop
Basket cells. *See* Cerebellum, basket cells
Billiards, 122
Biomechanical vocal tract model, 26–27, 126
Bite block. *See* Somatosensory perturbation, jaw
Blindness, 116n9, 124
Botulinum toxin, 286
Bradykinesia, 280
Brain-computer interface, 21, 69, 306–308, 311
Brain rhythms, 19, 303
Brain stem, 39–44. *See also* Cranial nuclei; Inferior olive; Periaqueductal gray matter; Pons; Red nucleus; Reticular formation; Superior olive
 and ataxic dysarthria, 277
 and auditory feedback control, 161–163, 169–170
 and feedforward control, 210
 and flaccid dysarthria, 275
 motor nuclei, 37, 39–42, 82, 94, 309
 and prosodic control, 258
 somatosensory nuclei, 39–42, 186
 and spastic dysarthria, 276
Broca's aphasia. *See* Aphasia, Broca's (motor)
Broca's area. *See* Inferior frontal gyrus, Broca's area
Brodmann areas, 59, 103
Buccofacial apraxia, 76

Carryover coarticulation. *See* Coarticulation, carryover
Categorical perception, 161
Caudate nucleus. *See* Basal ganglia, caudate nucleus
Central executive. *See* Baddeley working memory model
Central operculum. *See* Operculum, central (Rolandic)
Central sulcus, 19, 23, 57–59, 61–62, 64, 68, 82
Cerebellar cortex. *See* Cerebellum, cortex
Cerebellar loop. *See* Cortico-cerebellar loop
Cerebellar peduncles, 40, 51
Cerebellum, 49–56. *See also* Cortico-cerebellar loop
 activity during covert speech, 56, 241–242
 agenesis of, 56
 articulator representations in, 54, 56, 324–326, 328, 331–333
 basket cells, 51, 53
 climbing fiber, 51, 53, 211–214
 cortex, 45, 49, 51–56, 81, 83n8, 105, 167, 186, 210–214, 241–242, 276–278, 309, 328, 333
 cortical connectivity of, 46, 49, 51, 54–55, 62
 deep nuclei, 49, 51–55, 211–213, 217n11, 277–278. (*see also* Cerebellum, dentate nucleus; Cerebellum, fastigial nucleus; Cerebellum, interpositus nucleus)
 dentate nucleus, 53–55, 213, 217n11, 277–278, 309
 electrical stimulation of, 55
 fastigial nucleus, 53–54, 277
 Golgi cells, 51–52, 210
 granule cells, 51–53, 83n8, 210–211
 interpositus nucleus, 53–55, 277
 lateralization of processing in, 279
 and metrical structure, 267
 mossy fibers, 51–52, 210–211
 parallel fibers, 51–53, 83n8, 211–213, 216
 Purkinje cells, 51–53, 83n8, 211–213, 216
 and sensory prediction, 165, 185
 sparse coding in, 51–53, 210–211
 subcortical connectivity of, 45–46, 49, 51, 53
 and working memory, 167, 241–242
Cerebral cortex, 57–80. *See also individual cortical regions*
Chorea, 284–285, 288
Chorus reading, 298, 304
Chunking, 228–229, 237–239, 242. *See also* Sound chunk
Cineradiography, 11, 13
Cingulate cortex. *See* Anterior cingulate cortex; Cingulate motor area; Posterior cingulate cortex
Cingulate motor area 71–73. *See also* Anterior cingulate cortex
 anatomical definition of, 340–341
 cortical connectivity of, 72–73, 77, 356, 361–362
 in the DIVA model, 390–392
 electrical stimulation of, 72–73
 and motivation (will) to vocalize, 72, 82
 and mutism, 72–73, 82
 and non-speech articulator movements, 74–75, 325–326, 330–331
 and prosodic control, 262
 speech-related activity in, 60–61
Climbing fiber. *See* Cerebellum, climbing fiber
Closed-loop control. *See* Feedback control
Coarticulation, 131, 133, 145, 279, 293
 anticipatory, 106, 137, 204–205, 213, 215–216, 222
 carryover, 106, 135–137, 204, 215, 222
 window model of, 133
Cochlear implant, 155–156, 170, 182, 254
Cochlear nerve. *See* Cranial nerves, cochleovestibular (VIIIth)
Cochlear nucleus. *See* Cranial nuclei, cochlear
Cochleovestibular nerve. *See* Cranial nerves, cochleovestibular (VIIIth)
Competitive queuing model of serial order, 222–223, 233, 235–236, 242, 244, 300–301

Completion signal. *See* Cortico-basal ganglia motor loop, completion/termination signals in
Computational model, 26–30, 47, 114, 231, 273
Computed tomography, 22–23
Conceptual preparation, 4–5
Constriction reference frame, 7, 123
Control theory, 94–98
Convex region theory of the targets of speech, 127–128, 131–138, 145, 146n2. *See also* Target regions
Cortex. *See* Cerebellum, cortex; Cerebral cortex
Cortical parcellation
  cytoarchitectonic, 59, 103
  functional, 104
  landmark-based, 59, 103, 339–344
Corticobulbar tract, 7, 40, 42–43, 62, 81, 210
Corticomotoneurons, 66–68
Corticoreticular tract, 43
Corticorubral tract, 43
Corticospinal tract, 42–43, 61
Cortico-basal ganglia motor loop, 46–49
  and chunking/motor program learning, 107, 237–239, 244, 245n11
  completion/termination signals in, 209, 297–298, 301
  in the DIVA model, 103, 105, 107, 205–209, 216, 394–395
  and feedforward control, 105, 195, 206–209, 216
  in the GODIVA model, 232–233, 235, 237–239, 243–244
  and hyperkinetic dysarthria, 209, 274–275, 283–288, 309–310
  and hypokinetic dysarthria, 209, 274–275, 279–283, 309
  initiation signals in, 297–300, 304
  and motor program selection/initiation, 48, 81–82, 105, 206–209, 216, 237, 244
  and Parkinson's disease, 280–283, 309
  and prosodic control, 257, 263–264, 266–267
  and spasmodic dysphonia, 286–287
  and speech sequencing, 209, 231–232, 237, 239, 243–244, 245n11
  and stuttering, 209, 295–301, 304–306, 311
  timing signals in, 298 (*see also* Basal ganglia, completion/termination signals in; Basal ganglia, initiation signals in)
Cortico-basal ganglia planning loop, 232–233, 237, 243–244
Cortico-cerebellar loop, 46, 49–55, 81–82
  and anticipatory coarticulation, 213
  and apraxia of speech, 290–291
  and ataxic dysarthria, 55, 81, 210, 241, 267, 274–279, 309
  and auditory feedback control, 105, 164–167, 169, 171
  and chunking/motor program learning, 107, 237–239, 244
  in the DIVA model, 103, 105–107, 114, 205, 210–214, 216, 394–396
  and feedforward control, 105, 114, 205–206, 209–214, 216
  and inverse dynamics, 214
  and motor programs, 107, 235, 237–239
  in the primate vocalization circuit, 38–39, 257
  and prosodic control, 257–259, 263, 265, 267
  and somatosensory feedback control, 106, 167, 184–186, 189
  and speech sequencing, 237–239, 241–242, 244
  and speech targets, 143–146, 164–165, 171, 185
  and tuning of feedforward commands, 211–213
Covert speech, 21, 56, 62, 79, 230, 241–243
Cranial nerves, 39–43, 81
  accessory (XIth), 39–41, 257
  cochleovestibular (VIIIth), 39–41, 81
  facial (VIIth), 39–41, 81, 275
  and flaccid dysarthria, 275, 309
  glossopharyngeal (IXth), 39–41
  hypoglossal (XIIth), 39–41, 81, 83n2, 275
  trigeminal (Vth), 39–41, 81, 183, 275
  vagus (Xth), 39–41, 257
Cranial nuclei, 7, 39–43, 61–62, 81–82, 106, 183, 275, 309
  ambiguus, 40–41, 81, 83n4, 257
  cochlear, 40–41, 81, 163
  facial, 40–42, 81, 275
  hypoglossal, 40–42, 81, 83n2, 275
  trigeminal motor, 40–41, 81, 275
  trigeminal sensory, 40–41, 81, 83n2, 183
Cross-modal sensory interactions, 111, 162, 188, 203
Cue trading, 254, 267. *See also* Articulatory trading relations
Cytoarchitecture, 59, 70–72, 76–77, 103, 369

Darts, 122
Deafness, 41, 124, 155–156, 170, 273, 304
Deep brain stimulation, 280–281, 283, 286, 288, 298
Deep cerebellar nuclei. *See* Cerebellum, deep nuclei
Delayed auditory feedback. *See* Auditory perturbation, delayed auditory feedback
Dentate nucleus. *See* Cerebellum, dentate nucleus
Development. *See* Speech motor development
Diaschisis, 279
Diffusion-tensor imaging. *See* Magnetic resonance imaging, diffusion-weighted
Diffusion-weighted imaging. *See* Magnetic resonance imaging, diffusion-weighted
Direct pathway of descending motor commands, 42–43, 62
Directional mapping, 127, 154–155, 166, 179, 188, 203
Directions into velocities of articulators model. *See* DIVA model
Distinctive features, 9

DIVA model, 27, 69, 99–109
  articulator map, 103, 105–107, 112, 163, 166, 184, 186, 205, 210, 216, 274, 290, 308, 389, 390–394, 396
  auditory error map, 79–80, 103, 106, 112, 163, 165–167, 171, 260, 308, 392
  auditory feedback control subsystem, 153–155, 163–171
  auditory state map, 103, 106, 110, 115, 163–164, 308, 391–392
  auditory target map, 103, 105–106, 113, 115, 163–165, 308, 391–392
  babbling phase, 108
  brain activity simulations, 102–104, 167–168, 186–187, 389–396
  feedback control map, 103, 106, 108, 112, 163, 166, 168, 171, 184, 186, 189, 291, 308, 391–392, 396n1
  feedforward control system, 193–197, 205–216
  GO signal, 194–195
  imitation phase, 108–109
  initiation map, 103, 105, 108, 205–207, 216, 232, 234–235, 237–239, 244, 274, 282, 292–294, 296–298, 305–306, 308, 310, 391–392
  learning in, 107–116, 122–123, 130–134, 143–145, 154–155, 164, 171, 179, 182, 193–197, 204, 208, 210–214
  somatosensory error map, 74, 103, 106, 109, 112, 115–116, 163, 184–189, 391–392, 396n1
  somatosensory feedback control subsystem, 178–179, 183–189
  somatosensory state map, 103, 106, 111, 115, 184, 189
  somatosensory target map, 103, 106, 114–115, 184–185, 188, 391–392
  speech sound map, 76, 80, 99–103, 105–108, 113–115, 140, 143–146, 153–154, 163–165, 171, 178, 184–185, 188, 193–194, 205, 210, 216, 232, 235–236, 241, 244, 274, 289–293, 310, 391–392
Dopamine
  and hyperkinesia, 284–285, 309–310
  and hypokinesia/Parkinson's disease, 23, 280–281, 309
  imaging of, 23
  and the substantia nigra, 47, 208, 216
  and stuttering, 23, 295, 298–300, 311
  as a teaching/reinforcement signal, 47, 208, 216
Dual-route theory of word production, 292–293
Duration (as a prosodic cue), 251
Dynamic aphasia. See Aphasia, transcortical motor
Dynamic articulatory model, 27
Dynamic dual-pathway model of prosody, 264
Dysarthria, 275, 288, 295, 306, 309
  ataxic, 55, 81, 210–211, 241, 267, 274–279, 283, 309
  flaccid, 274–276, 309
  hyperkinetic, 209, 274–275, 283–288, 309–310
  hypokinetic, 209, 274–275, 279–283, 288, 309
  spastic, 274–276, 309

Dysgranular cortex, 77
Dyskinesia, 283–285
Dysmetria, 276
Dyssynergia, 279
Dystonia, 284–286, 288, 310, 312n8

Echolalia, 71
Economy of effort, 137–138
Efference copy, 79, 142, 165–166
Efficiency. See Motor efficiency; Economy of effort
Electrocorticography, 20–21, 24, 30
Electroencephalography, 18–21, 24, 30
Electromagnetic articulometry. See Articulometry
Electromyography, 13
Electrophysiology, 18–22, 30
Eloquent cortex, 62
Emotional prosody. See Prosody, affective (emotional)
Emphatic stress. See Stress, emphatic
Error signal, 95–97, 163, 211–214
  auditory, 95–97, 153–154, 158, 163, 167, 197, 254–256, 287, 305
  climbing fiber, 211–214
  inferior olive, 211
  somatosensory, 179, 256, 287, 305
Event-related potential, 20
Extrapyramidal system, 83n5, 279
Extreme capsule fasciculus, 164

F0. See Fundamental frequency
Facial nerve. See Cranial nerves, facial (VIIth)
Facial nucleus. See Cranial nuclei, facial
Fastigial nucleus. See Cerebellum, fastigial nucleus
Feedback control, 56, 93, 95–98. See also Auditory feedback control; Somatosensory feedback control
Feedback error learning control scheme, 97–98, 100
Feedforward control, 93–95, 193–217. See also DIVA model, feedforward control system
  and apraxia of speech, 290–291
  articulation circuit, 206, 210–214
  and ataxic dysarthria, 278
  behavioral studies of, 197–205
  in a brain-computer interface, 307–308
  and hypokinetic dysarthria, 282
  impact of damage to, 273–275
  initiation circuit, 206–209
  and motor programs, 107
  neural mechanisms of, 55, 71, 103, 105, 114, 169, 205–216, 241–242
  in the overall control system for speech, 27, 100–102, 114, 193–195
  and prosodic control, 252–254, 256, 258, 265, 267
  and segmental versus postural parameters, 156, 170
  and spasmodic dysphonia, 286–287
  and stuttering, 302
  tuning of, 97–98, 158, 193–197, 211–214
Final common pathway, 42

Fluency disorder. *See* Stuttering
Focus. *See* Stress, emphatic
Formant frequencies, 8–13
  and auditory feedback control, 101, 156, 158–162, 164, 167–170
  and feedforward control, 196–198, 200, 202
  perturbation of (*see* Auditory perturbation, formant)
  representation in motor cortex, 69–70, 307
  as segmental parameters, 156, 170, 253–254, 267
  and speech motor development, 110–112
  and speech targets, 127–130, 139–140
Formant synthesizer, 307–308
Forward model, 98, 112, 116, 144, 165, 185, 216n2
Frame. *See* Syllabic frame
Frame-slot neurons, 234
Friedreich's ataxia, 276
Frontal lobe, 1, 57–59
Frontal operculum. *See* Operculum, frontal
Frontal pole, 61, 324, 329, 334, 341
Functional magnetic resonance imaging. *See* Magnetic resonance imaging, functional
Functional mosaicism, 66
Functional neuroimaging techniques, 23–24
Fundamental frequency, 8–13, 101, 155–156, 158, 170, 251–253, 256, 260, 264, 266, 293. *See also* Pitch

Gain factor, 95–97
  in auditory feedback control, 154–155, 157, 162, 171, 389
  in feedforward control, 194–195, 389
  in somatosensory feedback control, 179, 389–390
Gamma motor neurons, 8, 42
Gestural score, 4–5, 27, 194
Gestural target model, 123, 126
Gesture. *See* Articulatory gesture
Gesture-selective neurons, 234
Glide, 139
Globus pallidus. *See* Basal ganglia, globus pallidus
Glossopharyngeal nerve. *See* Cranial nerves, glossopharyngeal (IXth)
GODIVA model, 207, 226, 231–239, 243–244, 307
  and apraxia of speech, 291–293, 310
  and autism, 306
  and Parkinson's disease, 282–283
  phonological content buffer, 235–236, 239, 244, 292–293, 310
  sequential structure buffer, 239, 244, 300
  and stuttering, 296, 298, 300–302
  and supplementary motor area syndrome, 293–294
  and transcortical motor aphasia, 294
Golf, 122
Golgi cells. *See* Cerebellum, Golgi cells
GO signal. *See* DIVA model, GO signal
Gradient order DIVA model. *See* GODIVA model
Grammar, 3, 14, 72, 74, 221
Granular cortex, 77

Granule cells. *See* Cerebellum, granule cells
Groping, 289–290
Guillain-Barré syndrome, 275

Handwriting, 125
Hand movements, 62, 67–68, 141, 185, 233–234
Hearing aids, 155
Hearing restoration, 155–156
Heschl's gyrus, 78–79. *See also* Auditory cortex
  anatomical definition of, 340–341
  and non-speech articulator movements, 325, 329–330
  cortical connectivity of, 78–79, 369, 378
  excision of, 78
  speech-related activity in, 60
  subcortical connectivity of, 40–41, 78, 81
Huntington's disease, 266, 285, 288
Hyperkinesia, 279, 283–285, 312n7
Hypoglossal nerve. *See* Cranial nerves, hypoglossal (XIIth)
Hypoglossal nucleus. *See* Cranial nuclei, hypoglossal
Hypokinesia, 279, 283
Hypophonia, 280–281
Hypotonia, 276

Imitation, 113–114, 141. *See also* DIVA model, imitation phase
Indirect pathway of descending motor commands, 42–43, 62, 81
Inferior colliculus, 40–41, 81, 163
Inferior frontal gyrus, 74–76. *See also* Premotor cortex
  anatomical definition of, 340–342
  and apraxia of speech, 78, 82, 289–292, 310
  Broca's area, 9, 14–15, 18, 21, 74, 76, 80, 141–142, 356, 369 (*see also* Aphasia, Broca's (motor))
  cortical connectivity of, 75–77, 79, 164, 356, 366–373
  electrical stimulation of, 76
  lateralization of function in, 80, 82, 143, 168–169, 189, 210, 235–236, 262–264
  mirror neurons in, 141–145, 171
  and mutism, 76
  and prosody, 260–262, 268
  speech-related activity in, 60–61
  and stuttering, 295, 297, 302–303, 305, 311
  and working memory, 227, 230
Inferior frontal sulcus. *See also* GODIVA model, phonological content buffer
  cortical connectivity of, 369, 374–375
  and speech sequencing, 230–233, 235–241, 244
  and working memory, 227, 230
Inferior olive, 51, 53, 211
Inferior parietal lobule, 16, 24, 26, 73, 141–144, 164, 227, 295, 356. *See also* Angular gyrus; Supramarginal gyrus
Inferior temporal-occipital junction, 61, 240, 340, 342

Initiation circuit. *See* Feedforward control, initiation circuit
Initiation signal. *See* Cortico-basal ganglia motor loop, initiation signals in
Inner speech. *See* Covert speech
Insula, 57–59, 76–78. *See also* DIVA model, feedback control map; DIVA model, speech sound map; Premotor cortex
  anatomical definition of, 339–340, 343
  and apraxia of speech, 78, 82, 289–290, 310
  cortical connectivity of, 72–74, 77, 79, 164, 369, 376–377
  and jaw-perturbed speech, 187
  lateralization of function in, 78, 80, 82
  and non-speech articulator movements, 74–75
  and prosody, 258, 261, 265
  speech-related activity in, 60–61
  and stuttering, 303
  superior precentral gyrus of, 78
  and working memory, 230–231, 240–241
Intensity. *See* Loudness (intensity)
Interhemispheric fissure, 57
Internal model, 98. *See also* Forward model; Inverse model
Internal speech. *See* Covert speech
Interpositus nucleus. *See* Cerebellum, interpositus nucleus
Interval-selective neurons, 234
Intonation, 10, 80, 93, 224, 251–253, 263–268
Intracranial stimulation, 16–18
Intraparietal sulcus, 73, 240, 261–262, 330
Inverse dynamics, 2, 214
Inverse kinematics, 2, 214
Inverse model, 98, 108, 112, 116, 154, 170, 179, 188
Isocortex. *See* Neocortex

Juncture, 180, 189n2, 198, 256

Laryngeal dystonia. *See* Spasmodic dysphonia
Lateralization of speech/language processing, 17, 62, 70–71, 78, 80, 82–83, 143, 169, 210, 235–236, 262–266, 279, 287, 302, 304
Lateral lemniscus, 41
Lateral sulcus. *See* Sylvian fissure
Lee Silverman voice treatment, 281–282
Left-to-right coarticulation. *See* Coarticulation, carryover
Lemma, 4–5
Lesion studies, 14–16. *See also* Aphasia; Apraxia of speech; Dysarthria
Leukoencephalitis, 276
Levelt model of word production, 4–5
Levodopa, 280, 284
Lexical selection, 4–5, 234
Lexical stress. *See* Stress, lexical
Lichtheim model of aphasia. *See* Wernicke-Lichtheim model of aphasia

Limbic system, 39, 47, 71, 257–258, 268
Lingual gyrus, 60–61, 330–331, 334, 340, 342
Linguistics, 3–5
Linguistic prosody. *See* Prosody, linguistic
Lip reading, 123–124
Local field potentials, 21
Locked-in syndrome, 69–70, 307–308, 311
Lombard effect, 156–157, 171, 252
Longitudinal fissure. *See* Interhemispheric fissure
Long-term depression, 211–214, 216
Long-term memory, 226
Long-term potentiation, 211–213, 216
Loudness (intensity), 39, 44, 156–157, 159, 161–162, 170–171, 251–256, 267, 276, 280–281, 309
Loudness perturbation. *See* Auditory perturbation, loudness
Lower motor neurons, 42, 275–276
L-dopa. *See* Levodopa

Magnetic resonance imaging
  diffusion-weighted, 22–23
  functional, 24, 30
  structural, 22–23
Magnetoencephalography, 19–20
Masking noise. *See* Auditory perturbation, masking noise
McGurk effect, 111, 124
Mechanoreceptors, 8, 61, 106, 177, 183, 189, 199
Medial geniculate nucleus. *See* Thalamus, medial geniculate nucleus
Mental syllabary. *See* Syllabary
Mesocortex, 57, 77
Metrical structure, 241, 251–252, 256, 265–268. *See also* Prosody; Rhythm
Microelectrode recording, 21–22
Middle frontal gyrus, 235, 240, 261–262, 294, 303, 340, 343
Middle temporal gyrus, 14, 60, 74–75, 79, 164, 305, 340, 343
Mink model of basal ganglia function, 46–48, 208–209
Mirror neurons, 140–145, 171. *See also* DIVA model, speech sound map
Morpheme, 3–5
Morphological encoding, 4–5
Morphology (linguistic), 3
Morphology (vocal tract), 98, 109, 114, 193
Mossy fibers. *See* Cerebellum, mossy fibers
Motivation (will) to vocalize, 38–39, 47, 49, 72, 257–258, 268n2
Motoneurons. *See* Motor neurons
Motor cortex, 7, 61–70, 82. *See also* Precentral gyrus
  and apraxia of speech, 290–291
  and ataxic dysarthria, 278–279
  and auditory feedback control, 114, 162–164, 166–167, 169, 171

Motor cortex (cont.)
  beta suppression during movement, 19
  cytoarchitecture of, 77, 59
  descending projections of, 40, 42–44, 61–62, 81, 83n4, 311n2
  in the DIVA model, 100, 103–107, 114, 116, 163, 166–167, 171, 186, 189, 205–206, 210–211, 213, 216, 390–393 (*see also* DIVA model, articulator map)
  and feedforward control, 114, 205–211, 213, 216
  formant representation in, 69–70
  and hyperkinetic dysarthria, 285
  and hypokinetic dysarthria, 282–283
  lesions to, 63, 276
  and locked-in syndrome, 307–308
  mirror neurons in, 142–143
  and prosodic control, 257–259, 265
  single neuron properties of, 66–70
  and somatosensory feedback control, 116, 181, 184, 186, 189
  and spasmodic dysphonia, 287
  and spastic dysarthria, 274, 276, 309
  and speech sequencing, 237–239
  and stuttering, 297, 305
  and vocal control, 38–39, 257–259
Motor development. *See* Speech motor development
Motor efficiency, 128–129, 133, 155, 170, 179, 188, 204, 214–215, 228. *See also* Economy of effort
Motor equivalence, 8, 106, 125–129, 145, 181, 203–204, 215
Motor evoked potential, 142–143
Motor neurons, 6, 38, 42–43, 61, 66–68, 82, 83n3, 94, 257–258, 275–276, 309
Motor program. *See also* Motor target
  and apraxia of speech, 291–292, 310
  articulation of, 105, 210, 216
  definition of, 106–107
  in the DIVA model, 9, 99, 101, 105–108, 194–197, 203–204, 310, 389
  and hypokinetic dysarthria, 282–283
  learning of, 108, 113–116, 145, 195–197, 203–204, 210–214, 237–239 (*see also* Chunking; DIVA model, learning in)
  phonemic, 9, 30, 99, 298
  and prosody, 253, 265, 267
  selection and initiation of, 48, 105, 206–209, 216, 310
  and speech sequencing, 236–239, 241, 243–244
  and stuttering, 296, 298–300, 304–305, 311
  syllabic, 30, 99, 203–204, 229, 244, 290–291, 296, 298
  word-level, 9, 30, 99
Motor redundancy, 125, 171n3
Motor reference frame, 6–8, 100, 166
Motor state, 100–102, 193–195, 253
Motor target, 99–102, 107–108, 126, 134, 193–197, 253. *See also* Motor program

Movement-selective neurons, 234
Multiple sclerosis, 276
Multi-unit recording, 21
Muscle activation reference frame, 6
Muscle length reference frame, 6
Muscle spindles, 7–8, 61, 101, 106, 177, 199
Muscular dystrophy, 275
Mutism, 44, 71–72, 76, 82, 206, 293–294, 310
Myasthenia gravis, 275
Myoclonus, 283–284

Neocortex, 57, 77–78
Neurological disorders of speech, 273–312. *See also* Aphasia; Apraxia of speech; Dysarthria; Spasmodic dysphonia; Stuttering; Supplementary motor area syndrome
Nigrostriatal pathway, 47, 207–208, 280–281, 285
Nucleus accumbens. *See* Basal ganglia, nucleus accumbens
Nucleus ambiguus. *See* Cranial nuclei, ambiguus

Occipital cortex, 60–61, 75, 227, 230–231, 261, 330–331, 334, 340, 342
Occipital lobe, 57–58
Open-loop control. *See* Feedforward control
Operculum, 57, 59, 78, 83n13
  central (Rolandic), 61, 78, 356–358
  frontal, 76, 78, 356, 368–369, 372
  parietal, 77, 167, 329, 356, 365
  temporal (*see* Supratemporal plane)
Oromandibular dystonia, 286, 310

PAG. *See* Periaqueductal gray matter
Paleocortex. *See* Mesocortex
Pallidum. *See* Basal ganglia, globus pallidus
Paralimbic system, 71, 78, 258, 268, 356, 369
Parallel fibers. *See* Cerebellum, parallel fibers
Parallel representation model of serial order, 222
Paraphasia, 71, 292–293
Parietal lobe, 57–58
Parietal operculum. *See* Operculum, parietal
Parkinson's disease, 23, 266–267, 275, 280–285, 288, 298, 300, 309, 311, 312n6. *See also* Dysarthria, hypokinetic
PD control. *See* Proportional-derivative control
Penfield cortical stimulation studies, 17–19, 23, 28, 30n2, 62–64, 70, 76
Perceptual magnet effect, 161
Periaqueductal gray matter, 38–39, 44, 73, 257–258, 268
Perseveratory coarticulation. *See* Coarticulation, carryover
Perturbation experiments. *See* Auditory perturbation; Somatosensory perturbation
Phoneme. *See also* Phonological content; Subsyllabic constituent
  and apraxia of speech, 289, 231–293

and auditory feedback control, 159–163
and hypokinetic dysarthria/Parkinson's disease, 282–283
as a phonological unit, 1, 3–4, 9, 29–30, 93, 224–226, 242–243, 251
and prosody, 251, 256
and sensory targets, 116, 121–139
and somatosensory feedback control, 182–183
and speech development, 108–114, 124
and speech errors, 11, 224–226
and speech motor programs, 9, 27, 30, 95, 99, 107, 116, 193, 203–204, 213, 215–216, 228
and speech sequencing, 4, 221–226, 233–239
and stuttering, 296–298, 301
and working memory, 228–229, 243
Phoneme-selective neurons, 234
Phoneme-transition-selective neurons, 234
Phonemic paraphasia. See Paraphasia
Phonetics, 3, 10
Phonetic encoding, 4–5
Phonological content, 83, 224–226, 232, 235–237, 242–244, 292. See also GODIVA model, phonological content buffer; Syllable frame, frame-content models
Phonological encoding, 4–5, 224
Phonological loop. See Baddeley working memory model
Phonological reference frame, 9
Phonological store. See Baddeley working memory model
Phonological word, 4–5
Phonology, 3, 76, 242
PID control. See Proportional-integral-derivative control
Pitch, 8, 10, 93, 126, 140, 155, 157–162, 166–167, 169–170, 200, 203, 251–256, 260–261, 263–265, 267–268, 276, 304, 309. See also Fundamental frequency
Pitch perturbation. See Auditory perturbation, pitch
Pitch shift response, 158–159
PI control. See Proportional-integral control
Plant, 94–96, 98
Planum polare, 78, 340–341, 383–384. See also Auditory cortex; Superior temporal gyrus; Supratemporal plane
Planum temporale, 14, 78, 106, 164, 167, 340, 343, 379, 383. See also Auditory cortex; Superior temporal gyrus; Supratemporal plane
Pneumoencephalography, 22
Pons, 39–40, 46, 49, 51–52, 62, 81, 83n5, 105–106, 144–145, 210, 216, 276, 278, 309
Population vector, 68
Positron emission tomography, 23–24, 30
Postcentral gyrus, 61–69. See also Somatosensory cortex
 anatomical definition of, 340, 342
articulator representations in, 63–66, 324–326, 329, 332, 334
cortical connectivity of, 62, 346, 353–355
cytoarchitecture of, 104
electrical stimulation of, 17–18, 62–64
excision of, 62–63
somatotopy of, 17–18 (see also Postcentral gyrus, articulator representations in)
speech-related activity in, 60
subcortical connectivity of, 40–41, 46, 48, 51–52, 54, 62
Posterior cingulate cortex, 326, 330–331
Posterior parietal cortex, 62, 331
Postural parameters, 156, 170, 254
Pragmatics, 3
Precentral gyrus, 61–70. See also Motor cortex; Premotor cortex
 anatomical definition of, 340, 343
articulator representations in, 63–66, 324–326, 329, 332, 334
cortical connectivity of, 46, 62, 71, 233, 346–352
electrical stimulation of, 17–18, 62–64
excision of, 62–63
formant representation in, 69–70
and locked-in syndrome, 307
mirror neurons in, 141–143, 145
single neuron properties of, 66–70
somatotopy of, 17–18 (see also Precentral gyrus, articulator representations in)
speech-related activity in, 60
subcortical connectivity of, 46–48, 51–52, 54–55, 61–62 (see also Cortico-basal ganglia motor loop; Cortico-cerebellar loop)
Precentral sulcus, 69, 82, 167, 231
Prefrontal cortex. See also Frontal pole; Inferior frontal gyrus; Inferior frontal sulcus; Middle frontal gyrus; Superior frontal gyrus
and apraxia of speech, 288, 292
cortical connectivity of, 71, 233
and prosody, 261, 264
and speech sequencing, 49, 231–232, 235–237, 244
and stuttering, 297, 301, 303
subcortical connectivity of, 47, 51, 54–55
and transcortical motor aphasia, 294, 310
and working memory, 226
Premotor cortex, 61–69, 82. See also Anterior insula; Inferior frontal gyrus; Precentral gyrus; Pre-supplementary motor area; Supplementary motor area
and apraxia of speech, 76, 274, 288–293, 310, 312n12
and ataxic dysarthria, 278, 280
and auditory feedback control, 163–166, 171
and autism, 306

Premotor cortex (cont.)
  in the DIVA model, 76, 103–107, 114, 163–167, 171, 183–186, 188–189, 205–206, 210–211, 216, 392–393 (*see also* DIVA model, feedback control map; DIVA model, speech sound map)
  and feedforward control, 205–211, 216
  in the GODIVA model, 231–232, 236–240, 244
  and hyperkinetic dysarthria, 283–284
  and hypokinetic dysarthria, 282–283, 309
  lateralization of function in, 80, 143, 169, 210, 235–236, 263, 279
  mirror neurons in, 140–145
  monkey area F5, 141
  and prosodic control, 254, 261–264, 266
  and somatosensory feedback control, 184–189
  and spasmodic dysphonia, 287
  and speech sequencing, 227, 230–233, 235–240, 244
  and speech targets, 140–145
  and stuttering, 297, 301–303
  and transcortical motor aphasia, 294
PreSMA. *See* Pre-supplementary motor area
Pre-supplementary motor area, 69–71. *See also* Cortico-basal ganglia planning loop; GODIVA model, sequential structure buffer
  anatomical definition of, 340, 343
  and chunking/ motor program learning, 237–239
  cortical connectivity of, 71, 77, 356, 360
  in the GODIVA model, 231–235, 237–240, 244
  and feedforward control, 206, 216
  and hypokinetic dysarthria, 282
  lateralization of function in, 70–71, 80
  and non-speech articulator movements, 74, 325, 330–331, 334
  and Parkinson's disease, 282
  and prosodic control, 261–263, 268
  speech-related activity in, 60
  and speech sequencing, 82, 230–235, 237–240, 244
  and stuttering, 296–297, 300, 303, 305
  subcortical connectivity of, 71
  and supplementary motor area syndrome, 293–294
  and transcortical motor aphasia, 292, 294
Primary auditory cortex. *See* Auditory cortex; Heschl's gyrus
Primary motor cortex. *See* Motor cortex; Precentral gyrus
Primary somatosensory cortex. *See* Postcentral gyrus; Somatosensory cortex
Primate vocalization circuit, 37–39, 44, 257
Principle of reafference, 165–166. *See also* Efference copy
Progressive bulbar palsy, 275
Proportional-derivative control, 96–97
Proportional-integral control, 97
Proportional-integral-derivative control, 97
Proprioceptive information, 6, 8, 29, 41, 61, 69, 81, 101, 106, 108, 116, 125, 177–178, 183, 199
Prosodic frame. *See* Syllabic frame

Prosody, 10, 93–94, 251–268. *See also* Intonation; Metrical structure; Rhythm; Sequential structure; Stress
  acoustic features of, 251–252
  affective (emotional), 43, 49, 73, 78, 93, 251–252, 258–264, 266, 278
  and apraxia of speech, 289, 291, 310
  generation of, 262–268
  hemispheric lateralization of, 76, 80, 169, 260, 264–267
  linguistic, 93, 251–252, 259–263, 267–268
  motor execution of, 252–260
  perception of, 260–262
  and postural parameters, 156, 170
  and speech development, 113
  in stroke patients, 264–267
  and structural frames, 224
Purkinje cells. *See* Cerebellum, Purkinje cells
Putamen. *See* Basal ganglia, putamen
Pyramidal cells, 19–20, 24
Pyramidal system, 42
Pyramidal tract, 42, 66, 83n5

/r/
  articulatory variability of, 126–129, 135, 203–204
  bunched and retroflexed, 126–130
  coarticulation of, 204
  Japanese versus English, 110–111, 131, 134, 142
Rank-order neurons, 234
Reaching, 2, 10, 82, 125, 170, 214
Reading, 1, 25, 49–50, 53–54, 60–62, 73–75, 231, 234, 240, 279, 330–331, 333
Reafference. *See* Principle of reafference
Receptive aphasia. *See* Aphasia, Wernicke's (sensory)
Recurrent slide and latch model of cerebellar function, 212–213
Reduced buffer capacity model of apraxia, 293
Redundancy. *See* Motor redundancy
Red nucleus, 40, 42–43, 50, 81, 83n3
Reference frames for speech motor control, 6–9, 29–30, 126, 129
Regional cerebral blood flow, 20, 23–24
Reticular formation, 37–40, 42–44, 53, 81, 83n3, 83n5, 257–258, 311n5
Reticulospinal tract, 43
Rhythm, 93, 224. *See also* Metrical structure; Prosody
Right-to-left coarticulation. *See* Coarticulation, anticipatory
Rolandic cortex, 61–70, 74, 82. *See also* Motor cortex; Postcentral gyrus; Precentral gyrus; Premotor cortex; Somatosensory cortex
Rolandic fissure. *See* Central sulcus
Rolandic operculum. *See* Operculum, central (Rolandic)
Rubrospinal tract, 43

Scanning speech, 23, 267, 276
Segmental component of speech, 93–94, 113, 156–157, 161, 170, 180, 224, 251–254, 258–260, 264–265, 267–268, 289
Segmental (versus postural) parameters, 156, 170, 253–254, 267
Semantics, 3
Sensorimotor adaptation
  to auditory perturbation, 112, 158, 160, 198, 200–203, 215, 217n8
  generalization to untrained sounds, 199, 203
  incomplete compensation, 202–203
  of prosodic cues, 254, 256, 267, 304
  and sensory acuity, 201–203
  to somatosensory perturbation, 112, 198–199, 215
  in stuttering, 304
Sensorimotor cortex. *See* Rolandic cortex
Sensory aphasia. *See* Aphasia, Wernicke's (sensory)
Sensory prediction, 112, 165, 185, 199
Sequence-selective neurons, 233–234
Sequencing. *See* Speech sequencing
Sequential structure, 232–234, 237, 239, 244, 300. *See also* GODIVA model, sequential structure buffer
Serial order, 221–224
Servosystem model of speech production, 27, 95–97
Short-term memory. *See* Working memory
Sidetone, 79, 153, 156–157
Silent speech. *See* Covert speech
Singing, 80, 264–265, 267, 298
Single photon emission computed tomography, 23
Single-unit recording, 21–22
SMA. *See* Supplementary motor area
Somato-auditory error map, 188
Somatosensory cortex, 7, 61–66, 69, 81–82. *See also* Operculum, parietal; Postcentral gyrus; Supramarginal gyrus
  and apraxia of speech, 290–291
  in the DIVA model, 101, 103–104, 106, 114, 183–186, 188–189, 392–393 (*see also* DIVA model, somatosensory error map; DIVA model, somatosensory state map; DIVA model, somatosensory target map)
  and feedforward control, 207
  movement-related suppression of, 185
  and somatosensory feedback control, 74, 106, 183–189
  and somatosensory targets, 146
  and spasmodic dysphonia, 286–287
  and stuttering, 297, 302–303, 305
Somatosensory deprivation, 180
Somatosensory error, 106, 109, 162, 179, 186–189, 256, 279, 287, 305. *See also* DIVA model, somatosensory error map
Somatosensory feedback control, 177–189. *See also* DIVA model, somatosensory feedback control subsystem
  and apraxia of speech, 291, 310
  behavioral studies of, 179–183
  neural mechanisms of, 74, 100–102, 105–106, 114, 183–189
  in the overall control system for speech, 27, 97, 109, 114, 116, 177–179, 188
  and prosodic control, 252–253, 256, 258, 267
  relation to auditory feedback control, 162, 181–182, 202
  and somatosensory deprivation, 124–125, 180
  and spasmodic dysphonia, 286–287
  and stuttering, 302
  and tuning of feedforward commands, 122–123, 193–195, 202–203, 213–215
Somatosensory perturbation, 56, 112, 145, 167, 188, 194, 203, 215, 279, 291. *See also* Sensorimotor adaptation
  anesthesia, 180, 188, 197–198, 215, 256
  combined with auditory perturbation, 202–203
  compensation generalization, 199
  facial, 199
  jaw, 74, 99, 121, 125–126, 180–183, 186–188, 198–199, 202–203, 217n7, 389
  lip, 99, 126, 181–183, 188
  negative after-effects, 198–199
  palate, 198
  unexpected versus sustained, 198
Somatosensory reference frame, 8, 29, 100, 178–179
Somatosensory state, 95, 101–102, 106, 112, 116, 145, 178–179, 185. *See also* DIVA model, somatosensory state map
Somatosensory target, 101–102, 106, 108–109, 112, 116, 121–125, 129, 131–132, 140, 144–146, 178–179, 181–183, 185, 188–189, 194, 197, 199, 202, 214–215, 253, 279. *See also* DIVA model, somatosensory target map
Sound chunk, 9, 30, 56, 99, 116, 228–229, 237, 242, 290. *See also* Chunking; Motor program, learning of
Source-filter theory, 26
Spasm, 283. *See also* Spasmodic dysphonia
Spasmodic dysphonia, 252, 286–287, 310, 312n9
Spectrally altered speech. *See* Auditory perturbation, formant; Auditory perturbation, pitch
Spectrogram, 11–13, 196
Speech arrest, 45, 47, 70, 76
Speech clarity, 131, 137–138, 145
Speech error patterns, 222, 225–226, 229, 236, 242
Speech-induced suppression, 79–80, 166, 186
Speech motor development, 107–116. *See also* DIVA model, learning in; Motor program, learning of
  effects of hearing impairment on, 124–125, 155–156
  effects of somatosensory impairment on, 124–125
  effects of visual impairment on, 124–125
  language specificity of, 130–134, 161–162

Speech prosthesis, 306–308
Speech rate
 and apraxia of speech, 289, 293
 and ataxic dysarthria, 276
 control of, 137–139, 195, 389
 effect of thalamic stimulation on, 45
 and feedback control, 27, 193
 and Parkinson's disease, 282
 and spastic dysarthria, 276
 and target regions, 137–139, 145
 and transcortical motor aphasia, 71
Speech recognition system, 108, 113
Speech restoration in locked-in syndrome, 306–308
Speech sequencing, 221–245. See also GODIVA model
 and basal ganglia, 49, 105, 206–207, 209, 216, 236–239, 244
 and cerebellum, 56, 167, 241–242, 244
 and inferior frontal sulcus, 231–233, 235–241
 lateralization of, 80, 235–236
 and premotor cortex, 231–233, 237–241
 and sensory feedback disruption, 180
 and stuttering, 296
 and supplementary motor areas, 71, 82, 206–207, 216, 231–235, 237–241, 244
 and working memory, 230–231, 240–241, 244
Speech sound (definition of), 99. See also Sound chunk
Speech sound map. See DIVA model, speech sound map
Speech targets. See Targets of speech
Spoonerisms, 224–225. See also Speech error patterns
Spt, 168
Stammering. See Stuttering
State feedback control, 216n2
Stellate cells, 19, 51, 53, 83n8
Stress, 93, 198, 224, 253–254, 267, 276, 289
 emphatic, 5, 251–252, 254–256, 263, 266–267, 268n1
 lexical, 10, 265, 268
 pattern, 113, 180, 232, 256, 266
 retraction, 266
Striatum. See Basal ganglia, striatum
Structural frame. See Syllabic frame
Structural neuroimaging, 22–23
Stuttering, 23, 294–306, 311
Substantia nigra. See Basal ganglia, substantia nigra
Subsyllabic constitutent, 225–226, 229, 234–236, 239
Subthalamic nucleus. See Basal ganglia, subthalamic nucleus
Superior frontal gyrus, 261–262, 303, 340, 343
Superior longitudinal fasciculus, 164, 185
Superior olive, 40–41, 81, 163
Superior parietal lobule, 61, 73, 230–231, 240, 261, 303, 326, 329–330, 340, 343
Superior temporal cortex, 14, 78–80, 166, 168, 243, 295, 330, 333, 369, 383. See also Auditory cortex; Heschl's gyrus; Planum polare; Planum temporale; Superior temporal gyrus; Superior temporal sulcus; Supratemporal plane
Superior temporal gyrus, 78–80. See also Auditory cortex
 anatomical definition of, 340–341, 343
 cortical connectivity of, 76–79, 380, 383, 385
 and non-speech articulator movements, 74–75, 325, 329–330,
 phonological store in, 227
 and reading, 73
 speech-related activity in, 60
 subcortical connectivity of, 46, 51–52, 54, 78, 81
Superior temporal sulcus, 73, 76, 78, 106, 141–143, 164, 386–387. See also Auditory cortex
Supplementary motor area, 69–71. See also Cortico-basal ganglia motor loop; DIVA model, initiation map
 anatomical definition of, 340, 344
 and aphasia, 71, 82, 292, 294, 310
 and autism, 306
 and chunking/motor program learning, 237–239
 cortical connectivity of, 62, 71, 356, 359
 in the DIVA model, 103, 105, 205–209, 216, 392
 electrical stimulation of, 70
 excision/resection of, 70–71, 293–294
 and feedforward control, 205–209
 in the GODIVA model, 231–235, 237–239, 244
 and hyperkinetic dysarthria, 284–285
 and hypokinetic dysarthria, 281–283
 lateralization of function in, 70–71, 80
 and movement initiation, 82, 205–209
 and mutism, 71–72, 294, 310
 and non-speech articulator movements, 70, 74–75, 324–326, 330–332, 334
 and Parkinson's disease, 281–283
 and prosodic control, 263
 speech-related activity in, 60
 and speech sound sequencing, 230–235, 237–239, 244
 and stuttering, 296–298, 301, 305, 311
 subcortical connectivity of, 48, 71
 and supplementary motor area syndrome, 274, 292–294, 310
Supplementary motor area aphasia. See Aphasia, supplementary motor area
Supplementary motor area syndrome, 274, 292–294, 310
Supramarginal gyrus, 73–74. See also Somatosensory cortex
 anatomical definition of, 340, 344
 cortical connectivity of, 73, 356, 363–364
 and non-speech articulator movements, 325–326, 329–330
 and prosody perception, 261
 speech-related activity in, 60
 and stuttering, 303, 305

in the Wernicke-Geschwind model, 16
and Wernicke's area, 14, 79
and working memory, 73–74, 227, 230–231, 240
Suprasegmental components of speech, 93, 159, 161–162, 180, 224, 232, 251, 265. See also Prosody
Supratemporal plane, 57–59, 77–78, 167. See also Auditory cortex; Heschl's gyrus; Planum polare; Planum temporale
Syllabary, 228
Syllabic frame, 224–226
  consonant-vowel, 224, 243
  frame-content models, 224–226, 242–243
  in the GODIVA model, 232–236, 244
  onset-nucleus-coda, 225–226, 243
  onset-rime, 226, 243
  representation in supplementary motor areas, 234
Syllable
  and apraxia of speech, 289–293, 310
  and auditory feedback control, 159–161
  brain activity during production of, 25, 48–50, 60–61, 76 103–104
  and hypokinetic dysarthria/Parkinson's disease, 282–283
  as a motor program unit, 9, 95, 99, 101, 116, 139, 178, 193, 203–205, 209, 212, 228–229, 231, 236, 239, 243–244, 293
  as a phonological unit, 1, 3–5, 9, 30, 93, 122, 221, 224–226, 242–243
  and the pitch shift response, 253–254
  representation in left ventral premotor cortex, 210, 236
  and sensory targets, 106, 116, 121–122, 139
  and somatosensory feedback control, 178
  and speech development, 108–114
  and speech errors, 224
  and speech sequencing, 55–56, 209, 221, 230–242, 242
  and stress/metrical structure, 5, 252–254, 256, 263, 265–267, 276
  structure of (see Syllabic frame)
  and stuttering, 296–298
Syllable-selective neurons, 234
Sylvian fissure, 14–15, 57–59, 63–64, 78, 83n13, 188
Syntax, 3, 369

Tactile information, 6, 8, 29, 41, 61, 81, 101, 106, 108–109, 116, 125, 177–178, 183, 185, 199
Talker normalization, 9, 140, 164
Tardive dyskinesia, 284–285
Targets of speech, 106, 121–146. See also Auditory target; Convex region theory; Somatosensory target
Task dynamic model of speech production, 27, 177
Teaching signal, 47, 97, 98, 208
Temporal lobe, 57–59
Temporal operculum. See Supratemporal plane
Termination signal. See Cortico-basal ganglia motor loop, completion/termination signals in

Thalamus, 44–50. See also Cortico-basal ganglia motor loop; Cortico-basal ganglia planning loop; Cortico-cerebellar loop
  and ataxic dysarthria, 276–278, 285
  cortical connectivity of, 45–46, 48, 165
  electrical stimulation of, 17–18, 45–47, 237, 312n10
  lateralization of function in, 17, 80, 83
  medial geniculate nucleus, 40–41, 46, 78, 81, 103, 105–106, 163, 274, 308
  and non-speech articulator movements, 49–50, 324, 326–327, 331–333
  in the primate vocalization circuit, 38–39, 257–258
  and prosody, 263, 266
  subcortical connectivity of, 40–41, 43, 45, 51, 53–54
  ventral anterior nucleus, 46, 103, 105, 205, 207–209, 232, 237, 244, 274, 281, 284–285, 292, 297, 308
  ventral lateral nucleus, 38, 45–46, 48–49, 103, 105–106, 163, 167, 184, 186, 205–209, 211, 216, 232, 237, 244, 257, 263, 274, 276, 278, 281, 284–285, 292, 297, 308–309, 327
  ventral posterior medial nucleus, 40–41, 45–46, 61, 81, 103, 106, 163, 183–184, 186, 274, 308
Tics, 284
Time normalization, 140
Timing signal. See Cortico-basal ganglia motor loop, timing signals in
Tinnitus, 41
Tonal language, 159, 161–162, 200, 203, 265, 268
Tourette's syndrome, 284–285, 288, 312n10
Tract variables, 178, 181–182
Trading relations. See Articulatory trading relations
Transcortical motor aphasia. See Aphasia, transcortical motor
Transcranial current stimulation, 26
Transcranial magnetic stimulation, 24, 26
Transverse temporal gyrus. See Heschl's gyrus
Tremor, 283, 286
Trigeminal nerve. See Cranial nerves, trigeminal (Vth)
Trigeminal motor nucleus. See Cranial nuclei, trigeminal motor
Trigeminal sensory nucleus. See Cranial nuclei, trigeminal sensory

Ultrasound, 13
Unstable behavior, 96, 162, 299–300
Upper motor neurons, 42, 83n3, 275–276

Vagus nerve. See Cranial nerves, vagus (Xth)
Ventral anterior nucleus. See Thalamus, ventral anterior nucleus
Ventral lateral nucleus. See Thalamus, ventral lateral nucleus
Ventral posterior medial nucleus. See Thalamus, ventral posterior medial nucleus
Ventral striatum. See Basal ganglia, ventral striatum
Verbal working memory. See Working memory

Viseme, 124
Vocal pitch. *See* Pitch
Voice onset time, 124, 139, 164, 213
Voice quality, 252, 257–258, 267–268, 276, 280, 287
Vowel reduction, 138

Wernicke-Lichtheim model of aphasia, 15–16
Wernicke's aphasia. *See* Aphasia, Wernicke's (sensory)
Wernicke's area, 14–15, 18, 21, 79, 383
Will to speak. *See* Motivation (will) to vocalize
Window model of coarticulation. *See* Coarticulation, window model of
Word-selective neurons, 234
Working memory, 55–56, 73–74, 167, 221, 223, 226–231, 235–236, 239–244, 279

X-ray microbeam, 11, 13